Progress in IS

Progress in IS encompasses the various areas of Information Systems in theory and practice, presenting cutting-edge advances in the field. It is aimed especially at researchers, doctoral students, and advanced practitioners. The series features both research monographs, edited volumes, and conference proceedings that make substantial contributions to our state of knowledge and handbooks and other edited volumes, in which a team of experts is organized by one or more leading authorities to write individual chapters on various aspects of the topic. Individual volumes in this series are supported by a minimum of two external reviews.

The Series is SCOPUS-indexed.

Piero Mella

The Combinatory Systems Theory

A Powerful Theory for Understanding, Modeling and Simulating Collective Phenomena

Second Edition

Springer

Piero Mella
Department of Economics and Management
University of Pavia
Pavia, Italy

ISSN 2196-8705 ISSN 2196-8713 (electronic)
Progress in IS
ISBN 978-3-031-86945-7 ISBN 978-3-031-86946-4 (eBook)
https://doi.org/10.1007/978-3-031-86946-4

This Springer imprint is published by the registered company Springer Nature Switzerland AG
The registered company address is: Gewerbestrasse 11, 6330 Cham, Switzerland

To my son Lorenzo who has always been a source of inspiration to me.

Preface

If in the midst of a crowdyou stare intently at the moon, all around you will suddenly do the same. It is inevitable (Anonymous)

THE TEXT—This study adopts the logic of Systems Thinking(Mella 2012) and Control Systems (Mella 2014a), presenting a simple and hopefully complete theory that I have called the *Theory of Combinatory Systems*, which is able to describe, interpret, explain, simulate, and control collective phenomena and their observable effects. Why do paths form in fields? Why does a fashion quickly arise, only to disappear just as quickly? Why are some park benches or walls covered in graffiti-while others nearby are spotless? How does a feud develop? Why, in Pavia, in the span of a few decades, have over 150 towers been built, all alike, without any apparent function, if not a symbolic one? How can we explain the birth, branching out, or extinction of a species? Why do dangerous wheel tracks form in certain sections of highway while in others the asphalt, which is of the same quality, appears to resist the stress of traffic? How do we explain urban and industrial settlements in circumscribed areas? What mechanism can we use to explain the maintenance of languages and dialects in limited areas? Why do mounds of garbage spontaneously form in certain areas? Why do industrial innovations quickly spread? What is the force behind the continual improvement in the quality of products? Why do individuals chase records? Why does a background buzz arise in crowded rooms, which causes everyone present to talk in a louder voice? Why are speedlimits and no standing rules invariably ignored despite rigid controls? How does spontaneous orderarise in a ballroom when a waltz is played? How does order in the flight of a flock of birdsarise and maintain itself, so that the flock is so compact as to appear a single organism with its own movement? And what about the movement of a herdof elephants in flight, which is an incredibly destructive mass? Or what about the mechanism followed by the Can-Can dancersat the Moulin Rouge to allow them to remain aligned? How did the Hoplites create a Macedonian phalanx and maintain its order?

Despite specific differences among these phenomena—many of which are "one way," non-repeatable or reproducible—they can all be *described* or *explained*, and thus *understood*, using the model, as simple as it is general, of "combinatory systems"; that is, systems formed by collectivities, or populations of non-connected and unorganized individuals of some species, which appear to be directed by an invisible handthat guides the analogous actions of similar individuals in order to produce an *emerging collective phenomenon*.

THE FUNDAMENTAL IDEA—Collectivities can be viewed as units formed by a plurality of *similar* elements or agents, each of which produces similar *micro behaviors*—and in many circumstances also observable *micro effects*—which, "in combining," produce a *macro behavior*that gives rise to collective macro phenomena—and noticeable *macro effects*—which do not refer back to the individual micro behaviors, even if they necessarily derive from the agent's micro behaviors. If, on the one hand, the macro phenomena are produced from a "combination"—hence the term "combinatory systems"—of the agents' micro behaviors, on the other hand, at the same time those phenomena condition the agents' behaviors, as part of a *micro–macro feedback* relationship that represents the "invisible hand" that seems to guide the individual behaviors and produce the collective phenomena.

The combinatory systems generally are set off "by chance," but when activated they maintain their dynamics "by necessity," due to the presence of *necessitating* and *recombining factors*. The action of the *micro-macro feedback*, which is guaranteed by the contemporaneous presence of *necessitating* and *recombining factors*, turns these collectivitiesinto true systems, which can be observed as a unit as well as a multiplicity of elements, and which I have termed combinatory systems.

The feedback arises from *necessitating* factors, which force the agents to adapt their micro behavior to the system's macro behavior, and is maintained by the action of *recombining* factors, which lead the collectivity to recombine the micro behaviors, or the micro effects, in order to produce and maintain the macro behavior, or the macro effect. Combinatory Systems Theory emphasizes and tries to identify those necessitating and recombining factors, since *recognizing the existence of a* micro-macro feedback *and understanding the nature of both the* necessitating *factors and their* combining *ones is indispensable for interpreting collective phenomena as deriving from a combinatory system.* In this sense, the path dependence may be considered as proof of the action of the *micro-macro feedback*, even if Path Dependence Theory does not explicitly include this mechanism in the explanation of the path dependence.

THE FIELD OF ENQUIRY—Combinatory systems are not easily recognizable; nevertheless, they are widely diffused and produce most of the social and economic collective phenomena involving the accumulation of objects, the spread of features or information, the overcoming of a limit, and the achievement of general progress as the consequence of the individual pursuit of particular interests. Combinatory systems, together with their effects, are observable in the biological, social, political, economic, business, and organizational fields. The theory appears to be of interest to the fields of biology, ecology, economics, and social research, since the category of combinatory systems includes the defined phenomena of self-organization, population dynamics, whose evolution is influenced by the history of the population itself, and, in general, the collective phenomena where individual behavior is, to a greater or lesser extent and in various forms, influenced by the collectivity's "general macrobehavior," which is produced by the micro behaviors of the members of the collectivity.

Combinatory systems do not lend themselves to being *represented* and *interpreted* by means of traditional systems logic—that is, the logic of *organized* systems—but can be studied according to the proposed new Combinatory System Theory. Many of these collective systems are usually studied in the context of the *theory of complex systems* and *synergetics*; however, these disciplines offer an *external* description of these systems and do not bring out the conditions, factors, and rules that produce the feedback between micro- and macro-behaviors. The simple theory of combinatory systems presented in this study offers an *internal* interpretation by emphasizing the operative mechanisms that justify the behavior of such systems. In particular, Combinatory System Theory interprets, includes, and generalizes the "allelomimesis" and "stigmergie" approaches to the ordered behaviors of collectivities of agents.

THE THEORY OF COMBINATORY SYSTEMS—In plain words, the typical operative logic of any "social" combinatory system can be summarized by the following simple decisional rules that direct the micro- and macro-behaviors:

1. A collectivity of $N \geq 2$ agents constitutes the "base" of the system.
2. Each agent is characterized by an individual variable of some kind (qualitative or quantitative) whose values—at any time t_h—represent the individual micro states.
3. The collectivity is characterized by a global variable (qualitative or quantitative) whose values—at any time t_h—represent the system's macro state, which in turn represents global information for the agents.
4. Due to the presence of an opportune set of *recombining* factors, the system state—at any time t_h—derives from the "combination" (to be specified for each situation) of the individual states, following macro or recombining rules.
5. Each agent can perceive a gap (positive or negative) between his individual state and the state of the collectivity.
6. Due to the presence of an opportune set of *necessitating* factors each agent—at time t_{h+1}—decides, or is forced, to attempt to expand or reduce the perceived gap following the micro or necessitating rules.
7. As long as the necessitating and recombining factors are maintained, the *micro-macro feedback* can operate.
8. The agents are characterized by an initial state at time t_0; in most cases, this initial state may be assumed to be "due to chance."
9. The *micro-macro feedback* operates between the limits of the minimum activation number and the maximum saturation number of the agents presenting the state that maintains the *micro-macro feedback*.
10. The sequences of state values over a period represent the macro and micro dynamics, or behaviors, of the system and the agents.

According to this logic, when the actions of the agents are simultaneous, the agents of the system appear to synchronize their micro behavior. However, each micro behavior updates the global information, and this recreates a divergence that exerts even more influence on the individuals to conform to this information. For this reason, combinatory systems can also be called "self-produced global informationsystems," in order to distinguish them from local informationsystems, whose typical model is represented by complex systemssimulated by cellular automata.

The behavior of the system can be conditioned by environmental actions that can *strengthen* or *weaken* its behavior. For this reason, the control of the system's behavior can be carried out by means of appropriate *strengthening* or *weakening* actions; it can directly operate on the macrobehavior—we will define this as *macro* or *external* control (or *exogenous* control)—or it can influence the micro behaviors: in this case, the control will be termed *micro* or *internal control*(or *endogenous* control).

Typology of combinatory systems—This outlined logic can be observed in five relevant classes of combinatory systems, which differ with regard to their macro behavior and/or their macro effect:

1. Systems of *accumulation*, whose macro behaviorleads to a macro effect which is perceived as the accumulation of objects, of behaviors, or of effects of some kind.
2. Systems of *diffusion*, whose macro effect is the diffusion of a trait or particularity, or of a "state," from a limited number to a higher number of elements of the system.
3. Systems of *pursuit*, which produce a behavior that consists in a gradual shifting of the system toward an objective, as if the system, as a single entity, were pursuing a goal or trying to move toward increasingly more advanced states.
4. Systems of *order*, which produce a macro behavior, or a macro effect, perceived as the attainment and maintenance of an ordered arrangement among the elements that form the system.
5. Systems of *improvement* and *progress*, whose effect is to produce *progress* in the overall state of a collectivityas a consequence of the agents' need for individual *improvement*. We can go so far as to state the existence of a true law (metaphysical) of "progress" that originates from the action of biological and, above all, social combinatory systems.

Control Systems in Combinatory Systems—From the above definition, it is easy to understand that "social" combinatory systems function due to the presence of *micro control systems* which, operating at the individual level, lead to uniform micro behavior by individuals in order to eliminate the (gap) with respect to the objective that is represented—or revealed—by the globalinformation (macro behavioror effect) (Mella 2014). This can be better understood by referring to two simple examples.

What causes this buzz to form? It arises from the voice level (micro effect) of those present when they speak to each other (micro behavior); in turn, the voice level is recombined by the shape of the room in order to produce the buzzing noise. But why do those present speak in a loud voice (macro behavior)? Because there is the buzzing noise (global information) which prevents them from being heard, this becomes the minimum constraint (objective) to exceed in order to be heard; if the individual voice level is not adequate (gap), it is not possible for each agent to communicate. Thus, if the buzz increases, those present, in order to be heard, must raise their voices. It seems they do this all together, as if the global information forces them to synchronize their micro behavior, but this causes the buzz to increase further, which obliges those present to raise their voices even more, which increases the buzz, which forces those present ... etc., as part of a reinforcing loopthat takes the buzz to the maximum level of tolerance (constraint). Once this is reached, the individual Control Systems induce the speakers to be quiet for a brief moment, as we have all witnessed firsthand.

Let us now consider the system that produces the process for "passing on a *language* within a population." All parents pass on their mother tongueto their children (micro behavior), who thus learn this language (micro effect). The population communicates (macro behavior) using the mother tongue(macro effect), which represents the global information that obliges families to teach that language to their children (constraint/objective) in order not to disadvantage them in their communication. The feedback is evident but—remembering the bad marks inflicted on us by our teachers because of syntax errors—we can imagine there are also external Control Systems that both reinforce the main combinatory system, by detecting deviations between the language of the group members and the codified mother tongue, and try to eliminate these deviations by using the traditional levers of academic teaching to correct the syntax. The existence of so many languages and equally numerous dialects shows how powerful this combinatory system is and how efficient the individual and external Control Systems are. Thus, combinatory systems for the spoken language are some of the most powerful systems operating in human society.

COMBINATORY AUTOMATA—The simplest models to represent combinatory systems are the *descriptive* ones that indicate in words, or with the aid of diagrams and figures, the fundamental elements necessary for understanding the operative logic of systems that produce observable collective phenomena and, above all, how the *micro-macro feedback* works (Chap. 1). More powerful still are the *heuristic* models that try to simulate the system's dynamics by stating—or constructing ad hoc—a set of rules specifying the fundamental elements of the combinatory systems behavior (Chap. 2). "Combinatory Automaton" is a more sophisticated tool to simulate combinatory systems (Chap. 3). This is composed of a lattice, each of whose cells contains a variable representing the state of an agent. The value of each cell at a particular time depends on a synthetic global variable whose values derive from some operation carried out on the values of the cells and represent the synthetic stateof the automaton. The micro-macro feedback connects the analytical values of the cells and the synthetic state of the automaton.

The automaton may be stochastic, if a probability is associated with the transition of state of each agent; in the opposite case, it is deterministic. In stochastic combinatory automata, when both probabilities and periods of transition of state are agent/time/state sensitive, the probabilistic micro behaviors are conditioned by the macro behaviorof the entire system, which makes the micro-macro feedback more evident.

The "social" combinatory systems that are most interesting and easiest to represent are the irreversible ones (build a tower or not, teach French or English to babies). In these systems, both the micro- and macro-behaviors produce permanent effects that may be viewed as increasing or decreasing cumulative processes in which the probabilities of micro behaviors depend on the macro behavior of the collectivityas a whole. "Chaos" arises in combinatory systems when the hypothesis of reversibilityis introduced (e.g., to speak or to keep quiet in the next minute, wear a skirt or miniskirt on different days, choose road A or B on different days). When reversibility in micro behaviors or in micro effects is possible, the combinatory system's macro behavior, or macro effect, can show a cyclical dynamic and, under certain conditions concerning the probability function regarding the transition of state of the elements, a chaotic one as well, when no cycles are recognizable in the time series of the system starting from random initial values.

THE LAYOUT OF THE TEXT—The text is composed of four chapters, in which I have tried to treat the subject matter at a level accessible to all. Chapter 1 presents the basic ideas behind the theory, which are analyzed in some detail. Chapter 2 describes the heuristic models of several relevant combinatory systems in the context of the five typical classes. Chapter 3, while not making particular use of sophisticated mathematical and statistical tools, presents the Theory of Combinatory Automataand builds models for simulating the operative logic of combinatory systems. Chapter 4 tries to answer three questions: Are combinatory systems "systems" in the true sense of the term? Why is this theory able to explain so many and so varied a number of phenomena, even though it is based on a very simple *modus operandi*? Are combinatory systems different than complex systems?

The book is "for everyone." There is no prerequisite required to read and understand it; in particular, math, statistics, and computer knowledge. The statistical tools used for the construction of Combinatory Automataare among the most basic. The simulations, where possible, are performed in Excel to facilitate the use of random numbers and to perform tests with different data. I have endeavored to make the text simple and enjoyable to read, thanks to the many examples taken from the most diverse fields, while also supplying food for thought for further in-depth analyses.

Pavia, Italy Piero Mella

Preface to the Second Edition

Purpose of the Book. This book proposes the new Combinatory System Theory by constructing a formal model to demonstrate how simple "collectivities" of non-interconnected, similar agents (individuals, operators, entities, etc.) producing analogous micro behaviors—which I have termed "combinatory systems"—reveal very interesting forms of micro- and macro-behaviors and effects attributable to a cybernetic mechanism defined as "*micro-macro feedback*." On the one hand, the macro-behavior of the system as a whole derives from the "combination" of the analogous micro behaviors or effects of the agents, and on the other, the macro-behavior determines, conditions, or directs the subsequent micro behavior, thereby creating observable effects and patterns in the collectivity, in most cases producing self-organization, synchronization, path dependence, or chaos. Combinatory systems are not easily recognizable; nevertheless, they are widely diffused and produce most of the social and economic collective phenomena involving the accumulation of objects, the spread of features or information, the pursuit of a limit, the formation of orderly behaviors, and the achievement of general progress as the consequence of the individual pursuit of particular interests.

Seven years have passed since the first edition, during which numerous papers have been published on various aspects of the Theory. During this period, I have received several comments, including expressions of appreciation, requests for clarification, and constructive criticism. The most frequent requests concerned both the desire to expand on certain topics covered in Chapters 2 and 3 and the request to apply the Theory to interpret new collective phenomena. In light of these comments and requests, I have prepared additions to the First Edition and asked the publisher to consider a second edition, a proposal which was positively evaluated and accepted.

To allow the Reader to immediately grasp the changes compared to the First Edition, it is useful to present, for each of the four chapters, the main differences from the First Edition.

In general, the structure of four chapters has been maintained; the figures in the text have been revised and improved in appearance, with the addition of colors to allow for easier reading and understanding.

Chapter 1: *The Theory of Combinatory Systems*

This is the most relevant chapter, because it formalizes the "Theory" of Combinatory Systems. Apart from a few minor changes, the structure of the chapter is comprehensive in its treatment of the subject matter, and therefore there is no need for significant modifications; only several minimal changes and additions were necessary to further enhance the completeness of the theoretical treatment.

Chapter 2: *The Observable Variety. Heuristic Models of Combinatory Systems*

This is a pivotal chapter as it includes significant examples of Combinatory Systems whose effects can be observed in social, biological, and physical reality, thereby confirming the theory. The examples are ordered based on a typology of the effects produced by the five types of Combinatory Systems: Systems of Accumulation, Diffusion, and Pursuit; Systems of Order and Conservation; Systems of Improvement and Progress. This chapter has been significantly modified and expanded; the Combinatory Systems presented in the first edition have been updated and, when necessary, extended. In addition, other significant examples have been added, in particular:

1. In the section titled "*Systems of Accumulation*," the following two systems have undergone relevant expansion: 2.2.4 *"Urban-Settlement" System*; 2.2.5 *Industrial Districts and Clusters.*
2. Regarding the "*Systems of Diffusion*," two new sections have been added: *2.3.2 "Epidemics" System*; *2.3.8 "Nudging" System. Diffusion of Useful and Virtuous Behavior.*
3. Among the *Systems of Pursuit*, the following system has been added: 2.4.6 *Human and Animal "Stampedes."*
4. In "*Systems of Order and Conservation*," Section 2.5.1 "*Waltz-Spin System*" has been expanded to interpret the process involving the formation of "rotary motion" in the oceans; the system in Section 2.5.6, "Highway-Ruts" and System and "Chains of Potholes" on the Road, has also been expanded with an analysis of the combinatory process of the formation of annoying "Chains of Potholes" on the Road.
5. The section "*Systems of Improvement and Progress*" has been expanded with the presentation of five new Combinatory Systems: *2.6.5 "Survival and Expansion" System. One Population; 2.6.6 "Survival and Expansion" with Limits to Available Resources; 2.6.7 "Survival and Expansion" for a Coexisting Population; 2.6.8 "Survival and Evolution" System; 2.6.9 The Diffusion and Evolution of Non-Biological "Species." Improvement and Progress in Robotics.*

A new section has been added: "2.7 *Some Phenomena Produced by the Joint Action of Multiple Combinatorial Systems*," which examines the following two systems: 2.7.1 *Enterprises and Entrepreneurial Dynamics. Co-Localization of Firms and the Formation of Productive Clusters, Business and Industrial Districts, and Colonies; 2.7.2 Joint Action of Combinatory Systems Producing Evolution in IA and Robotics Interaction.*

Chapter 3: Simulation Models. The Combinatory Automaton

This chapter introduces the concept of a Combinatory Automaton, a tool for simulating, mainly in probabilistic terms, the different classes of Combinatory Systems. Combinatory Automata represent a significant innovation in the modeling of Combinatory Systems as they allow for the generation of numerical simulations consistent with the theory, offering consistent *added value* to Combinatory Systems Theory. All Combinatory Automata are presented with specifications of simulation programs in Excel and Powersim, and the calculations they produce allow for direct verification by the reader of the dynamics produced by the simulated systems. The Automata included in the first edition of the book have been retained, with marginal modifications. Section *3.6., The Theoretical Behavior of Combinatory Automata Simulating Systems of Improvement and Progress,* has been expanded with two new Combinatory Systems: *3.6.5 Combinatory Automata Simulating Evolutionary Dynamics under Various Assumptions; 3.6.6 Coevolution of Two Interacting Species, Such as Prey and Predators.*

Chapter 4: The Heuristic Value of Combinatory Systems Theory

The final chapter considers the "explanatory power" of Combinatory Systems Theory for understanding the logic of collective dynamics, understood as a unitary (macro) object of observation, even though collectivities are composed of a "plurality" of individual entities capable of autonomous (micro) behavior. In particular, Combinatory Systems Theory is compared with Systems Theory in general and with the theory of Complex Systems. Finally, an attempt is made to answer the question of whether Combinatory Systems Theory is a "scientific theory" in the sense of Popper or a "procedural theory" in the sense of Darwin. Apart from some minor additions, the chapter has remained the same with regard to the First Edition.

I would like to express my deepest gratitude to the publisher, for encouraging me to tackle this second edition, and to the editorial staff for their invaluable support, without which this book could not have been published in such an accurate form.

Pavia, Italy
November 2024

Piero Mella

Contents

Chapter 1
The Theory of Combinatory Systems

Abstract We often observe phenomena connected to the actions of collectivities or populations of similar unorganized elements, or the actions of individuals (agents) of some species, which seem directed by an invisible hand guiding the analogous behaviors of the individuals to produce an emerging collective (macro) phenomenon. Such phenomenon cannot be reduced to the simple sum of the effects from the individual (micro) actions; rather, it is the result of the "combination" of these individual actions that conform to the collective behavior, which in turn conditions the behavior of the individual, in a *macro-micro feedback*. The awareness of the existence of traits common to collective phenomena leads to the formulation of the Combinatory System Theory to explain the genesis and maintenance of these macro phenomena. This chapter formally presents the basic ideas behind the theory, analyzing in some detail the logic of combinatory systems and their constitutive characteristics.

Keywords Micro-macro feedback · Necessitating and recombining factors · "Chance" and "Necessity" · Incomplete and limited information · Minimum and maximum density · Control of combinatory systems strengthening and weakening actions · Robustness and persistence

> First, you know, a new theory is attacked as absurd; then it is admitted to be true, but obvious and insignificant; finally it is seen to be so important that its adversaries claim that they themselves discovered it. (James, 1907, p. 76)

> Yet observation and theory are woven together, and it is futile to attempt their complete separation. Observations always involve theory. (Hubble, 1936, p. 35)

We often observe phenomena, connected to the actions of collectivities or populations of individuals of some species, that seem to be caused by an invisible hand that guides the analogous actions of similar individuals in order to produce an *emerging phenomenon* that cannot be reduced to the simple sum of the effects deriving from individual behavior. These are diverse phenomena, all of which, however, can be tied to a single interpretative scheme entailing a chance component that gives rise to ineluctable dynamics. This first chapter begins by examining these

P. Mella, *The Combinatory Systems Theory*, Progress in IS,
https://doi.org/10.1007/978-3-031-86946-4_1

phenomena which, though completely different from each other, can all be considered the result of the action of particular systems composed of populations of individuals acting as a unit. The analysis of these differing phenomena will allow us to determine the common operative traits of the system. In particular, the observable phenomena will be shown to be the result of the *combination* of individual micro actions that conform to a collective behavior that itself conditions individual behavior. The awareness of the existence of traits common to all the collective phenomena leads to the formulation of a theory to explain the genesis and maintenance of these phenomena, many of which will be described in greater detail in Chap. 2.

After this informative examination, the chapter introduces the concept of a *combinatory system*, setting out the guidelines for constructing a Theory of Combinatory Systems required to interpret, understand, and explain the phenomena that derive from the action of a collectivity. The macro effects that refer to the macro behavior of a plurality of individuals can be caused by micro effects resulting from individual behavior; at the same time, however, these macro effects condition the micro effects themselves as part of a *micro-macro feedback relationship* that represents the "invisible hand" that appears to guide the individual micro behavior in the direction needed to maintain the macro behavior. In particular, the focus will be directed at the notions of *recombining* and *necessitating* factors, which are fundamental for starting and maintaining over time the *micro-macro feedback* that sets *combinatory systems* apart from every other type of collective behavior.

If we accept the traditional definition of *self-organization* as the macro behavior of a collectivity of agents in which the micro behaviors appear to be directed, or organized, by an Invisible Hand, Supreme Authority, or Internal Organizer in order to produce the emerging phenomenon represented by the formation of ordered structures, or recognizable patterns, then it is easy to recognize the "synergetic effect" of the *micro-macro feedback* action (or circular causality) between the agents' micro behaviors and the global system's macro behavior or effect.

The most interesting combinatory systems are those defined here as "social," in that they are made up of agents operating in a "rational manner" to decide on their micro behavior based on *global information* (of varying types) that defines a *global objective* individuals must or desire to achieve. The system's macro behavior, or macro effect, represents (or conditions or suggests) the *objectives* behind the individual choices. This objective guides the choices of individual agents who, noting a distance between their own micro behavior or micro state and that of the overall state of the system, try to eliminate the gap by adjusting their individual and collective behavior. The *micro-macro feedback* starts its action by producing the micro and macro dynamics of the social combinatory system.

1.1 Defining Combinatory Systems

1.1.1 Strange but Simple Phenomena

Both the ordinary man and the scientist come face to face with many phenomena and processes which, because of the simplicity of their effects, often cannot be easily or convincingly explained by the analytical tools of traditional system theory—which views systems as black boxes that operate according to an [input-Δstate-output] logic—or by the logic of complex systems, whose dynamics is produced by local rules directing the agents' behavior (see below, Sect. 1.2.2). Here is a brief sample of such phenomena (a large catalog will be presented in Chap. 2).

Why do paths form in fields? Why does something suddenly become fashionable and then just as suddenly disappear? Why do industrial innovations spread rapidly? What is the force behind the continual improvement in the quality of products? Why do athletes chase records? Why are some park benches or walls full of graffiti, while others nearby are immaculately clean? How is a feud born? Can it be interrupted? Why in 200 years have over 150 towers been built in medieval Pavia, all similar and without any apparent function save a symbolic one? How can we justify the birth of a new species? Why do dangerous wheel tracks form in certain stretches of highway, while in others the asphalt, which is of the same quality, appears to resist the stress of traffic?

How many times have we witnessed the breaking out of applause after an initial hesitation, and why, at other times and in similar circumstances, has applause not arisen? How do we explain the settlement of urban and industrial centers in limited areas (Valenza for jewelry, Vicenza for gold-work, Silicon Valley for chips, Prato for knitwear, Cantù for furniture, Como for silk, etc.)?

Why are languages and dialectical inflections maintained even in circumscribed areas? Why in England does everyone line up in an orderly queue while waiting their turn, while in other "Latin" countries people wait their turn in a disorderly fashion? Why do garbage dumps spontaneously arise in certain areas? Why are records sought after (the highest skyscraper, the fastest plane, the most complex watch will soon be surpassed in the record books)? Why, in rooms full of people, does a background murmur develop that causes everyone to speak in an increasingly louder voice? Why are speed limits and no-standing laws invariably ignored despite rigid controls and heavy fines? How is order spontaneously created in ballrooms when a waltz is played? Why, after some initial moments of disorientation and without any apparent direction, does the entire dance floor seem animated by a circling motion?

How does a "flock of birds" form and maintain flight order, appearing to form a single organism propelled by its own motion? And what about the movement of a herd of elephants in flight? And the almost miraculous behavior of ants that form long orderly files, or termites that, each carrying a grain of sand, can build perfect arches? How do cancan dancers at the Moulin Rouge remain in a straight line? How could the Hoplites create and maintain in orderly fashion a phalanx of Macedonians?

It is apparently easier to explain the "wave" at the stadium, which transforms a disorderly group of spectators into a wave that circles the stadium several times.

How can we interpret and explain the phenomena that occur in competitive markets? How do we interpret the "invisible hand" that leads to the formation of a stable price, or justify the dynamics that we observe within certain productive sectors or in the entire economic system? What causes the irreversible dynamics involving the increase in productivity? How do we take account of the cyclical fluctuation of economic trends? Why in some cases do new models of products replace others, while in other cases the new products fail to overcome the competition?

We could continue with this list, but this short sample should be sufficient to begin the search for a common model of explanation based on an appropriate "systems thinking" capable of explaining the operative mechanisms that regulate these collective phenomena: "*In one way or another we are forced to deal with complexities, with 'wholes' or 'systems', in all fields of knowledge. This implies a basic re-orientation in scientific thinking across almost all disciplines*" (Von Bertalanffy, 1968, p. 5).

Despite specific differences, the phenomena mentioned above—many of which are "one-way" and cannot be repeated or reproduced—can be described and explained using the simple and general model of *combinatory systems*; these are systems constituted by a group of independent elements which, while acting individually and freely, behave in a uniform way, as if they constituted a single entity, in order to produce a given phenomenon, process, or effect, so that the behavior of the system as a unit derives from the "combination" of the analogous behavior of its similar elements.

In order to observe, interpret, and explain combinatory systems, one must first be able to apply the First Rule of Systems Thinking (Appendix 1), as introduced by Peter Senge in his book *The Fifth Discipline: The Art and Practice of the Learning Organization* (Senge, 1990; Senge et. al. 1994); this rule can be summarized as follows:

> First rule: To understand reality we must not limit ourselves to observing only individual objects, elements, or entities; it is necessary to "see" even the larger groupings that these compose, attributing to them an autonomous meaning. The converse process is also true: we cannot limit ourselves only to considering an object in its unity but must force ourselves "to see" its component parts. (Mella, 2012, p. 9)

In order to interpret and explain the operative mechanisms that regulate these phenomena, a particular kind of system is proposed here—which will be defined as a "combinatory system"—which has not yet been specifically studied in the field of Systems Thinking.

1.1.2 Macro (Collective) and Micro (Individual) Behavior: Collectivities

To construct a general model of a combinatory system, let us try to recognize three relevant common features of the phenomena mentioned in the previous sections. Above all, we immediately observe that these derive from, or are caused by, a collectivity or a plurality of similar agents (who are thus unorganized, following Maturana and Varela's (1980) concept of organization) that produce an analogous micro behavior over time—or similar micro effects—but, considered together, are capable of developing a macro behavior—and/or macro effects—which is attributed to the collectivity as a whole.

The graffiti is left by a multiplicity of people who, day after day, write or carve names, sayings, initials and messages; and a garbage pile is also the result of numerous acts involving people leaving garbage behind. The Macedonian phalanx is made up of a large number of Hoplites; highway carriageways are the result of the passing of countless numbers of trucks; records are achieved and surpassed by the action of thousands of athletes who, in different places and at different times, compete in the same event; a herd of fleeing elephants, a school of fish, a flock of birds are formed by many specimens; applause does not break out, and the murmur in a room does not arise if there are not enough people present. Even fashion trends and epidemics can only spread within a collectivity.

We therefore recognize that these phenomena derive from the action of a *collectivity* (population, plurality, matrix, set, etc.), no matter how it came to be formed, whose elements we shall call the *base* of the system. The *collective action* attributed to the *base* (walking down the same path or driving along the same stretch of highway, the wave at the stadium, the circular dance of the ballroom dancers, sitting on the same bench and carving initials in it, etc.) will be defined as the *macro behavior* of the system. Its observable effects (path, carriageways, the wave circling the stadium, rotating motion, graffiti, the catching on of a fashion trend, etc.) are defined as the *macro effects* of the system. From these examples the following definition is proposed:

Definition 1 Collectivity. A *collectivity* is defined as a plurality of similar unorganized elements, or agents—that is, agents not organized according to hierarchical relations or interconnected in network or tree relations—that, acting individually and freely, produce analogous *micro* behaviors over time (which lead to similar *micro* effects), but considered together are capable of developing a *macro* behavior, and at times a *macro effect* or a recognizable pattern, which is not included in advance in the operating program of the agents' behavior, but is instead attributed to the collectivity as a whole.

In this sense a collectivity differs from organizational and social systems ("social machines" according to Maturana & Varela, 1980, p. 77), which are composed of *organized agents* where the collective and individual behavior is determined by a network of stable relations (i.e., the organization).

Collectivities can be *observed* (e.g., swarms, flocks, crowds, spectators at a stadium, students in a classroom, persons that are talking in a crowded room, dancers doing the Can Can), or simply *imagined* (e.g., trailer-trucks traveling along a stretch of highway in a month, the noble families of Pavia who erected the 100 towers in the span of two centuries, a group of scientists who dedicate themselves to a branch of research, the consumers of a particular product during its entire life-cycle, stockbrokers working on a certain day in world or European stock markets).

Collectivities can be composed of persons (social collectivities or populations), animals and active biological organisms (herds, schools, swarms, etc.), or reactive ones (plants, micro-organisms), even other inanimate objects; in any case, if considered from a certain distance, collectivities appear distinct with respect to the individuals they are composed of, and thus seem able to show an autonomous macro behavior due to the joint action of the micro behavior of the agents.

It is generally possible to also circumscribe an *environment* in which the *base* operates; that is, a real or imaginary place where the *micro behavior* arises and the *macro behavior* is formed. A set of bases can be represented by 50 pairs of dancers, by a herd of 500 elephants, 100 members of a chorus, or 5000 trucks going down a stretch of highway; but the action of these elements depends on the type and size of the environment. The dancers will behave in a different way depending on whether or not they are in a small or large dance hall; the elephants will behave differently depending on whether or not they are in a vast savanna or a narrow valley; the trucks will produce different effects if the highway is straight as opposed to winding.

But how does a murmur arise in a room if the individuals do not speak to each other? How does applause break out if the individual spectators do not applaud? And how does a fashion spread if those belonging to the collectivity do not develop imitating behavior? If the mass of graffiti comes from a plurality of people, it is nevertheless formed by individual graffiti writings. Thus a *second observation*: the *macro* behavior and correlated macro effects result from the collective action of the base, in that each of its elements produces a distinct behavior, a *micro behavior*, that can produce a particular effect, a *micro effect*. Let us look again at several phenomena using the terminology we have just introduced.

A pair of individuals (elements) are speaking (micro behavior) at a given voice level (micro effect), and the crowded room (base) of talking people (macro behavior) creates a background noise (macro effect). A carrier of influenza (element) comes into contact with other persons (micro behavior) and spreads the virus (micro effect), and in a population (base) of persons who interact (macro behavior) an epidemic quickly develops (macro effect). Many individuals (base) cross a field (macro behavior); each individual (element), while crossing (micro behavior), steps on the grass (micro effect) and a path is thus formed (macro effect). It is a useful exercise to also consider other phenomena in terms of micro and macro behavior.

Note. It is important to note explicitly that from Definition 1 it follows that "collectivities" cannot be identified with the organized systems and, in particular, with organizations and societies. In both cases the agents produce micro behaviors delimited by shared or imposed links and rules that transform them into structured (even hierarchically) members or organs that cooperate to achieve a common goal or the common good.

1.1.3 The Micro-Macro Feedback

In order to fully explain collective phenomena we must, however, take a *third step*: determine the mechanism that connects their *micro* and *macro behaviors* or their *micro* and *macro effects*. To easily understand this mechanism, let us again consider the phenomenon of a murmur arising in a crowded room. What is the murmur made up of? From the voice level of those present while they are talking to one another. But why do they talk out loud? Because there is a murmur. If the murmur increases, those present, in order to be heard by the person they are talking to, must raise their voices. But this increases the level of the murmur, which causes those present to raise their voices even higher, which increases the murmur and forces those present, and so on and on. The murmur is formed by the voice level of those talking, but this depends, in turn, on the murmur.

In other words, the macro *effect* (murmur) from the macro *behavior* (collective chatter) depends on the micro *effect* (voice level) produced by the micro *behavior* (speaking in pairs). Reciprocally, the macro *behavior* (exchange of information among those present) and the macro *effect* (murmur) influence the micro *behavior* (attempts at communicating) and the micro *effects* (voice level).

Let us consider another clear example. During the process whereby a language spreads within a population, each parent transmits the mother tongue (micro behavior) to the children, who thus learn it (micro effect). The population speaks (macro behavior) the mother tongue (macro effect) and forces the families to teach this to their children. The feedback is clear.

Definition 2 Micro-macro feedback. In a collectivity, the mutual dependence between the agents' *micro* behavior and the collectivity's *macro* behavior (or their effects) can be defined as *micro-macro feedback*, as represented in the model in Fig. 1.1.

Recognizing the existence of a micro-macro feedback is indispensable for interpreting collective phenomena as deriving from a combinatory system: the state of the system at a given time *must* depend on the state of its elements; but, in turn, the subsequent state of the latter *must* depend on the state of the system (for a more precise formulation see Sect. 1.3.3).

The *micro-macro feedback* generates a "synergetic effect" that produces self-organization and emerging macro behavior that is attributable only to the collectivity (Sect. 1.2.1).

In other words, in order to identify a combinatory system, it is not sufficient to describe a collectivity in terms of micro and macro behaviors (or effects); it is also *necessary* to recognize the mutual dependence among these types of behavior and

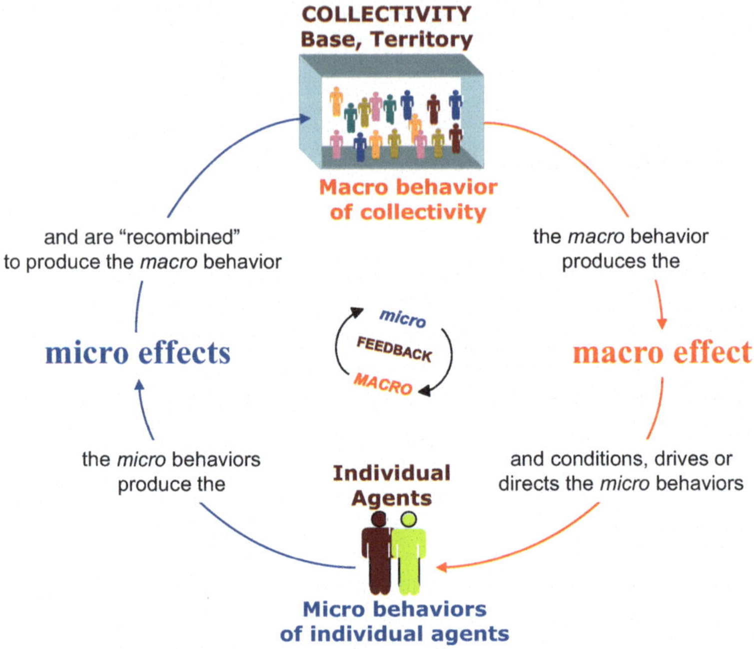

Fig. 1.1 *Micro-macro feedback*. The invisible hand

the effects produced by a *micro-macro feedback* action. Without the *micro-macro feedback* no combinatory system can be observed or conjectured.

Here is an example: "It is forbidden to put your feet on the wall" was written on a sign in clear view in every room in a museum. And yet … a set of footprints defaced the entire wall, a sign of the incivility of the visitors who rested their feet on the walls in order to look at the paintings and, in order to be more comfortable, pressed their soles against the wall. Is this a case of a combinatory system? The answer is "No!," since the reprehensible micro behavior of each visitor can be said to be completely independent of the number of dirt marks on the wall. The footprints simply derive from the need for each individual to better observe the works by leaning on the wall. That mass of footprints is not the macro effect of any combinatory system, due to the lack of feedback between the individual micro behavior and the collective effect.

A second example: if an initial nucleus of families by chance begins to settle in a fertile plain, other settlements will be attracted to this nucleus, and a larger city will develop due to the evident action of a combinatory system; however, if a population should settle in a narrow fertile area between two steep rock faces, a city would develop, but without the action of any combinatory system; the new settlements would not depend on the preceding ones but on the shape of the environment.

We immediately understand how all the phenomena mentioned in the first section can be represented by the same schema indicated in Fig. 1.1. Figure 1.2 shows the phenomenon of the buzzing that originates in a crowded room.

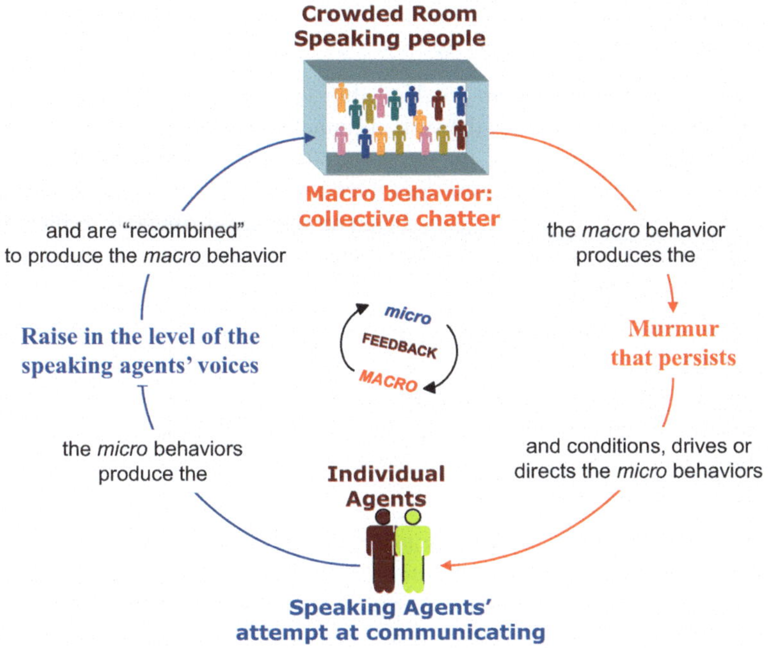

Fig. 1.2 Voice and noise feedback

1.1.4 The Central Idea of Combinatory System Theory (CST)

The central idea of the Theory is that in combinatory systems the collective phenomena are produced by a combination of the micro behavior of the agents; but at the same time they condition this behavior as the result of a *micro-macro feedback* relationship that seems to guide the individual behavior and produce the collective phenomena.

It is assumed here that a *necessary* and *sufficient condition* for a collectivity (observable or assumed) to be considered a combinatory system *is the existence of a feedback* between the micro behavior of the individuals and the macro behavior of the collectivity constituting the system, and/or the connected micro and macro effects. This assumption easily leads to the following definition:

Definition 3 (Structural) Combinatory system. A "combinatory system" is defined as any *collectivity* (see Def. 1) made up of a plurality of unorganized *similar* agents (or elements) producing *analogous* micro behaviors and showing, as a whole, a macro behavior and/or a macro effect, whose dynamics are created by a *micro-macro feedback* action. If, on the one hand, the macro behavior of the system, as a whole, *derives* from the *combination*, appropriately specified, of the analogous behavior (or effects) of its similar agents (hence the name combinatory system), on the other hand the macro behavior (or the macro effect) *determines*, *conditions*, or *directs* the subsequent micro behavior, *according to a feedback relation between the micro and macro behavior or effects.*

The existence of *micro-macro feedback* thus leads to an essential consequence: the macro behavior (or effect) of the system cannot be considered a mere sum of the micro behavior (or effects) of its elements; the *micro-macro feedback* causes "emerging" types of macro behavior (or effects) which are not included in advance in the operating program of the agents' behavior and are attributable to the unit.

The operation of *combination* may be defined in different ways in different combinatory systems; in particular as:

- The sum of the values produced by, or associated with, the effects of the agents' micro behavior
- The mean value of the agents' performance
- The maximum value produced by all the agents at each instant
- The minimum value
- The modal value, and so on

In this sense, combinatory systems are *recursively closed systems*; their dynamics can be attributed only to the joint action of "chance and "necessity" by adopting, though with a different meaning, the same terminology used by Monod (1970), who, in his famous *Chance and Necessity*, examined a very powerful combinatory system: that leading to a dynamic evolution in a population due to random mutations produced in the DNA, which "by necessity" spread as a result of the invariant reproductive mechanism of cells. Haken also speaks of "chance" and "necessity" when he proposes constructing models of complex systems. Here Haken considers chance as the unpredictable fluctuation from an unstable equilibrium state and necessity as the movement toward a new, more stable state (Haken, 1977a, 1983; Prigogine & Stengers, 1984; Klir, 2013), Chance will not only set under way the macro behavior but will also determine its direction; that is, the direction of the "winning" fluctuation. Prigogine bases his theory on the emergence of order in complex systems on the consequences of fluctuations (see Sect. 1.3.5).

The analysis and understanding of many phenomena mentioned below is even more difficult because they often are "one way" and cannot be repeated or reproduced, as if they were only due to one-way phenomena and processes. While the phenomenon of urban settlements appears to repeat itself many times, even with particular variations, the same cannot be said for the construction of towers in medieval Pavia, which is an amazing event because it is unique.

If the perception of combinatory systems appears to be difficult, in any case it becomes possible as soon as we apply the First Rule of Systems Thinking in this most complete formulation:

> To understand reality we must not limit ourselves to observing only individual objects, elements, or entities; it is necessary to "see" even the larger groupings that these compose, attributing to them an autonomous meaning. The converse process is also true: we cannot limit ourselves only to considering an object in its unity but must force ourselves "to see" its component parts. This rule, which is at the basis of Systems Thinking, can be translated as follows: if we want to broaden our intelligence we must develop the capacity to "zoom" between parts and wholes and between wholes and components. (Mella, 2012, p. 9)

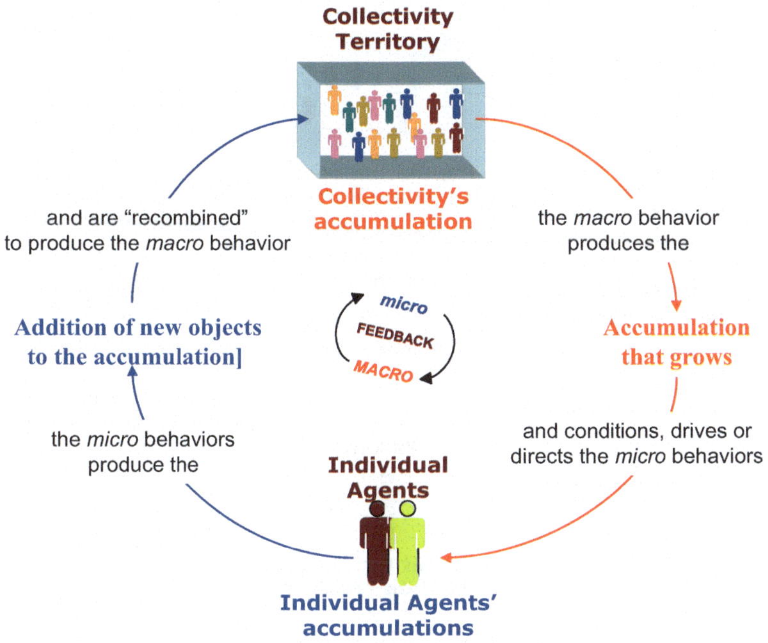

Fig. 1.3 Accumulation feedback

This rule forces us to observe collective phenomena from a "height" sufficient enough to allow us to observe unitary behavior, which is a macro behavior, when a close-up observation would lead us to think of isolated and distinct behaviors, which are micro behaviors.

The *micro-macro feedback* giving rise to the phenomenon represented by the accumulation of objects is presented in Fig. 1.3. Figure 1.4 illustrates the *micro-macro feedback* that maintains the mother tongue in a collectivity (this combinatory system will be examined in detail in Sect. 2.2.3).

1.1.5 The Invisible Hand Producing Self-Organization and Path Dependence

The macro behaviors of the collectivity can produce many important phenomena or effects, four of which are: the accumulation of objects, the spread of features or information, the pursuit or exceeding of a limit, and the attainment and maintenance of an order among the micro behaviors. A very relevant fifth effect, which includes the others, is the interdependent dynamics of individual *improvement* and collective *progress* in the overall state of a collectivity (defined in opportune ways).

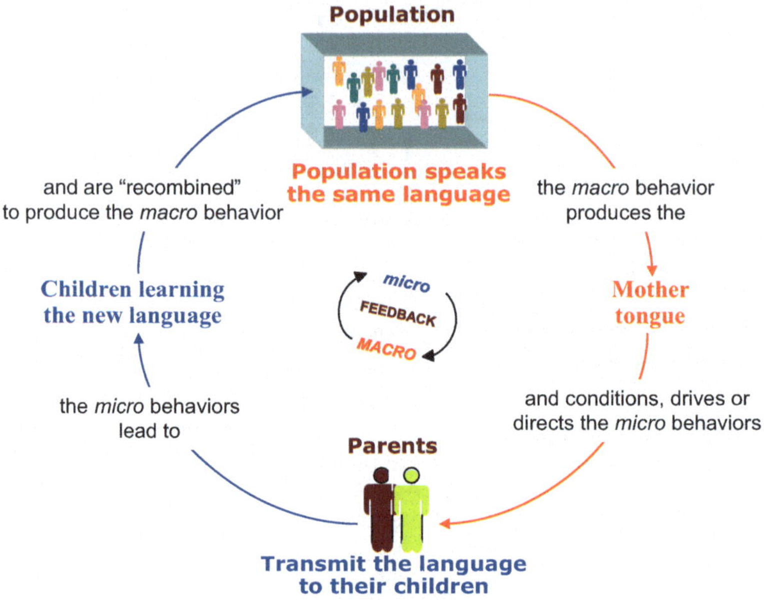

Fig. 1.4 Feedback from which a mother tongue is maintained

The internal feedback between *micro* and *macro* behavior plays a crucial role because it guarantees the maintenance over time of the system's *macro* behavior and directs the *micro* behaviors. When the system starts up—"by chance" or by "external decision"—it then normally maintains its behavior over time, as if an "Invisible Hand," a Supreme Deity or Higher Authority, regulated its time path and produced the observable effects and patterns.

There is nothing strange or mysterious about this: the "invisible hand" is nothing other than the *micro-macro feedback* action. Adam Smith's "invisible hand" naturally comes to mind, which "mysteriously" appeared to provide equilibrium to the chaotic competitive markets, leading to the emergence of a single price which every actor had to accept. Adam Smith used the term "invisible hand" only once in his *The Wealth of Nations*. He argued that the dynamic trends in the economy (the macro effect of the collective behavior) are caused by the production organizations which, due to their constant self-interested effort to gain the most advantages for themselves (agent's micro behavior), behave as if directed by an "invisible hand" in order to reach increasingly higher standards:

> It is not from the benevolence of the butcher, the brewer, or the baker, that we can expect our dinner, but from their regard to their own interest. By directing that industry in such a manner as its produce may be of the greatest value, he intends only his own gain, and he is in this, as in many other cases, led by an invisible hand to promote an end which was no part of his intention. Nor is it always the worse for the society that it was not part of it. (Smith, 1776, pp. 13–14)

Hermann Haken, the founder of Synergetics, speaks of *order parameters*; this time the object is a complex system formed by a multiplicity of elements whose micro behavior cannot be individually observed but can give rise to "macro phenomena" that is observable and describable.

> To describe collective behavior we need entirely new concepts compared to the microscopic description. ... In more complicated systems quite other "modes" may appropriately describe spatio-temporal patterns or functionings. ... It will turn out that equations governing self-organization are intrinsically nonlinear. From those equations we shall find in the following that, "modes" may either compete, so that only one "survives", or coexist by stabilizing each other. Apparently the mode concept has an enormous advantage over the microscopic description. Instead of the need to know all "atomic" coordinates of very many degrees of freedom we need to know only a single or very few parameters, e.g., the mode amplitude. As we will see later, the mode amplitudes determine the kind and degree of order. We will thus call them order parameters and establish a connection with the idea of order parameters in phase transition theory. (Haken, 1983, p. 14)

The *micro-macro feedback* may be thought of as an internal dynamic director or, better yet, as an internal dynamic organizer *which produces and uses global information* (Sect. 1.3.3) as an order parameter and, following the slaving principle, directs or organizes the individual behavior and produces the self-organization of the agents, and hence the collective phenomena (Haken, 1983; Kauffman, 1993; Martelli, 1999; Prigogine, 1985; von Foerster, 1960).

If we accept the traditional definition of *self-organization* as a macro behavior in which the micro behavior appears to be directed, or organized, in order to produce some *emerging phenomenon* represented by the formation of ordered structures, of recognizable patterns (Aziz-Alaoui & Bertelle, 2007) then all the collective phenomena mentioned in Sect. 1.1.1 can be defined as the result of self-organization.

> Self-organization is a process in which pattern emerges at the global (collective) level by means of interactions among components of the system at the individual level. What makes a system self-organized is that the collective patterns and structures arise without: (i) the guidance of well-informed leaders; (ii) any set of predetermined blueprints; (iii) recipes of templates that explicitly specify the pattern. Instead, structure is as an emergent property of the dynamic interactions among components in the system. (Aziz-Alaoui & Bertelle, 2007, p. 113)

> Self-organization may be defined as a spontaneous (i.e. not steered or directed by an external system) process of organization, i.e. of the development of an organized structure. The spontaneous creation of an "organized whole" out of a "disordered" collection of interacting parts as witnessed in self-organizing systems in physics, chemistry, biology, sociology ..., is a basic part of dynamical emergence. (Heylighen, 1989, p. 1)

> Self-organization is basically a process of evolution where the effect of the environment is minimal, i.e. where the development of new, complex structures takes place primarily in and through the system itself. (Heylighen, 2009, online)

> Self-organization is a process in which pattern at the global level of a system emerges solely from numerous interactions among the lower-level components of the system. Moreover, the rules specifying interactions among the system's components are executed using only local information, without reference to the global pattern. In short, the pattern is an emergent property of the system, rather than a property imposed on the system by an external ordering influence. (Camazine et al., 2001, p. 9)

Figure 1.5 shows instead the feedback loop that gives rise to the formation of a path where many individuals (collectivity imagined as a succession of passers-by over a period of time) crossing a field (macro behavior) form a permanent record of trampled grass. Each subsequent passer-by (agent) crosses (micro behavior) where he observes the track of already-trampled grass, treading again (micro effect) on this track and making it even more noticeable and permanent (macro effect) (Sect. 2.5.4).

The cancan dancers spin around individually in a coordinated manner, forming a line that rotates as one. Figure 1.6 shows how this phenomenon derives from the *micro-macro feedback* between the rotation of the individual dancer and that of the line of dancers (Sect. 2.5.1).

The Combinatory Systems Theory also considers *reversible systems*; that is, systems whose elements may again show a state that occurred in the past, so that they may present a cyclical behavior and, under certain conditions concerning the probability function regarding the transition of state of the elements, a chaotic one as well (Gleick, 1987; Kellert, 1994). Examples of reversible systems are those of diffusion and dissemination (fashion and contagion), whose elements may present a state chosen in a "repertoire" (Lustick, 2000).

The combinatory systems which are most interesting and easiest to represent are the irreversible ones, where both the micro and macro behavior produce permanent effects (residential or industrial settlements, the maintenance of the language, the spread of epidemics) as irreversible systems explain almost all the cases of "path dependence."

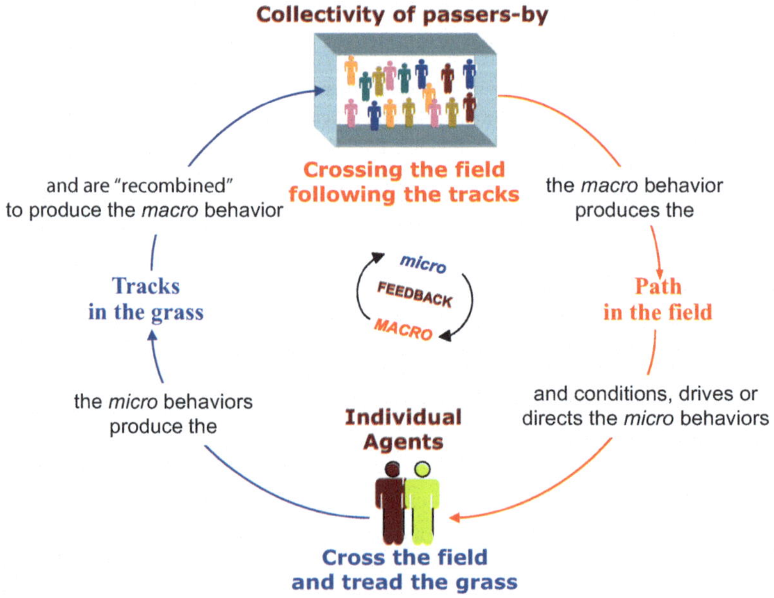

Fig. 1.5 Feedback from which a path derives

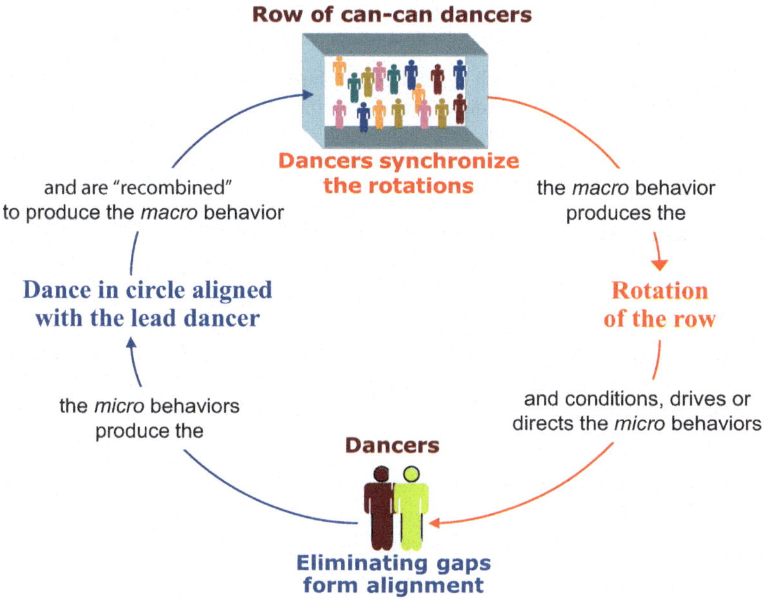

Fig. 1.6 Feedback that gives rise to a rotational motion

There are various definitions of path dependence. A comprehensive definition concerns the dependence of the path of a system on its initial conditions.

> Sensitivity to initial conditions: when a small change in the initial conditions produces major and unpredictable qualitative changes. Traditional approaches implicitly assume that events occur at an average rate (there are exceptions, and Robust Planning for example does not make that assumption) and that they can be adjusted if they deviate from the desired plan by employing the appropriate adjustment mechanism. But events do not unfold with average regularity and adjustments rarely produce the desired effect. No planning mechanism can take all initial and influencing conditions into account, and at times a small change in the initial conditions produces major and unpredictable qualitative changes. This coupled with positive feedback or increasing returns (Arthur, 1988, 1990, 1994) makes accurate forecasting and the planning of specific outcomes extremely difficult (LSE, online).

In their analysis, Liebowitz and Margolis showed three types of path dependence with different implications (Liebowitz & Margolis, 1995, 1998).

> There are three possible efficiency outcomes where a dynamic process exhibits sensitive dependence on initial conditions. First, this sensitivity might do no harm. That is to say, initial actions, perhaps insignificant ones, do put us on a path that cannot be left without some cost, but that path happens to be optimal (although not necessarily uniquely optimal). … We will call instances in which sensitivity to starting points exists, but with no implied inefficiency first-degree path dependence.
>
> Where information is imperfect, a second possibility arises. Where this is the case, it is possible that efficient decisions may not always appear to be efficient in retrospect. … In such a situation, which we will call second-degree path dependence, sensitive dependence on

initial conditions leads to outcomes that are regrettable and costly to change. They are not, however, inefficient in any meaningful sense, given the assumed limitations on knowledge.

Related to this second type of path dependence is third-degree path dependence. In third-degree path dependence, sensitive dependence on initial conditions leads to an outcome that is inefficient—but in this case the outcome is also "remediable." That is, there exists or existed some feasible arrangement for recognizing and achieving a preferred outcome, but that outcome is not obtained. (Liebowitz & Margolis, 1995, p. 2)

The conclusion is:

The three types of path dependence make progressively stronger claims. First-degree path dependence is a simple assertion of an intertemporal relationship, with no implied error of prediction or claim of inefficiency. Second-degree path dependence stipulates that intertemporal effects together with imperfect prediction result in actions that are regrettable, though not inefficient. Third-degree path dependence requires not only that the intertemporal effects propagate error, but also that the error was avoidable. (Liebowitz & Margolis, 1995, p. 2)

In regard to Combinatory System Theory, path dependence takes on a new, twofold significance. From a macro point of view, recognizing the phenomenon of path dependence is not a theory but simply the observation that the macro dynamic of a combinatory system—its macro behavior or its macro effect—can be thought to depend on initial *chance* (dependence from initial conditions) and on the *recombining rules* of the micro behaviors of the agents. From a micro point of view, the individual choices of the agents at any moment depend on the historical trends, since they derive from past history: that is, from the macro behavior (historical dependence).

In this sense the *path dependence* is the proof of the action of the *micro-macro feedback*, even if path dependence theory does not include this mechanism in the explanation of the path dependence. The deterministic action of path dependence, the *necessity*, is not a consequence of the past evolution of the path of the system, but of the *micro-macro feedback*, and then of the necessitating and recombining factors. Ignoring *micro-macro feedback* leads, moreover, to a second consequence: the path dependence theory focuses particularly on the micro behavior, considering the macro behavior as a constraint to the individual freedom to decide.

In order to understand the path dependence in combinatory systems, we must remember that the macro behavior of many combinatory systems is characterized by a particular "direction." When the macro behavior can develop in a particular *direction* or in opposite *directions*, toward one general state rather than another (see, e.g., the "waltz-spin" systems in Sect. 2.5.1 or the "stadium-wave" system in Sect. 2.5.3 that can have a right-directional or left-directional dynamics), then this behavior can be determined by *random fluctuations* in the initial micro behavior from one to the other state. "Chance"—that is to say, the uncontrollable and unknowable elements (with respect to the outside observer) which represent the base and the environment of the system—will not only set under way the macro behavior but also determine the *direction*, that is the direction of the "winning" fluctuation. Once the macro direction of the collectivity's macro behavior is stabilized, the micro behaviors are guided toward that direction, and the *micro-macro feedback*, thanks to the recombining factors, guarantees that the macro direction is maintained. Path dependence at the macro level is observed in the maintenance of the winning direction; at the micro level, it is this phenomenon which conditions the micro behaviors of the agents over time. When the direction of the collective behavior is defined, the agents are no longer free to follow it. The problem of individual rationality and collective behavior will be dealt with in Sect. 4.3.2.

1.1.6 Behavior Due to "Chance" and "Necessity"

Section 1.1 above presented the three characteristics that are needed in order for a combinatory system to be observed or conceived of: (1) the existence of individual micro behavior, (2) the development of collective macro behavior, and (3) the existence of feedback between the micro and macro behavior (or between their micro and macro effects). Although they are fundamental, these features are not sufficient for understanding the formation of collective phenomena. We must recognize another essential element: the joint action of "chance" and "necessity."

Let us return to the example of the murmur phenomenon in a crowded room. Why does a murmur arise in a closed space? Because someone "by chance" begins to speak. We cannot know who among the crowd will speak first, but in order for the murmur to be produced it is indispensable that a certain minimum number of people (minimum density) begin "by chance" to speak. The initial murmur that is thus produced triggers the *micro-macro feedback*, which will then do the rest. The murmuring noise will increase and those present will speak in a louder voice, thereby increasing the murmur until a deafening noise makes it impossible for many to hear and be heard (at this point we have reached the maximum density of the speakers); some will stop talking and the murmuring will diminish (the feedback is still operating) until, with some quiet having returned, those present start again to speak.

Most collective phenomena begin when "by chance" a minimum density is reached (i.e., a certain activation level, specific to each phenomenon: at times the action of a single element is enough; at other times all the elements in the collectivity must act) and end when the maximum density is reached (i.e., a given saturation level). Let us consider again the combinatory system of the epidemic, as shown in the model in Fig. 1.7. It seems clear that if the first infected individual does not come into contact with anybody, then the combinatory system will not be activated; but if "by chance" contacts with other individuals are frequent, then the system has a high probability of becoming activated. When the number of infected individuals exceeds a minimum density the epidemic, thanks to the *micro-macro feedback*, occurs "by necessity."

Let us suppose that in a village in Libya with white houses an inhabitant decides by chance to paint his house blue. If the color is pleasing, another inhabitant will imitate the first one, then another and another. The initial state of the system formed by the painted houses is represented by N = 1 white houses. The state is modified by an impulse, which owes to chance, represented by the first inhabitant who paints his

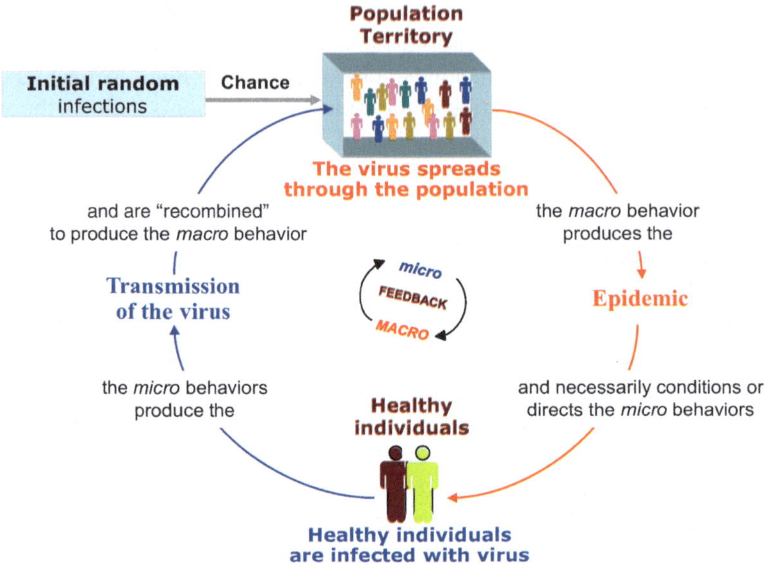

Fig. 1.7 Feedback that gives rise to the spread of influenza

house blue. An internal feedback begins: the higher the number of blue houses, the stronger the incentive for people to paint their white houses blue, and the faster the change in color of the houses. We can assume that the state of the system is quickly transformed from houses that are all white to houses that are all blue. The macro behavior thus produces a macro effect (gradual change of color) which, in turn, influences the micro behavior as part of an internal feedback.

Once the minimum density is reached, the *micro-macro feedback* guarantees that the macro behavior "by necessity" initiates and grows, feeding on the succeeding micro behavior while at the same time conditioning them. At times we observe an expansive dynamic in the macro behavior (or in its effects) and at other times a gradual weakening until it ceases. Sometimes there is even a cyclical fluctuation.

Since the activity of most combinatory systems is thus produced by the *joint action* of initial "chance" and "necessity," these systems can therefore also be called "chance-necessity systems" and, as such, are distinguished from "operative systems," which are usually "cause-effect systems."

Let us now consider the combinatory system that explains how graffiti comes to appear on a wall or a bench. It is absolutely clear that the initial graffiti is placed there "by chance" by someone who wishes to leave behind a trace of himself. If it is removed the system does not start up; if it is left there, and if "by chance" others leave behind new graffiti, then the system will produce the macro effect that enters into feedback with the subsequent micro behaviors. The system "by necessity" leads to the formation of a greater mass of graffiti.

A simple way to observe the inflow of the random fluctuations in orientating the *direction* of the *macro* dynamics of combinatory systems—even if it is not sufficient to describe the effect of *chance* on the overall dynamics of a combinatory system—is offered by a particular combinatory system represented by the so-called Polya Urns, a random elementary process consisting in the random extraction of balls from an urn and the doubling of the extracted balls which are reinserted in the urn. In its simplest form, let us consider an urn at time t = 0 that contains 50 balls of color "w" and 50 of color "b." The state of the system is S(0) = [50w, 50b]. For successive discrete moments we apply the following *micro rule*: extract a ball, check the color, and insert two balls of the same color. At moment t = 1, assuming a color "b" ball was extracted, the system moves to S(1) = [50w, 51b]. It is now clear that the probability of extracting a "b" ball is a bit greater than that of extracting a ball of a different color. To get the system to evolve, the first extraction is usually not sufficient; a certain number of other extractions are needed during which the number of white balls may increase, so that we have a fluctuation in the number of balls around the value of "w/b = 1." After a given number of extractions, a certain colored ball becomes prevalent, so that the fluctuations toward that color are "winning." Chance has decided how the system should evolve; the system inevitably has a state that tends to favor the balls of the emerging color from the prevalent fluctuation. This shows that the system normally tends to take on a stable state after a certain number of extractions. The combinatory system produces a change in the *macro effect*—mix of different colored balls—deriving from the *combination* of the single colored balls; the macro effect in turn influences the probabilities of subsequent draws, which in turn change the state of the system, according to a typical *micro-macro feedback*.

A variant of this model is the Ehrenfest Urns in which the overall number of balls remains constant but, upon extracting a ball of a certain color, say "w," a ball of another color is eliminated. Also in this case, each random shift in the number of balls influences the probabilities of the succeeding extractions. The models offered by the Polya and Ehrenfest Urns, though similar, are not adequate to describe the dynamics of combinatory systems, since they only reveal the effect of *chance* but do not consider the *necessitating factor* that guides the individual choices of agents.

Note. I have only non-formally touched on Polya and Ehrenfest's two Urn Models, even if these models underlie a broad statistical treatment (Johnson & Kotz, 1977).

1.1.7 Necessitating and Recombining Factors

This brings up a new and very important aspect: which mechanisms push the elements of the system to initiate their micro behavior and how can these be influenced by the macro behavior. In other words, we must investigate the logic that allows the *micro-macro feedback* to act after the initial "random impulse." In effect, "chance" by itself is never enough to maintain the macro behavior, only to set it off; some *necessitating factor* must operate.

Necessitating factors are all the factors that *force* the agents to *adapt* their micro behavior to the system's *global information* (macro behavior or effect). They may be a constraint, a rule, a condition, a law, a conviction, an imitative act, a biological or social impulse, and so on. The stronger the necessitating factor is, the more powerful the micro behavior will be. In combinatory systems formed by people, the *necessitating factor* often originates from a *conscious* motivation: necessity, convenience, opportunity, the desire not to be inferior, and so on, and depends on the culture, education, distinctive mental and emotional conditions, and so on, of the individual agents. At other times this factor can also be *natural* and act in an *unconscious* way, in that it is derived from the genetic or operative program of the individuals making up the base of the combinatory system.

The existence of one or more *necessitating factors* is indispensable but not sufficient. It is also necessary for the system to be able to recombine the micro behaviors (or the micro effects) in order to produce the macro behavior (or the macro effect) dynamics; some kind of *recombining factors* must operate on the system so that, through the *micro-macro feedback*, the *necessitating factors* can act. *Recombining factors* are the "elements" or causes that *allow* or force the system to *notice* and *recombine* the micro behavior (or the *micro* effects) in order to produce and maintain the *macro* behavior (or the *macro* effect). They may derive from a rule, a convention, an algorithm, and so on, and may also simply follow on from the conditions of the environment, or result from the social condition or the culture of the collectivity constituting the system. Considering also the necessitating and recombining factors, the model in Fig. 1.1 has been modified to look like that in Fig. 1.8.

Several *necessitating factors* are strong at times and can be included in the vital operating program of the living agents. Think of the power of the survival instinct that, in the presence of a predatory fish, creates in a school of small fish the need to hide within the group. This combinatory system can be easily observed in every sea, but above all in those rich in small-sized fish. Since they can be food for larger fish, the smaller fish, when there is danger, gather together in large schools. If, by chance, some fish in flight come together and form an initial grouping, the instinct to hide (necessity) attracts other fish, and very soon we can observe the formation of a compact mass; this continues as long as there is still the threat from the predatory fish, which represents the recombining factor. If the danger subsides, the mass of fish soon disappears (Fig. 1.9).

The school is formed by the micro behavior of the individual fish that, reaching the boundary of the school, with a dart and following their instinct (innate operative program), slip into the school once again in order to hide from the fish predators, leaving on the outskirts other fish that will behave in the same way. The school thus appears to be an autonomous, pulsating entity that offers fascinating spectacles for underwater photography lovers. Nevertheless, it is a true combinatory system, equipped with an autonomous movement, composed of micro behaviors but able to condition these, since each micro behavior is motivated by a voluntary necessitating factor: to remain hidden in the school in order not to be attacked by predators. The presence of a predator fish represents the recombining factor, since the school

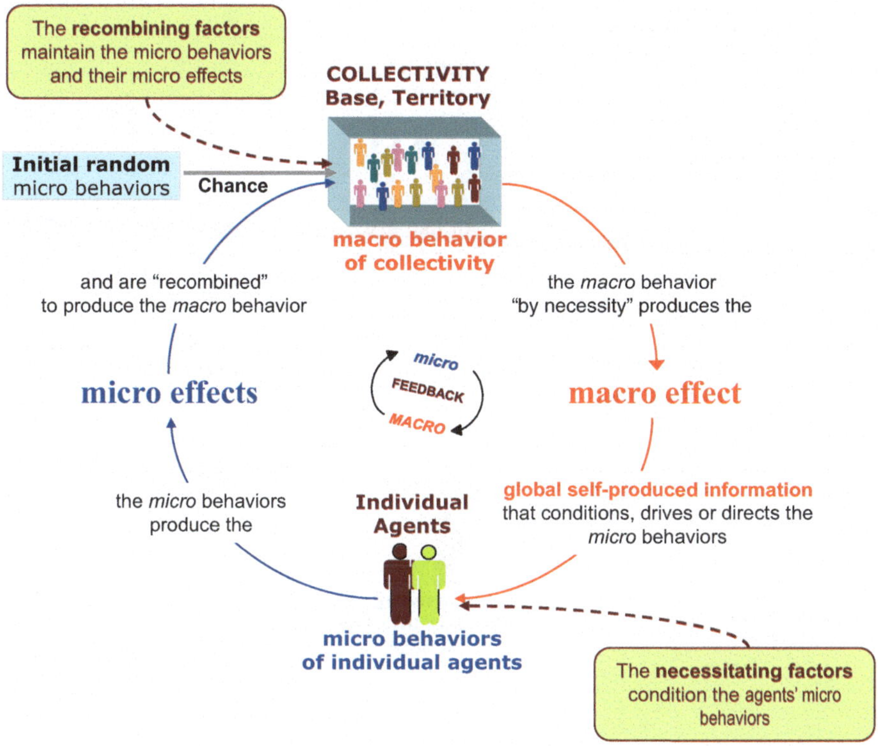

Fig. 1.8 The action of necessitating and recombining factors

remains compact in order to hide from view the fish inside it. It is an autonomous entity with its own macro behavior: it can navigate in a particular direction or remain in the same place; it can become oblong, reform into a sphere, or break up into two smaller schools, according to how the individual fish develop their micro behavior as they re-enter the school.

Why do the cancan dancers need to dance in a row? If the disapproval of the public toward a row that rotated in a disorderly way was not enough, the director or even the theater impresario would think of removing the dancer that could not learn to stay in sync. And what about the need to carve graffiti into a bench? Here the necessitating factor is weaker, but who among us, seated in good company one evening in the park and seeing a mass of graffiti, would not be tempted to loosen the reins imposed by education and leave our mark behind?

The one-hundred-towers-of-Pavia system started up when, after the construction of the first tower, other families by chance, and for reasons of prestige, felt the need or convenience to imitate the initial behavior. Consider, on the other hand, the much higher degree of need a family has in passing on its language (conscious need) and its dialectal inflection (unconscious need) to its children.

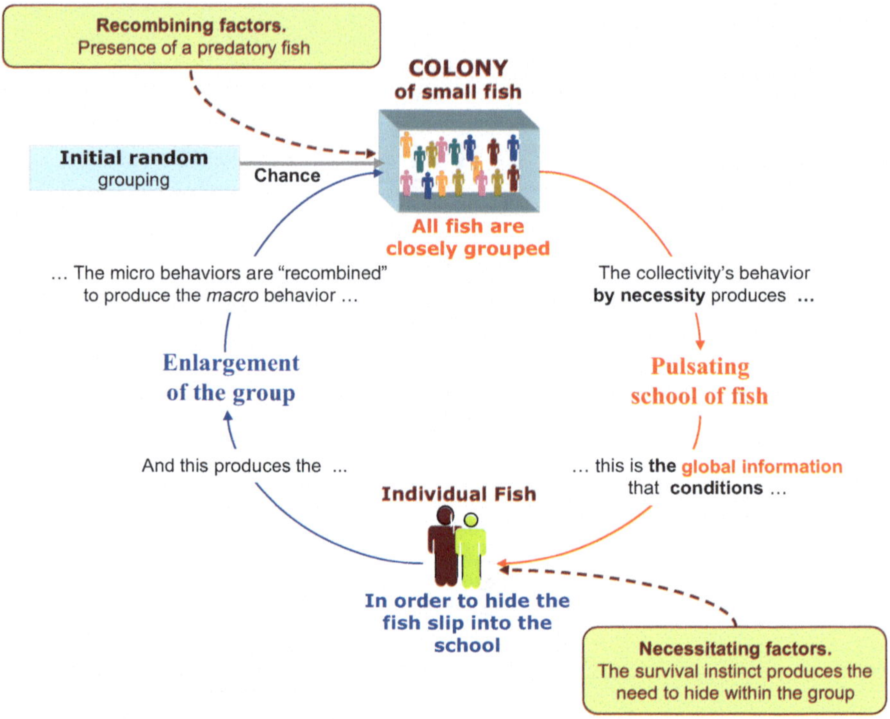

Fig. 1.9 *Necessitating* and *recombining* factors in the formation of a school of fish

As mentioned above, the necessitating factor can even be natural and act in an *unconscious* way. In the system characterized by the spread of an epidemic, the necessitating action does not depend on the desire of the carrier of the pathogenic agent to infect other individuals, as much as on the operative aggression mechanism of the virus. Even the working of a combinatory system that gives rise to the spread of an animal or vegetal species depends on a very strong necessitating factor (reproductive instinct) of which individuals are probably not aware. Similarly, the system that leads to the formation of grooves in highway roadbeds depends on the need felt by the individual truck drivers not to run the risk of skidding by driving outside the lanes, certainly not on the convenience of strengthening the wheel ruts.

A particular *involuntary* combinatory system that derives from an *involuntary necessitating factor* is the "information contagion," which causes individual behavior based on preferences which are formed as the result of "limited information" obtained from subsystems of individuals from a population. At the beginning of the process the information is equally distributed throughout the population; however, if some elements in the system choose a given alternative, the information can be distributed in an unsymmetrical way. There may be some initial fluctuations in the preferences, and thus in the information, which causes the system to randomly assume a given state following a fluctuation which is a winning one with respect to the others. The system is nevertheless combinatory, since the micro behavior depends on the decisions of the agents in the system, but these in turn influence the decisions of other agents (Sect. 1.3.5).

If *necessitating* factors are fundamental for the subsistence of any combinatory system, the *recombining* factors also play a fundamental role in its creation. If the collectivity in some way is not able to "join together" their individual micro behavior (or their effects), then the *micro-macro feedback* cannot occur.

Consider how the recombining effect from the passing of trucks to form the grooves in highway roadbeds is influenced by whether or not the stretch of road is straight or winding; how, in a large environment, it is much simpler to scatter one's garbage than to accumulate it; or how it is easier for an epidemic to break out in an area with a low population density. Even prevention by means of widespread vaccinations and the possibility of a rapid identification of the infection, together with the abundance of appropriate medicines, makes the recombining effect of such a system very weak.

Here is a subtle but necessary point to consider: *in order to explain the activity of combinatory systems, the theory behind such systems must recognize and understand the nature of both the recombining factors as well as the necessitating ones, since without the joint action of these factors there would be no micro-macro feedback and the collective phenomena we wish to explain would not occur.*

Note. All the combinatory systems presented as examples in the preceding sections are analyzed and formalized in Chap. 2.

1.2 The Study of Collectivities: A Literature Review

Collectivities have always been a very complex subject of study, and for this reason both fascinating and interesting. Conway's discovery of the fantastic world of *Life* (Gardner, 1970), Schelling's (1971) model of neighborhood segregation, and Sakoda's model of group formation (Sakoda, 1971) are fundamental milestones in the study and simulation of the behavior of collectivities. The following, like Schelling's (2006) famous *Micromotives and Macrobehaviour*, is an attempt to offer a logical explanation for the *collective macro behavior* shown by intelligent agents which, acting on behalf of their own interest, produce emergent collective dynamics. Although it is rather difficult to provide a list of available theories, models, and instruments that are useful for exploring this research field, we can nevertheless recognize two main approaches:

(A) The *macro* or *synthetic approaches*, which aim to build models of systems capable above all of justifying the *macro behavior*. The *micro behavior* is considered unobservable or unimportant because the relations that link the elements are too complex and numerous. Since the abundance of connections makes the construction of meaningful models based on elements very difficult, the macro approaches produce a macro description of the behavior of collectivities, which includes the following typologies:

A. The cybernetic approach
B. The system dynamics and systems thinking approach
C. The synergetic approach
D. The "allelomimesis" and "stigmergie" approaches
E. The autopoietic approach
F. The holonic approach

(B) The *micro* or *analytic approaches*, whose models are built exclusively (or prevalently) by studying the *micro behavior* and the *micro rules* which connect them (Gilbert, 1995, 1999; Gilbert & Doran, 1994). The macro behavior is a consequence—often unexpected—of the action of these connections. Forming part of this typology are the models worked out by the *cellular automata theory*, which allow us to explore the systems by simulating Artificial Life. We can consider, in particular:

A. The *complex systems* approach
B. The *complex adaptive systems* approach
C. The *cellular automata* approach
D. The *Alife* approach
E. The *Ants* approach
F. The *Swarm* approach
G. The *Floys* approach
H. The *Recursive* approaches
I. The *Genetic Algorithms* approach

It is clear that progress in the computer simulation of the behavior and evolution mechanisms and the first technological outcomes (Artificial Life environments, robots, intelligent toys, self-reproducing machines, agents on the web) are creating the basis for a new age: the coming of *artificial beings* and artificial societies (for a general survey: http://www.generativeart.com/).

> Our ability to apply a range of algorithmic tools in building evolutionary models has expanded enormously under the stimulus provided by large expansions in computing power. Cellular automata, neural nets and genetic algorithms are all devices that can be used to simulate the behaviour of agents in response to environmentally conditioned interactions with other agents (Foster & Metcalfe, 2001a, p. 3; see also Axelrod, 1997; Drogoul et al., 1994; Drogoul & Ferber, 1994).

Note. The literature on the topics I will discuss is vast. Generally, I will give priority to the authors that first presented and developed such themes, rather than citing authors that dealt with them more recently.

1.2.1 The Macro Approaches: A Short Survey

A.—Wiener's Cybernetic approach (Wiener, 1961; Ashby, 1957; von Foerster, 1960; Kauffman, 1993; Sandquist, 1985), in particular evolutionary cybernetics (Campbell, 1960; Sean Gould, 2000; Stephen Jay Gould, 2002), aims at developing a theory to explain the process of arranging components based on a particular criterion in order to form a pattern different from and improving on what could occur by chance. This approach attempts to provide generalizations about how cells or organisms interact, or how and why people work together to form larger unities. Evolutionary cybernetics develops a complete philosophy or "world-view" based on fundamental Darwinian principles:

> Blind Variation and Selective Retention' (BVSR) is a phrase introduced by Campbell (1960), as a way of describing the most fundamental principle underlying Darwinian evolution. (Campbell only applied it to the evolution of knowledge, but we here apply it in the most general context). The BVSR formula can be understood as a summary of three independent principles: blind variation, asymmetric transitions, and selective retention. (Heylighen, 1993, online)

We can include in the *cybernetic approach* the models of *population dynamics*, which try to represent population dynamics (increase, evolution, coevolution, and competition) in terms of the dynamics of the number of their elements, as shown in Malthusian models and in Volterra (1926, 1931) and Lotka's (1925) well-known equations, in various forms (Sect. 2.6.7). These equations show how two prey and predator populations can have a dynamic equilibrium that takes various forms in relation to the variation rates for the populations, which are a function of the number of individuals in each population (Mella, 2021; Ardeni & Gallegati, 1999).

B.—Forrester's *System* (Industrial) *Dynamics approach*—which may be viewed as an extension of the cybernetic one—and Senge's *Systems Thinking approach* (Senge, 1990; Mella, 2012) are methods and techniques that can be applied to understand how the behavior of concrete collectivities, arise and change over time. *System Dynamics* was developed initially from the work of Forrester (1961) on Industrial Dynamics. System Dynamics is connected to Systems Thinking, which looks at exactly the same kind of systems from the same perspective. It constructs the same causal loop diagrams but rarely takes the additional steps of constructing and testing a computer simulation model and of testing alternative policies in the model. Internal feedback loops within the structure of the system influence the entire behavior of the system; however, these loops cannot be interpreted as the *micro-macro feedback* that operates in combinatory systems.

> "System dynamics" is a professional field that deals with the complexity of systems. System dynamics is the necessary foundation underlying effective thinking about systems. System dynamics deals with how things change through time, which covers most of what most people find important. System dynamics involves interpreting real life systems into computer simulation models that allow one to see how the structure and decision-making policies in a system create its behavior. (Forrester, 1999, p. 1)

> Systems thinking is a discipline for seeing wholes, recognizing patterns and interrelationships, and learning how to structure those interrelationships in more effective, efficient ways. (Senge & Lannon-Kim, 1991, p. 24)

C.—Haken's *Synergetics approach* is the theory of *cooperative behavior* in active systems. *"The word synergetics is composed of Greek words meaning 'working together'."* (Haken, 1982, p. 2). Following Haken, Synergetics is an interdisciplinary field of research. It deals with open systems that are composed of many individual parts or units that interact with each other and that can form spatial, temporal, or functional structures by self-organization. The research goal of Synergetics is threefold: (1) Are there general principles of self-organization? (2) Are there analogies in the behavior of self-organizing systems? (3) Can new devices be constructed because of the results in (1) and (2)? From a mathematical point of view, Synergetics deals with nonlinear partial stochastic differential equations and studies their solutions close to those points where the solutions change their behavior qualitatively.

The *synergetics approach* provides an exogenous description of collectivities of individual parts without entering into internal operative mechanisms and without examining the micro and macro rules from which the behavior originates. This approach instead highlights the *macro* results of the *cooperative behavior* of the "unobservable" elements of collectivities (Haken, 1977b, 1982). In other words, in the *synergetics* approach *systems* are composed of a number of elements that is so high it can only be analyzed and described in terms of a limited number of parameters defined as order parameters (Serra & Zanarini, 1990; Corning, 1995). As we can see, in the *Synergetics approach* the concept of *micro-macro feedback* is never explicitly considered.

Note. I want to specify that I am speaking of Haken's *synergetics,* to avoid confusion with the language used by R. Buckminster Fuller (1977) and E. Applewhite to construct a metaphysical description of the world. What distinguishes *Fuller's synergetics* from more traditional metaphysical discussions of the temporal versus the eternal, and from most contemporary philosophy, is Fuller's consistent use of geometric concepts to express such ancient dichotomies. Fuller's hope for the future lies in doing more with less. Again and again he discovers that there is no such thing as continuity and the infinitesimal: with sufficient resolution we find that we look at a very large, but finite number of very small, but finite multiples, put together with very great, but attainable ingenuity. There always appears to be a structure (Loeb, 1977).

D.—The "allelomimesis" and "stigmergie" approaches are important for the study of systems made up of animals and social animals, such as bees, ants, and termites, which behave like agents of combinatory systems.

> The most widely-observed social interaction concerns allelomimesis in its many forms (roughly speaking, do what my neighbour is doing; see e.g. Sudd 1963, Scott 1972, Altmann 1985). For example, one bird takes off, those near it also take off, and very quickly the whole flock has taken off. Recruitment in social insects is another classical example, in which one forager discovers an important food source, recruits inactive foragers in the nest to go to it, which in turn recruit still more foragers. Allelomimesis is by definition autocatalytic, in that if I do as others, then others do as I, and we all end up doing the same thing. Another term for this is positive feedback, and we shall use the three terms rather indiscriminantly. ... they all refer to the idea that the probability of an individual adopting a particular behaviour or state is an increasing function of the number of individuals already exhibiting that behaviour or state. (Deneubourg & Goss, 1989, p. 296)

These approaches attempt to interpret *systems of diffusion* as well as of *order*. Specifically, Grassé's "stigmergie" approach interprets the coordination we observe in social animals, termites, and ants in particular, as the result of their own micro behaviors, which lead to the release of pheromones as micro effects. The combination of the micro effects, understood as the sum (accumulation) of the amount of pheromones released (after subtracting the part that has vanished over time) represents the self-produced global information that guides the insects' subsequent micro behaviors, which together produce the relevant macro effects.

> Nest building in termites is the typical example of stigmergy, and is also the original example used by Grassé to introduce the concept. Termite workers use soil pellets, which they impregnate with pheromone (i.e., a diffusing chemical substance) to build pillars. Two successive phases take place during nest reconstruction. First, a non-coordinated phase occurs which is characterized by a random deposition of pellets. This phase lasts until one of the deposits reaches a critical size (…). Then, a coordination phase starts if the group of builders is sufficiently large and pillars emerge. The existence of an initial deposit of soil pellets stimulates workers to accumulate more material through a positive feedback mechanism, since the accumulation of material reinforces the attractivity of deposits through the diffusing pheromone emitted by the pellets. (Dorigo et al., 2000, p. 852)

The combinatory system that describes the behavior produced by stigmergy will be analyzed in Sect. 2.5.5.

E.—Humberto Maturana's and Francisco Varela's *Autopoiesis approach* (Maturana & Varela, 1980, 1987; Varela, 1979, 1981; Maturana & Guiloff, 1980; Zeleny, 1981; Jantsch, 1980) offers powerful conceptual frameworks and practical tools for building models of the behavior of collectivities considered as organized *self-maintaining* systems, based on Maturana and Varela's interpretation (1980), which is also clearly described by Stafford Beer in the Preface to their work.

> The relations between components that define a composite unity (system) as a composite unity of a particular kind, constitute its organization. In this definition of organization the components are viewed only in relation to their participation in the constitution of the unity (whole) that they integrate. This is why nothing is said in it about the properties that the components of a particular unity may have other than those required by the realization of the organization of the unity (Beer, 1987, p. XIX).

> A unit realized through a closed organization of production processes such that (a) the same organization of processes is generated through the interaction of their own products (components), and (b) a topological boundary emerges as a result of the same constitutive processes. (Zeleny, 1981, p. 6)

The *autopoietic approach* to collectivities conceived of as closed organizations is based on two fundamental concepts: the idea of *structural* and *behavioral coupling* and the idea of *operational closure* of the system. As the behavior of each agent may be thought of as depending on the behavior of other agents, we can argue that

> the autopoietic conduct of an organism A becomes a source of deformation for an organism B, and the compensatory behaviour of organism B acts, in turn, as a source of deformation for organism A, whose compensatory behavior acts again as a source of deformation for B, and so on recursively until the coupling is interrupted. (Maturana & Varela, 1980, p. 120)

> In general, when two or more plastic dynamic systems interact recursively under conditions in which their identities are maintained, the process of structural coupling takes place as a process of reciprocal selection of congruent paths of structural changes in the interacting

systems which result in the continuous selection in them of congruent dynamics of state. (Maturana & Guiloff, 1980, p. 139)

Therefore, during the course of behavioral coupling each participating autopoietic agent is, with respect to the other(s), a source (and a target) of compensatory perturbations for each other. These are 'compensatory' in the sense that (a) there is a range of 'compensation' bounded by the limit beyond which each system ceases to be a functional whole, and (b) each iteration of the reciprocal interaction is affected by the one(s) before. (Whitaker, 2003: Enolagaia, online)

F.—A different approach to agent-based systems is Koestler's holonic systems approach (Koestler, 1967, 1978; Shimizu, 1987; Wilber, 2000, 2001), which is particularly useful for studying the behavior of collectivities of living organisms and social organizations, as these are composed of self-reliant units, called holons, that are capable of flexible behavior and which form holarchies, defined as a hierarchically organized structure of holons. Koestler (1967) viewed the holon as a Janus-faced entity: if it observes its own interior it considers itself a whole formed by (containing) subordinate parts; if it observes its exterior it considers itself a part or element of (contained in) a vaster whole. In any event, it sees itself as a self-reliant and unique entity that tries both to survive (it is a viable system) and to integrate with other holons:

These sub-wholes—or 'holons', as I have proposed to call them—are Janus-faced entities which display both the independent properties of wholes and the dependent properties of parts. Each holon must preserve and assert its autonomy, otherwise the organism would lose its articulation and dissolve into an amorphous mass—but at the same time the holon must remain subordinate to the demands of the (existing or evolving) whole. (Koestler, 1972, pp. 111–112)

Each holon includes those from lower levels, but it cannot be reduced to these; it transcends them at the same time that it includes them, and it has emerging properties (Edwards, 2003). Therefore, a holarchy is a whole that is also a structure of parts that are in themselves wholes; but it is not a collectivity (for a more detailed description of the holonic view, see Mella, 2009).

1.2.2 The Micro Approaches: A Short Survey

The *micro* or *analytic approaches* study and simulate the behavior of collectivities by observing the micro behavior of the component agents in order to develop different kinds of *agent-based models*. The common fundamental idea of the *micro approaches* is that the macro behavior of collective systems is determined by the interaction of the micro behavior of the agents which form the system; these agents act in an isolated manner or in teams, and for this reason the macro behavior is *complex*. The only reasonable approach to complexity in such systems is analytic: to *recognize* or to *define* the micro rules which produce or direct the micro behavior. In other words: not to describe a complex system with complex equations, but to let the complexity emerge from the interaction of simple individuals following simple rules (Waldrop, 1993).

> Many natural systems (e.g., brains, immune systems, ecologies, societies) and increasingly, many artificial systems (parallel and distributed computing systems, artificial intelligence systems, artificial neural networks, evolutionary programs) are characterized by apparently complex behaviours that emerge as a result of often nonlinear spatio-temporal interactions among a large number of component systems at different levels of organization. Consequently, researchers in a number of disparate areas including computer science, artificial intelligence, neural networks, cognitive science, computational economics, mathematics, optimization, complexity theory, control systems, biology, neuroscience, psychology, engineering, etc., have begun to address, through a combination of basic as well as applied, theoretical as well as experimental research, the analysis and synthesis of such systems. (Honavar, 2006)

The analytic approach is based on three main ideas:

> I am fascinated by these three ideas about Evolution, Artificial Life, and Computers. The idea that complex, sophisticated, adaptive solutions can be generated by automatic, blind, knowledge-lacking mechanisms (Evolution). The idea that complex systems, such as life, are actually the emergent behaviors of systems with many elements that interact according to simple, local rules (Artificial Life). The idea that a personal desktop or notebook can be used as a laboratory tool for conducting scientific experiments from one's home (Computers). (Dolan, 1998, online)

The model-creator aims to accurately describe a system's components and plausible interactions, and to then use a realization of that description as an empirical basis for the study of the system's global dynamics. The emphasis is on finding appropriate abstractions for describing components and interactions rather than on finding abstractions that are useful for reasoning about global dynamics. This bottom-up approach is called agent-based modeling (ABM).

A.—The *complex systems approach* is a relatively new field of science studying the behavior of collectivities made up of heterogeneous elements that interact to form dynamic wholes, which are emerging and not predictable a priori and shall be defined as complex systems (EMK, Complexity Group, 2014).

To be more precise, a general definition is that complexity is the study of the behavior of macroscopic collections of units that are endowed with the potential to evolve over time. Their interactions lead to coherent collective phenomena, so-called *emergent properties* that can be described only at higher levels than those of the individual units (Coveney & Highfield, 1995).

Murray Gell-Mann (1995) traces the meaning to the root of the word *complexity*. *Plexus* is a Latin word that means braided or entwined, from which *complexus* is derived, meaning braided together; the English word "complex" is derived from the Latin. Complexity is therefore associated with the intricate intertwining or interconnectivity of elements within a system and between a system and its environment. For a survey of different meanings and distinctions of the complexity idea see, for example, Delorme (2001) (Foster & Metcalfe (2001b), who gives a formal distinction between complexity and complication.

> There is no single unified Theory of Complexity, but several theories arising from various natural sciences studying complex systems, such as biology, chemistry, computer simulation, evolution, mathematics, and physics. This includes the work undertaken over the past four decades by scientists associated with the Santa Fe Institute (SFI) in New Mexico, USA, and particularly ... by Murray Gell-Mann (1994) on complex adaptive systems (CAS) (Mitleton-Kelly, 2003, p. 24)

There are three interrelated approaches to the modern study of complex systems: (1) how interactions give rise to patterns of behavior; (2) understanding how to describe complex systems; and (3) the formation process of complex systems through *pattern* formation and evolution. A *pattern* is a property of the system as a whole but not a property of small parts of the system. It is a property of a system that allows its description to be shortened as compared to a list of the descriptions of its parts (NECSI online; Stacey, 1995).

B.—Included in the area of complex systems is the *complex adaptive systems* (CAS) approach (Kauffman, 1993, 1996; Castelfranchi, 1998; Gilbert, 1995; Gilbert & Troitzsch, 1999; Flake, 1998; Coveney & Highfield, 1995; Lewin, 1999; Mitleton-Kelly, 1997a, 1997b, 2003; Henderson (2000); Allen, 2001; Foster & Metcalfe (2001b); Axelrod, 1997; Goldspink, 2000) and the *complex evolving systems* approach.

> Coherent Systems are Systems whose structure does not remain rigid, fixed, and unchanged, but changes in coherent ways. Organisms of all different kind, and Ecosystems are coherent systems, in the same way as cities, communities, economies, financial markets and a host of other forms of social organization .... Human Beings, the entities within which they organize themselves, and other living systems, therefore, are Self-Organizing, Dynamic, Coherent and Evolving Systems. (Weis, 2008, p. 12)

> All biological Organisms, communities, and ecosystems, are dynamic, self-organizing, autopoietic, coherent, or complex, evolving systems—in the very same way as the complex anthropogenic systems so beloved in economics, sociology, or political science: Households, firms, governments, oligopolies, networks, markets, economic systems (regimes), and other niceties. (*ibidem*, pp. 18–19)

In particular, *complex evolving systems* study how complex systems interact and exchange information with their environment (Conte & Castelfranchi, 1994) to maintain their internal processes over time and to develop some form of "cognition," as described by Maturana and Varela (1980):

> A cognitive system is a system whose organization defines a domain of interactions in which it can act with relevance to the maintenance of itself. (Maturana & Varela, 1980, p. 13)

> The term Complex Adaptive System is used here to also refer to Complex Evolving Systems. The following extended definition will be adopted:

> Definition (1): A complex adaptive system (CAS) consists of non-homogeneous, interacting adaptive agents. Adaptive means capable of learning.

> Definition (2): An emergent property of a CAS is a property of the system as a whole which does not exist at the individual elements (agents) level. Typical examples are the brain, the immune system, the economy, social systems, the ecology, insect swarms, etc.

> Therefore, to understand a complex system one has to study the system as a whole and not to decompose it into its constituents. This totalistic approach is against the standard reduc-

tionist one, which tries to decompose any system to its constituents and hopes that by understanding the elements one can understand the whole system. (Ahmed et al., 2005, pp. 1–2)

Complex systems are usually systems which have been created by evolution or an evolutionary process. Evolved systems which have a long historical background are nearly always complex. Complexity can be found everywhere where evolution is at work:

- in all living organisms which are subject to evolution
- in all evolving complex adaptive systems which have a long history
- in systems that have grown over a long period of time (CASG, 2013, online)

According to Murray Gell-Mann (1992, 1994, 1995), the category of *complex adaptive systems* should also include all the basic components of this system as well as individuals who are capable of *surviving* by *adapting* their behavior and producing new *schema of interaction* and coexistence that allow this behavior to be predicted and adapted to (Stonier & Yu, 1994).

Now how does a complex adaptive system operate? How does it engage in passive learning about its environment, in prediction of the future impacts of the environment, and in prediction of how the environment will react to its behavior?

[...] The answer lies in the way the information about the environment is recorded. In complex adaptive systems, it is not merely listed in what computer scientists would call a look-up table. Instead, the regularities of the experience are encapsulated in highly compressed form as a model or theory or schema. Such a schema is usually approximate, sometimes wrong, but it may be adaptive if it can make useful predictions including interpolation and extrapolation and sometimes generalized to situations very different from those previously encountered. In the presence of new information from the environment, the compressed schema unfolds to give prediction or behavior or both. (Gell-Mann, 1992, p. 10)

In "social collectivities" (Sect. 1.5.3) in particular, mutual adaptation among the individuals-agents is not only due to adaptation, that is a reaction to the disturbances, but derives from *forecasts*, *prospects*, and *expectations* regarding the behavior of the other individuals-agents, as is well stated by John Holland (1975, 1992, 1995, 1998), the pioneer of CAS studies:

Here we confront directly the issues, and the questions, that distinguish CAS from other kinds of systems. One of the most obvious of these distinctions is the diversity of the agents that form CAS. Is this diversity the product of similar mechanisms in different CAS? Another distinction is more subtle, though equally pervasive and important. The interactions of agents in CAS are governed by anticipations engendered by learning and long-term adaptation. (Holland, 1995, p. 93)

In particular, in complex systems theory the feedback is considered between agents and not as a determining feature of the system. In effect, the changes in agents' states are often the result of internal infra agents micro feedback that the agents receive.

C.—The *cellular automata approach* may be considered the most general approach to simulating behavior in collectivities, especially those which act as CAS.

The underlying rules of the system are changing over time, which means that different agents behave according to different rules at different times. [...] Because of these difficul-

> ties, a class of models, variously called "artificial worlds", "particle-based", and "agent-based", have been a popular approach to studying CAS. This style of modelling is quite different from the differential equation style of models used most frequently to model non-linear dynamical systems. In agent-based models, each "actor" and each interaction among actors (that is, not just each type of interaction) is represented (simulated) explicitly. Individuals are capable of quite different kinds of behaviours (the agents in the system are heterogeneous). Agent-based models are discrete in most dimensions, typically time, state, and update rules. Thus, the standard approximations for infinite-sized systems and the techniques developed for studying asymptotic behaviour of continuous nonlinear dynamical systems often do not directly apply. As a result, these systems tend to be more difficult to analyse. (Forrest & Jones, 1994, p. 4)

The theory of cellular automata (CA) builds mathematical models of a system which consists of an array of cells (possibly in more than one dimension), and the theory can be traced back to John von Neumann in the late 1940s (edited by Von Neumann & Burks, 1966).

> A cellular automaton can be thought of as a stylised universe. Space is represented by a uniform grid, with each site or cell containing a few bits of data; time advances in discrete steps; and the laws of the universe are expressed by a single recipe—say, a small look-up table—through which at each step each cell computes its new state from that of its close neighbours. Thus, the system's laws are local and uniform. (Toffoli & Margolus, 1987, p. 6)

The basic element of a CA is the cell. A cell is a kind of a memory element and, simply put, stores states. In the simplest case, each cell can have the binary states 1 or 0. In more complex simulation the cells can have several different states (it is even imaginable that each cell has more than one property or attribute, and each of these properties or attributes can have two or more states) (Wolfram, 1984).

These cells are arranged in a spatial web—a lattice. The simplest one is the one-dimensional "lattice," meaning that all cells are arranged in a line like a string of pearls. The most common CAs are built in one or two dimensions. Up to now, these cells arranged in a lattice represent a static state. To introduce dynamics into the system it is necessary to add rules to define the state of the cells for the next time step. In cellular automata a rule defines the state of a cell as a function of the neighborhood of the cell (Schatten, 2004; Berto & Tagliabue, 2002).

It is important to note that a set of rules defines the transition from one state to another from one step in the time frame and that the rules that define the micro behavior of a cell are only local rules, in the sense that the state of the cell depends only on one of a specified number of neighbors and not on the state of the array (Wolfram, 1984, 2004; Dewdney, 1990; Ulam, 1986, 1991).

D.—The *Alife approach* may be considered a specific case of the cellular automata approach (Langton, 1989; Levy, 1992; Reynolds, 1987), because it refers to cells simulating simple living autonomous reactive agents which do not (in general) learn or "think" but simply react to messages received in a stereotyped way (blind behavior) (Liekens, 2000). This approach aims to show how interactions among neighboring agents, following local rules, lead at a higher level to complex patterns of self-organization.

> The term "Artificial Life" is used to describe research into man-made systems that possess some of the essential properties of life. Artificial Life is often described as an attempt to understand high-level behavior from low-level rules; for example, how the simple rules of Darwinian evolution lead to high-level structure, or the way in which the simple interactions between ants and their environment lead to complex trail-following behaviour. Understanding this relationship in particular systems provides novel solutions to complex real-world problems, such as disease prevention, stock-market prediction, and data-mining on the Internet. (ISAL, 2016, online)

The *Alife approach* presents many interesting variations, depending on the supposed nature of the agents. The most well known are: *Ants*, *Swarm*, and *Floys*. A few of these variations will be mentioned below.

E.—The *Ants* approach (Dorigo et al., 1999, 2000; Hölldobler & Wilson, 1990), which derives from Grassé's stigmergy approach (Sect. 1.2.1-D), considers ant colonies as collections of reactive agents in order to study the self-organization and sociogenesis in ant and in wasp colonies; that is, the process by which a single individual, a gravid queen, gives rise to an entire society of insects. Although each ant is characterized by limited capabilities (limited local movement, recognizing food or ants, marking territory with chemical traces, etc.) and acts blindly according to local rules, ant colonies can perform collective tasks which are far beyond the capacities of their constituent components.

> Ants occupy a central place in artificial life due to their relative individual simplicity combined with their relatively complex group behavior. Ant colonies have evolved means of performing collective tasks which are far beyond the capacities of their constituent components. They do so without being wired together in any specific architectural pattern, without central control, and in the presence of strong intrinsic noise. Ants can create architectural structures dynamically when and where they are needed, such as trails between nest and food sources, or "living bridges" when swarms of ants migrate in the rain-forest. (Cox, 2004, online)

F.—The *Swarm approach* shows a logic and patterns of behavior similar to those of ants but differs from them because the basic architecture of the swarm is the simulation of collections of concurrent agents. The swarm program and the swarm software were launched in 1994 by Burkhart (1994), with Chris Langton and Nelson Minar, at Santa Fe Institute in New Mexico (for more go to: SWARM online).

G.—The *Floys approach* is similar to the Ants and Swarm approaches; however, it considers flocking creatures characterized by collective flying or flocking and territorial instinct that acts following simple local rules.

Floys belong to the flocking Alife creatures variety, sharing with them the social tendency to stick together, and the lifelike emergent behaviour which is based on a few simple, local rules. (http://www.psicopolis.com/fisikepsic/descrifloys.htm)

> Dolan's floys follow just two rules: stick close to your fellows but not too close, and when you spot an intruder, move toward it and attack. These instructions also generate a flock. But the assembly benefits individuals in an unexpected way. Avoiding too much togetherness causes the floys to move in an ensemble that spreads out over a substantial region. They thus patrol airspace that extends far beyond the sensory range of any one floy. (Carlson, 2000, pp. 112–113)

H.—The *Recursive approaches* (mentioned only in passing here) consider many phenomena observed in populations (growth and diffusion) that give rise to unexpected patterns as the result of a recursive application of *simple local syntactical rules* (alphabet and syntax), often defined in a qualitative way. These approaches are often an application of the cellular automata approach. As an example, we can refer to L-systems (Lindenmayer & Rozenberg, 1972; Prusinkiewicz & Lindenmayer, 1990; Green & Green, 1993) that model growth processes which arise from the application of sets of rules and symbols, also known as Formal Grammars (Fedoriaka, 2010), to Fuzzy Systems (Zadeh, 1978; Yager & Zadeh, 1991; Negoita, 1981; Negoita et al., 1978; Cox, 1994), and to the theory of Self-Organized Criticality:

> The idea provides a possible 'explanation' of the omnipresent multi-scale structures throughout the natural world, ranging from the fractal structure of mountains, to the power law spectra of earthquake sizes. Recent applications include such diverse topics as punctuated evolution and traffic flow. The concept has even been invoked to explain the unpredictable nature of economic systems, i.e. why you can't beat the stock market. (Creutz, 1996, p. 1)

I.—The *Genetic Algorithms* approach—invented by John Holland in his book *Adaption in Natural and Artificial Systems* (1975)—represents a model of *machine learning* which simulates its behavior following the metaphor of the *processes of evolution* in nature. The machine works with a *population* of *individuals* represented by a set of character strings (or chromosomes). A recursive process of *crossover* operations—generally stochastic—simulate the reproductive behavior; a defined environment generates the selection process as a function of the *fitness measure* of the individual that is supposed to compete with other individuals in their environment. Some *genetic algorithms* use a function of the *fitness measure* to select individuals (probabilistically) to undergo genetic operations such as crossover or *reproduction*, and this leads to the propagation of unaltered genetic material (Bak, 1994, 1996; Schatten, 2004). As Holland attempts to demonstrate, the most powerful approach to understanding and showing the hidden order in collective behavior is the genetic algorithms approach (Holland, 1995, 1998). In 1992, Koza (1992) used genetic algorithms to develop programs to perform certain tasks. He called his method Genetic Programming (GP) (Goldberg, 1989). The *Evolutionary Economics* approach is derived from the genetic algorithms approach (see: Anderson et al., 1988; Arthur et al., 1997; Fogel, 1995; Allen, 2001; Potts, 2000; Dopfer & Potts, 2007).

> The birth of modern evolutionary economics can be traced back to the beginning of the 1980s. In 1981, Kenneth Boulding published a remarkable little book appropriately titled Evolutionary Economics. This was quickly followed by Richard Nelson and Sidney Winter's *An Evolutionary Theory of Economic Change* in 1982. (Foster & Metcalfe, 2001a, p. 1)

1.2.3 The Third Approach: The Micro-Macro Feedback Approach

A *third approach* for exploring collectivities is provided by the simple *Combinatory System Theory*, as outlined in this book. The combinatory systems view is neither a *macro* approach, since it also refers to local rules by considering micro behavior, nor a *micro* approach, since it also includes the macro behavior in the model of the system. It is rather a *micro-macro feedback approach*, precisely in that the operating rules, describing the behavior of the system, must in some way include not only local rules but also the feedback between the micro and macro behavior, which acts out over many cycles. Of course, the fundamental *micro-macro feedback* may also be accompanied by several other loops which make the system's dynamics nonlinear. It has been decided here to mention only the *micro-macro* fundamental feedback that may be specified in appropriate micro-macro loops.

For this reason, we can consider the *micro-macro feedback* approach as a particular form of Rousseau's *meso approach* (Rousseau, 1985; House et al., 1995), in that it applies to non-organized collectivities, whereas Rousseau's considerations mainly refer to organized collectivities that act with a specific end in mind.

> Formally defined, meso theory and research concerns the simultaneous study of at least two levels of analysis wherein (a) one or more levels concern individual or group behavioral processes or variables, (b) one or more levels concern organizational processes or variables, and (c) the processes by which the levels of analysis are related are articulated in the form of bridging, or linking, propositions. (House et al., 1995, p. 73)

> Organizations affect behavior and behavior affects organizations. The process is likely reciprocal, sometimes simultaneous, and sometimes serial. (*ibidem*, p. 83)

> In essence we are asserting that the meso perspective is capable of generating unique insights and knowledge about organizational behavior, over and above that contributed by micro and macro level research and theory. (*ibidem*, p. 85)

It should be noted, however, that these authors do not consider the *micro-macro feedback* as a "construct" in the sense of Morgeson and Hofmann (1999) that is useful for investigating organizational behavior.

> Guideline 5: To understand the structure of a collective construct, it may be helpful to identify the role the outcome plays in the collective, particularly in terms of how it facilitates goal accomplishment. This can help provide insight into why the construct exists and why it persists (or fails to persist) over time. (Morgeson & Hofmann, 1999, p. 259)

Combinatory System Theory follows the bottom-up approach of Epstein and Axtell (Epstein & Axtell, 1996), but unlike that approach it considers the *micro-macro feedback* as the origin of self-organization in combinatory systems.

If, on the one hand, it is easy to explain (perhaps properly speaking, to describe), assuming only *local information* and *rules*, the behavior of a flock of birds, a school of fish, or a herd of elephants when these collectivities have already formed, or the spread of information, the imitation of choices (information contagion), or the percolation effects in probabilistic diffusion systems (Grimmett, 1999a, 1999b), on the other hand it is not so easy to apply this micro approach to describe, for example, the grouping of flocks (a bird is attracted by the flock and not by its neighbors), swarms, herds, and other collectivities, the formation of graffiti on walls (people are attracted by the cloud of graffiti and not by the behavior of other observed people), the breaking out of applause (many people applaud until the applause dies down), or the phenomenon of a rising murmur in a crowded room (Sect. 3.4.2).

It is clear that the noble families of ancient Pavia built a new tower not only after it had observed the neighborhood but the whole swarm of towers in the town as well (Sect. 3.4.1). It is also clear that a person who is talking raises his voice to go beyond the increasing murmur of the crowded locale only for individual necessity, and not because his neighbors are raising their voices; or that the applause begins, rises, and is maintained because the clapping itself directs the clapping people; or that a fish joins a school of fish because of the presence of a predator, and only if it can perceive the school, not because it sees other fish join the school. Similarly, it is hard to explain, by exclusively using local rules, the exceeding of limits (all people park or drive fast even in the presence of parking limits and speed limits), the pursuit of records, the eternal maintenance of feuds, and the phenomenon of urban settlements.

In many cases, moreover, agents cannot observe the collectivity, and thus their neighbors, and must act based only on individual necessity, as in the case of the formation of piles of garbage (if I need to throw away garbage and I see a garbage pile, I prefer to leave my garbage behind), annoying and dangerous wheel ruts on the highway (passing trucks need to maintain their trajectory on the carriageways, and this is reinforced by such micro behavior), or paths in fields (people prefer to cross a field where a path is visible), and so on.

How do we explain the formation of paths in fields? What is the force behind the continual improvement in the quality of products? How does a feud develop? Why are some park benches or walls covered by graffiti while others nearby are spotless? Why are records continually overcome? What mechanism can we use to explain the maintenance of languages and dialects in limited areas?

In all these circumstances, the agents' micro behavior seems to follow some necessitating *global information* represented (or derived) by some macro variable(s) deriving from the collectivity (the cloud of graffiti, the pile of garbage, the applause, the carriageway, the feud, etc.), rather than derive from the agent obeying a set of local rules or information (Sect. 1.2.2). These and many other interesting collective phenomena, or effects, might be properly attributed to the basic behavior of combinatory systems, since they represent a particular class of systems acting in a "combinatory" way based on the *micro-macro feedback* action. Combinatory systems are recursively closed systems whose dynamics are prevalently due to the joint action of "chance" and "necessity"; while chance acts in an exogenous and inscrutable manner, necessity is the consequence of *micro-macro feedback* that connects the collectivity and its agents over a period.

A question arises: are agents conscious of the *micro-macro feedback*? Normally, they are not! But this is not in contrast with the logic of the combinatory system approach based on *micro-macro feedback*. It is necessary to specify the point of view: from an *external point of view*, the observer must recognize the macro behavior/effect and the *micro-macro feedback* action in order to define and build a model of the combinatory system; from an *internal point of view*, the agents normally are unaware of the *micro-macro feedback* action, and in many cases also of the *necessitating* and *recombining* factors that produce it.

1.3 Peculiarities of Combinatory Systems

Combinatory systems have been defined above as a plurality of *similar* units or agents that produce relatively analogous, independently defined *micro* behavior—often simple and predictable—which are distinctly observable or, in any case, refer to each single unit. The *micro* behavior is *analogous* to each other but, when *combined together* (whether or not in a conforming or non-conforming manner), leads to an overall joint *macro* behavior. Combinatory systems include all *self-organizing collective phenomena* and population dynamics, whose evolution is normally influenced by the history itself of the collectivity. The dynamics of combinatory systems is characterized by the action of *micro-macro feedback,* since individual micro behavior is, to a greater or lesser degree and in various ways, influenced, conditioned, or directed by the macro behavior of the collectivity in which the individual agents are included. In order to better clarify and integrate the logic presented in the previous sections, it is necessary to specify the most relevant characteristics that permit us to define or recognize a combinatory system.

1.3.1 Base and Environment

The first relevant feature that should always be specified is that combinatory systems can only be observed when we can perceive or hypothesize a plurality of *similar* entities—the collectivity—which can be conceived of as a single unit. The *base set*—or, more simply, the *base*—of a combinatory system, indicated by S(N), is defined as a plurality of N agents, A_n, n = 1, 2, … N—so that S(N) = [A_1, A_2, … A_N]—which can be considered as a *single unit* for an observer. We can assume that N is predetermined, but this assumption can be eliminated, especially when we observe "expanding systems" (Sect. 1.4.2). Every combinatory system will be characterized by its own *base*, which must be identified or specified in the system's descriptive model.

Observing the population of a city, for example, we can even identify different combinatory systems, each of which is composed of a specific *base*: the combinatory system of those that live in a certain district, or those that have a dog, or the system of those that live in white houses, and so on. Observing a woods, we can assume as the base of a combinatory system the winged predators of small animals; the population of small animals that can represent prey make up the base of a different combinatory system. The consumers of a certain product represent a possible combinatory system; we can interface with this the combinatory system whose base is made up of the producers of that product.

It is usually possible—even if unnecessary in many instances—to circumscribe an *environment* in which the *base set* operates. By *environment* I mean a real, supposed, or imaginary place where the *micro* behavior occurs and in which the system's *macro* behavior is shaped. The environment may or may not be limited: in the first case the environment's *boundaries* can be observed or defined. However, we must not confuse *environment* with *base set*. The latter represents a structure of agents that can give life to the system, a structure that can produce the macro behavior. The *environment*, on the other hand, is simply a way of circumscribing the action of the elements that make up the *base* of the system, so that the same base can, in different environments, produce different micro and macro behavior.

A *base set* can be represented by 50 pairs of dancers, a herd of 100 elephants, 150 members of a chorus, or by 5000 tractor-trailers going along a stretch of highway. But the action of these elements differs according to the *type*, *size*, and shape of the environment. The dancers behave differently when they are in a narrow room as opposed to a large one, or in a circular or square ballroom; the elephants' behavior will differ depending on whether they are in a wide savannah or a narrow valley; the tractor-trailers will produce a different effect depending on whether the stretch of highway is straight or curved.

The environment can influence or condition the *shape* of the combinatory system; that is, the arrangement of the elements that form the base. A combinatory system formed by 1000 persons behaves differently when the latter are in single file as opposed to side by side or in a square arrangement.

1.3.2 Micro and Macro Behaviors and Effects

A *collectivity* may constitute a combinatory system only if *each agent* of the *base*, A_n, n = 1, 2, … N, can show a plurality of different *states* (values, quantity, quality, etc.) whose dynamics over time defines the *micro behavior*, or *micro dynamics*. Assuming observations at discrete times, t_h, h = 0, 1, 2, …, let us suppose we observe the *same single individual variable*, *X*, characterizing each agent in the base, and that the value $x_n(t_h)$ of *X* represents the micro state of agent A_n for a given instant, t_h, in a period T (to be specified), or also the *instant micro behavior*. In what follows, and for simplicity's sake, unless stated otherwise it is assumed that the *micro behavior* of each A_n can be identified with the dynamics in its state over a period T:

$$x_n(\mathrm{T}) = [x_n(t_1),\ x_n(t_2), \ldots] \tag{1.1}$$

Let us assume that for each agent, A_n, a *micro function* "g_n" exists which transforms each *micro state*, $x_n(t_h)$, into an individual micro *effect*, $e_n(t_h)$, which can be conceived of as the *micro output* of the agent.

$$e_n(t_h) = g_n\left[x_n(t_h)\right] \tag{1.2}$$

The dynamics of the micro effects:

$$e_n(T) = \left[e_n(t_1),\ e_n(t_2),\ldots\right] \tag{1.3}$$

may be defined as the *micro effect—or output*—of each A_n over period T.

In many cases it is easier and useful to consider the micro effects (1.3), instead of the micro behavior (1.1), as the *micro state* of the agent.

A mass of graffiti forms because individuals carve their name on a bench or write words on a wall. The *micro behavior* of each agent is identified with one of the two states: carve/don't carve, or write/don't write in subsequent instants. The carved or written name is the *micro effect* of the individual behavior. Individual birds fly by following a given route and form a flock. The position occupied in the flock represents the micro effect of the individual behavior. Animals in flight, people not obeying a prohibition, athletes competing to win an event: these are all examples of micro behavior brought about by single elements. The micro effects can be easily worked out.

Generally, we can also assume that the agents change their state within a finite interval, $\Delta_n t_h$, which we can call the *transition period of the state*. For some systems the change in state of the elements can be a continuous function of time. At times we can also hypothesize that a *probability for the transition in the state* is associated with the change in state of each agent; in this case the combinatory system is defined as stochastic (Sect. 3.2.2). The combinatory system "as a whole," at each t_h in T, assumes a *macro state* that can be *analytical* or *synthetic*. The *analytical state* of the system, at each t_h, $X(t_h)$, is defined as the configuration of the states assumed by its elements at that instant; that is:

$$X(t_h) = \left[x_1(t_h),\ x_2(t_h),\ldots x_n(t_h),\ldots x_N(t_h)\right] \tag{1.4}$$

The *analytical macro behavior* of the system is the dynamics of its *analytical states* in $T = [t_0, t_1, \ldots t_h, \ldots t_N]$:

$$X(t_0)=[x_1(t_0),\ x_2(t_0),\ldots x_N(t_0)] \tag{1.4/0}$$
$$X(t_1)=[x_1(t_1),\ x_2(t_1),\ldots x_N(t_1)] \tag{1.4/1}$$
$$\ldots$$
$$X(t_h)=[x_1(t_h),\ x_2(t_h),\ldots x_N(t_h)] \tag{1.4/h}$$
$$\ldots$$

Let us suppose that, due to the action of *recombining factors*, there is a *macro function* "*F*" which, at each t_h, allows the system to "recombine" the *analytical state* of the systems, $X(t_h)$—representing the micro states of its agents—in order to produce a *synthetic macro state*, $Y(t_h)$:

$$Y(t_h) = F[X(t_h)] = F[x_1(t_h), x_2(t_h), \ldots x_N(t_h)] \tag{1.5}$$

The (macro) dynamics of the synthetic states (1.5) over a period, T: $\{Y(t_0), Y(t_1), Y(t_2), \ldots Y(t_h), \ldots\}$, can be termed the (synthetic) *macro behavior* of the combinatory system. Finally, we assume that a *macro function* "*G*" exists that, at each t_h, transforms the synthetic *macro states* (1.5) into a *macro effect*, which can be conceived of as the (macro) *output* of the combinatory system:

$$E(t_h) = G[Y(t_h)] \tag{1.6}$$

Note. In most cases it is easier and useful to consider the dynamics of the *macro effect* (1.6), instead of the dynamics of the *synthetic macro state* (1.5), as the *macro behavior* of the system.

Figure 1.10 shows the micro and macro dynamics and behaviors in a combinatory system, highlighting the previous definitions.

1.3.3 The Micro-Macro Feedback Action

As clearly shown in Fig. 1.10, for a combinatory system to exist it is necessary that between the *synthetic state of* the system, $Y(t_h)$, and the *states* of its units, $x_n(t_h)$, a mutual interdependence *must* be observed, or assumed, provided by the *micro-macro feedback* action; as a consequence, the *micro* behavior produces the *macro* behavior, but this in turn directs the subsequent *micro* behavior.

However, the functions *F*, *G*, and g_n, defined in the previous section, are not sufficient to define the *micro-macro feedback* action; it is also necessary to specify how the *micro states* of the agents at t_{h+1} derive from the *macro state* of the system at t_h. We must then hypothesize a set of functions "f_n" (assumed independent from t_h) that allows for the following derivation:

$$x_n(t_{h+1}) = f_n[Y(t_h)] \tag{1.7}$$

In many combinatory systems the agents' *micro* states at t_{h+1} not only are a function of the *macro* state of the system at t_h but also depend on the *micro states* (1.5) at t_h and/or the *macro effect* (1.6). Function (1.7) must be integrated as follows:

$$x_n(t_{h+1}) = f_n[x_n(t_h),\ Y(t_h)\, or\, E(t_h)] \tag{1.8}$$

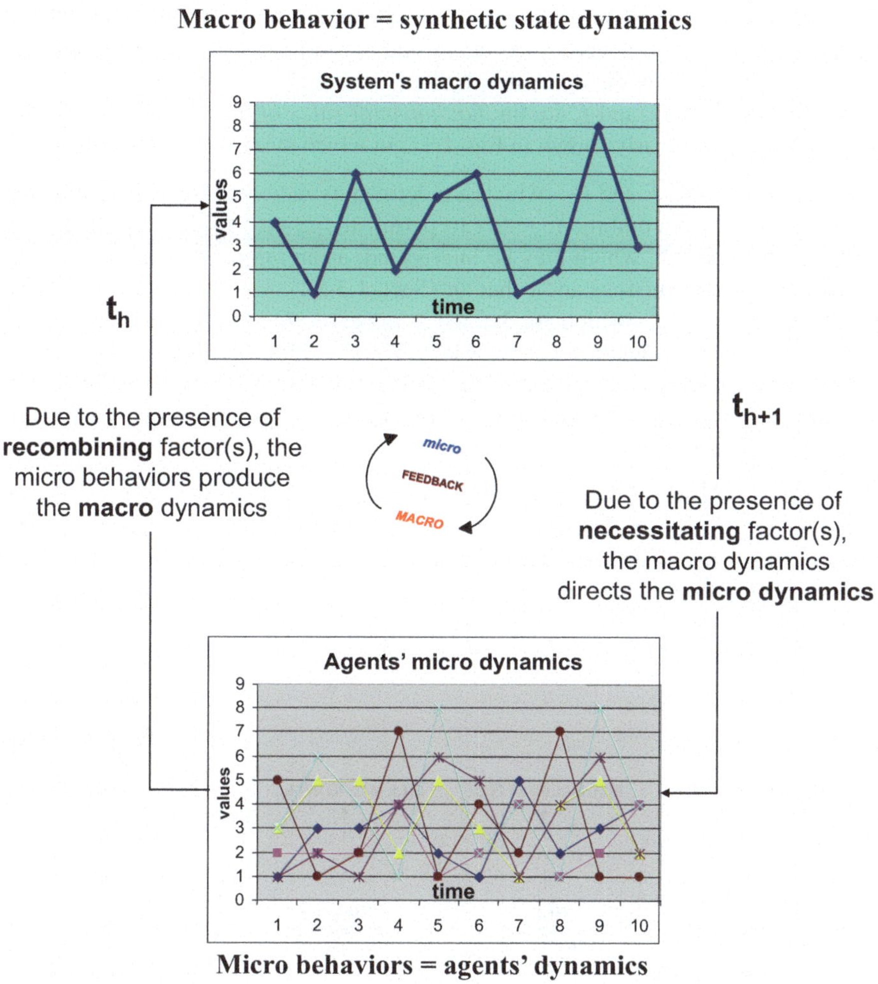

Fig. 1.10 Micro and macro dynamics in a combinatory system

The *micro-macro feedback* between the micro and macro behavior is thus described by the dynamics of the following system:

$$\begin{cases} Y(t_h) = F[X(t_h)] = F[x_1(t_h),, x_2(t_h),, \ldots x_N(t_h)] \\ x_n(t_{h+1}) = f_n[x_n(t_h),\ Y(t_h)] \end{cases} \tag{1.9}$$

The operative logic of the *micro-macro feedback* action can be simply summarized as follows: the synthetic macro state of the system at a given time *must* depend on the state of its elements; but, in turn, the subsequent state of the latter *must* depend on the macro state of the system, following the logic shown in Fig. 1.11, which represents a combinatory system as a dynamic model that links the macro and micro dynamics produced by the *micro-macro feedback* based on the micro and macro rules from which the dynamics derive.

The functions *F*, *G*, and f_n are the *fundamental rules of the system* and account for its behavior, as clearly shown in Fig. 1.11, in which system (1.9) is applied:

(a) The *macro* rules *F* and *G*—which are defined as *recombining*—show how the *micro* behaviors combine together to permit the system to develop the *macro* behavior or effect, which may be interpreted, as an *informative input* for individual agent's decisions and behavior (Sect. 1.3.5).
(b) The *micro* rules f_n—which are defined as *necessitating*—determine the agents' *micro states* as a function of the system's *macro state* (the *informative input*), which recursively changes the value assumed by the *macro variable,* which in turn modifies the perceived positive or negative gaps, driving the agents to adapt their *states* through new decisions.

Figure 1.11 clearly shows that the first initial analytical state (1.4/0) at t_0, $X(t_0)$, is usually due to an *initial casual input* that triggers the modification of the micro states and gives rise to the macro behavior; in many combinatory systems, however, the first input, rather than resulting from *chance*, can be produced voluntarily by an

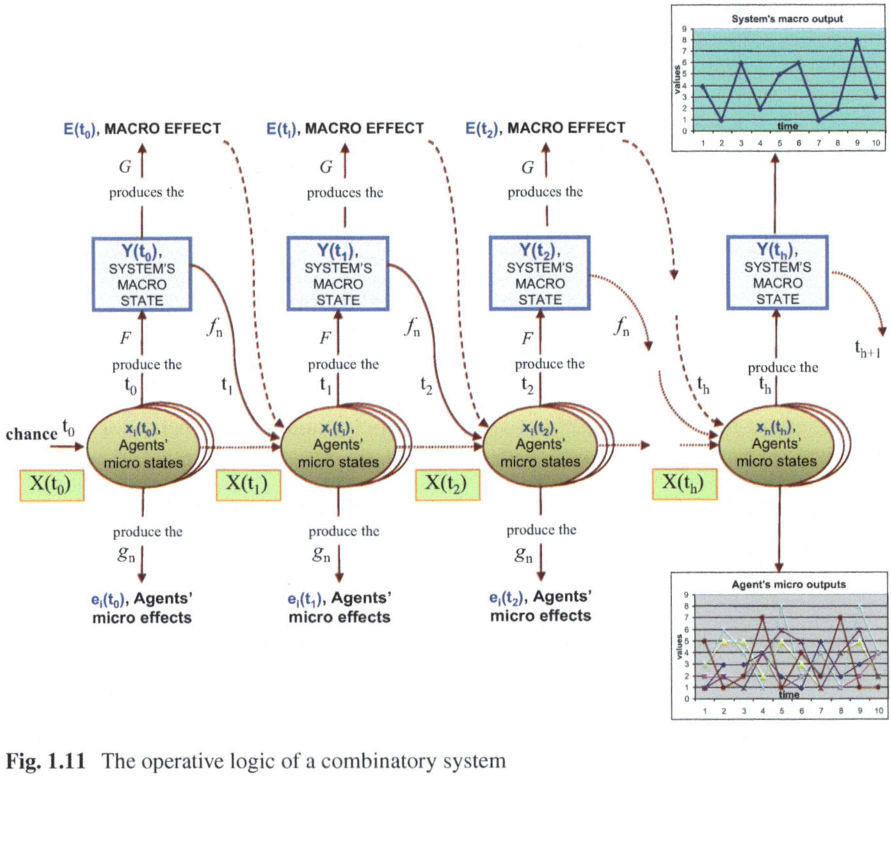

Fig. 1.11 The operative logic of a combinatory system

external agent or by an established program, as happens, for example, in the spread of fads.

Note. It is important to remember that the feedback between the micro and macro behavior in most cases occurs indirectly by means of the *effects* of this behavior, as shown by (1.8). The *micro behavior* produces *micro* effects—the *outputs* of the single agent's action—which, "combined together," produce the *macro effect*, the *output*, of the entire system. From this observation it follows that the observer must choose the best variable—*behavior* or *effects*—capable of explaining the *micro-macro feedback*. Thus, generalizing, the *micro-macro feedback* can be described by the following more explicit system:

$$\begin{cases} Y(t_h) = F[X(t_h)] = F[x_1(t_h), x_2(t_h), \ldots x_N(t_h)] \\ E(t_h) = G[Y(t_h)] \\ x_n(t_{h+1}) = f_n[x_n(t_h),\ Y(t_h) \text{ or } E(t_h)] \end{cases} \tag{1.10}$$

Definition 4 Global information. The system above (1.10)—which is illustrated in Fig. 1.11—allows us to define in general the macro behavior, $Y(t_h)$, or the macro effect, $E(t_h)$, as *synthetic global information* that conditions (consciously or unconsciously) the micro behavior of the agents. This synthetic *global information* is continually updated by the *micro-macro feedback.*

The *micro-macro feedback* generates a "synergetic effect" that produces "emerging" macro behavior or effects which are only attributable to the unit. When we cannot observe or assume a *micro-macro feedback* action between the micro and macro behaviors or effects, we cannot find any combinatory system.

The *micro-macro feedback* is often positive, in the sense that it amplifies the initial casual impulse. At other times it is negative and tends to order the behavior of the system as a whole by pegging the micro behavior to the macro behavior or, vice versa, by eliminating the micro behavior that is deviant with respect to the macro behavior. In the example of troops parading by or the dance group at the Moulin Rouge, the negative feedback between the micro and macro behaviors is completely evident: the overall movement of the base derives from the movement of the single individuals, but these in turn are synchronized with the overall movement. The feedback is negative, since it eliminates the errors in individual positioning.

Let us consider a combinatory system that we observe in the process of transmitting a language within a population. Each parent passes on his or her mother tongue (micro behavior) and the children learn it (micro effect). The children form the population that speaks the mother tongue (macro behavior and macro effect of the combinatory system). The *micro-macro feedback* operates between the macro effect (existence of a mother tongue) and the micro behavior; the mother tongue obliges the families to teach this language to their children. The negative feedback that makes the language stable is clear.

1.3.4 State and Output, Macro Behavior and Macro Effect

We have observed above how combinatory systems produce a macro behavior that derives from a *micro-macro feedback* and that the macro *behavior* produces a macro *effect*. At times it is easy to distinguish between the *internal states* of the system—that is, the macro *behavior*—and the system's *output*, that is, the macro *effect* in the environment, caused by that behavior. In other cases, the macro *behavior* is perceived as the sole output of the system, because the macro effect coincides with the macro behavior. A few examples will help clarify this important feature.

First of all let us consider the "paths-in-fields" system, which forms when a certain number of persons crosses a grassy field, thereby forming tracks which are then followed by other passers-by (see Sect. 2.5.4). Here the crossing of the field is the micro behavior and the trampled grass the micro effect of this individual behavior. If other individuals should subsequently cross the field in the same direction, they can decide to follow the tracks in the grass left by the first individual. In this case a macro behavior is produced which corresponds to the crossing of the base of the system in the same direction and along the same tracks. But it is also easy to observe the macro *effect* of this macro *behavior*, represented by the formation of a path, which derives from the agents' micro effects: the trampling of the ground and the drying of the grass. It is thus easier to think that "the path"—that is, the macro effect—conditions the subsequent micro behavior—crossing the field *following* the path—even if it is the micro behavior itself that consolidates the path thanks to the micro effects, which lead to the gradual disappearance of the grass. Therefore, micro and macro behavior and micro and macro effects can be easily distinguished. Whenever we see a path in any environment, it is necessary to investigate the *necessitating* and *recombining* factors from which the micro-macro feedback originates.

Let us now consider the "stadium wave" system we observe when spectators stand up and sit down in rhythmic fashion, thereby creating a "wave" that runs around the stadium (see Sect. 2.5.3). We can thus synthetically express the micro rule as follows: "If the wave arrives, stand up, count to three (for example), and sit down." The micro behavior thus corresponds to each spectator standing up and sitting down. The "combination" of the micro behaviors produces a macro behavior that refers to the entire stadium and is represented by the wave that circularly runs around the stadium several times until the stop signal intervenes. In this example, the state of the system—represented by a mix of standing and seated spectators, which form the wave that is observed with a continuous motion—can also be considered, at least in observational terms, as the output of the system; the macro effect and the macro behavior coincide.

At other times it is difficult to make out the macro behavior, so that it is easier to consider a feedback between the micro behavior of the component agents and the macro effects produced by the system itself. Thus the distinction between state and output is at times real, while at other times it represents a useful tool for the analysis of the combinatory system.

Similar considerations lead us to distinguish the *micro behavior* of the units of the system from the *micro effects* produced in the environment by such individual behavior. At times it is even useful to describe the individual behavior in terms of micro effects rather than in terms of micro behavior.

Let us now examine the combinatory system represented for simplicity's sake by a column of seats in the theater: that is, a series of seats set in single file that go from the rear of the theater toward the stage. If "by chance" the spectator(s) seated in front of the stage stands up to applaud and blocks the view of the spectator sitting behind him, who in turn stands up, thereby causing the third to do so also, then very soon all the spectators in the column of seats will "by necessity" be on their feet. The micro rule is simple: "Try to always see the stage; if you cannot because the person in front of you is standing, then stand up yourself." The micro behavior is easy to observe, since it consists in standing up. The number of persons standing represents the state of the system, the macro behavior, and there is no doubt that the state of the system conditions the micro behavior, which in turn modifies the state itself. In this case the macro behavior corresponds to the gradual standing up of the column of spectators and coincides with the state of the system: in fact the macro behavior is: "if a spectator stands up, the number of spectators who are standing necessarily increases." This rule enters into feedback with the micro behavior until all the spectators are on their feet; whoever has been to the theater can testify to this combinatory system, which also acts at conferences, in political meetings, and wherever there are agents arranged in rows and columns. The macro effect, at time t_h, is instead identified with the state of the system at t_h; that is, with the fact that a certain number of spectators in the row are standing up. The final macro effect/state of the system corresponds to the observation that all the spectators in the row are standing up. This combinatory system is well known and brings to mind the description of the crowd in Chapter XIII of Alessandro Manzoni's (nineteenth-century Italian novelist) *The Betrothed* (I Promessi Sposi): "*And everyone, rising on their toes, turns to look to where the unexpected arrival* [of the grand chancellor Antonio Ferrer] *was announced. With everyone rising, they could see, some more some less, as if they had remained with their feet on the ground; in any event [pushed by the micro-macro feedback] everyone stood up*" (free translation from Manzoni, 1985, Chap. XIII, p. 240).

Recalling the "graffiti" system, let us assume that two lovers are sitting on a park bench and, "by chance," decide to carve their names into its dark wood by using a hair pin. The micro *behavior* is the act of leaving behind a "mark of oneself," while the mark itself is the micro *effect.* If other lovers or other frequenters of the park should decide to leave a mark on the same bench, a mass of graffiti becomes very visible that would probably increase the likelihood that others will leave behind marks of themselves. However, the *macro behavior* is truly difficult to define, since the group of those who leave carvings in the bench is not observable but can only be postulated, while we can immediately observe the mass of graffiti as the *macro effect* that derives from the macro behavior. It is thus easy to verify that the graffiti (macro effect) represents the result of past micro behavior and guides future micro behavior, precisely according to the logic of combinatory systems.

Note. In the previous combinatory system it should be clear, or understandable, that the mass of graffiti (macro effect) influences the agents' micro behavior, in the sense that it changes the probability that a passer-by will deposit new graffiti. In other words, the macro effect acts on the *necessitating factors*, which can be expressed in terms "of probability of action" by the agents. All of Chap. 3 will consider the role of probabilities in simulating the behavior of combinatory systems, and the reader is thus referred to this chapter.

1.3.5 Incomplete and Limited Information

As seen in previous sections, in agent-based models, and in particular in complex systems and CAS, collectivities normally act on the basis of limited information and local rules. In general, the behavior of a collectivity can be defined as *local*—or based on *limited information*—or *global*—or based on *complete information*—depending on whether or not the agent's micro behavior derives from *local information* possessed by the agents (a person acquires a good because he observes that at least N friends have bought it; an elephant in a herd runs to the left because the elephants on its right push it in this direction) or from *global information* (over time and/or space) possessed by all the agents (all the students rise because the teacher orders them to; all the animals flee because they see the fire advancing).

Global information may derive from the outside (external director, starting traffic lights or the starting gun in races, trumpeting by the leader of the herd, etc.) or may be "self-produced" by the joint micro behavior of the agents considered as a whole. However, there is no doubt that *self-produced global information* plays an essential role in many relevant phenomena created by collectivities, "social" collectivities in particular; such phenomena cannot simply be explained on the basis of local rules based on *limited information*. Depending on the importance of the *self-produced global information* in determining the agents' micro behavior, different types of combinatory systems can be distinguished.

When agents possess complete information about the behavior of the whole system, and the micro behavior of the agents is determined *exclusively* by the macro behavior/effect, in that it is perceived as *global information* that is *sufficient* to change their state, then the collectivity is a "typical" or "perfect" combinatory system.

We must nevertheless recognize that in many, or most, cases each agent is normally blind to the *macro* behavior of the collectivity as a whole, while being aware of the *micro* behavior of some other agents; if the agent's micro behavior depends on information about the state of an opportune neighborhood (as well as, when possible, on the macro behavior), the combinatory system is characterized by *incomplete and limited information*:

- It is an *incomplete information combinatory system* in that the agents produce their own micro behavior without considering the macro behavior of the unitary system *as the only information* (except as an extreme case of a completely observable macro effect).

- It is a *limited information combinatory system* in that the agents' micro behavior depends on information about the micro behavior (which occurs or is only expected or foreseen) of a limited number of other *neighborhoods* (defined in an opportune way), exactly as in a cellular automaton, and a pattern forms, which is the effect of the phenomenon defined as *spontaneous order* (Ashford, 1999, 2001; Sugden, 1989; Swenson, 2000).

- The second point relates to the principle of order in a complex system, and, as Kauffman argues, a system achieves spontaneous order from a process of localization of interactions about a low number of inputs. Each element's behaviour becomes conditional upon a low number of other elements. There is no necessary specification of which other elements this may be: that is, the general result holds for sets of random interactions provided only that the input parameter is suitably low (…). Thus the mystery of a spontaneous order in a completely decentralized market system, for instance, hangs only upon the behaviour of each element (agent) being conditional not upon everything, but only upon a few things. It is to be noted that from Adam Smith to Hayek and to Hahn (…), this basic question on the workings of a market system has remained unsolved in terms of the extraction of the underlying principle that leads to the emergence of the spontaneous order. And it is simple: incompleteness. (Potts, 2001, p. 97)

Note. In cellular automata the term "neighborhood" can be defined in different ways, including:

(a) *von Neumann Neighborhood*: four cells, the cell above and below and to the right and left of each cell are called the von Neumann neighborhood of this cell. The radius is 1 in this definition, as only the next layer is considered.
(b) *Moore Neighborhood*: the Moore neighborhood is an enlargement of the von Neumann neighborhood, containing the diagonal cells as well. In this case, the radius is also r = 1.
(c) *Extended Moore Neighborhood*: equivalent to the description of the Moore neighborhood above, but the neighborhood extends beyond the distance of the adjacent cells. Hence r = 2 (or larger).
(d) *Margolus Neighborhood*: considers 2 × 2 the cells of a lattice at the same time.

If the agents' behavior depends *only* on *local rules* acting on a defined neighborhood, without considering any *micro-macro feedback*, collectivities can be conceived of as *local information systems*, whose typical model is represented by "complex systems" simulated by cellular automata. Local information systems *lose the characteristics of combinatory systems.*

For this reason we cannot in general consider the ants, swarms, and, more generally, artificial collectivities which exclusively act according to the cellular automata approaches as examples of combinatory systems, except in the case where the *macro* behavior may affect the *micro* behavior of the agents in some way, as occurs, for example, when the agents create an "aromatic potential field" by spreading pheromones or other permanent messages, thus producing the *stigmergy effect* (Grassé, 1959, 1960, 1967) and activating a "typical" combinatory system (see Sect. 2.5.5) with their micro behavior. The agents, in fact, spread pheromone across one site (local or micro information); the increasing concentration of pheromone (global or macro information) increases the probability that each agent will move in the direction of that site even when they do not perceive the behavior of a neighborhood. The *micro-macro feedback* in this case is quite evident (Zollo et al., 2001; Deneubourg & Goss, 1989).

> It appears to leave human organisations and institutions little different in principle from wasp's nests or even piles of sand. They can all be said to emerge from the actions of the individuals. The difference is that while we assume that, for instance, wasps have no ability to reason—they just go about their business and in doing so construct a nest—people do have the ability to recognise, reason about and react to human institutions, that is, to emergent features. Behaviour which takes into account such emergent features might be called second order emergence. (Gilbert, 1995, online)

Finally, when we observe a group of agents that produce micro behaviors independently of any type of self-produced global information, or of behavior from some *neighborhood*, then those agents do not constitute a collectivity, since they do not possess the requisites of either a combinatory or a complex system. If, in acting freely and independently, they produce some form of regular behavior this can be viewed as the result of the action purely of chance or of some unobservable constraint set by the micro behavior.

The following simple redefinition is now presented:

Definition 5 Combinatory systems. I define as a "combinatory system" (see Def. 3) any collectivity (see Def. 1) whose agents, consciously or unconsciously, act (exclusively or prevalently) on the basis of *global information* (see Def. 4) which they directly produce and update as the consequence of their micro behavior and the *micro-macro feedback* (see Def. 2).

Definition 5 emphasizes the cognitive activity of the agents: the *macro effects* produced by the macro behavior of the system do not necessarily lead all by themselves to self-organization; they become factors in self-organization only when these effects are *interpreted* by the agents as *information* they can base their decisions on.

Many authors have studied the relation between the agents' micro behavior and the collectivity's macro behavior; but they have not explicitly recognized the combinatory effect produced by *micro-macro feedback*. For example, von Foerster theorized that there existed a particular class of systems formed by agents in which the actions of individual agents determine the very state of the system, which in turn serves as the reference point for these same actions. This theory clearly embraces combinatory systems, even if von Foerster never described the latter in detail, and thus never adopted the notions of *micro-macro feedback* and necessitating and recombining factors (in the following quote, the comments in brackets have been added):

> What you are trying to describe is the relationship of circular causality [*micro-macro feedback*] between the whole (a human community for example) and its parts (the individuals from which it is comprised). On the one hand, individuals are related to each other, and on the other hand they are related to the whole. The bonds between individuals can be more or less "rigid"—the technical term I use is "trivial". The more trivial they are, by definition the less the behaviour of one of them provides information to the observer who already knows the behaviour of the others. I will make the following conjecture: the more trivial the inter-individual relationships, the more the behaviour of the whole will appear to the individual elements from which it is made up as having its own dynamics which escape their control. [...] The whole appears to become autonomous with respect to its conditions of emergence, and its development to be immobilised as its destiny. (von Foerster, 1972, p. 38)

> By "trivially" connected, von Foerster means that the influence of the state of the system on the action of the elements takes the form of a rigid, univocal determination. (Atlan et al., 1986, p. 38)

In trying to interpret the behavior of a crowd in panic situations, Elias Khalil and Kenneth Boulding distinguish between external information (external leader) and self-produced information by describing the modus operandi of a typical incomplete and limited information combinatory system (comments in brackets have been added):

> To say that the leader (or power) is an endogenous fixed point is to claim that the human group in fact uses an element within the group as its external point of reference [information that is exogenous to the system]. [...] In a systems perspective, the transformation of the crowd [collectivity] during a situation of panic raises no logical problems since it simply consists of substituting one endogenous fixed point for another. Although the leader has disappeared in the panic, another fixed point representing the collectivity takes his place and appears to transcend the members. This fixed point is nothing other than the movement of the group itself [the macro behavior is the self produced global information], a movement that acquires distance and autonomy in relation to the individual movements [the micro behaviors of agents update the macro behavior of the collectivity]. (Khalil & Boulding, 1996, p. 75)

A final *note*: let us ask ourselves why the individual elements of the system produce their micro behavior even though they know that at times this leads to undesired, harmful, even catastrophic global effects? Why do other people continue to speak their own language when English is so much more useful? What moves elephants fleeing in a herd to head toward the crevasse so that all fall in and a predictable disastrous outcome occurs? Why do companies continue to increase productivity knowing this will lead to fewer jobs and thus social tensions? The answer must be found in the incomplete and limited information mechanism according to which the individual elements in the system produce their micro behavior; in fact, each element does not usually know the macro behavior of the system. It acts only on the basis of information about the micro behavior of a few other elements that it can directly observe. Each individual acts on the basis of *incomplete* information concerning the collective macro behavior and on information *limited* to the micro behavior of a small number of other elements, as we have seen above.

1.3.6 Minimum and Maximum Density

Figure 1.11 shows clearly that the initial state at t_0 is usually due to an initial casual *impulse* that triggers the modification of the internal state and gives rise to the macro behavior. In many combinatory systems, however, the first input, rather than due to *chance*, can be produced voluntarily by an external agent or by an established program, as happens, for example, in the spread of fads or in the combinatory system that produces the stadium wave.

Another relevant characteristic of the combinatory system logic is the fact that *every* combinatory system begins to operate when there is an initial change in the state of a minimum number of elements, and it ceases to develop its macro behavior when there is a change in the state of a maximum number of elements. This means that the *micro-macro feedback* is manifested only if the number of the A_n that develop their micro behavior at t_0 exceeds a *minimum* number, N_{min}, or *critical activation mass*, and remains below a *maximum* number N_{max}, or *critical saturation mass*, which is defined each time for each specific system.

In equivalent terms we can write:

$$N_{min} / N = D_{min} > 0,$$
$$N_{max} / N = D_{max} \leq 1,$$

where D_{min} and D_{max} indicate, respectively, the *critical activation* and *saturation density* necessary for the *micro-macro feedback* to reveal and maintain itself over time, thereby allowing the system to produce its own emerging effects. In extreme cases $N_{min} = 1$ (the applause frequently comes when only one agent starts to clap) or $N_{min} = N$ (cancan dancers show their alignment only if all the dancers try to align to the rotating row).

Let us look at the example of the Combinatorial System concerning the Village of North Africa North Africa—presented in Section 1.3.6—whose houses, originally white, are gradually but inevitably colored blue. The initial state of the system was N white houses. This state is gradually changed due to micro behaviors of individual inhabitants who paint their houses blue; but these micro behaviors combined together give the idea of a macro behavior of the village, that is, of the community of its inhabitants, which produces a macro effect (gradual change of color) which, in turn, represents a global information of the opportunity to change the color of the houses; This is the power of feedback between micro and macro behaviors that inevitably changes the color of the village. The final state becomes N blue houses.

An interesting quote in this context is an excerpt from the "Prelude" to Chap. 11, in *Gödel, Esher, Bach* (1979), in which Douglas Hofstadter describes the ant colony as an entity distinct from the ants which produces emerging phenomena, clearly bringing out the need for a sufficient base and a minimum density:

ANTEATER: You could put it that way if you want to insist on seeing the trees but missing the forest, Achilles. In fact, ant colonies, seen as wholes, are quite well-defined units, with their own qualities, at times including the mastery of language. (Hofstadter, 1979, p. 165)

[...]

ACHILLES: There must be some amazingly smart ants in that colony, I'll say that.

ANTEATER: I think you are still having some difficulty realizing the difference in levels here. Just as you would never confuse an individual tree with a forest, so here you must not take an ant for the colony. You see, all the ants in Aunt Hillary are as dumb as can be. They couldn't converse to save their little thoraxes! [*ibidem*, p. 166]

[...]

ANTEATER: I knew there was something parallel going on in the two very different systems. Now I understand it much better. It seems that group phenomena which have coherence—trail building, for example—will take place only when a certain threshold number of ants get involved. If an effort is initiated, perhaps at random, by a few ants in some locale, one of two things can happen: either it will fizzle out after a brief sputtering start. (ibidem, p. 167)

ACHILLES: When there aren't enough ants to keep the thing rolling?

ANTEATER: Exactly. The other thing that can happen is that a critical mass of ants is present, and the thing will snowball, bringing more and more ants into the picture. In the latter case, a whole "team" is brought into being which works on a single project. That project might be trail making, or food gathering, or it might involve nest keeping. Despite the extreme simplicity of this scheme on a small scale, it can give rise to very complex consequences on a larger scale. (*ibidem*, p. 168)

1.3.7 Energy Inputs

In order to produce the micro behavior/effects and the macro behavior/effect it is usually necessary to supply energy to the system. In addition to the "initial" *micro states* input we must also consider the *energy* inputs, which, however, must be kept distinct from the *initial impulse* necessary for the production of the micro and macro behaviors. Let us consider the last example of the system of the white village that becomes changed into a blue village (Sects. 1.3.6 and 3.4.3). It is obvious that an input of paint and painting material is also needed to produce the micro behavior.

Since the main objective of the theory of combinatory systems is to bring out the operative logic typical of such systems, in order to simplify its description energy inputs, when not essential, are usually not represented, considering that the representation of such inputs is usually implied and superfluous, if not impossible. If we consider the following systems: graffiti on a park bench, dancers at the Moulin Rouge, a herd of elephants in flight, soldiers parading, and most of the other combinatory systems which will be presented in Chap. 2, it is clear that energy inputs are not necessary to understand the operative logic of these systems. Nevertheless, we must keep in mind the fact that in order to give a "technical explanation" of the action of such systems, in particular for purposes of designing them, knowledge of the energy input *constraints* can be indispensable.

We can make similar considerations for any *instructive* inputs that may be necessary to set off the combinatory system. These inputs are necessary in all *artificial* combinatory systems in which the micro rules for the development of the micro behavior are explicitly decided by the agents or by an external coordinator, for example, in the "stadium wave" system. In other cases the instructive inputs are represented by the same micro behavior which owes to chance, such as in the "applause" system.

1.3.8 Control of Combinatory Systems: Strengthening and Weakening Actions and Effects

Thanks to the presence of *necessitating* and *recombining* factors and the action of the *micro-macro feedback*, a combinatory system tends to survive for a period of time. If the system is considered useful, the environment in which it operates strengthens the factors on which the system's dynamics depend. If not considered useful, the environment weakens those factors. In effect, combinatory systems often act in an environment—in which other individuals or other systems, both operative and combinatory, operate—that can condition its behavior through interventions (actions, provisions, constraints, obligations, etc.) of a *strengthening* or *weakening* nature. Thus, under similar conditions (base, environment, elements, necessitating and recombining factors) similar combinatory systems can have quite different developments. Borrowing the concept of teleonomy presented by Jacques Monod in his famous work, *Le Hazard et la Nécessité* (1970, p. 22), *exogenous teleonomy* will be defined here as the aptitude of a combinatory system to be preserved in the environment with suitable *strengthening actions* because it is held to be useful (advantageous, convenient, indispensable, etc.).

If a combinatory system has a high exogenous teleonomy it can also *institutionalize* itself; if it has scarce exogenous teleonomy it suffers *weakening actions*, so that the macro behavior weakens and tends to disappear. Thus, in order to fully understand the nature and dynamics of any combinatory system it is necessary to examine its environment and determine the *strengthening* and *weakening* actions it receives from this.

Let us consider the combinatory system that leads industries to locate in a certain area, or to the formation of industrial zones (see Sect. 2.2.5). When a certain area offers a *positive economic differential* over other areas [necessitating factor], a certain number of entrepreneurs may decide to locate [micro behavior] a productive and commercial facility in that area [micro effect]. If the *minimum activation number* is reached—which depends on the intensity of the necessitating factor—then the combinatory system can start up and produce the collective phenomenon of joint location [macro behavior], with the creation of typical industrial and commercial settlements [macro effect]. The presence of a group of enterprises nevertheless brings economic advantages for the settlement [recombining factor], and this attracts new businesses that may push for *strengthening* actions by territorial authorities as part of a typical *micro-macro feedback* process. As a result, the dynamics of the settlement can be favored by certain exogenous strengthening actions aimed at creating the conditions for producing the economic differential. These strengthening actions must become permanent and easily perceivable. Particularly evident are actions aimed at creating logistic infrastructures (highways, ports, equipped areas, etc.) and the "conscious" recognition of cost advantages (lowering of the cost of labor, and tax and financial advantages). The system ceases when weakening actions occur (e.g., urban constraints, taxes) that reduce the economic advantages.

The strengthening and weakening actions can act either on the macro behavior or the macro effect—that is, on the combinatory system at the *macro level*—and either on the micro behaviors or the micro effects of the elements that make up the system itself. In this case we observe strengthening and weakening actions at the *micro level.*

1.3.9 Exogenous and Endogenous Control

We can now examine the possibility of the *control* of a combinatory system, where by *control* we mean any action, exogenous to the system, aimed at making the macro behavior conform to the desired macro behavior. The control is carried out through suitable *strengthening* or *weakening* actions, which operate by acting directly on the *macro* behavior—we will define this as *macro* or *external* control—or by influencing the *micro* behaviors; in this case the control will be called *micro* or *internal* control, as indicated in Fig. 1.12 (taken from Fig. 1.8), which makes immediately clear how the macro-level control is achieved through strengthenings and weakenings aimed at modifying some *recombining* factor. Instead, the micro-level control acts on the *necessitating* factors.

The micro control must therefore not be confused with the *feedback action* between the macro and micro behavior. In fact, the *micro-macro feedback* always causes the internal self-control of the elements of the system, which adapt their micro behavior to the macro behavior; this is a "self-coordination" effect. The micro control derives instead from actions by the environment that influences the agents' behavior.

Let us consider a typical combinatory system that leads to the spread of new businesses, generated endogenously in a given area, whose modus operandi (similar to the example of co-localization, described in Sect. 2.7.1) can be summed up as follows: when, in a given area, successful businesses locate by chance and are able to professionally train their personnel (employees, managers, professionals), some

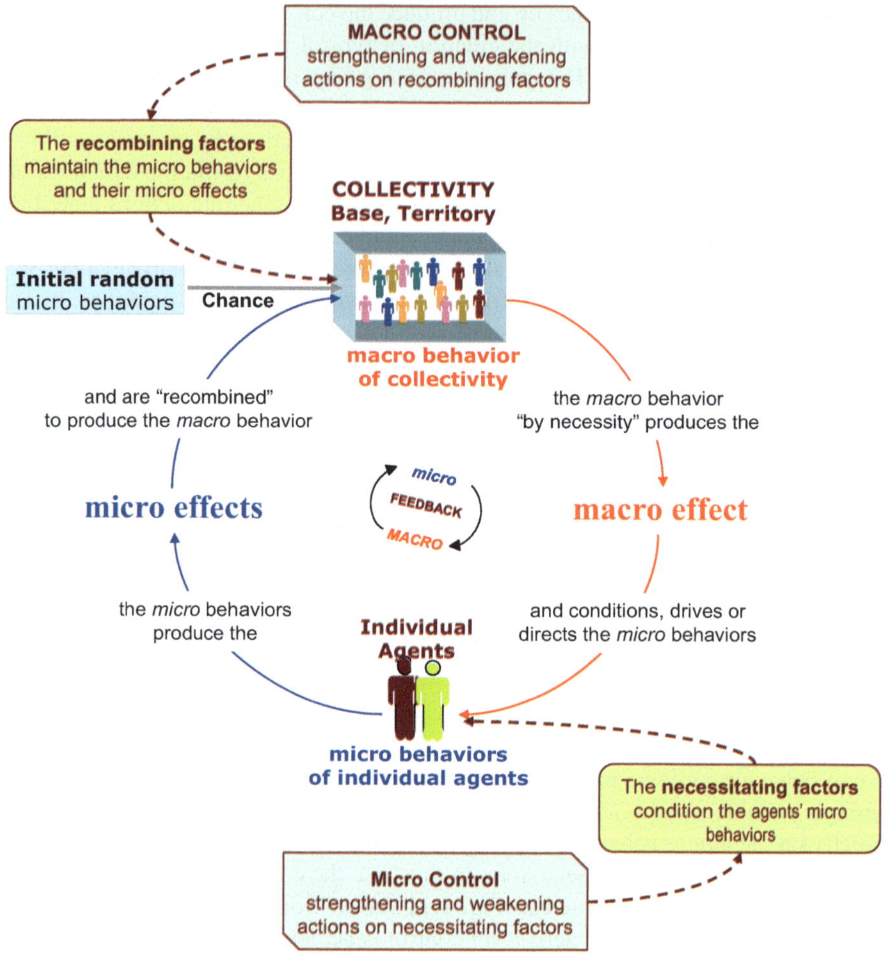

Fig. 1.12 Macro and micro control in combinatory systems

of the latter may, "by chance," and after having acquired the necessary skills, decide to undertake their own activities in order to take advantage of these acquired skills for their own benefit (micro behavior). New businesses then develop (micro effect). If those who are successful in these undertakings attain the minimum activation density, then the combinatory system can start up and the area will become full of endogenously created enterprises (macro effect). This represents an incentive for other workers, who possess the same skills, to risk going into business on their own and starting up new enterprises. The process spreads and the group of workers is gradually transformed into a group of businessmen (macro behavior).

The enterprises become ever more numerous (macro effect), which motivates individuals to undertake new business activities (micro behavior), in a typical *micro-macro feedback* action that characterizes "systems of diffusion." An area of workers gradually becomes an area of entrepreneurs, who quickly will have to bring in subordinate work from other areas. Since the spread of businesses is considered a source of wealth for the entire collectivity, several forms of macro and micro controls can be carried out, such as: the creation of professional training schools that guarantee the initial start of work (macro control); the availability of venture capital and loans (micro control); the incentive to form new businesses by means of particular legislation to favor, for example, young entrepreneurs (macro control); and the creation of protective measures against business failure (micro control). The exogenous creation by public authorities of enterprises with appropriate characteristics (small-size businesses, the need for small-size collateral production, business training programs) could artificially trigger the system, on the condition that the critical mass of new businesses starting up in the area is soon reached, so that the necessitating and recombining factors can emerge to assure the *micro-macro feedback* process occurs.

The harmful effects caused by the micro behavior/effects of combinatory systems that operate without an adequate number of internal and external *micro* and *macro* controls have been studied as *counter-productivity* phenomena (self-deregulation and self-disorganization) of a system, when the latter "becomes foreign to the elements from which it is made up."

> In the case of a society, counter-productivity characterises a system that escapes the control of those who contribute to it, and is destroyed by the same means which are intended to serve it, as in the case of an organism affected by an auto-immune disease: "medical science corrupts health, school makes one mindless, transportation immobilises, communications make one deaf and dumb, information flow destroys the senses (…) industrial food converts to poison (Dupuy, 2006)." (Chavalarias, 2016, p. 10)

As the above quote illustrates, the *counter-productivity* mainly refers to organized systems (societies or processes) and not to combinatory systems. However, we can observe even in the latter the effect of the *self-deregulation* produced because of both a slowdown in the *necessitating* factors and the *micro-macro feedback* on the detrimental micro behavior, which, in recombining, produces harmful macro effects that cause the agents to "worsen" even more their micro behavior, leading to a downward spiral that changes the quality of the effects produced by the system.

1.4 Self-Organization and the Evolution of Combinatory Systems

1.4.1 Natural and Artificial Combinatory Systems. Spontaneous Genesis and Design

Because the combinatory system is formed and functions by means of micro and macro *rules* (Sect. 1.3.2), we can also assume that it can be designed by suitably defining the two types of rules. Many combinatory systems, as we have seen in the above examples, apparently have a spontaneous formation. This means that the micro and macro *rules* are already "naturally" inserted into the elements of the system.

Spontaneous systems can be defined as *natural* (in the relative and conventional meaning of the term), producing ordered micro behavior that can be viewed as forms of *self-organization*. In other cases we can observe ad hoc rules to form combinatory systems, and as such we can define these as *artificial*.

Let us consider the phenomenon of *self-organization*, which is typical of many combinatory systems and seems to produce and maintain a spontaneous order in the micro behavior of the component elements ("systems of order" will be examined in Sect. 2.5). A typical example of an ordered system is the circular ordered dance that can occur in dance halls when a rotating dance (e.g., a waltz) is played; after a disordered start the dancers appear to self-organize themselves and maintain a rotating dynamics, so that they all dance in a rotating fashion (Sect. 2.5.1).

Another typical example of *self-organization* is the division of a disorderly two-directional flow into two well-ordered separate directional flows. We invariably observe this system when many individuals (large base) must enter or exit a door (necessitating factor) at the same time from opposite directions, or walk in opposite directions on a sidewalk where there is an obstacle to the flow of traffic. If, after several disordered passages there *by chance* should form an initial ordered flow—for example, if the people who must exit do so in a row on the right-hand side of the door, or if the passers-by on the sidewalk walking in one direction line up to pass on the right side of the obstacle—then by *necessity* even those who must enter or walk will tend to arrange themselves in an ordered flow on the right-hand side of the other group. The disordered flow is transformed into two ordered and persistent flows (macro effect); the initial disorder has turned into an ordered phenomenon thanks to the *self-organization* caused by the *micro-macro feedback*, which makes it necessary to join the ordered arrangement when, by chance, this is formed.

Combinatory systems are not always natural; at times they are "designed" or "programmed" with micro and macro rules so that the base can manifest the desired macro behavior. The design itself of combinatory systems can even lead to the concrete starting up of the system. In this case "the design" can substitute "chance" in starting up the system. All systems formed during epidemics are natural systems, as are those systems which are produced through *micro rules* inserted into the genetic patrimony of living beings who make up their biological or social behaviors, as occurs, for example, in social systems that are based on the tendency for imitation among the component elements. All those systems that are formed due to the action of "ad hoc" structural rules, that permit the desired macro behavior to be achieved, are *artificial*, as occurs clearly in the case of dance or parade troops.

In principle, even *natural* systems can be "designed" by means of sufficiently *simple rules*. Let us assume that a *living form*—for example, a leaf, a tree, or even our nose—can be considered to be the macro effect of a combinatory system whose base is formed by cells that reproduce (micro behavior) by transmitting the information to reproduce to contiguous cells. The information that each cell receives is limited to the number of cells that have already received the information and is gradually modified as the living form acquires its *form* (also considering, of course, the function for which the cell is programmed). The macro effect of the macro behavior is the production of the living form, but this conditions the micro behavior of the new cells being formed.

Let us imagine a possible process leading to the formation of a leaf (this is just an intellectual experiment). The first cell of the leaf must be able to reproduce and transmit to the offspring information on how they in turn must reproduce and transmit information to the new cells. However, such information depends on how much of the leaf has already been formed, and thus the new cells modify the state of the leaf, which in turn directs the reproduction of new cells, in a typical *micro-macro feedback*.

To understand the mechanisms for designing *artificial* systems, suppose that a scientist, through the genetic manipulation of an individual in a population of bacteria, succeeds in causing a mutation. If the mutation spreads to a sufficient number of individuals, and the minimum activation density is reached, then the combinatory system that changes the previous species or creates a new species can begin to operate (Sect. 2.6.8). In this case it was the scientist himself who substituted for "chance"; nevertheless, for purposes of studying combinatory systems, it makes generally no difference whether the first casual input is natural or desired by some agent. The growing resistance of bacteria to antibiotics is a consequence of the action of the above-mentioned combinatory system.

Here are other examples of artificial combinatory systems. A band of criminals forces a group of individuals to become addicted to a drug. If the minimum activation density is reached, the combinatory system involving the spread of the drug will begin. A large pharmaceutical company supports research in a given sector. If the first efforts are successful, the combinatory system of the researchers who produce further improvements in that area of research will begin. A large car manufacturer undertakes research that produces a useful innovation. If successful, the innovation will be used and appreciated. The combinatory system involving the adoption of that innovation may begin. If the innovation is an electric engine that solves the pollution problem, then it is likely that important strengthening actions will occur: the elimination of gas pumps, the installation of energy refueling stations, tax rebates for buying new automobiles, and so on.

1.4.2 Expansion, Organization, and Ramification of Natural Combinatory Systems

Natural systems show several interesting characteristics: on the one hand, they can *expand* beyond the environment in which they are originally observed; on the other, they can *organize* themselves into specialized subsystems, thereby transforming themselves into *operative* systems. Over time combinatory systems can also show *ramifications*.

Regarding the first characteristic—*expansion*—let us recall that combinatory systems are defined in the context of a given *environment*, in which both the micro and macro *rules* can act on the N elements of the system base. The base of the system is thus defined as the closed set of elements in the originally observed environment. However, natural biological systems are able to develop their effects on a limited *base*, in a defined *environment*, but they are also able to *expand* their effects on elements belonging to a vaster environment. With the *expansion* effect, the micro and macro *rules* that initially operated on a limited base of N elements are extended to an open set of elements.

A fashion arises in a city but spreads all over the country. A disease explodes into an epidemic in a specific area but expands in others. Human life is typical of the earth, but procedures are under way for its expansion by looking for other habitable planets.

Organization is a characteristic typical of operative systems, but it can also be observed in many natural biological combinatory systems in which the individual elements—for example, cells—can take on different states. Each cell specializes in its own function by taking on the appropriate state in relation to the position it occupies in the system. The form once again appears to guide the system toward a given evolution according to the micro rules contained in the genetic code.

A desert area is colonized when after a certain number of settlements the "urban settlement" (Sect. 2.2.4) combinatory system starts to operate: the presence of settlements attracts other settlements and a town or city is formed. At the beginning all the settlements are of the same type (e.g., settlements of farmers), but when the density increases some colonies become animal breeding or industrial settlements and so on, and specialized partial combinatory systems are formed (farmers, breeders, industrialists, warriors, priests), giving rise to the typical operative system of a city or even a nation.

The characteristic of *ramification* appears in combinatory systems, typically of the diffusion variety, that have a temporal dynamics during which part of the *base* is transformed into another combinatory system that can subsequently expand, and whose elements have certain features in common with the other system and others that are different. We can thus imagine that a new branch takes off from the original system that in time can be reabsorbed or independently maintained. In the latter case, other ramifications can subsequently occur.

There are many systems which permit ramifications: the history of evolution testifies to the progressive ramification of combinatory systems which, beginning from the primordial organisms, has produced the variety of all living species.

On the description of the branching of the living beings, Humberto Maturana and Francisco Varela, in their work *The Tree of Knowledge* (1987), refer to branching as the consequence of *genetic drift*; that is, the change in the frequency of a gene variant in a population due to random diffusion of the allele in new organisms. This typical phenomenon produced by a combinatory system of diffusion is similar to that of the diversification of languages in a population (Sect. 2.3.3).

Monod, in *Chance and Necessity* (1970), considers the combinatory system of evolution and justifies the branching in terms of random mutations in the genotype and of uniform replications of the new genotype; the resulting phenotype change spreads if it has advantages for life; otherwise it tends to disappear. Let us not forget that Darwin believed in the common origin of species and saw the production and spread of varieties of species as proof of the environmental mechanism for maintaining branching in combinatory systems represented by populations of living beings: in other words, the appearance, spread, and extinction of the species.

> Owing to this struggle for life, any variation, however slight and from whatever cause proceeding, if it be in any degree profitable to an individual of any species, in its infinitely complex relations to other organic beings and to external nature, will tend to the preservation of that individual, and will generally be inherited by its offspring. The offspring, also, will thus have a better chance of surviving, for, of the many individuals of any species which are periodically born, but a small number can survive. I have called this principle, by which each slight variation, if useful, is preserved, by the term of Natural Selection, in order to mark its relation to man's power of selection. (Darwin, 1859, p. 61)

It would be profitable to meditate on the vision of evolution proposed by Dawkins (1989, 2004), which places at the center of evolution the search for stability in the *genes* that transform the phenotypes of organisms (their *survival machines*) in order to ensure greater *chances of survival* for their offspring in subsequent generations.

> The gene is defined as a piece of chromosome which is sufficiently short for it to last, potentially for long enough for it to function as a significant unit of natural selection. Exactly how long is 'long enough'? [...] It is its potential immortality that makes a gene a good candidate as the basic unit of natural selection. (Dawkins, 1989, pp. 35–36)

> We are survival machines—robot vehicles blindly programmed to preserve the selfish molecules known as genes. This is a truth which still fills me with astonishment. (*ibidem*, p. xxi)

To better understand the diffusion of ramified combinatory systems, let us also consider all those systems that transmit language; after several generations these are transformed into branches with various dialects that, after further generations have passed, become autonomous languages (Sect. 2.3.3). Even the spread of scientific and technical innovations proceeds in stages or by ramifications (Sect. 2.6.4).

1.4.3 Robustness and Persistence in Combinatory Systems

It is easy to verify that some combinatory systems are very fragile; all that is needed is a modest weakening in order for the action of the system to cease or even for the system to break up. At other times the systems resist any form of weakening and, even without particular strengthening actions, they operate and develop, even to the point where they organize themselves.

Robustness, or resilience, is defined here as the characteristic of a combinatory system to resist perturbations and weakenings that, in some way, limit the production of the macro behavior or the macro effect. The most robust combinatory system is the one that equally produces the macro behavior or the macro effects even when part of the *base* is eliminated or when the *environment* modifies its *form*.

> Resilience (from the Latin etymology resilire, to rebound) is literally the act or action of springing back. As a property, two strands can historically be identified: a) in social psychology [...], where it is about elasticity, spirit, resource and good mood, and b) and in material science, where it is about robustness and elasticity. The notion of resilience has then been elaborated [...] A common point to the above senses of the notion of resilience is the ability to successfully accommodate unforeseen environmental perturbations or disturbances. (Laprie, 2008)

One of the most robust systems is that of the spread of drugs. We only need a few drug addicts and pushers for the system to form again. Another robust system involves the spread of a religion. At the start of Christianity even the arrest of N − 1 believers allowed the Nth to form the system once again. Even the system of the lanes in a highway is very robust; it is enough for one lane to form over a short stretch in order for it to then go on for kilometers.

A *weak* system is the one called "voice-noise." A mere change in the environment or a reduction in the base is enough for the background noise not to appear when speaking arises among the elements in the base (Sect. 2.4.1).

Persistence is defined as the characteristic of a combinatory system to be able to resist over time and to operate for a long period of time. The most persistent is the one that over time reproduces the macro behavior even when the base activity is reduced to the minimum density, even to a single element of the entire base. The most *volatile* is the system that, though achieving the maximum density in a short time, ceases just as suddenly to produce its macro behavior. Among the most persistent combinatory systems are those involving language and religion. As we can easily verify, many languages and religions have persisted for hundreds, even thousands of years. The most volatile system is that of applause, or the look-at-the-moon system (Sect. 2.3.7), even if such systems should be considered robust. In fact, even if the base is halved, applause can just as well break out.

Robust systems are not always persistent, but usually *robustness* and *persistence* are characteristics that occur together. In fact they depend on *necessitating* factors; that is, on the micro rules that trigger the feedback between the micro and macro behavior. The more this rule makes it necessary, convenient, or useful for the individual to adapt his or her micro behavior to the system's macro behavior, the more the latter will be robust and persistent.

In order to thoroughly understand the chance-necessity mechanism that characterizes combinatory systems, we must examine the reasons for the *micro-macro feedback*, trying to determine the nature and form of the micro rules and their dependence on the "will" of the individual.

1.5 Typology of Combinatory Systems

1.5.1 Five Relevant Classes of Combinatory Systems

If we classify combinatory systems according to their macro behavior or effects, despite the variety of phenomena produced we can determine five fundamental types of combinatory systems (Mella, 2005):

1. Systems of Accumulation, whose macro behavior leads to a *macro effect* which is perceived as the accumulation of objects, behavior, or effects of some kind; the accumulation represents the *global information* that directs the choices of the agents to accumulate or disperse the objects. If an agent has to accumulate some object with others similar in nature (micro behavior), he normally looks for already-made accumulations, since this gives him an advantage or reduces some disadvantage (necessitating factor). This logic applies to quite a diverse range of phenomena, among which the formation of urban or industrial settlements of the same kind or of industrial districts, and the accumulation of garbage, graffiti, and writings on walls. However, it can also be applied to phenomena such as the breaking out of applause, the formation of lines in fashion shows, the grouping of stores of the same type in the same street (Sect. 2.7.1), and the formation and maintenance of colonies, forests, herds, and schools.
2. Systems of Diffusion, whose *macro effect* is the diffusion of a trait or particularity, or of a "state," from a limited number to a higher number of agents of the system; the *global information* is represented by the observed or hypothesized diffusion of objects or features among the collectivity. A greater diffusion (macro effect) implies a greater desire to acquire the object (micro effect); the single acquisition (micro behavior) widens the collective diffusion (macro behavior). Systems of diffusion explain quite a diverse range of phenomena: from the spread of a fashion to that of epidemics and drugs, from the appearance of monuments of the same type in the same place (e.g., the towers of Pavia) to the spread and maintenance of a mother tongue or customs.
3. Systems of Pursuit, which produce a gradual shifting of the system toward an objective, as if the system, as a single entity, were pursuing a goal or trying to move toward increasingly more advanced states. The *global information* is represented by the distance of the agents' states from the objective to be reached; this model can represent quite a different array of combinatory systems: from the pursuit of records of all kinds to the formation of a buzzing in crowded locales; from the start of feuds and tribal wars in all ages to the overcoming of various types of limits.
4. Systems of Order, and Conservation, which produce macro behavior, or a macro effect, perceived as the attainment and maintenance of an ordered arrangement among the agents that form the system. The *global information* is represented by the difference of the agents' position from that of other ordered agents. Systems of order can be used to interpret a large number of phenomena: from the sponta-

neous formation of ordered dynamics (to an observer) in crowded places (dance halls, pools, city streets, etc.) to that of groups that proceed in a united manner (herds in flight, flocks of birds, crowds, etc.); from the creation of paths in fields, wheel-ruts on paved roads and successions of holes in unpaved roads, to the ordered, and often artificial, arrangement of individuals (stadium wave, cancan dancers, Macedonian phalanx).

5. Systems of Improvement and Progress, whose effect is to produce progress, understood as an improvement in the overall state of a collectivity that is attained through individual improvement. Individual improvements raise the parameter that measures collective progress, which represents global information; this leads to the formation of positive and negative gaps that push the individuals to improve in order to increase the gaps (if positive) or eliminate them (if negative). The system must be able to notice the individual improvement and adjust the progress parameter to the average (or, more generally, to the combination) of the individual improvement measures. In particular these systems offer a simple interpretation of the operational logic of the continuous and irreversible increase in productivity, quality, well-being, and scientific progress (Sect. 2.6).

1.5.2 Models for Representing Combinatory Systems

To understand the collective phenomena described in the above categories, we must try to build models that represent, in a clear and simple way, the operative mechanisms of the combinatory systems these phenomena produce. Combinatory systems can be represented by different models of increasing complexity.

The simplest models are the *descriptive* ones that indicate in words (as done so far in this chapter), or by patterns analogous to Fig. 1.12 (energy inputs are not included), the fundamental elements necessary for understanding the operative logic of systems that produce observable collective phenomena. Descriptive models are useful but not sufficient to specify the modus operandi of different classes of combinatory systems.

More precise are the *heuristic models* that the observers build to understand the logic of specific combinatory systems, or a class of combinatory systems. Such models try to simulate the system's dynamics by stating—or constructing ad hoc—a set of rules necessary, and perhaps sufficient, to produce observable collective phenomena. These rules must specify with as much detail as possible:

(1) The micro, or *necessitating rules* producing the micro behavior of agents; these rules interpret the action of the necessitating factors.
(2) The macro, or *recombining rules*, which explain the action of the recombining factors in producing the system's macro behavior.
(3) The modus operandi of the *micro-macro feedback*, which allows the system to generate the observed phenomena.
(4) The strengthening, weakening, and *control actions*, when possible or admitted, and their effects on the micro and macro behavior.

The heuristic model (like the verbal one) can also include a graphic model that describes all the typical elements.

> Heuristic methods for system modelling constitute all those methods and techniques that, although unconfirmed, may be valuable for discovering, revealing, or suggesting the qualitative and quantitative nature and characteristics of actual systems. Such methods, although unproven or even incapable of verification, can be useful, nevertheless, in the construction of both practical and accurate system models. What is probably most distinctive about heuristic methods is that they generally lack firm theoretical bases and are, therefore, subject to intuition and opinion. (Sandquist, 1985, p. 269)

Finally, we can build mathematical and statistical *simulation models*, which make it possible to determine the quantitative dynamics, associated with micro and macro behavior. Among the *mathematical-statistical models*, particularly effective are the Combinatory Automata which, in their simplest form, represent combinatory systems, numerically quantifying rules (1.1) to (1.10) in Sects. 1.3.1–1.3.3. The Combinatory Automata show the system arranged as a matrix whose N cells, for a given instant t_h, represent the micro states of the N agents, $x_n(t_h)$, n = 1 to N, and/or the micro *effects*, $e_n(t_h)$, that correspond to those states. The "combination" (based on the *recombining rules*) of the $x_n(t_h)$ allows us to quantify the macro state, $Y(t_h)$ of the collectivity, which is considered as a whole, and/or the global macro effect $E(t_h)$, which, as we know, in most cases can be interpreted as *self-produced global information*, based on which the agents determine their new micro state $x_n(t_{h+1})$, which produces the new macro state $Y(t_{h+1})$ together with the micro and macro *effects*.

Chapter 2 will construct heuristic models from a significant sample of combinatory systems. Chapter 3 will present the Combinatory Automata approach, together with several examples of applications.

1.5.3 Social Combinatory Systems: The Modus Operandi

Combinatory System Theory considers combinatory systems composed of agents of all kinds. However, the most interesting combinatory systems are the social ones, which are made up of people or organizations and entail the following definition, which is a more specific version of the general Definition 3, in Sect. 1.1.4.

Definition 6 Social combinatory system. A "social combinatory system" is defined as any combinatory system made up of people, groups, or organizations which can compare their micro states with the *global information* associated with the macro state of the system as a whole. If the system components perceive a gap, they try to eliminate it, if judged as being negative, or increase it, if judged as being positive. As a consequence of their micro behavior, they update the variables from which the global information is produced, in a typical *micro-macro feedback* action.

The basic characteristic that distinguishes social combinatory systems based on the general definition is the fact the micro state of the agents represents a value that, in the presence of *necessitating* factors, the agents try to optimize after comparing them with the macro state of the system. In fact, through a succession of decisions the agents try to harmonize their state in order to eliminate or increase the perceived gap with the macro state. Due to the presence of *recombining* factors, with each decision the agents independently modify the state of the system, so that an adaptive decision is once again needed, as part of a typical *micro-macro feedback* process. Therefore, the action of the *social combinatory system* derives from the decisions taken by the agents to eliminate or increase the gap between the individual and global states of the system; the "combination" of the effects of these decisions produces a new macro state for the collectivity, so that over time the gap is, in theory, continually modified but never eliminated.

The typical modus operandi of generic combinatory systems can be specified for the social class of system by introducing the *control systems* that direct the agents' micro behavior—according to the logic of control systems described in Appendix 2—and by modifying the general model in Fig. 1.8 to produce the model in Fig. 1.13.

Following Fig. 1.13, the action of social combinatory systems occurs because the macro behavior—and the associated macro effect, ME—translates into *global information* (of varying types) that determines a *common objective*, me* = f(ME), which individuals must or desire to achieve. This objective guides the choices of each agent who, noting a gap (or error), E(me), between his or her own micro behavior or micro state and that of the overall state of the system, due to the presence of an opportune set of *necessitating factors*, at time t_{h+1} decides, or is forced, to identify the most appropriate strategies to expand or reduce the perceived gap, E(me), following the micro or *necessitating rules*. This logic evidences that the system's macro behavior, or macro effect, represents (or conditions) the *objectives* behind the individual choices. Agents' decisions recursively change the value assumed by the *global information variable*, and this modifies the perceived positive or negative gaps, driving the agents to adapt their behavior through new decisions. The overall behavior of the system is determined by the *micro-macro feedback* action; as long as the *necessitating* and *recombining* factors are maintained, the *micro-macro feedback* can operate.

The mutual dependence between the *micro* and *macro* behavior (or their states or effects) represents the *micro-macro feedback* [R], which is maintained thanks to the individual control systems represented by the loops [B] in the lower portion of the model in Fig. 1.13. As a result, the combinatory system in principle repeats its cycles until the individual actions can update the self-produced global information, which makes clear to each individual the distance, E(me), separating the individual state (me) from the state representing the individual objective (me*).

As in the general model, the agents are characterized by an *initial state* at time t_0; in most cases this initial state may be assumed to be due to *chance*; the *micro-macro feedback* operates between the limits of the *minimum activation number* and the *maximum saturation number* of the agents, who reveal the state by maintaining the *micro-macro feedback*.

The sequences of state values over a period represent the macro and micro dynamics, or behavior, of the *social* combinatory system and the agents, as represented by Fig. 1.11.

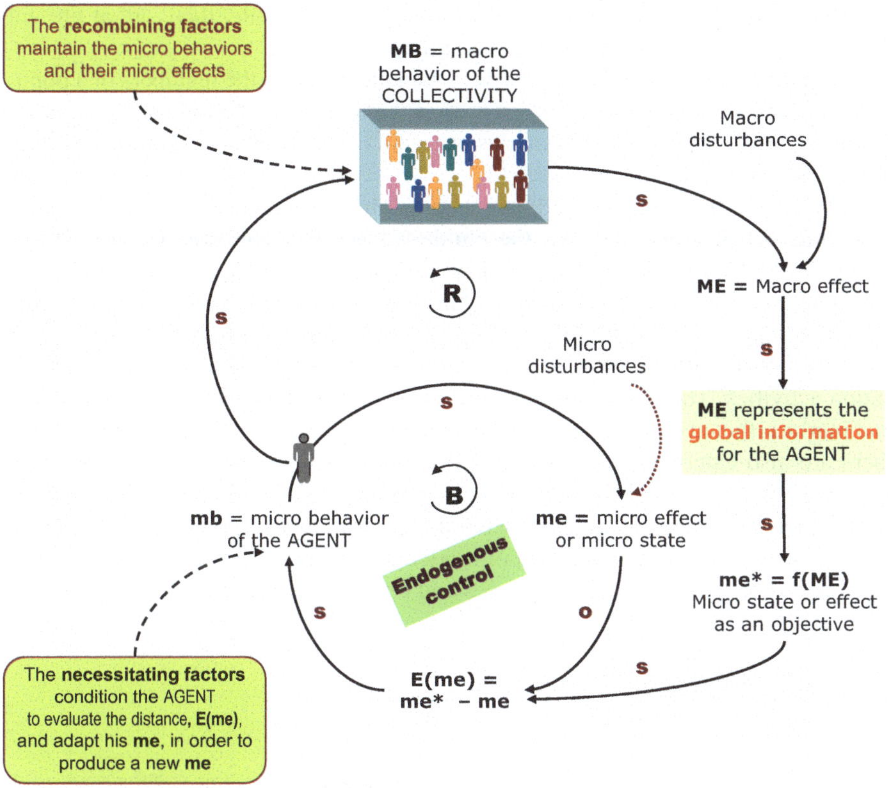

Fig. 1.13 Schematic general model of a social combinatory system (Legend: mb = micro behavior; me = micro effect; MB = macro behavior; ME = macro effect; me* agent's objective at time "t"; E(me) = gap (error) between objective and actual state) (source: Mella, 2014a, p. 343)

It clearly follows from this that *social* combinatory systems function due to the presence of *micro control systems* which, operating at the individual level, lead to uniform micro behavior by individuals in order to eliminate the gap, E(me), with respect to the "objective" that is represented—or revealed—by the *global information* (macro behavior or effect). The above description is represented in the model in Fig. 1.13. For this reason, social combinatory systems are standard *Self-Produced Global Information Systems* that must be distinguished from *local information systems*, whose typical model is represented by "complex systems" simulated by cellular automata (Sect. 1.2.2).

In order to better understand the previous definitions it is useful to consider the social combinatory system that produces the process for "passing on a *language* within a population" and maintaining it for a long time (see Sect. 2.3.3). Referring to Mella (2014a, Ch. 7), it is easy to see that the most deeply rooted tradition in any population is the learning and maintenance of the *mother tongue* by the inhabitants of an area of varying size. This is a very powerful control system given that, as reported by the SIL International and Ethnologue (2013) site, there are at present 7105 languages, in the world, without counting dialects.

The maintenance of spoken and written language among the members of a population thus appears as one of the most powerful and *persistent* combinatory systems operating in human society. The modus operandi of this combinatory system is easy to understand with the aid of Fig. 1.14.

Let us begin by stating that all parents pass on their mother tongue to their children (micro behavior), and that the children learn this language (micro effect). Generation after generation, the population communicates (macro behavior) using the same language, which becomes the mother tongue (macro effect) representing the *global information* that obliges families to transmit this mother tongue to their children (constraint/objective) in order not to disadvantage them in their communication activities. The *micro-macro feedback* is evident; however, remembering the bad marks inflicted on us by our teachers because of syntax errors, we can imagine there are also external control systems that both reinforce the main combinatory system, by detecting deviations between the language of the group members and the codified mother tongue (Eccles & Robinson, 1984, p. 141), and try to eliminate these by using the traditional levers of academic teaching to assure the correct syntax. The existence of so many languages and equally numerous dialects shows how

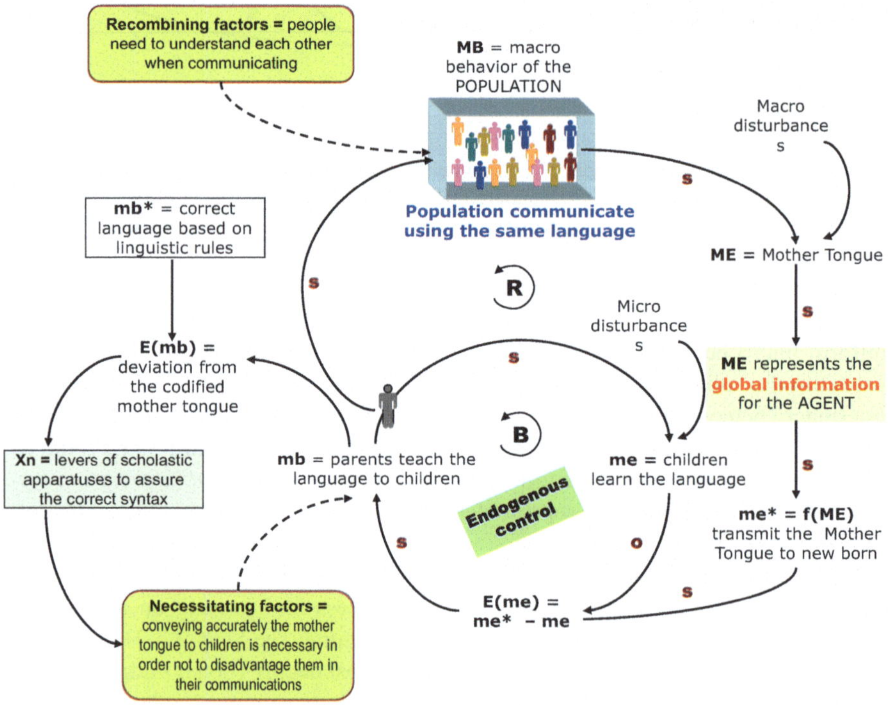

Fig. 1.14 Schematic general model of the combinatory system that produces the process for "passing on a *language* within a population," with external control of the micro behavior

powerful this combinatory system is and how efficient the individual and external control systems are. Thus, combinatory systems for the spoken language—to the same extent as those for the written language, with its more strictly codified syntax, which is less affected by accepted practice (consider all the syntax and spelling errors that have driven scholars crazy!)—are some of the most powerful systems operating in human society.

Since by definition the agents are similar and have analogous behavior, it follows that we can assume that the same *global information* produces similar decisions regarding the change in state of the agents, who, obeying their *micro control systems*, thus appear to conform or even *synchronize* their micro behavior. However, each change in state of the agents updates the *global information*, and this recreates a gap that exerts even more influence on the individuals to conform to this information.

To more easily understand the modus operandi of combinatory systems and the role of global information in producing the synchronization effect, let us return to the phenomenon of a *buzzing noise* arising in a crowded locale, whose modus operandi is shown in Fig. 1.15.

We know that the buzz arises from the voice levels (micro effects) of those present when they speak to each other (micro behavior), which are recombined by the shape of the room. But why does the collectivity speak in a loud voice (macro

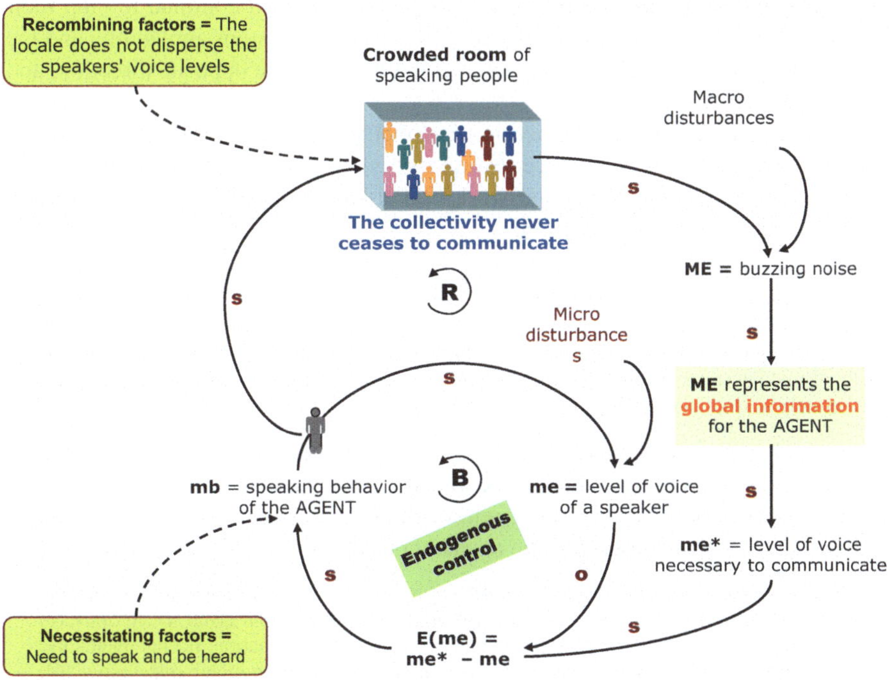

Fig. 1.15 Schematic general model of the combinatory system that produces the *buzzing noise* arising in a crowded locale

behavior)? Because there is a buzzing noise (global information) which prevents them from being heard. This becomes the minimum constraint (objective) to exceed in order to be heard; if the voice level is not adequate (gap), it is not possible to communicate. Thus, if the buzz increases, those present, in order to be heard, must unconsciously raise their voices. It seems they do this all together, as if the *global information* forces them to *synchronize* their micro behavior; but this causes the buzz to increase further, which obliges those present to raise their voices even more, which increases the buzz, which forces those present, and so on, as part of a reinforcing loop that takes the buzz to the maximum bearable level (constraint). Once this is reached, the individual control systems induce the speakers to be quiet, as we have all witnessed firsthand.

Another typical example of evident *synchronization* is that of applause. How many times have we experienced this! A certain number of persons attend an event. Suddenly someone—by chance or directed by someone—claps (micro behavior), thereby producing a typical sound (micro effect). If the number of those that begin to clap does not reach the minimum activation number, then the applause does not begin. But if the initial clapping does not die down, others will join in to produce thundering applause. The micro behavior translates into macro behavior (everyone applauding), of which the applause, understood as a typical sound, represents the macro effect and the global information, according to which the subsequent micro behavior is synchronized. The individual spectators produce the applause by clapping their hands, but this obliges everyone by necessity to continue to clap their hands in order to sustain the applause itself, until someone stops applauding and the macro effect fades away. The feedback inevitably acts in the opposite direction, and the applause slowly dies out.

More in general, due to the action of global self-produced information, social combinatory systems may produce several forms of *self-organization*, as Foster and Metcalfe (2001a), considering the work of Pelikan (2001), tried to demonstrate, starting from this simple definition:

> Processes during which disorderly sets of agents spontaneously form orderly organizations have been described many times, but to my knowledge (and taste) not yet fully explained in clear operational terms. The perspective on organization is an effective remedy to this lack of clarity. Its contribution begins with a simple clear definition: self-organizing is organizing for which some or all of the information required is supplied by the constituent agents themselves. (Pelikan, 2001, p. 130)

On the basis of this definition, Pelikan also attempted to explain the appearance of self-organization by introducing the idea of a feedback between a flow of (organizational) information, circulating within the organizational structure of agents, and the capability of learning of the agents themselves:

> The true problem with flexible self-organizing is that it can generate intricate organizing feedbacks: (a) agents start forming an organization using their initial given selectivities, (b) the organization leads to the production of certain current information, (c) the agents use their learning abilities to respond to this information by modifying their selectivities, (d) these in turn modify the organization; and so on. Much like feedbacks in automatic control, organizing feedbacks also may generate complex adjustment processes, during which both

> the organizations formed and their agents may undergo a long series of mutually caused modifications, possibly involving path dependency, or cycles which may or may not converge to stable states (attractors, equilibria). What may confuse a superficial observer is that the feedback effects of an organization upon its agents are often more conspicuous than the forward effects of individual agents upon their organization. These effects may often be only infinitesimal, and thus easy to overlook, but without them the organization could not have been formed and maintained in the first place. (Pelikan, 2001, pp. 132–133)

Combinatory System Theory represents a theory which is much simpler and more complete, in that it demonstrates the importance of the *micro-macro feedback* action in generating self-organization, order, and synchronization even in collectivities of unorganized, similar agents.

1.5.4 The External Control of Social Combinatory Systems

As shown in Fig. 1.13, and in the subsequent figures, combinatory systems can operate thanks to the *endogenous control* implemented through the micro control systems that allow agents to adapt their individual states to the global information arising from the macro state/effect of the system as a whole. Nevertheless, as we know from Sect. 1.5.4, combinatory systems can, in turn, also be controlled to favor or inhibit the attainment of certain individual and collective results considered to be harmful or useful. Without an *external control* on the combinatory system, the effects of collective action based on self-produced global information could lead to harmful, even disastrous effects. When, on the contrary, the combinatory system produces a useful macro effect, then the external control could accelerate the formation of the macro effect; we have seen this clearly, for example, in the formation of industrial districts (macro effect), which are formed when new firms are enticed to locate in a particular district (micro behavior), which makes the area even more advantageous by offering more employment and increasing the local wealth.

What happens if the combinatory system acts freely, without any external control? In some cases a type of self-control is produced: no one can continue raising his voice to overcome the buzz in the crowded room; such a volume is reached that people spontaneously stop talking, and then the buzz diminishes (only to soon begin to rise again). In other cases the combinatory system finishes operating when the maximum admissible density (or maximum level of the macro effect) is reached, making any further individual action impossible.

Leaving aside the spontaneous cessation of combinatory systems, there is no doubt that very often *external controls* are carried out on combinatory systems by using *external* control systems constructed expressly for that purpose. In order to understand how these external control systems for combinatory systems operate, we must always remember that the *micro-macro feedback* that produces the micro and macro behavior requires the contemporaneous presence of *necessitating* and *recombining* factors, by acting with the appropriate levers on the *necessitating* and *recombining factors*.

By using the proper *necessitating* and *recombining* factors, it is possible to activate two forms of *external* controls: an external *macro control* which, by acting directly on the *recombining* factors, tries to modify the *macro behavior* of the collectivity as a whole, and an external *micro control* which, by acting on the *necessitating* factors, seeks to modify the agents' *micro behavior*.

By adding these two forms of external control to Fig. 1.13 we obtain the complete model in Fig. 1.16, which clearly shows that when the *macro* behavior (MB)—or, in many cases, directly the *macro* effect (ME)—of the combinatory system must be guided toward a desired objective or limit (MB*), set by some policymaker in the socio-economic environment within which the combinatory system operates, specific external control systems can be activated whose control levers, [Xn], are represented by *reinforcing* or *weakening measures* (actions, provisions, constraints, limits, obligations, etc.) which, operating at a macro and/or micro level, modify the *recombining* and *necessitating* factors, influencing the *macro* and/or *micro* behavior and thus directing the *macro* behavior of the combinatory system (see Appendix 1).

As an example, let us apply all this to the combinatory system of *accumulation* that gives rise to industrial settlements or the formation of industrial districts (Mella, 2006, 2014a). When a certain area offers a positive differential in terms of economic efficiency compared to others (necessitating factor), then a certain number of entrepreneurs may decide to locate there (micro behavior) and set up production and commercial units (micro effect). The presence of a group of firms (global

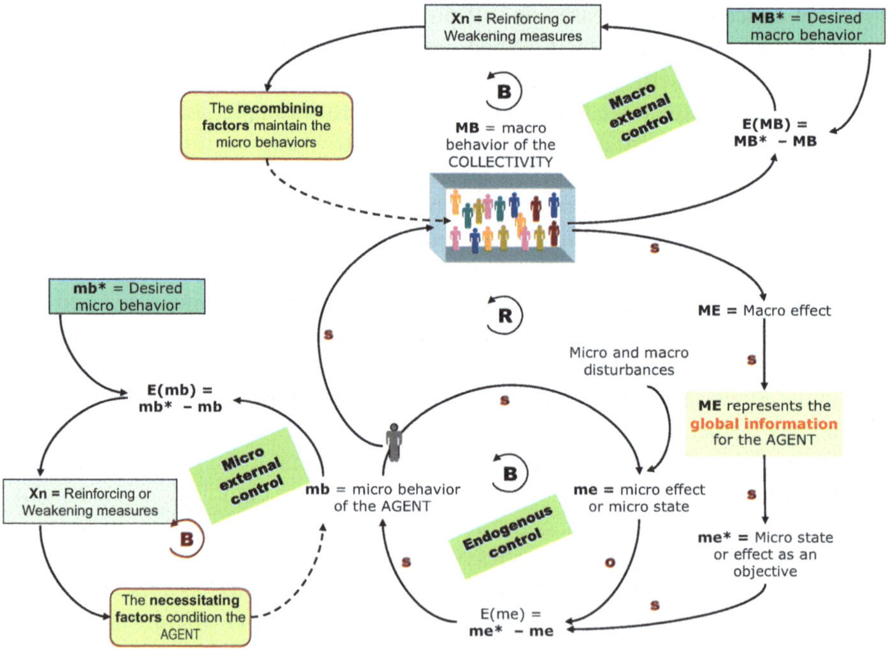

Fig. 1.16 External macro and micro controls in combinatory systems (source: Mella, 2014a, p. 350)

information) attracts new entrepreneurs that weigh the economic advantages of locating in the industrialized area against the disadvantages of locating in non-industrialized areas, in so doing activating a typical *micro-macro feedback* process.

When the *minimum activation density* is reached (number of firms needed to produce a positive differential in economic efficiency), the combinatory system starts up and, in subsequent cycles, produces the collective inflow of other production units (macro behavior) and a rise and expansion in the typical industrial and commercial districts (macro effect). The system can also be controlled by both *reinforcing actions* (outright grants to locate in an area, the creation of infrastructures, etc.), which create new economic efficiency differentials, and *weakening actions* (e.g., urban regulations, higher taxes and charges). Reinforcements and weakenings are the levers [Xn] that control the combinatory system, and these can be applied both directly to the *macro* behavior—*macro* control—or to the *micro* behavior, thereby activating the *micro* control.

Without an adequate number of external *macro* and *micro* control systems the combinatory systems can also produce harmful macro effects, since they are particularly sensitive to aspects of counter-productivity: self-deregulation and self-disorganization, which have been examined in Sect. 1.3.9.

Appendix 1: The Language of Systems Thinking (the Basics)

Systems Thinking is one of the most powerful tools of knowledge and understanding because it teaches us to devise coherent and sense-making models of the world (Sterman, 2001). Systems Thinking was presented by Peter Senge in a way that could be understood by a broad readership in his important work, *The Fifth Discipline: The Art and Practice of the Learning Organization* (1990). Senge's work introduced a powerful logical approach linked to an effective *formal language* in order to construct interpretative and operational models to simulate the dynamics of a wide range of phenomena (Senge, 1990; Senge & Lannon-Kim, 1991).

In his masterly book, Senge presents Systems Thinking in an intuitive way, but he does not provide the logical principles behind it. My recent book, *Systems Thinking. Intelligence in Action* (Mella, 2012), attempts to recognize the fundamental rules and principles as well as the cultural background of this discipline. The logical structure of Systems Thinking can be summarized in five fundamental rules which the systems thinker must follow at all times.

First Rule: *If we want to describe and understand the world we must be able to* "see the trees and the forest"; *we must develop the capacity to* "zoom" *from the whole to the parts, from systems to components, and vice versa.*

Second Rule: *We must not limit our observation to that which appears constant but* "search for what varies over time"; it is *the* variables over time that *interest the systems thinker.*

Third Rule: *If we truly wish to understand reality and change we must make an effort* "to understand the *cause* of the variations in the variables we observe" *by forming chains of causal relationships among the connected variables* while identifying and specifying:

1. The *processes* that "produce" the dynamics in the variables, and the *machines* (or systemic structures) that "produce" those *processes*
2. The *variables* that "carry out" those processes (causes or inputs), and those that "derive" from the processes (effects or outputs)

Systems Thinking admits that the processes that produce variations can be conceived of as a *black box* whose internal structure and functioning may not even be known. What is truly indispensable is to understand the rules (laws, functions, operations) by which the variations in the input variables cause variations in the output variables.

Fourth Rule: *It is not enough to search for the causes of the variations we observe; we must also link together the variables in order to* "specify the loops among all the variations."

In brief, we must see the world in terms of circular processes, or feedback loops, abandoning "linear thinking" ("laundry list thinking"), which only considers chains of causes and effects, and becoming accustomed to "circular thinking" (loops and Causal Loop Diagrams), identifying the loops that interconnect the variables.

The Systems Thinking paradigm offers alternatives to each of these assumptions. First, according to this paradigm, each of the causes is linked in a circular process to both the effect and to each of the other causes. Systems thinkers refer to such circular processes as feedback loops (Richmond, 1993, p. 117).

We can conclude by stating that systems thinker must strive to link the variables until we find the circular connections that unite them; that is, we need to identify the loops between their variations. Loops are the fundamental elements of the "systems" in Systems Thinking.

Fifth Rule: *When we observe the world we must always* "specify the boundaries of the system" *we wish to investigate.*

Systems Thinking employs formal but simple *language*. The relations of cause and effect between variables can be simply represented using arrows that unequivocally correlate their variations. To represent the causal connection between variables, a simple directional arrow of any shape can be used to link the "cause" variable to the "effect" variable. The cause (input) variables are written in the tail of the arrow; the effect (output) variables at the head of the arrow. By connecting, in succession, several variables between which a causal relationship is observed or hypothesized, an open causal chain is formed, which can be very long due to the linear concatenation of numerous variables.

Two variables have the same direction of variation ("**s**") if increases or decreases in the former result in corresponding increases or decreases in the latter (first arrow of the model shown below). They have the opposite direction ("**o**") if increases or decreases in the former result in corresponding decreases or increases in the latter (second arrow).

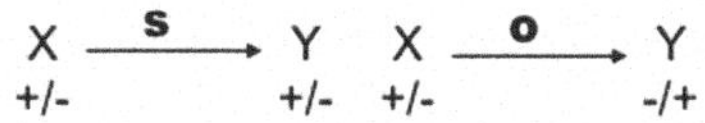

A *basic loop* is a circular link between two variables which can be interconnected in two opposite directions, forming a *closed causal chain*; a loop is *compound* when more than two variables are joined in a circular link. There is no limit to the number of interconnected variables.

There are only two basic types of loop:

1. *Reinforcing* loops [**R**], which produce a *reciprocal increase* or *reduction*—in successive repetitions of the system's cycle—in the values of the two variables, which have an identical *direction of variation*: "**s** and **s** "or "**o** and **o.**"

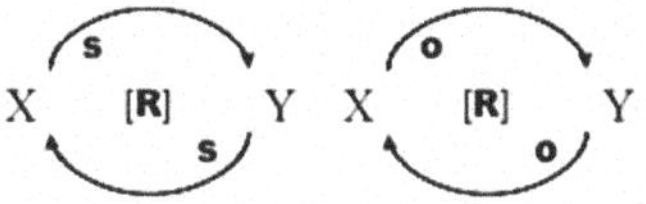

2. *Balancing* loops [**B**], which maintain relatively stable the values of the connected variables, which are connected by a different direction of variation: "**s** and **o"** or "**o** and **s.**"

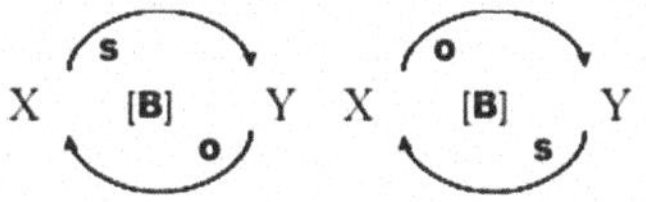

A system of loops in which all variables are linked by arrows, without there being an initial and final variable, is defined as a *Causal Loop Diagram* (Sterman, 2000).

Systems Thinking, precisely because it is a tool for developing our intelligence, must be learned gradually through practice and continual improvement. It is a "discipline" that requires the systems thinker to have a deep knowledge and to constantly apply its rules, as well as to have the willingness to continually improve: *A discipline is a developmental path for acquiring certain skills or competencies. […] To practice a discipline is to be a lifelong learner. You "never arrive"; you spend your life mastering disciplines* (Senge, 2006, p. 10).

This objective can be achieved by developing the necessary competencies (Richmond, 1993) in order to:

- Perceive and recognize the circularity of phenomena
- See how systems really function
- Intuit the effects of actions over time
- Feel responsible for the system's performance
- Predict the future
- Simulate dynamics of variables and systems (Forrester, 1961; Sterman, 2000)
- Control dynamic processes

Appendix 2: Control Systems (the Basics)

A variable Y_t is defined here as "controllable" if, on a temporal, discrete, or continuous scale, t = 1, 2, …, we can assign it a given value Y* (set-point) which can represent an objective, goal, constraint, or limit of Y_t. If $Y_t \neq Y^*$, we can measure a distance, gap, variance or *error*, which we denote by $E(Y)_t = Y^* - Y_t$. X_t is defined here as a control variable (or a control lever) which determines the values of Y_t according to a *causal relation* (defined by some process or apparatus), so that, by acting on X_t, we can produce dynamics for Y_t that tends toward Y*.

A control system is defined here as any set of apparatuses, logical or technical (algorithm or machine, rule or structure, etc.) that, for a set of instants, perceives $E(Y)_t$, calculates, and assigns the values X_t, thereby producing the appropriate Y_t to gradually annul, when possible, the error $E(Y)_t = Y^* - Y_t^*$ at instant t*.

Without going into detail about the history of Systems Thinking, here is how Norbert Wiener elegantly presents the idea and the function of both "distance"—that is, error—and the technical structure subject to the control.

> Now, suppose that I pick up a lead pencil. To do this, I have to move certain muscles. However, for all of us but a few expert anatomists, we do not know what these muscles are; and even among the anatomists, there are few, if any, who can perform the act by a conscious willing of the contraction in succession of each muscle concerned. On the contrary, what we will is to pick the pencil up. Once we have determined this, our motion proceeds in such a way that we may say roughly that the amount by which the pencil is not yet picked up is decreased at each stage. This part of the action is not in full consciousness. To perform an action in such a manner, there must be a report to the nervous system, conscious or unconscious, of the amount by which we have failed to pick up the pencil at each instant. (Wiener, 1961, p. 7)

The control system is repetitive and functions by means of action (X acts on Y) and reaction (E(Y) acts on X through Y); with a certain number of iterations on the control lever, it tries to achieve the objective (goal-seeking systems) or to respect the constraints or limits (constraint-keeping systems).

Among the various possible models, we chose to represent (in more detailed form) the simple logic of control according to the standard model of the control system in Fig. 1.17, which clearly shows that in Systems Thinking terminology a control system is nothing other than a balancing loop where the values of X_t—which determine Y_t in the direction "**s**" ("same")—do not depend on Y itself but on the error, $E(Y)_t$, in the direction "**s**" (the higher the error, the greater the corrective intervention of the X). It is clear that the higher the Y_t is, the lesser the *error* is; for this reason, there is a relation in direction "**o**" ("opposite") between Y and E(Y).

The variable X_t (or, if more than one, the vector [X]) is also defined as the action variable, the active variable, or the *control lever*. If [X] is composed of N action variables, the system is called a *multi-lever* control system. The variable Y_t (or vector [Y]) is also defined as the controlled or passive variable. If [Y] is composed of M variables, the system is called a *multi-objective* control system. The manager of the control system is defined (in the broadest sense of the term) as the subject (individual, group, organ, or organization) that, through a series of decisions—based on

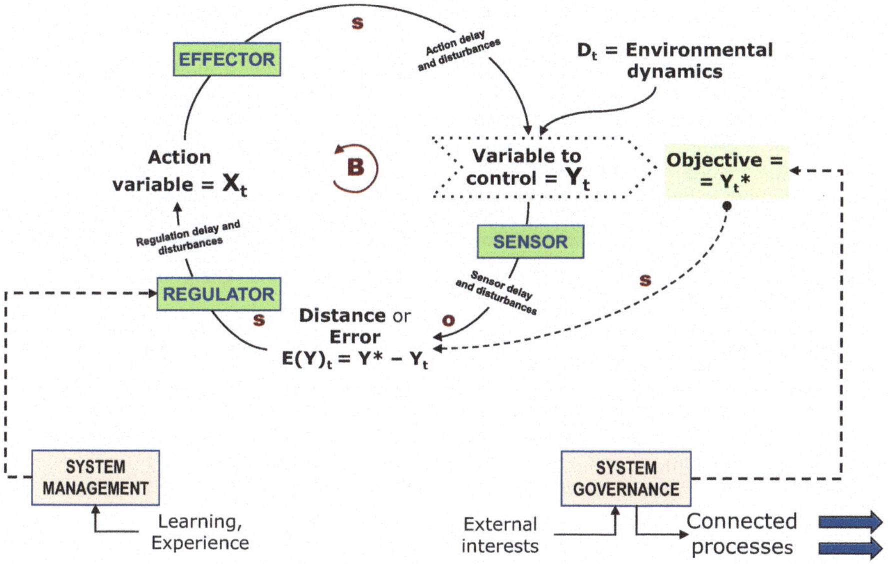

Fig. 1.17 Standard model of a One-Lever Control System (source: adapted from Mella, 2021, p. 77)

his/its particular culture, experience, and preferences—can regulate the X_t in order to change the Y_t. The governance of the system is defined here as the process by which the objective Y*, or the vector [Y*], is determined.

In multi-lever control systems, especially those where the control levers can be activated independently of one another, the manager must choose the activation order for the various levers in order to define the *control strategy* to be adopted. In multi-objective control systems the control requires that the manager determine the order of priorities of the objectives by first deciding on a *control policy*.

The structure and typology of control systems, and their ubiquitous presence in all observed contexts, have been analyzed in the book entitled *The Magic Ring* (Mella, 2014a).

References

Ahmed, E., Elgazzar, A. S., & Hegazi, A. S. (2005). *An overview of complex adaptive systems*. arXiv preprint nlin/0506059.

Algorithmic Botany. Website of the Biological Modeling and Visualization research group. Website: http://algorithmicbotany.org

Allen, P. M. (2001). In J. Foster & J. S. Metcalfe (Eds.). *Knowledge, ignorance, and the evolution of complex systems. Frontiers of evolutionary economics: Competition and self-organization and innovation policy*.

Anderson, P. W., Arrow, K. J., & Pines, D. (1988). *The economy as an evolving complex system*. Addison-Wesley.

Ardeni, P. G., & Gallegati, M. (1999). In Gallegati & Kirman (Eds.), *Fluctuations and growth due to technological innovation and diffusion*.

Arthur, W. B., Durlauf, S. N., & Lane, D. A. (Eds.). (1997). *The economy as an evolving complex system II* (Vol. 28). Perseus Books.

Ashby, R. W. (1957). *An introduction to cybernetics* (2nd ed.). Chapman & Hall.

Ashford, N. (1999). Spontaneous order. *The Freeman: Ideas on Liberty, 49*(7). Retrieved from: http://www.fee.org/freeman/99/9907/ashford

Ashford, N. (2001). *Principles for a free society*. Jarl Hjalmarson Foundation. Retrieved from: http://www.opportunitiesforafricans.com/wp-content/uploads/2013/04/1.Principles-of-a-FreeSociety.pdf

Atlan, H., Dupuy, J. P., & Kiippel, M. (1986). *Individual alienation and systems intelligence*. IFAC Economics and Artificial Intelligence.

Axelrod, R. (1997). *The complexity of cooperation. Agent-based models of competition and collaboration*. Princeton University Press.

Aziz-Alaoui, M., & Bertelle, C. (Eds.). (2007). *Emergent properties in natural and artificial dynamical systems*. Springer Science & Business Media.

Bak, P. (1994). In: Cowan, Pines, & Meltzer (Eds.), *Self-organized criticality: A holistic view of nature* (pp. 477–496).

Bak, P. (1996). *How nature works: The science of self-organized criticality*. Copernicus, ed. Retrieved from: http://jasss.soc.surrey.ac.uk/4/4/reviews/bak.html

Beer, S. (1987). In H. R. Maturana & F. J. Varela (Eds.), *Preface*.

Berto, F., & Tagliabue, J. (2002, revised in 2023). Cellular Automata. *Stanford Encyclopedia of Philosophy*. http://plato.stanford.edu/entries/cellular-automata/

Burkhart, R. (1994). The swarm multi-agent simulation system. In: OOPSLA, *94 Workshop on "The Object Engine"*.

Camazine, S., et al. (2001). *Self-organization in biological systems*. Princeton University Press.

Campbell, D. T. (1960). Blind variation and selective retentions in creative thought as in other knowledge processes. *Psychological Review, 67*(6), 380.

Carlson, S. (2000). Boids of a feather flock together. *Scientific American, 283*(5), 112–114.

CASG, Complex Adaptive Systems Group. (2013). Website: http://www.cas-group.net

Castelfranchi, C. (1998). Through the minds of the agents. *Journal of Artificial Societies and Social Simulation, 1*(1). Retrieved from: http://jasss.soc.surrey.ac.uk/1/1/5.html

Chavalarias, D. (2016). The unlikely encounter between von Foerster and Snowden: When second-order cybernetics sheds light on societal impacts of Big Data. *Big Data & Society, 1*(11). Retrieved from: http://bds.sagepub.com/content/3/1/2053951715621086

Corning, P. A. (1995). Synergy and self-organization in the evolution of complex systems. *Systems Research and Behavioral Science, 12*(2), 89–121.

Coveney, P., & Highfield, R. (1995). *Frontiers of complexity: The search for order in a complex world*. Faber and Faber.

Cowan, G., Pines, D., & Elliott Meltzer, D. (Eds.). (1994). *Complexity: Metaphors, models, and reality* (pp. 477–496). Perseus Books.

Cox, B. (2004). *Swarm. Virtual School*. Retrieved from: http://virtualschool.edu/mon/Bionomics/Swarm.html

Cox, E. (1994). *The fuzzy systems handbook: A practitioner's guide to building, using, and maintaining fuzzy systems*. Academic Press.

Creutz, M. (1996). *Self-organized criticality*. Retrieved from: https://arxiv.org/pdf/hep--lat/9611018.pdf

Darwin, C. (1859). *On the origins of species by means of natural selection*. Murray. Retrieved from: http://darwin-online.org.uk/content/frameset?itemID=F373&viewtype=text&pageseq=1

Dawkins, R. (1989). *The selfish gene*. Oxford University Press (1st ed., 1976). Online at: http://pratclif.com/biologie-moleculaire/dawkins.htm

Dawkins, R. (2004). *The ancestor's tale: A pilgrimage to the dawn of evolution*. Houghton Mifflin.

Delorme, R. (2001). *Theorizing complexity*, in Foster, J. and Metcalfe, J. S. (2001b).

Deneubourg, J. L., & Goss, S. (1989). Collective patterns and decision-making. *Ethology Ecology & Evolution, 1*(4), 295–311.

Dewdney, A. K. (1990). *The magic machine: A handbook of computer sorcery*. WH Freeman & Co..

Dolan, A. (1998). *Floys. New members in the Artificial Life zoo*. Retrieved from: http://www.aridolan.com/eFloys.html [the site is no longer online].

Dopfer, K., & Potts, J. (2007). *The general theory of economic evolution*. Routledge.

Dorigo, M., Bonabeau, E., & Theraulaz, G. (2000). Ant algorithms and stigmergy. *Future Generation Computer Systems, 16*(8), 851–871.

Dorigo, M., Di Caro, G., & Gambardella, L. M. (1999). Ant algorithms for discrete optimization. *Artificial Life, 5*(2), 137–172.

Drogoul, A., Ferber, J., & Cambier, C. (1994). Multi-agent simulation as a tool for analysing emergent processes in societies. In *Proc. Simulating Societies Symposium: The Computer Simulation of Social Phenomena* (pp. 127–142).

Edwards, M. (2003). *A brief history of Holons*. Retrieved from: http://www.integralworld.net/edwards13.html

EMK, Complexity Group. (2014). *Complexity Lexicon*. Retrieved from: http://emk-complexity.org/guide/lexicon.html

Epstein, M. J., & Axtell, R. (1996). *Growing artificial societies, social science from the bottom up*. Brookings Institution Press.

Fedoriaka, D. (2010). *L-systems list*. Retrieved from: https://fedimser.github.io/l-systems.html

Flake, G. W. (1998). *The computational beauty of nature: Computer explorations of fractals, chaos, complex systems, and adaptation*. The MIT Press.

Fogel, D. B. (1995). *Evolutionary computation: Toward a new philosophy of machine intelligence*. IEEE Press.

Forrest, S., & Jones, T. (1994). Modeling complex adaptive systems with Echo. *Complex Systems: Mechanism of Adaptation, 2*, 3–21.

Forrester, J. W. (1961). *Industrial dynamics*. Pegasus Communications.

Forrester, J. W. (1999). *System dynamics: The foundation under systems thinking*. Sloan School of Management, MIT. Retrieved from: http://static.clexchange.org/ftp/documents/system-dynamics/SD2011-01SDFoundationunderST.pdf

Foster, J., & Metcalfe, J. S. (2001a). In J. Foster & J. S. Metcalfe (Eds.), *Modern evolutionary economic perspectives: An overview* (pp. 1–16).

Foster, J., & Metcalfe, J. S. (Eds.). (2001b). *Frontiers of evolutionary economics: Competition, self-organization, and innovation policy*. Edward Elgar Publishing.

Gardner, M. (1970). Mathematical games: The fantastic combinations of John Conway's new solitaire game "life". *Scientific American, 223*(4), 120–123.

Gell-Mann, M. (1992). Complexity and complex adaptive systems. In *Santa Fe Institute Studies in the Sciences of Complexity-Proceedings* (Vol. 11, pp. 177–177). Addison-Wesley Publishing Co..

Gell-Mann, M. (1994). *The Jaguar and the Quark: Adventures in the simple and the complex*. Freeman & Co.

Gell-Mann, M. (1995). What is complexity? *Complexity, 1*(5). Retrieved from: http://complexity.martinsewell.com/Gell95.pdf

Gilbert, N. (1995). Simulation: An emergent perspective. In: Conference on *New Technologies in the Social Sciences*, October, 27–29.

Gilbert, N. (1999). Simulation: A new way of doing social science. *The American Behavioral Scientist, 42*(10), 1485.

Gilbert, N., & Doran, J. (1994). *Simulating societies. The computer simulation of social phenomena*. UCL press.

Gilbert, N., & Troitzsch, K. G. (1999). *Simulation for the social scientist*. Open University Press.

Gleick, J. (1987). *Chaos theory: Making a new science*. Viking.

Goldberg, D. E. (1989). *Genetic algorithms in search, optimization, and machine learning*. Addison-Wesley.

Goldspink, C. (2000). Modelling social systems as complex: Towards a social simulation meta-model. *Journal of Artificial Societies and Social Simulation, 3*(2), 1–23.
Gould, S. (2000). *The theory of options: A new theory of the evolution of human behavior.* Universal Publishers/Upublish.com (2001).
Gould, S. J. (2002). *The structure of evolutionary theory.* Harvard University Press.
Grassé, P. P. (1959). La reconstrucion du nid et les coordinations inteindividuelles chez Bellicositermes Natalensis et Cubitermes. La théorie de la stigmergie: essai d'interpretation du comportement des termites constructeurs. *Insect Sociaux, 6*(1), 41–80.
Grassé, P. P. (1960). The automatic regulations of collective behavior of social insect and "stigmergy". *Journal de psychologie normale et pathologique, 57*, 1–10.
Grassé, P. P. (1967). Nouvelles experiences sur le termite de Müller (Macrotermes mülleri) et considerations sur la théorie de la stigmergie. *Insectes Sociaux, 14*(1), 73–101.
Green, D. G., & Green, D. (1993). L-SYSTEMS. *Environmental and Information Sciences*. Charles Sturt University. See: https://www-archiv.fdm.uni-hamburg.de/b-online/e28_3/lsys.html
Grimmett, G. R. (1999a). Inequalities and entanglements for percolation and random-cluster models. In *Perplexing problems in probability* (pp. 91–105). Birkhäuser.
Grimmett, G. R. (1999b). *Percolation*. Springer-Verlag.
Haken, H. (1977a). *Synergetics. An introduction.* Springer-Verlag.
Haken, H. (1977b). Some aspects of synergetics. In *Synergetics* (pp. 2–17). Springer-Verlag.
Haken, H. (Ed.). (1982). *Evolution of order and chaos: In physics, chemistry, and biology.* Springer-Verlag.
Haken, H. (1983). *Synergetics. Introduction and advanced topics*. Springer-Verlag.
Henderson, P. (Ed.). (2000). *Systems engineering for business process change: Collected papers from the EPSRC research programme.* Springer Science & Business Media.
Heylighen, F. (1989). Self-organization, emergence and the architecture of complexity. In *Proceedings of the 1st European conference on System Science* (Vol. 18, pp. 23–32). AFCET.
Heylighen, F. (1993). Blind variation and selective retention. *Principia cybernetica web.* Retrieved from: http://pespmc1.vub.ac.be/bvsr.html
Heylighen, F. (2009 on line). Self-organization. Principia Cybernetica Web. Retrieved from: http://pespmc1.vub.ac.be/SELFORG.html
Hofstadter, D. (1979). *Godel, Escher, Bach: An Eternal Golden Braid.* Basic Books, Inc.. Retrieved from: http://themindi.blogspot.it/2007/02/chapter-11-prelude-ant-fugue.html
Holland, J. H. (1975). *Adaptation in natural and artificial systems: An introductory analysis with applications to biology, control, and artificial intelligence.* University of Michigan Press.
Holland, J. H. (1992). Complex adaptive systems. *Daedalus, 121*(1), 17–30.
Holland, J. H. (1995). *Hidden order: How adaptation builds complexity.* Perseus Books.
Holland, J. H. (1998). *Emergence: From chaos to order.* Addison-Wesley.
Hölldobler, B., & Wilson, E. O. (1990). *The ants.* Harvard University Press.
Honavar, V. (2006). Complex Adaptive Systems Group at Iowa State University. Cited at: https://web.mit.edu/esd.83/www/notebook/Complex%20Adaptive%20Systems.pdf
House, R., Rousseau, D. M., & Thomas-Hunt, M. (1995). The Meso Paradigm: A framework for the integration of micro and macro organizational behavior. *Research in Organizational Behavior, 17*, 71–114.
Hubble, E. P. (1936). *The realm of the nebulae* (Vol. 25). Yale University Press.
ISAL, International Society for Artificial Life. (2016). At: http://alife.org/index.php?page=alife&context=alife
James, W. (1907). Pragmatism's conception of truth. Lecture 6 in *Pragmatism: A new name for some old ways of thinking.* Longman Green and Co, pp. 76–91. Retrieved from: https://brocku.ca/MeadProject/James/James_1907/James_1907_06.html
Jantsch, E. (1980). *The self-organizing universe: Scientific and human implications of the emerging paradigm of evolution.* Pergamon Press.
Johnson, N. L., & Kotz, S. (1977). *Urn models and their application; an approach to modern discrete probability theory.* John Wiley & Sons.

Kauffman, S. A. (1993). *The origins of order: Self organization and selection in evolution*. Oxford University Press.
Kellert, S. H. (1994). *In the wake of chaos: Unpredictable order in dynamical systems*. University of Chicago press.
Khalil, E. L., & Boulding, K. E. (Eds.). (1996). *Evolution, order, and complexity*. Routledge.
Klir, G. J. (2013). *Facets of systems science* (Vol. 7). Springer Science & Business Media.
Koestler, A. (1967). *The ghost in the machine*. Arkana.
Koestler, A. (1972). *The roots of coincidence* (2nd ed.). Hutchinson (Paperback, 1973).
Koestler, A. (1978). *Janus; a summing up*. Random House.
Koza, J. R. (1992). *Genetic programming: On the programming of computers by means of natural selection* (Vol. 1). The MIT Press.
Langton, C. G. (1989). *Artificial life* (pp. 1–48). Addison-Wesley Publishing Company.
Laprie, J. C. (2008). From dependability to resilience. In: 38th IEEE/IFIP *Int. Conf. On Dependable Systems and Networks* (pp. G8–G9). Retrieved from: https://users.ece.cmu.edu/~koopman/dsn08/fastabs/dsn08fastabs_laprie.pdf
Levy, S. (1992). *Artificial life: The quest for a new creation*. Penguin Books.
Lewin, R. (1999). *Complexity: Life at the edge of chaos*. University of Chicago Press.
Lewis, M. P., Simons, G. F., & Fennig, C. D. (Eds.). (2013). *Ethnologue: Languages of the world*. SIL International. Retrieved from: http://www.ethnologue.com/ethno_docs/distribution.asp?by
Liebowitz, S. J., & Margolis, S. E. (1995). Path dependence, lock-in, and history. *Journal of Law, Economics and Organization, 11*(1), 205–226.
Liebowitz, S. J., & Margolis, S. E. (1998 online). *Path dependence*. Online: Retrieved from: https://www.utdallas.edu/~liebowit/palgrave/palpd.html
Liekens, A. (2000). From individuals to populations, approaches to the study of biological emergent phenomena. In: Artificial Life. *Proceedings of the Seventh International Conference on Artificial Life*. The MIT Press.
Lindenmayer, A., & Rozenberg, G. (1972). Developmental systems and languages. In: *Proceedings of the fourth annual ACM symposium on Theory of computing* (Denver, Colorado, May, 1972). ACM, pp. 214–221.
Loeb, A. L. (1977). *Preface* to Fuller & Applewhite. (1977, 1979). Synergetics.
Lotka, A. J. (1925). *Elements of physical biology*. Williams & Wilkins.
Lustick, S. (2000). Agent-based modelling of collective identity: Testing constructivist theory. *Journal of Artificial Societies and Social Simulation, 3*(1), 1.
Manzoni, A. (1985). *The Betrothed* [I promessi sposi, about 1840]. Mondadori, Milano Italy. At: https://moodle2.units.it/pluginfile.php/441330/mod_resource/content/1/MANZONI-PROMESSI%20SPOSI.pdf
Martelli, M. (1999). *Introduction to discrete dynamical systems and chaos*. J. Wiley & Sons.
Maturana, H. R., & Guiloff, G. D. (1980). The quest for the intelligence of intelligence. *Journal of Social and Biological Structures, 3*(2), 135–148.
Maturana, H. R., & Varela, F. J. (1980 1st ed. 1972). *Autopoiesis and cognition. The realization of living*. Reidel Publishing.
Maturana, H. R., & Varela, F. J. (1987). *The tree of knowledge: The biological roots of human understanding*. New Science Library/Shambhala Publications.
Mella, P. (2006). Spatial co-localisation of firms and entrepreneurial dynamics. The combinatory system view. *The International Entrepreneurship and Management Journal, 2*, 391–412.
Mella, P. (2009). *The Holonic Revolution. Holons, Holarchies and Holonic Networks. The ghost in the production machine*. University Press.
Mella, P. (2012). *Systems thinking. Intelligence in action*. Springer Verlag.
Mella, P. (2021, 1st Ed. 2014). *The Magic Ring. Systems thinking approach to control systems*. Contemporary systems thinking series. Springer International Publishing.
Mitleton-Kelly, E. (1997a). In: P. Henderson (Ed.), *Complexity: Partial support for BPR* (2000).
Mitleton-Kelly, E. (1997b). *Complex adaptive systems in an organisational context: Organisations as co-evolving complex adaptive systems*. ESRC Business Processes Resource Centre, University of Warwick.

Mitleton-Kelly, E. (Ed.). (2003). *Complex systems and evolutionary perspectives on organisations: The application of complexity theory to organisations. Advanced series in management.* Elsevier.

Monod, J. (1970). *Chance and necessity: Essay on the natural philosophy of modern biology.* Alfred A. Knopf, 1971 (Original, *Le hazard et la nécéssité: Essai sur la philosophie naturelle de la biologie moderne*. Paris: Seuil).

Morgeson, F. P., & Hofmann, D. A. (1999). The structure and function of collective constructs: Implications for multilevel research and theory development. *Academy of Management Review, 24*(2), 249–265.

NECSI. (online). New England Complex Systems Institute. Cambridge, MA USA. Web Site: http://www.necsi.org

Negoita, C., Zadeh, L. A., & Zimmermann, H. (1978). Fuzzy sets as a basis for a theory of possibility. *Fuzzy Sets and Systems, 1*(3–28), 61–72.

Negoita, C. V. (1981). *Fuzzy systems (cybernetics and systems).* Abacus Press.

Pelikan, P. (2001). In J. Foster & J. S. Metcalfe (Eds.), *Self-organizing and Darwinian selection in economic and biological evolutions: An enquiry into the sources of organizing information* (pp. 121–151).

Potts, J. (2001). Knowledge and markets. *Journal of Evolutionary Economics, 11*(4), 413–431.

Prigogine, I. (1985). *New perspectives on complexity.* United Nations University. Reprinted in: Klir G. J. (2013), pp. 483–492.

Prigogine, I., & Stengers, I. (1984). *Order out of chaos. Man's new dialogue with nature.* Bantam Books (Original: La Nouvelle Alliance. Paris: Gallimard 1989).

Prusinkiewicz, P., & Lindenmayer, A. (1990). *The algorithmic beauty of plants.* Springer-Verlag.

Reynolds, C. W. (1987). Flocks, herds and schools: A distributed behavioral model. In *Proceedings of the 14th annual conference on Computer graphics and interactive techniques* (pp. 25–34). Retrieved from: http://www.macs.hw.ac.uk/~dwcorne/Teaching/Craig%20Reynolds%20Flocks,%20Herds,%20and%20Schools%20A%20Distributed%20Behavioral%20Model.htm

Richmond, B. (1993). Systems thinking: Critical thinking skills for the 1990s and beyond. *System Dynamics Review, 9*(2), 113–133.

Rousseau, D. M. (1985). Issues of level in organizational research: Multi-level and cross-level perspectives. *Research in Organizational Behavior, 7*(1), 1–37.

Sakoda, J. M. (1971). The checkerboard model of social interaction. *The Journal of Mathematical Sociology, 1*(1), 119–132.

Sandquist, G. M. (1985). *Introduction to system science.* Prentice-Hall, Inc.

Schatten, A. (2004). *Modelling on the edge of chaos: Cellular automata and agents.* Retrieved from: http://www.schatten.info/publications/presentations/modell_edge_chaos-ca.pdf

Schelling, T. C. (1971). Dynamic models of segregation. *Journal of Mathematical Sociology, 1*(2), 143–186.

Schelling, T. C. (2006 first ed. 1978). *Micromotives and macrobehavior.* WW Norton & Company.

Senge, P. M. (1990). *The fifth discipline: The art & practice of the learning organization.* Currency Doubleday.

Senge, P. M., Kleiner, A., Roberts, C., Ross, R., & Smith, B. (1994). *The fifth discipline fieldbook.* Boubleday, Random House Inc.

Senge, P. M., & Lannon-Kim, C. (1991). The systems thinking approach. *The Systems Thinker Newsletter, 2*(5). Cambridge MA: Kendall Square.

Serra, R., & Zanarini, G. (1990). *Complex systems and cognitive processes.* Springer-Verlag.

Shimizu, H. (1987). A general approach to complex systems in bioholonics. In *Lasers and synergetics: A colloquium on coherence and self-organization in nature* (pp. 204–223). Springer Berlin Heidelberg.

Smith, A. (1776). *An inquiry into the nature and causes of the wealth of nations.* Edited with an Introduction, Notes, Marginal Summary and an Enlarged Index by Edwin Cannan. London: Methuen, 1904). Vol. 1.

Stacey, R. D. (1995). The science of complexity: An alternative perspective for strategic change processes. *Strategic Management Journal, 16*(6), 477–495.

Sterman, J. D. (2001). System dynamics modeling: Tools for learning in a complex world. *California Management Review, 43*(4), 8–25.
Stonier, R. J., & Yu, X. H. (1994). *Complex systems: Mechanism of adaptation*. IOS Press.
Sugden, R. (1989). Spontaneous order. *Journal of Economic Perspectives, 3*(4), 85–97.
Swarm online. (2024). Main page at: www.swarm.org/wiki/Main_Page
Swenson, R. (2000). Spontaneous order, autocatakinetic closure, and the development of space-time. *Annals of the New York Academy of Sciences, 901*(1), 311–319.
Toffoli, T., & Margolus, N. (1987). *Cellular automata machines: A new environment for modeling*. The MIT Press.
Ulam, S. M. (1986). *Science, computers, and people: From the tree of mathematics*. Birkhauser.
Ulam, S. M. (1991). *Adventures of a mathematician*. Univ of California Press.
Varela, F. (1979). *Principles of biological autonomy*. Elsevier.
Varela, F. (1981). *Describing the logic of the Living. The adequacy and limitations of the idea of Autopoiesis,* in Zeleny (1981).
Volterra, V. (1926). Variations and fluctuations of the number of individuals in animal species living together. *Animal Ecology*, New York, NY: McGraw-Hill (1931) [Original: Variazioni e fluttuazioni del numero d'individui in specie animali conviventi. Atti dell'Accademia Nazionale dei Lincei, Italy], 31–113].
Volterra, V. (1931). *Lessons on the mathematical theory of struggle for* life [original: Leçons sur la théorie mathématique de la Lutte pour la vie. Paris, France: Gauthier-Villars].
von Bertalanffy, L. (1968). *General system theory: Foundations, development, applications*. Braziller.
von Foerster, H. (1960). On self-organising systems and their environments. In M. C. Yovits & S. Cameron (Eds.), *Self-organising systems* (pp. 31–50). Pergamon Press.
von Foerster, H. (1972). Perception of the future and the future of perception. *Instructional Science, 1*(1), 31–43.
Von Neumann, J., & Burks, A. W. (1966). Theory of self-reproducing automata. *IEEE Transactions on Neural Networks, 5*(1), 3–14. Retrieved from: https://cba.mit.edu/events/03.11.ASE/docs/VonNeumann.pdf
Waldrop, M. M. (1993). *Complexity: The emerging science at the edge of order and chaos*. Simon and Schuster.
Weis, E. (2008). *Fundamentals of complex evolving systems: A primer*. Inst. of Social Ecology, IFF-Fac. for Interdisciplinary Studies, Klagenfurt Univ.
Whitaker, R. (2003). *The Observer Web: Autopoiesis and Enaction*. ENCYCLOPAEDIA AUTOPOIETICA WEB.
Wiener, N. (1961). *Cybernetics or control and communication in the animal and the machine*. MIT Press. Google Books (1st ed. 1948).
Wilber, K. (2000). *Sex, ecology, spirituality: The spirit of evolution*. Shambhala (1st ed., 1995).
Wilber, K. (2001). *A brief history of everything*. Shambhala (1st ed., 1996).
Wolfram, S. (1984). Cellular automata as models of complexity. *Nature, 311*(5985), 419–424.
Wolfram, S. (2004). *A new kind of science and the future of mathematics*. Joint Mathematics Meetings, Phoenix, Arizona. Retrieved from: http://www.stephenwolfram.com/publications/a-new-kind-of-science-future-mathematics/
Yager, R. R., & Zadeh, L. A. (Eds.). (1991). *An introduction to fuzzy logic applications in intelligent systems*. Kluwer Academic Publishers.
Zadeh, L. A. (1978). Fuzzy sets as a basis for a theory of possibility. *Fuzzy Sets and Systems, 1*(1), 3–28.
Zeleny, M. (1981). *Autopoiesis, a theory of living organization*. Elsevier.
Zollo, G., Iandoli, L., & De Maio, L. (2001). *An application of an ant colony system to the analysis of organizational learning processes*. In: VIII SIGEF Congress Proceedings, New Logics for the new economy, Naples.

Chapter 2
The Observable Variety: Heuristic Models of Combinatory Systems

Abstract This is a pivotal chapter as it includes clear examples of Combinatory Systems whose effects can be observed in social, biological, and physical reality, thereby confirming the theory. The chapter examines the most relevant classes of combinatory systems based on the nature of the macro-behavior or macro-effect they produce—Systems of Accumulation, Systems of Diffusion, Systems of Pursuit, Systems of Order, Systems of Improvement and Progress—presenting for each class a sample of very interesting phenomena that can be interpreted in terms of combinatory systems. To make immediately clear the modus operandi of the various classes of systems, the respective *heuristic* and *graphical* models will be presented to explain the rules that allow the system to function and that produce its dynamics.

Keywords Heuristic models · Systems of accumulation · Systems of diffusion · Systems of pursuit · Systems of order and conservation · Systems of improvement and progress

> Intelligent persons are those who understand (and comprehend) quickly and effectively; who are not content to "look at the world with their eyes" (objects, facts, phenomena and processes) but who are able "to see the world with their minds" by constructing models to "understand" how the world is (description), how it functions (simulation), and how we can act as part of it (decision and planning), even without having the need, or possibility, of "looking at everything." (Mella, 2012, p. 3)

> In the history of science, we have discovered a sequence of better and better theories or models, from Plato to the classical theory of Newton to modern quantum theories. It is natural to ask: Will this sequence eventually reach an end point, an ultimate theory of the universe, that will include all forces and predict every observation we can make, or will we continue forever finding better theories, but never one that cannot be improved upon? We do not yet have a definitive answer to this question … (Hawking & Mlodinow, 2010, p. 8)

The previous chapter presented a simple but coherent and complete theory for understanding and becoming aware of the operative logic of combinatory systems. This chapter examines the most relevant classes of combinatory systems based on the nature of the macro behavior or macro effect they produce, presenting a small sample of very interesting phenomena that can be interpreted in terms of

P. Mella, *The Combinatory Systems Theory*, Progress in IS,
https://doi.org/10.1007/978-3-031-86946-4_2

combinatory systems. To readily understand the *modus operandi* of the systems described in the various categories, the relative *heuristic model* will be presented, which explains in an easily understandable manner the logic and the rules that produce the system's operation and dynamics. The aim in presenting these examples of combinatory systems—others are described in the examples in Chap. 1 and in other parts of this work—is simply to suggest to the reader how to observe common phenomena from a "sufficient height" (Senge, 1990), in order that he may form his own system perspective that will allow him to familiarize himself with this new way of observing reality. The reader will easily be able to relate the descriptive characteristics of combinatory systems in Chap. 1 to the examples presented and arrive at complete descriptions of the combinatory systems. The subsequent chapter will construct some "Combinatory Automata" to be used in the simulation of the quantitative dynamics of several fundamental Combinatory Systems.

2.1 A Bit of Order: Heuristic Models of Five Classes of Combinatory Systems

This chapter proposes a classification of combinatory systems based on the nature of the macro behavior or macro effect it produces; in addition, a small sample of phenomena for each class will be presented, which can be interpreted in terms of combinatory systems. If we classify combinatory systems according to their macro behavior (or their macro effect) (Fig. 2.1) we can, despite the variety of phenomena produced, determine five fundamental types of combinatory systems (Sect. 1.5.1):

1. Systems of accumulation.
2. Systems of diffusion.
3. Systems of pursuit.
4. Systems of order and conservation.
5. Systems of improvement and progress.

In order to make immediately clear the *modus operandi* of the various classes of systems, the respective *heuristic models* will be presented, which explain in a nontechnical manner the rules that allow the system to function and that produce its dynamics.

2.2 Systems of Accumulation

We can define "accumulation" systems as those combinatory systems whose macro behavior leads to a macro effect that can be perceived as an accumulation of objects, types of behavior, or effects of some kind, understood in the broadest sense. In these systems, the feedback is usually between the micro and macro effects more so than

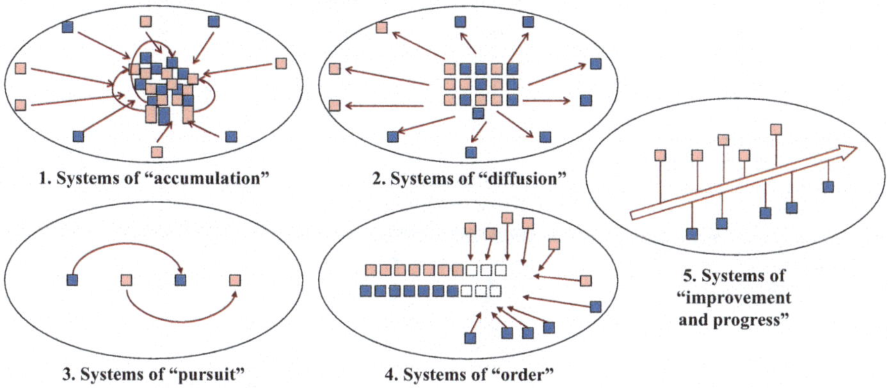

Fig. 2.1 Effects of combinatory systems

between the various types of behavior. The individual elements of the base often act with complete information; the accumulation, or a part of this, represents the informational input on which the micro behavior is based. At other times, each individual can be informed only through the observable or predictable micro behavior of some other element; in this case, we have the typical behavior based on "limited information." These systems can be described in the graphical model in Fig. 2.2 and by the following *heuristic* model:

Micro behavior and necessitating rule: if you have to accumulate some "object" with others similar in nature (micro behavior), look for already-made accumulations, since this gives you an advantage or reduces some disadvantage (necessitating factor).

Macro behavior and recombining rule: the environment preserves the accumulated objects or is not able to eliminate them, while maintaining the advantages of the accumulation; everyone accumulates (macro behavior), and an accumulation of some kind is created (macro effect).

Micro-macro feedback. Chance and necessity: the larger the accumulation (macro effect), the more incentive (facility, probability) there is to accumulate (micro behavior) objects (micro effects); the collective accumulation (macro behavior) leads to the maintenance or the increase of the accumulation. The system activates when, "by chance," an initial accumulation is produced; the *micro-macro feedback* inevitably makes it larger.

Strengthening, weakening, and control actions: a rule or constraint that prohibits the accumulation, the prompt removal of the initial accumulated objects and a careful vigilance represent weakening factors. The creation of sites for accumulation (e.g., residential or industrial areas) represents an effective strengthening action.

This logic is applied to combinatory systems that are quite different, some of which we have already considered as examples in the preceding chapter. We will examine others in detail and provide an easily recognizable name which will be useful for subsequent reference.

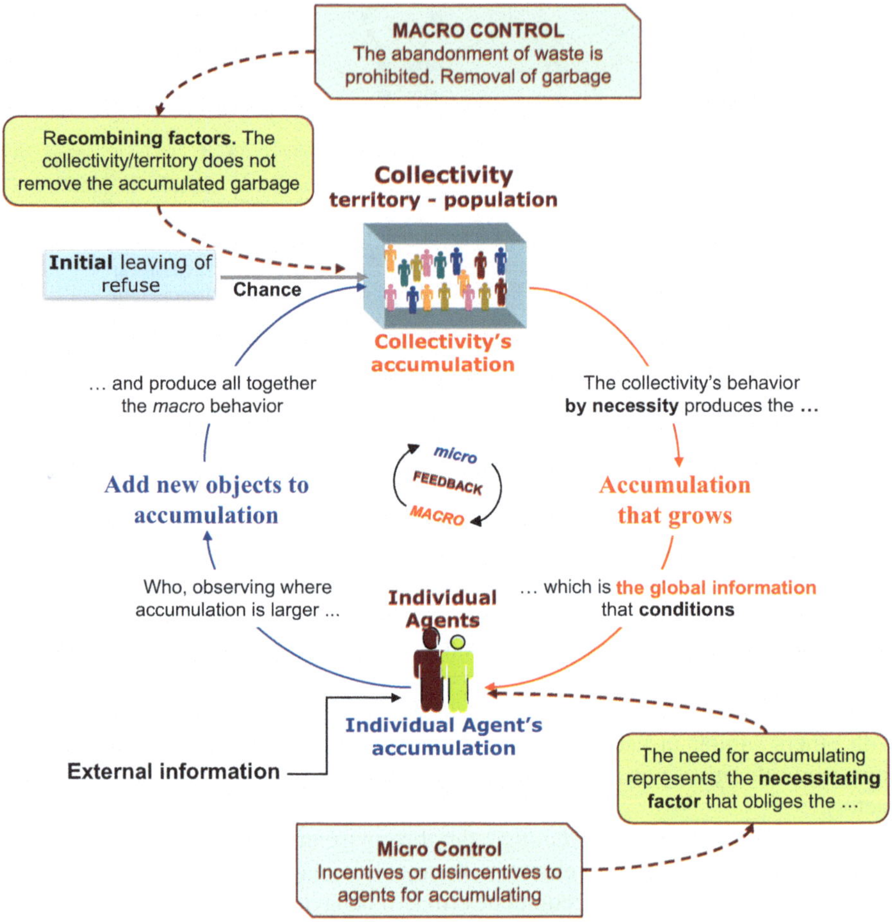

Fig. 2.2 Graphical model of a system producing "accumulation"

2.2.1 "Pile-of-Garbage" System

How often have we observed squares, streets, courtyards, fields, woods, and other frequented places (environment) with garbage spread over them, and how many times have we found garbage in the most improbable sites! Piles of garbage can be considered as the macro effect of a combinatory system that acts according to the following logic.

Assume there is a clean square (we can easily extend this example to other places), walking along which someone "by chance" feels the need (necessitating factor) to throw some garbage (micro effect) somewhere in the square (micro behavior). If the episode remains an isolated one or some sanitation worker removes the garbage, then the square will be clean until someone else comes along to leave some more garbage. If "by chance" other individuals (N_{min}, minimum number for activation) should leave garbage, at the same time or successively, in the same spot (micro behavior), the square would have a visible amount of garbage in one place, usually an area where people walk, and thus an initial pile of garbage would have formed (macro effect).

Garbage thrown away by some individual does not always activate a combinatory system; the macro behavior occurs only if the number of individuals who leave garbage in one spot in the environment exceeds a minimum number, which must be determined by considering the personal characteristics of the individuals that pass through that area as well as the environment. In a small square of a town without sanitation workers, N_{min} will presumably be less than that needed for a pile of garbage to form in the well-controlled Piazza San Marco, in Venice, Italy, unless there is a rapid succession of student groups passing through this Piazza.

As long as the garbage pile remains, a true combinatory system is in operation: the micro behavior—the leaving behind of the pile of garbage by the individuals—together with the combined micro effects (discarded refuse) produce the macro effect, the pile of garbage, which represents the "global" information that refuse can be left in that place without the risk of general reproach; this information in turn conditions the successive micro behavior (*micro-macro feedback*). Therefore, if "by chance" the pile forms and remains for a period of time, then "by necessity" this is fed—as we all have had a chance to observe—according to the typical operation of a combinatory system.

The system operates on complete information, since the decision to abandon refuse depends on the already-formed pile of garbage, which represents the global information that guides the subsequent decisions of passers-by; in many circumstances, the individual can also take into account the limited information regarding the behavior of the other people he directly observes leaving their garbage on the already-formed pile without any obstacle.

The heuristic model can be formulated as shown in Fig. 2.3.

Micro behavior and necessitating rule: if you need to throw away refuse without being reproached by others, look for a garbage dump; if you cannot find one, throw it where you see a garbage pile that is already formed (global information), if you want to make it easier for the sanitation worker, avoid criticism by others, or reduce pollution.

Macro behavior and recombining rule: if the environment is favorable to the formation of a pile of garbage, there are enough passers-by to throw away garbage, and, above all, there are no other ways to throw away refuse, then there will be an increase in the number of people who decide to throw away their garbage on already-formed piles, thereby conserving and increasing them.

Micro-macro feedback. Chance and necessity: the pile is the macro effect of the thrown away refuse, but it conditions the number of subsequent occurrences of littering. The system activates when "by chance" an initial pile of garbage is formed; the *micro-macro feedback* inevitably makes it larger.

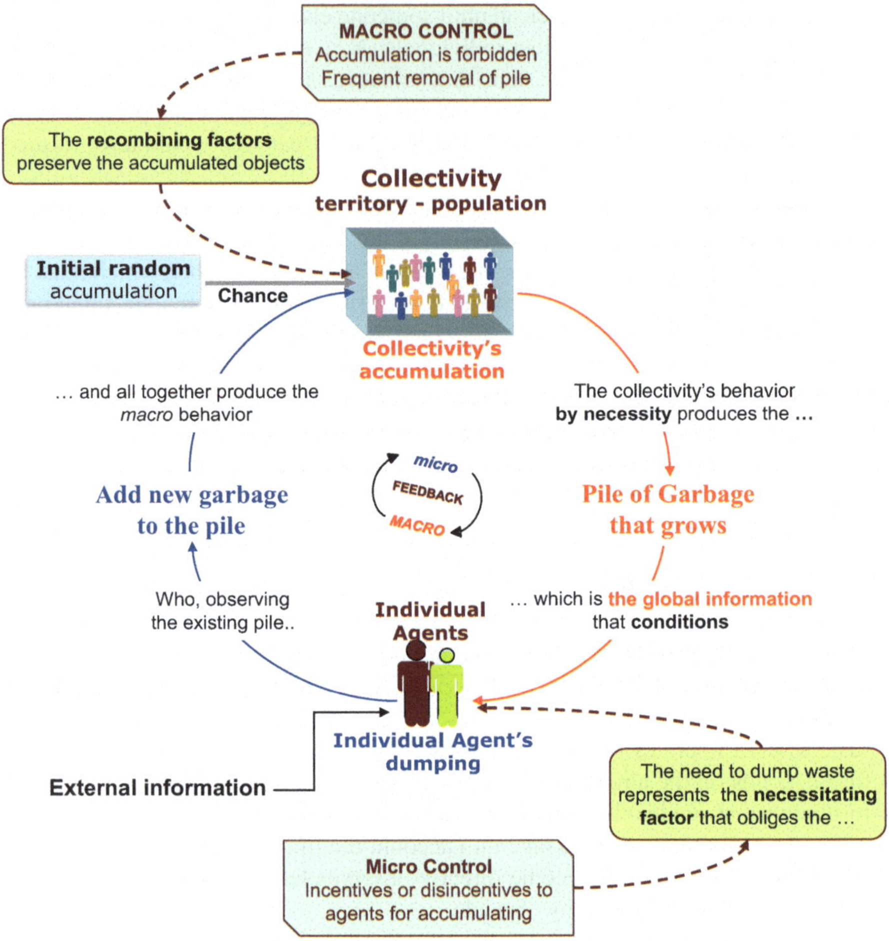

Fig. 2.3 Graphic model of the "pile-of-garbage" system

Strengthening, weakening, and control actions: a sign reading "garbage dump" or "garbage disposal prohibited" represents effective strengthening or weakening actions. Another such action is an overflowing garbage can, at the base of which a pile of garbage will inevitably form. The prompt removal of the initial refuse and a careful vigilance represent weakening factors. The external control acts on the accumulation and on the site where this forms; the internal control acts on the desire to throw away garbage and on the individuals' manners.

Figure 2.4 provides a clear example of the power of the "Pile-of-Garbage" combinatory system. In the photo, we can see a bicycle basket full of garbage left by the collectivity of pedestrians walking along a central street in Tokyo. The basket on the second bike is in the process of being filled.

Fig. 2.4 A concrete example of the pile-of-garbage system in action (photo taken in Tokyo by the author)

2.2.2 "Planet-Formation" System

Systems of Accumulation are not always made up of biological agents. We can identify the logic of Systems of Accumulation even in a variety of phenomenon whose effect is the physical aggregation of material. One of these physical systems, whose logic is similar to a *"Pile-of-Garbage" System*, is the one that explains the formation of planets. Planet formation is a complex process that takes place over millions of years, starting from a cloud of gas and dust in space. Even if there are several theories to explain the formation of the solar system, it is by now accepted that the solar system began from a solar nebula, a spinning disc of dust that formed the Sun. The solar nebula was formed from particles that, through their rotational motion, were randomly attracted to each other and collided. We can presume that, given their microscopic size and the reduced force of gravity that attracted them to each other, the particles joined together to form billions of larger particles, which, in turn, collided when joining together through the same mechanism, thus giving rise to billions of even larger objects (Weidenschilling, 1991, 2000). Those that reach a diameter of about 1000 m up to 100 km (Ringwood, 1979, p. 55) are called planetesimals. "*A planetesimal is a rock-type object formed in the early solar system from collisions with other objects in the solar system. The collisions eventually formed larger objects that led to the formation of planets*" (Fennell, 2016, online).

With the increase in their size, a significant gravitational force was created, with the result that the collision among planetesimals and between planetesimals and other smaller particles was guided by the same gravity. Thanks to the increased gravity, the collisions occurred at a very high speed. We can imagine that at times these collisions led to the breaking apart of the colliding objects, while at other times the objects once again joined together, giving rise to objects with an even greater mass. During the millions of years in which the planetesimals continued to collide—and to attract other objects of lesser mass—they did so not because their orbits crosses "by chance," but due to the increasingly greater reciprocal gravitational pull. This process of growth in the new objects, called accretion, ended up by forming the larger objects, called protoplanets, whose size and mass reached the levels of our moon "*Protoplanets are small celestial objects that are the size of a moon or a bit bigger. They are small planets, like an even smaller version of a dwarf planet. Astronomers believe that these objects form during the creation of a solar system*" (Cessna, 2015, online).

The process deriving from the accumulation of material continued until the planets were formed. This formation made the objects that could be attracted rarer, so to speak; however, the process of accretion has not yet ended. Numerous small-to-medium-sized objects, the asteroids, which could increase the size of the planets, are orbiting around the sun to form the asteroid belt. Nevertheless, many of these continue to collide with the planets and their satellites, in addition to colliding with each other. We can observe even with the naked eye the craters left by the impacts on the moon, and the earth also collides with thousands of particles each year, which, fortunately, due to their small size, are pulverized on contact with the atmosphere. We know, however, the damage that followed the collision of the asteroid that caused the extinction of the dinosaur.

These brief descriptions immediately show that the formation of the planets in our solar system was guided by a combinatory system of accumulation in which the agents—if we can call them as such—are the particles that rotate around the sun: the small particles, the planetesimals, and the protoplanets. We can interpret the macro particles that gradually form as collectivities whose micro behavior is the continuous "absorption" of mass and whose macro effect is the formation of planets (or other forms) due to the increase in gravity. Mass and gravity represent the *global information* that conditions the micro behaviors of the agents-particles, which direct their trajectory toward the bodies of higher mass. Such micro behaviors produce the micro effects of the collisions and the unions-mergers over time. The recombining effect of the gravitational pull from the body of greater mass preserves the absorbed mass (if the attracted object does not first become pulverized), thereby increasing the mass of the asteroids of the planetesimals and protoplanets, as part of a process that has lasted millions of years and which has led to the formation of planets.

Perhaps there are even other models to describe the formation of planets; however, the combinatory systems theory seems quite simple, and in our view convincing, as shown in the model in Fig. 2.5.

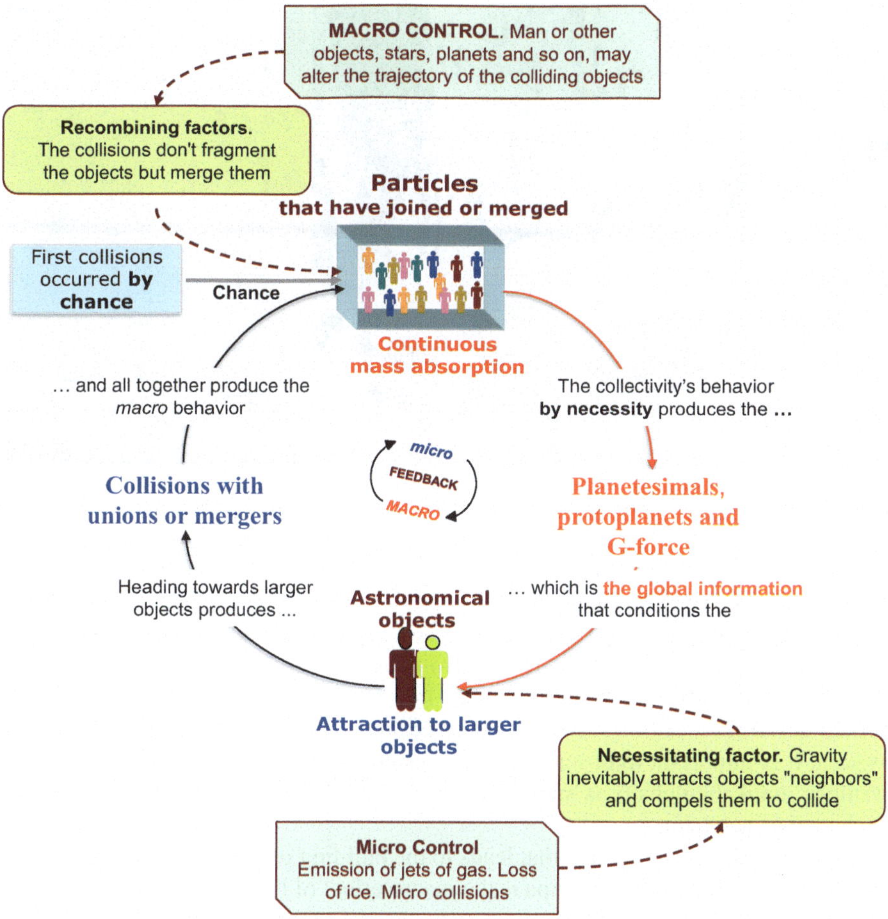

Fig. 2.5 Graphic model of the "planet formation" system

2.2.3 "Graffiti-on-Wall" System

We have all observed in certain places (environment)—walls, backs of doors, school desks, tree trunks, monuments, and so on—a concentration of writings or carvings, more or less permanent, done by individuals moved by the desire to leave a mark of their presence in that place (Fig. 2.6). Graffiti represent a complex and variable form of visual expression that develops in urban and other spaces. For this reason, Graffiti can be considered both art and vandalism, depending on the context and the way it is created. This phenomenon occurs when "by chance" someone leaves the first writing (micro behavior); if, again by chance, the initial writing (micro effect) is not quickly removed and is read by other individuals moved by the same desire for personal expression (necessitating factor), then other writings will be added.

Fig. 2.6 A collection of "Graffiti-on-Wall" Systems (photo taken by Google images)

The act of leaving graffiti represents the *micro behavior* and the trace, written or carved, the micro effect; the spread of the desire to leave graffiti can be considered as the macro behavior of a specific combinatory system for which the mass of graffiti represents the macro effect. The macro behavior is fed by past micro behavior but ends up conditioning future behavior in a typical *micro-macro feedback*. The formal description of this system is entirely similar to that of the "pile-of-garbage" system. It is important to observe that the system begins to operate attracting other writings, until the density is such as to make it no longer worthwhile to continue such uncivil behavior.

Similar to this is the system that leads to the building of "stupas" in places where the Buddhist cult exists. The stupa is the macro effect of the laying of stones (micro effects) on which the faithful carve or write a prayer in the hope that, as long as the prayer is preserved on the stone, the hoped-for effects will be produced.

2.2.4 "Urban-Settlement" System

When one or more individuals need a house (necessitating factor) and feel they have found a place (environment) which represents a favorable place to live—for example, due to the presence of water (the great cities of antiquity grew up along the great rivers), or roads (many cities arose along important routes), or a hill that is useful for defense (how many fortified cities have been built on hills that overlook valleys or the sea?), or even simply a beautiful view—then the first dwellings (micro effects) are built (micro behavior).

If those dwellings are soon abandoned, then no combinatory system will be formed; but if "by chance" the chosen place is thought to be favorable for community life (recombining factor), then the first dwellings are strengthened or increase, either because of new births or new arrivals. The presence of the settlement leads to the maintenance and growth—in line with the environmental constraints—of the settlement itself (macro effect), thereby conditioning the construction of new houses on the same site.

If an urban settlement held to be satisfying arises "by chance" in an uninhabited area, "by necessity" this is maintained, since the dwelling is passed on from parents to children and new residents are attracted to the place; a combinatory system is thus formed. The impossibility of knowing all the potential settlement alternatives and the perception or even only information regarding the advantages of living in a community represent limited information on which the decision to build new dwellings is based.

The macro behavior reveals itself only if the number of dwellings concentrated in a single site exceeds N_{min}, which depends on the type of favorable conditions sought by the individuals. N_{max} instead depends on the degree of saturation of the favorable conditions; in an agricultural area with fertile land, we might have $N_{min} = 1$ and $N_{max} = 1$. In this case, the area would be divided into farmhouses equidistant from one another, and no cities would arise.

A city thus appears as the result of individual decisions to build a house in a favorable place. However, the presence of a city provides *information* that favorable living conditions have been found, and this gives rise to powerful necessitating factors that influence the individual *micro* behavior; the city itself indicates the favorable conditions (recombining effect).

If the *necessitating factors* are very strong (hope of employment, fun, social encounters, etc.) and the area is favorable for recombining factors, then the number of people settling there will continue to grow. In 1950, there were only two megacities: London and New York. In 1975, Mexico City and Shanghai joined this group, and today there are more than 50 (Megacity online; Wikipedia, 2024d, Megacity). As Fig. 2.7 shows, the largest cities are found mainly in Southeast Asia.

The heuristic model of this system can be formulated as follows (see Fig. 2.8).

Micro behavior and necessitating rule: if you need to build a house, look for favorable conditions; if there is a city there already, it is assumed that favorable conditions exist. Leave your house to your descendants. The house represents the micro-effect deriving from the individual micro-behavior.

Macro behavior and recombining rule: an urban settlement of multiple families who have built their homes, forming a city, represents the macro-effect produced by collective macro-behavior. The construction of new houses strengthens the urban settlement; the strengthening and growth of the city are signs that favorable conditions exist (opportunities, services, protection, etc.), and this influences the micro behavior, whereby new houses are added to the existing settlement. The older and larger the city is, the greater the incentive, or necessity, for new arrivals to the area to locate there.

Micro-macro feedback. Chance and necessity: the city is the result of individual decisions to build a house in a favorable place; but the presence of a city gives *information* that favorable living conditions have been found, and this influences the

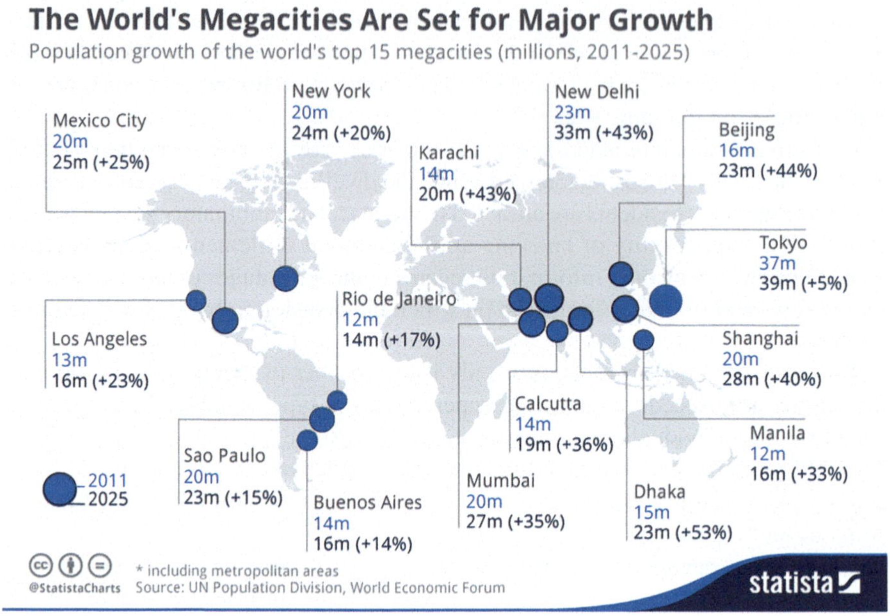

Fig. 2.7 Megacities in the world (source: Megacities online n.d.)

individual micro behavior (the city itself indicates the favorable conditions). A city arises "by chance" but, once begun, the phenomenon is "by necessity" maintained over time as long as the necessitating factor operates.

Strengthening, weakening, and control actions: overcrowding, increase in the time needed to cross a town, the desire for solitude, and the need to preserve the surrounding areas: these all represent *weakening* factors, and where these prevail over the need for a community life, the "urban-settlement" combinatory system will not operate; instead a different system would be activated that can be referred to as "maintain-the-territorial-division." Examples of *strengthening* factors are the danger of invasion, tourist attractions, tax incentives for construction, and the supply of attractive urban services. The macro control can act by means of urban planning; the micro control can influence the desire to live in a town or the opposing desire to flee the crowds.

If cities are the *macro-effect* of the need for individuals to establish their homes within an already existing city, driven by the desire for increasingly larger communities, and if no maximum population limit (N_{MAX}) is established, so that urban settlement continues to expand with the arrival of new inhabitants, we must ask ourselves how a city might expand, both in terms of the number of buildings and the number of inhabitants. In fact, the expansion of cities can occur in two main "directions": horizontal and vertical.

Horizontal expansion results from the "physical" growth of a city through the addition of new "low-rise" housing, with the increasing occupation of new

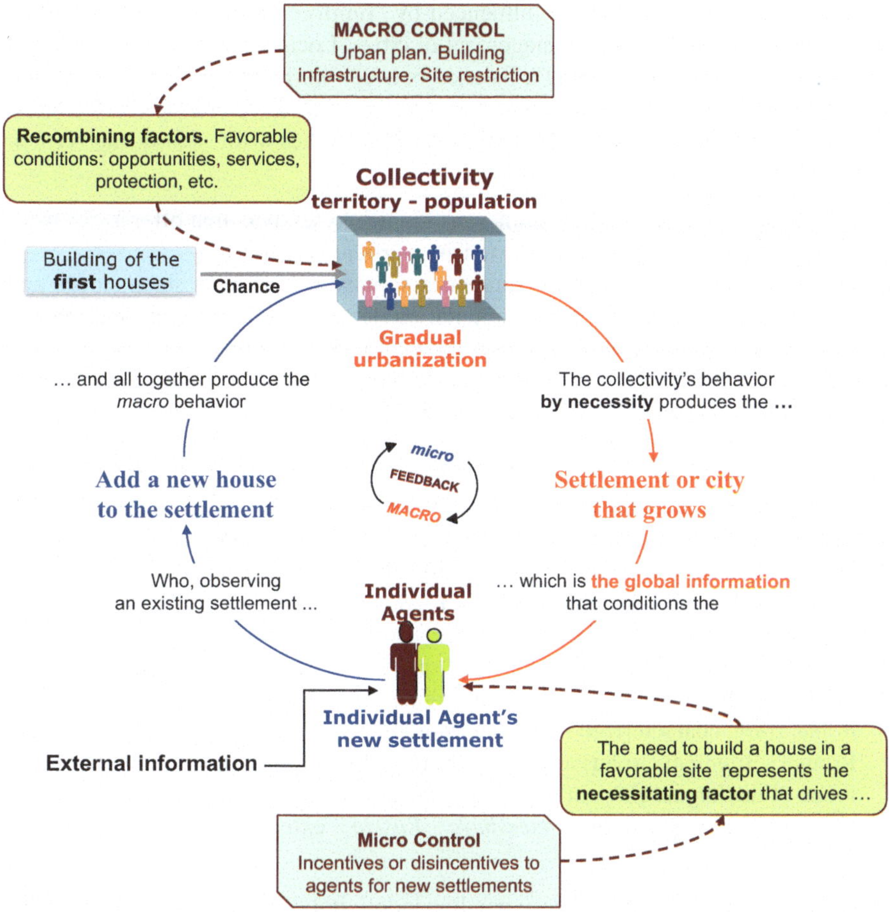

Fig. 2.8 Graphic model of the "urban settlement" system

territories for the development of infrastructure, housing, and commercial and industrial areas. This type of expansion can lead to urban sprawl in a disorganized and scattered manner, consuming vast amounts of land and leading to problems such as traffic congestion, pollution, and the reduction of green spaces.

Vertical expansion, on the other hand, involves the growth in height of buildings within the existing urban core, with the construction of increasingly taller and denser skyscrapers to provide housing without expanding the occupation of territorial space. This form of expansion results in greater urban density, reducing the need to occupy more land and promoting the sustainable development of cities. However, vertical expansion, through numerous skyscrapers, can lead to issues such as the overshadowing of surrounding areas, the concentration of high population densities with potential risks related to cohabitation and the creation of urban microclimates.

Both forms of expansion are influenced by "reinforcing factors" such as urban planning and the sustainable management of cities. Local administrators and urban planners are tasked with finding a balance between horizontal and vertical expansion to ensure urban development that is compatible with acceptable living standards.

The following are among the numerous mega-cities with significant *vertical expansion*:

1. *New York City, USA*, with its numerous iconic skyscrapers that offer a spectacular skyline, reflecting the city's extensive vertical expansion.
2. *Hong Kong, China*, where skyscrapers rise upward, often on steep slopes and hilly terrain, allowing for a high population density in relatively limited spaces.
3. *Dubai, United Arab Emirates*, characterized by breathtaking skyscrapers such as the Burj Khalifa, the tallest building in the world, symbolizing the city's remarkable vertical expansion.

The following are among the large cities with notable *horizontal expansion*:

1. *Los Angeles, USA*, characterized by developments across a vast urban area that extends over a large surface, with widespread low-rise and scattered neighborhoods. To facilitate population mobility, the city has extensive and complex highways that complement traditional road networks.
2. *Brasilia, Brazil*, which, despite being relatively recent (constructed in the 1950s and designed by the famous architect Oscar Niemeyer), is known for its horizontally developed urban structure, featuring large open spaces and a rational design.
3. *Rome, Italy*, characterized by its rich monumental heritage derived from ancient Roman settlements, spans an area of about 1285 km^2 (496 square miles), making it one of the largest cities in Europe in terms of surface area (*London, UK*, covers about 1572 km^2; *Saint Petersburg, Russia*, extends over approximately 1439 km^2).

Are cities destined to continue expanding indefinitely? The answer is not straightforward; however, the expansion of large cities can encounter various "limits" that may affect urban sustainability and quality of life, which differ from city to city.

In general, the most evident limits to expansion are the "physical" ones, including the increasingly limited availability of land and geographical constraints such as mountains, rivers, or oceans, which can restrict a city's physical growth. These very limits have allowed many cities to remain "small," giving rise to picturesque towns that are abundant in European countries. Among the "physical" limits are the risks of compromising ecosystems, reducing natural habitats, making deforestation unsustainable, causing unacceptable air pollution, and altering hydrological cycles, thereby endangering biodiversity and the quality of the environment.

The significant increase in population and high rates of urbanization can create serious problems for existing infrastructure, such as roads, public transportation, water supply, and waste disposal systems, leading to congestion and issues with city accessibility, which gradually diminishes the city's attractiveness.

One must not underestimate the limitations associated with uncoordinated urban growth, the emergence of socioeconomic disparities, the risk of urban segregation, the lack of access to basic services such as education and health care, and the potential for social tensions.

Macro and *micro*-control factors are linked to both political and administrative interventions aimed at adopting sustainable urban planning. This planning should be compatible with increasing population density, urban densification, urban renewal, public transportation efficiency, and the preservation of green spaces.

2.2.5 "Industrial Districts" and "Colonies"

The logic of the previous combinatory system also applies to different collective phenomena: the formation of cemeteries on the boundaries of towns and the industrial settlements of firms engaged in the same type of production or "industrial districts" (Fig. 2.9). Why is Valenza, Italy, the capital of jewelry and Silicon Valley (USA) the capital of microchips? The answer is similar to the one we have seen in the case of urban settlements: when the number of enterprises that produce the same type of good (or develop the same processes), and which "by chance" are settled in the same area, exceeds a certain minimum, that number "by necessity" will grow until it reaches the maximum number compatible with the urban and logistical resources.

It is not difficult to identify the necessitating and recombining factors that guide these Combinatory Systems. Alfred Marshall (1890, 1919) examined the concept of *districts* and *production clusters*, proposing that these local production systems are motivated by "economies external to the firm" but "internal to the district." Such systems enhance overall *productivity* because the joint location of numerous firms engaged in similar processes or segments of processes generates a distinct form of *efficiency* within the broader production system.

Building on Marshall's perspective regarding districts as sources of efficiency (Bellandi, 1996; Marshall, 1890) and Williamson's transaction cost approach (Dyer, 1997; Williamson, 1993), improved performance measures can stem from *reduced costs*, *increased revenue*, and effective *knowledge exploitation* and retention (Beccattini et al., 2009). *Cost savings* result from the *benefits of specialized processes* provided by location and can be linked to improved production and logistical conditions (Albu, 1997). *Revenue advantages* are linked to market benefits, which are determined by the market "fertility" of the site; in other words, the relative abundance of potential clients, particularly in commercial areas. The advantages can also be shaped by *pricing strategies* and are linked to the quality of production or to effective marketing processes (Chandler, 1990). One approach to exogenously attracting new companies and businesses is through the development of benchmarking processes (Cook, 1995), where policymakers aim to draw investments to a particular location (Riva & Pilotti, 2019), thereby promoting cost reduction and boosting the overall attractiveness of the area (Rondo-Brovetto & Saliterer, 2007). In this framework, industrial clusters and districts serve as geographical expressions of a *learning economy* (Lundvall & Johnson, 1994) since districts embody the

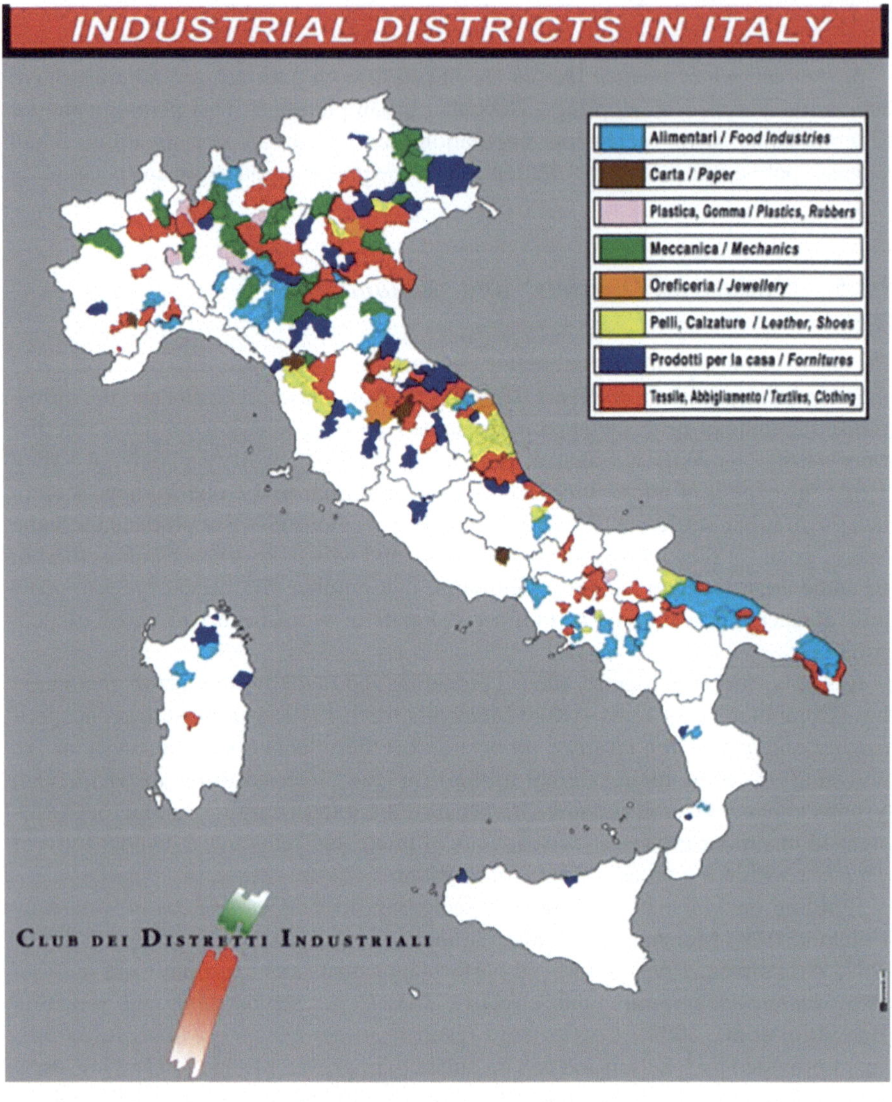

Fig. 2.9 Industrial districts in Italy (photo taken by Google images)

principle that "*knowledge is the most important resource and learning the most important process*" (Lundvall, 1992), thus defining them as *learning regions* (Belussi & Pilotti, 2002).

Even if these systems have been described by Arthur as positive feedback phenomena, considered as phenomena of *increasing yield*, and although location behavior can be explained using an agent-based model, which "*allows us to study human behavior at the microlevel in a spatial context*" (Otter et al., 2001), the explanation provided by Combinatory System Theory is equally convincing and

Fig. 2.10 A collection of "accumulations" in natural Systems (photo taken by Google images)

complete (Sect. 3.4.4). It has been demonstrated that the location of firms in the same area and the formation of settlements of firms derive from the joint action of systems of accumulation and diffusion as will be shown in the heuristic models in Sect. 2.7.1. These systems can be simulated by the probabilistic stochastic combinatory automata described in Sect. 3.4.6.

The same logic of diffusion and accumulation systems can be applied to explain the formation and maintenance of clusters of natural biological systems like those shown in Fig. 2.10. Why are there many mono-valve shells on some cliffs while on others they are scarce? Why do sponges concentrate in certain marine areas and not in others, which nevertheless have similar conditions? Why, where by chance a seed takes root and a plant grows, do woods of similar plants arise? All of these questions have the same answer: when *by chance* an initial biological settlement is formed, the necessitating factor of reproduction combines with the recombining factor of the presence of colonies to activate the *micro-macro feedback*: the colony is formed by the individual elements, but this facilitates reproduction and the consequent expansion. The action of "accumulation" systems leads to the formation of a colony, while the action of "diffusion" systems drives its expansion, regardless of the presence of suitable sites for the colony's reproduction, the availability of resources for nourishment, the abundance or scarcity of predators, and finally, human activities related to predation or cultivation.

2.3 Systems of Diffusion

This term can be used for all systems having as a macro effect the diffusion of an object, a trait, a feature, a peculiarity, or a "state," from a limited to a high number of elements of the system in a given environment. In these systems, the feedback is

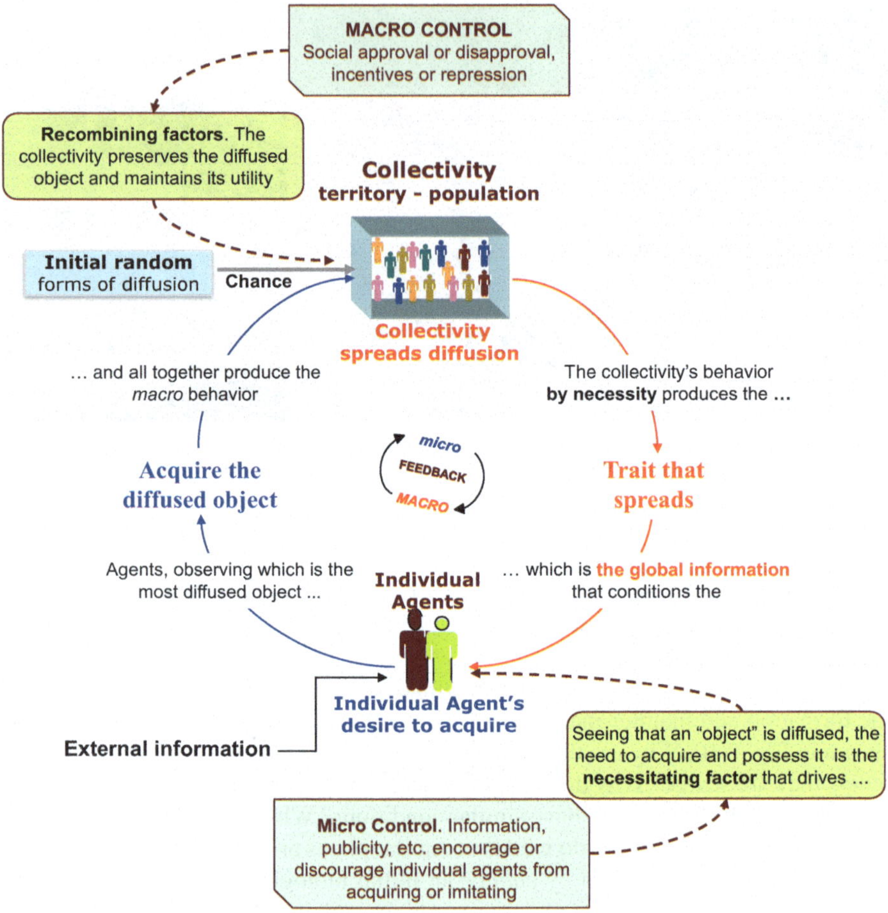

Fig. 2.11 Graphical model of a system producing "diffusion"

usually between the micro and macro behavior more so than between the effects. The individual elements of the base normally act with incomplete information.

Figure 2.11 shows the logic behind the following *heuristic* model:

Micro behavior and necessitating rule: if you see that an "object" is diffused, then it is "useful" for you to possess it or harmful not to possess it (necessitating factor), and you must try to "acquire" it.

Macro behavior and recombining rule: the environment or the collectivity preserves the diffused objects and maintains the utility of possessing the object; the higher the utility or need to acquire the object is for the individuals, the more the object will spread throughout the collectivity.

Micro-macro feedback. Chance and necessity: a greater diffusion (macro effect) implies a greater desire to acquire the object (micro effect); the single acquisition (micro behavior) widens the collective diffusion (macro behavior). Once the imitative phenomenon has begun "by chance," it is maintained "by necessity" until the saturation density is reached.

Strengthening, weakening, and control actions: information, publicity, word of mouth, and social gratification represent strengthening factors for the system; social disapproval (the fashion of driving at break-neck speed in cars, for example) and repression represent weakening ones.

Systems of diffusion explain quite a diverse range of phenomena, a few of which are described in the following sub-sections.

2.3.1 "A-Fashion-Is-Born"

How can we explain the fashion phenomenon? How do fashions arise? How do they cease? Why do some return while others disappear for good? A fashion—whether this concerns clothing, clocks, cars, toys, a particular linguistic form, and so on—arises in a given environment from a novelty that is introduced (micro behavior) "by chance" by a given initiator for the purpose of getting away from the usual routine. If the novelty was not pleasing, cost too much, or had other negative features, the episode would remain an isolated one and no combinatory system would develop. If the number of individuals that imitate the innovation reaches the necessary activation density, then that innovation becomes a "fashion."

A new trend can originate from an innovation introduced into a product by designers, stylists, artists, designers, influencers, or even ordinary people who invent something new and different. The first individuals to spread a new trend—called trendsetters—are usually people possessing a strong social impact, followed by a larger group of early adopters who help make the trend known to a wider audience. Alternatively, they may be individuals with many social contacts who make the adoption of the item visible, leading it to become fashionable.

The macro behavior referring to the fashion system leads to an increase in the desire to imitate (necessitating factor) and to the macro effect of the diffusion of the good that is the object of imitation and the advantage for those who possess it (recombining effect). The *micro behavior* is represented by the imitative actions regarding a novelty, conditioned by the individual need to set oneself apart from the norm, that is, to join in with the fashion. The micro effects consist in the acquisition of the good which is the object of imitation. Once a fashion has been adopted by a wider public, it becomes mainstream.

The system operates under *limited information*, which is represented by the degree of imitation that a certain individual observes, or estimates, in a sampling of other individuals and by the degree of consensus-dissent of that individual concerning the innovation, which in turn relates to the economic and social conditions of the base group. Traditional media—television, magazines, newspapers—along with social media, celebrities, and influencers can play a crucial role in spreading trends, as they disseminate information about the new "object" of imitation, thereby broadening and accelerating the limited information.

The *macro behavior* (usually not directly observable) can be seen as an increase in the number of individuals imitating the innovation; the *macro effect* consists in the increase in the average density of the good in the environment and in the base. This macro behavior is fed by the micro behavior, but in turn conditions future micro behavior until the density of imitation becomes so high that it no longer provides satisfaction. Indeed, when a trend becomes too common, it may start to lose its appeal; the novelty fades, and people begin to seek out other new things, making it beneficial for producers to focus on searching for the "new." Therefore, once the saturation density is reached, the system ceases to operate, although some trends do not disappear entirely but can be revived cyclically, as happens in the fashion industry: what once went out of style may become trendy again in the near or distant future. Once the saturation density is reached, the system ceases to operate. The spread of the innovation is the result of the imitative micro behavior of the single individuals, and it conditions the future imitative micro behaviors in the typical *micro-macro feedback.*

The heuristic model that describes this system can be represented in Fig. 2.12.

Micro behavior and necessitating rule: if you need to stand out, abandon the traditional way and follow an innovation; if you do not want to be considered as just a "somebody" but as someone different, then join in with a fashion and undertake an imitative micro behavior.

Macro behavior and recombining rule: if an individual distinguishes himself, he is appreciated and well-considered; the diffusion of goods in the environment, which represents the effect of the imitative micro behavior, increases the desire to adhere to a fashion. The greater the number of imitative behaviors, the more powerful is the fashion and the more pressing the desire to follow it, until the saturation point is reached.

Micro-macro feedback. Chance and necessity: the diffusion of an innovation is the result of the imitative micro behavior of the single individuals, but it also conditions future behavior. The innovative idea arises "by chance" and conditions the initial imitations. Once the imitative phenomenon has begun, it is maintained "by necessity," until the saturation density is reached.

Strengthening, weakening, and control actions: publicity and social gratification represent strengthening factors for the system; social disapproval (the fashion of driving at break-neck speed in cars, for example) and repression represent weakening factors. The macro control must act to limit the availability of purchasable goods, whereas the micro control must act on individual preferences.

Therefore, the spread of and changes in a fashion are produced by a typical *incomplete and limited information* combinatory system (Sect. 1.3.5), in which the agents carry out imitative micro-behaviors based on information obtained not through direct transmission but by observing the behavior of the other agents. Bikhchandani et al. (1992) use game theory to construct a model to interpret the diffusion mechanism arising from the imitation of the behavior of other agents as a behavioral phenomenon based on limited information: the behavior of individuals is imitated by other agents, thereby producing an informational cascade. Nevertheless, they do not highlight the role of *micro-macro feedback* in producing this phenomenon

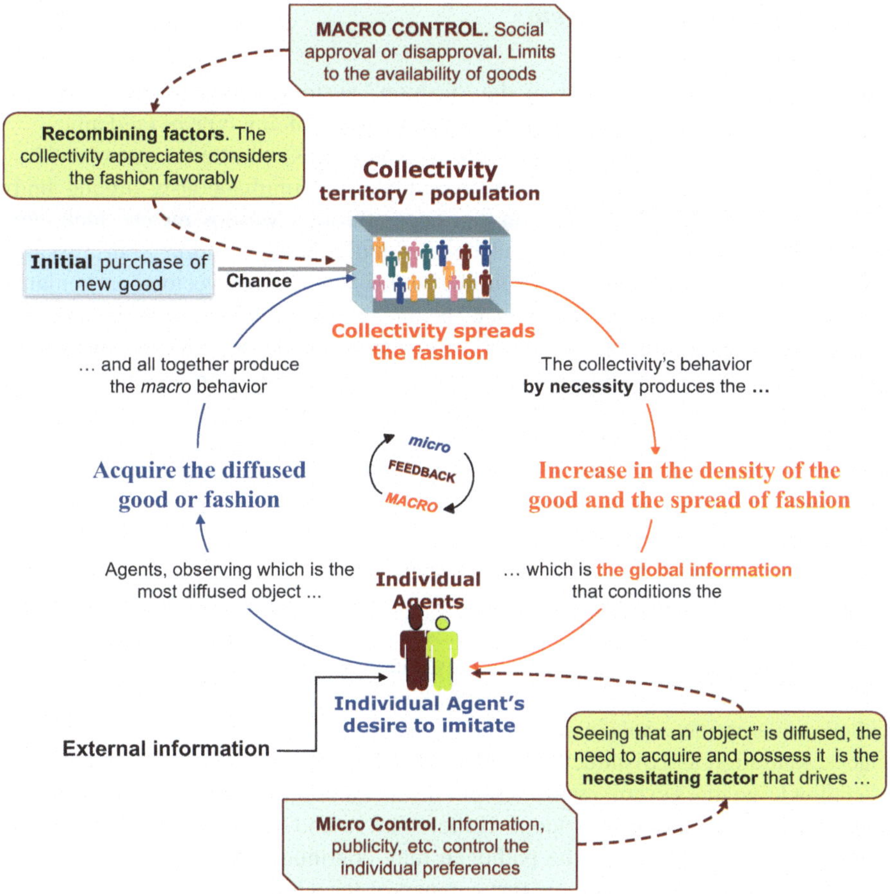

Fig. 2.12 Graphic model of the "fashion" system

In our model, individuals rapidly converge on one action on the basis of some but very little information. If even a little new information arrives, suggesting that a different course of action is optimal, or if people even suspect that underlying circumstances have changed (whether or not they really have), the social equilibrium may radically shift. Our model, which is based on what we call "informational cascades," explains not only conformity but also rapid and short-lived fluctuations such as fads, fashions, booms, and crashes. (Bikhchandani et al., 1992, p. 994)

An informational cascade occurs when it is optimal for an individual having observed the actions of those ahead of him, to follow the behavior of the preceding individual without regard to his own information. (*ibidem*, p. 994)

Information transmission among individuals can take many forms. For example, individuals may observe all other individuals' information, only the signals of predecessors, or only the actions of predecessors. Our analysis concentrates on the least informative case in which individuals observe only the actions of previous individuals. Since "actions speak louder than words," the information conveyed by actions may also be the most credible. (*ibidem*, p. 996)

2.3.2 "Epidemics" System

The heuristic model representing the "fashion" system can be applied to many forms of rapid diffusion; in particular, to the diffusion of any type of epidemic.

An epidemic begins when a pathogen, such as a virus or bacteria, spreads within a human population or among captive animals, or simultaneously among both humans and animals. The diffusion can occur through various means, including direct *contact* between infected and healthy individuals, *airborne* transmission, *contamination* of water or food, or through *bites from animal* vectors, particularly insects. When the pathogen "by chance" begins to infect individuals in the targeted population, and a minimum threshold of infected population density is surpassed, the epidemic can expand through a process of transmission from one individual to another. A Combinatory System clearly starts to operate. Individuals become infected because others have transmitted the pathogen to them; these newly infected individuals then increase the density of the infected population, which, in turn, raises the probability of further infections, creating a typical *micro-macro feedback* loop.

The size and degree of concentration of the population, limited territory, rapid spatial movement, and lack of information are *reinforcing factors* that increase the likelihood and rate of spread. In epidemics affecting human populations and livestock communities, *weakening factors* may come into play—these form the basis of the *micro* and *macro-controls*. Such factors include the isolation of infected individuals, quarantines, vaccination campaigns, the closure of schools and public places, sanitation and disinfection of sites, travel restrictions, bans on organizing or attending large mass events, as well as public information and awareness campaigns about preventive measures to adopt. If these *macro* and *micro-controls* are insufficient to weaken the spread, the pathogen may continue to expand exponentially, potentially leading to a pandemic that can persist for a long time.

To make clearer the action of the Combinatory System that spreads a pandemic, we can use a flu epidemic as an example. The epidemic begins when the carrier of a flu virus infects a certain number of individuals (limited information), thereby transmitting the virus (micro-behavior) and passing on the infection to other individuals (micro effects). The higher the number of sick people (macro-effect), the more likely it is that other people will become sick (macro-behavior). To produce an epidemic-like spread, the carrier must transmit the flu to other individuals, who must in turn pass it on to others, and so on. If we assume that the first carrier of the flu, having contracted the virus during a trip abroad, returns home and becomes sick in his isolated country house, the virus will probably not spread among the population. In other words, there must be a *minimum number* of infected persons to have an epidemic-like spread of the disease. If the first infected individual does not come into contact with anyone, then the combinatory system will obviously not be triggered. However, if *by chance* there are frequent contacts with other individuals, the system has a high probability of being triggered. When the number of infected persons exceeds a minimum activation number, the epidemic develops *by necessity*.

If the controls implement effective containment measures, the rate of new infections decreases, and the epidemic reaches its peak and begins to decline. This can also happen through a form of "biological control" when the majority of the population has already been infected and has developed immunity or when prevention and control practices are rigorously followed. When the number of new cases drops to very low levels and the number of recovered or immune individuals increases, the epidemic is no longer able to spread significantly, leading to the extinction of the epidemic. However, in some cases, the pathogen may remain latent and re-emerge in the future if conditions are favorable.

A *deterministic automaton* to simulate the dynamics of a pandemic is constructed in paragraph 2.6.7; reference can also be made to the *combinatory automata* which will be presented in paragraphs 3.6.5 and 3.6.6 below.

2.3.3 "The-Hundred-Towers" System

Pavia is a picturesque Northern Italian town along the Ticino River, famous not only for having the oldest university in the world, founded by King Lotario in A. D. 825, but also for its many towers, which is why it is called the *Civitas turrigera*, *Civitas centum turrium*, "the city of the hundred towers" as depicted in Fig. 2.13.

The tower phenomenon began around 1000, perhaps some decades before, and rapidly developed in 1100 and the following century, so that by around the year 1300 many had already been ruined (Anonymous Ticinese about 1320). In 1570, the historian Breventano (1570) mentioned more than 170 towers, Spelta (1603) around 100, and Zuradelli (1888) counted 76, a very impressive number if we consider that within its walls Pavia had a surface area of only 900 × 900 m (Fig. 2.14).

The map of the town in 1525, found in a fresco in the basilica of San Teodoro of Pavia, shows some 50 towers still intact. The inventory drawn up by the mayor of Pavia in 1965 revealed 71 traceable towers, 7 of them almost intact (Fig. 2.15), almost all built during the communal strife between the Ghelphs and Ghibellines and incorporated into the palaces of the rich families. Those that are still standing and the still-visible remains of the others allow us to imagine a truly fascinating panorama—one even mentioned by the Italian poet Petrarch (Francesco Petrarca 1304–1374)—testified to by the ancient prints of the city.

The towers of Pavia, some even 50–60 m in height, are very simple and fairly "modest" if compared with the more elaborate ones of other towered cities. Square and with each side around six meters wide, they are built in brick (the province of Pavia has always been rich in high-quality clay), windowless and without an interior, at times containing only a very narrow spiral stairway. The towers are truncated, with a flat or slightly slanted roof, just barely covering the tower. These features make it very compact but slender, with a typical clay color. They usually have deep foundations filled with inexpensive material. How can we explain this unique phenomenon? Once again, we can turn to the theory of combinatory systems. The towers of Pavia were built at the corners of the town streets, on the

Fig. 2.13 Pavia and its many towers in an ancient print

Fig. 2.14 The fresco of Pavia in AD 1525 in St. Theodore's church

outsides of the palaces of important families. They were built along with the construction of the palace to celebrate the subsequent birth of a male heir.

Their construction without openings, their narrow shape, and their location indicate they were not used as a dwelling or for defense; their only role was as a mark of distinction for the family (usually a noble one and certainly wealthy) that had

Fig. 2.15 The towers of Pavia as indicated in the inventory drawn up by the mayor of Pavia in 1965. I have indicated two clear concentrations of towers (source: official document)

built the palace and that, in order to set itself apart from the others, built its own tower. For this reason, during the struggles between the Ghelphs and Ghibellines, the faction that was winning at the time would pillage the palace of its adversary and knock down or break off the tower as a sign of contempt.

We can thus assume that the first tower was built "by chance," to celebrate the birth of the first-born son, by a rich and noble citizen of Pavia who saw the tower as a way of testifying to his wealth, power, nobility, and pride from the event. The tower rose up clear against the Po Valley sky and from afar indicated to visitors where the palace was located.

If the first tower had collapsed after a short period of time, had been prohibitively expensive to build, or had been taken as a sign of haughtiness, and thus not worthy of imitation, then Pavia would probably not have had its hundred towers. But the idea was instead appreciated. It is easy to imagine that another powerful noble, in order not to be left behind, was attracted by the idea of building a second tower. As this was also pleasing, the combinatory system involving the towers of Pavia began to develop due to the presence of recombining and necessitating factors, which fanned the desire of the noble and rich families not to be left behind their neighbors.

The swarm of towers outwardly represents the macro effect that is fed by the individual micro behavior, represented here by the construction of the individual towers by the families of Pavia. The macro behavior is difficult to imagine, but we can consider it to be the "race to build towers." Once set off "by chance," the combinatory system exhausts itself when the density of towers is such that a new one would no longer play a distinguishing role but, on the contrary, would be taken for "unfashionable" social climbing.

Thus, on the one hand, the macro-behavior of the system, as a collective unit (Pavia becomes increasingly towered), derives from the "combination" of the similar micro-behaviors of the individuals (every family builds a tower). On the other hand, the macro-behavior (the town rich in towers) determines, conditions, or directs the subsequent micro-behaviors, thereby conditioning the agents' choices and decisions: "if there are so many towers, our family must have one as well." Once again, this reciprocal "agent–collectivity relationship" represents the *micro-macro feedback* which constitutes the necessary and sufficient condition for the Combinatory System to exist; for this reason, it can be regarded as the fundamental engine of the Combinatory System giving rise to the *Civitas centum turrium.*

We know that a combinatorial system is activated and maintained over time thanks to the presence of appropriate sets of *necessitating* and *recombining* factors. Envy, the thirst for power, and the desire not to appear inferior to nobles who owned towers were key *necessitating factors* that drove the families of Pavia to build their own towers, in the hope of not appearing less wealthy and powerful compared to their neighbors. The townspeople valued these towers, viewing them as symbols of power, and supported their construction. This led to the emergence of construction contractors who specialized in building these structures, which, taken together, *reinforced* the *necessitating factor*: the desire of each family to own a tower (a "will-for-tower"). This desire impacted the individual *micro-behaviors* (building towers) and the resulting *micro-effects* (the addition of new towers), opening the way for the collective phenomenon: a cluster of towers in the town. The crucial presence of these *necessary* and *recombining factors* shows that the *micro-macro feedback*s starts functioning once a certain minimum number of initial micro-behaviors and effects—the towers—is reached, known as the *minimum activation number.* This loop is halted when the construction of more towers becomes unsustainable, marking the *maximum saturation point.*

Let us now consider a second, related question: What activates the micro-macro feedback? In general, the activation of the micro-macro feedback requires an initial casual input: a convenient number of agents—greater than a minimum activation number—must initially build a tower. Combinatory Systems function within a broader *environment.* The environment in which these systems operate must often *manage* both the micro- and macro-behaviors, as well as their effects, to steer the system toward a specific goal or limit. This is done through *control levers*, which may involve actions that either *strengthen* or *weaken* certain factors (such as measures, provisions, constraints, limits, etc.). These mechanisms can influence the system at both the macro and micro levels, thereby altering the recombining and necessitating factors. The environmental controls carried out by the strengthening and weakening actions are known as the *fourth pillar* of Combinatory Systems Theory.

Historical Note. Cristiano Zuradelli (1888) links the construction of the towers in Pavia to a legend about rivalry in the ranks of the nobility. According to the legend, a sorceress, whom the nobles consulted to learn their fate, told them: "*Whoever builds the highest tower next to or attached to their residence will rule over Pavia and all its people.*" As a result, Zuradelli explains, "*Some worked by night, others by day, some secretly, others openly, but all began building towers in every part of the town.*"

The heuristic model for this system is composed of the following rules and can be represented as in Fig. 2.16.

Micro behavior and necessitating rule: if there are a lot of towers and you want to distinguish yourself, to demonstrate that your family is noble and rich, build "your own" tower and try to make it taller than those of your rivals.

Macro behavior and recombining rule: each new tower built increases the stock of towers, and the towers that are built take on such a charm and a distinguishing

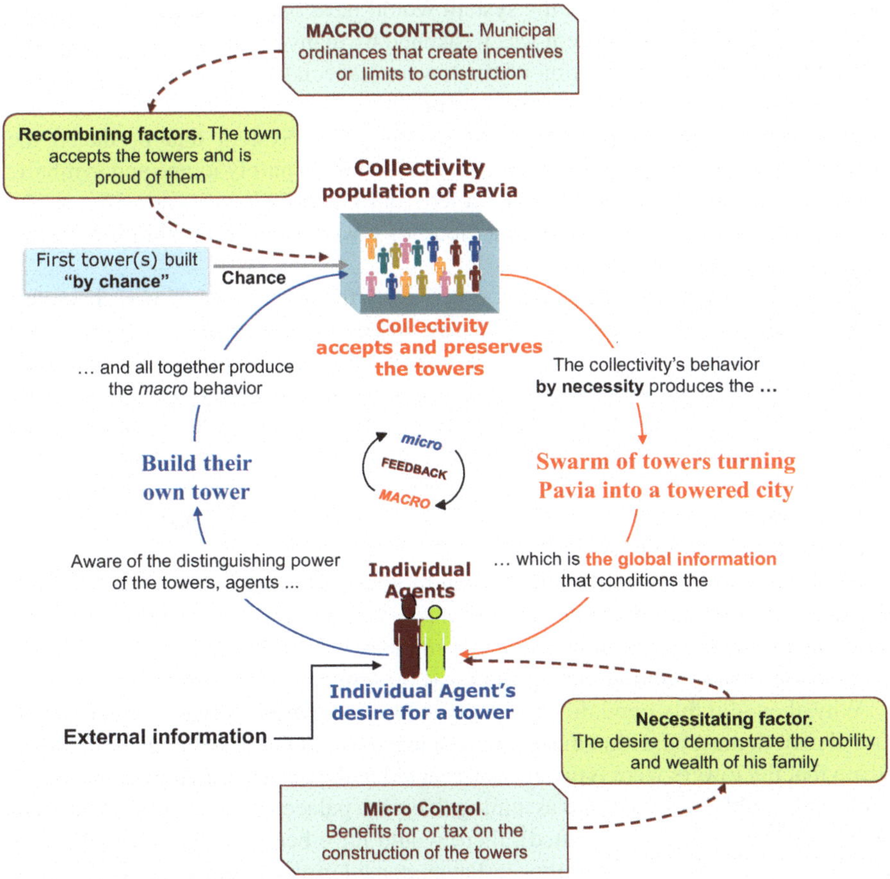

Fig. 2.16 Graphic model of "hundred-towers" system

power as to influence those who, in turn, wish to distinguish themselves. The greater the number of towers, the more likely it is that the only way to distinguish oneself is to build a new tower that is taller (until the saturation density is reached). The town accepts the towers and is proud of them, and this maintains the desire to stand out at a high level.

Micro-macro feedback. Chance and necessity: the mass of towers is the *macro-effect* (the town full of towers) of past micro-behavior, which in turn determines, conditions, or directs the subsequent micro-behavior (building a new tower), thereby conditioning the agents' choices and decisions (if there are so many towers, our family must have one as well). The idea of distinguishing the family by building a tower arose "by chance" but, once begun, the phenomenon is maintained "by necessity." If instead of building an "economical" but clearly visible tower, the first family had built a costly church with frescoes, then probably this would have been imitated, and Pavia would now be referred to as the town of the hundred churches.

Strengthening, weakening, and control actions: if the town had given money for the construction of the towers, the system would have been considerably strengthened; on the contrary, the system was weakened by many municipal ordinances that created limits to the construction of the towers (especially regarding their height), whose density created clear risks for the citizens.

To complete the description of the Combinatory System that produced the incredible phenomenon of the spread of towers having a purely iconic and symbolic significance, it must be noted that the micro and macro-behaviors and effects (the building of single towers leading to a cloud of towers) develop within the limited territory of the town of Pavia.

Before the Combinatory System was activated, the town's "state" was characterized by "palaces-without-towers." There are no historical records specifying which family was the first to construct its own tower or which families followed, ultimately reaching the *minimum activation number* required to activate the Combinatory System. Following this initial chance-driven impetus, the system's initial state shifted to "palaces-without-towers-but-one-with-tower." The behavior of each nobleman started to be affected by the growing number of towers (a macro-effect), and within less than two centuries, Pavia's state had shifted to "palaces-almost-all-with-tower."

The spread of tower construction came to a halt after nearly two centuries, not because there were no longer any noble palaces without towers, nor due to concerns about the towers posing a danger (Pavia is in an area with very low seismic risk), or because the cost of building new towers was prohibitive.

Why, then, did this trend die out? A plausible explanation is that during that historical period, the area within the town's defensive walls covered roughly 900/900 m. Given that the towers were typically constructed at the corners of palaces owned by noble and wealthy families, and assuming that each palace averaged around 30/30 m (which can be objectively verified), there would have been no more than 300 palaces for the rich and noble families in Pavia, considering the space taken up by the less affluent residents, as well as by the squares, roads, churches, and public buildings in the town. This leads to two possible explanations: first, the towers had become so numerous (with a density of one tower for every two palaces) that they lost their *allure*. The *global information* had shifted from "if you don't have a tower, you're not considered wealthy" to "if you build a new tower, you're seen as a parvenu." Furthermore, as the number of towers increased, the citizens' fear of the risk of collapse increased accordingly.

The tower-clustering had come to an end because the main necessitating factor had weakened to the point of exhaustion. In fact, once set off *by chance*, the Combinatory System exhausted itself when the *maximum saturation number* was reached; that is, the density of towers was such that a new one would no longer play a *distinguishing role* but, on the contrary, be taken for *unfashionable* social climbing.

In conclusion, the "reinforcing loop" typical of Combinatory Systems was accompanied by two "balancing loops" that gradually weakened the *necessitating factors*, as can be seen in the Causal Loop Diagram in Fig. 2.17, which uses the symbols typical of Peter Senge's Systems Thinking approach (Mella, 2012).

A great teaching. The phenomenon of the building of a large number of towers is not limited to Pavia. The phenomenon of tower construction also appeared in other

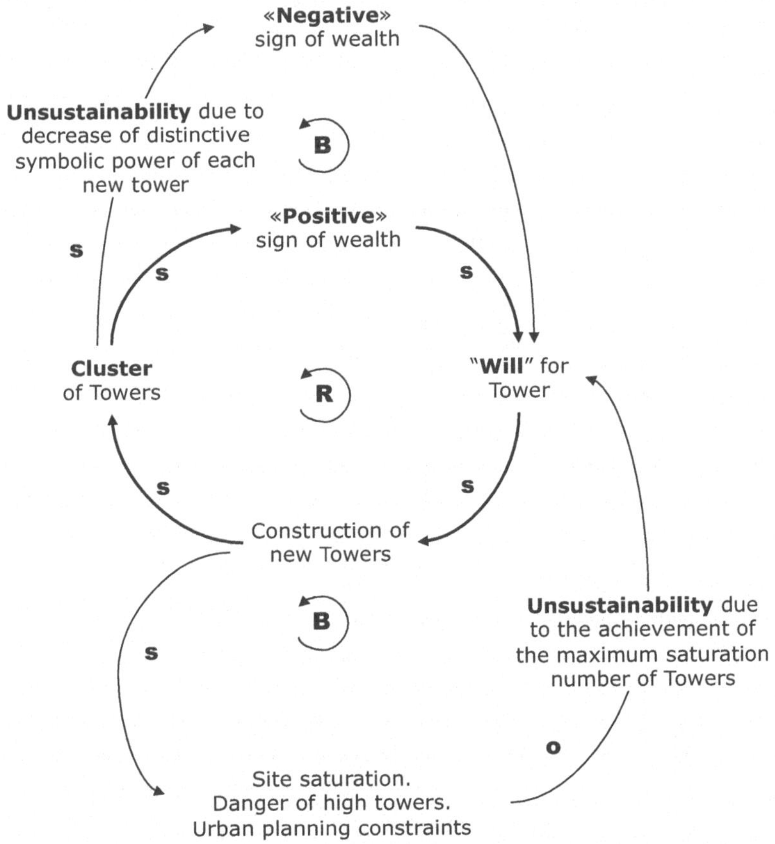

Fig. 2.17 Graphic model of the unsustainability of the "hundred-towers" combinatory system

Fig. 2.18 The towers of several Italian cities built a few centuries after those in Pavia

Italian cities during the same medieval period the towers of Pavia were built: from the eleventh to the thirteenth centuries, as depicted in Fig. 2.18. For instance, *Asti*, a city in the Piedmont region of northwestern Italy, is famously known as the "city of a hundred towers," which were constructed by noble families who fortified their houses by adding a prominent tower, either attached to the house or enclosed within it. Although the same model used to explain the tower construction in Pavia can be applied to Asti, there are three key differences between the two cases. The *first* lies in the time periods: in Pavia, towers were built from the tenth to the twelfth centuries, whereas in Asti, the construction occurred between the twelfth and thirteenth centuries. The *second* difference, more pronounced and significant, lies in the function of the towers, highlighting the uniqueness of the combinatory system in Pavia. In Pavia, the towers served a primarily symbolic purpose, while in Asti they had a practical defensive role, acting as the last stronghold to protect family members during attacks or sieges. Consequently, the interiors of the towers in Asti were not empty but equipped with essentials such as a furnace, a well, and storage for provisions. The *third* difference pertains to the design of the towers. In Pavia, the towers were built with a uniform exterior, distinguished only by their height. In contrast, the towers in Asti featured distinctive crenelations, which identified the family that owned them.

Similar to the development of the towers in Asti was the formation of the tower cluster in *San Giminiano* and *Pisa*, in Tuscany, a region in central Italy, and *Bologna*, in Emilia Romagna, in northern Italy. In these cases as well, the towers were erected many centuries after those in Pavia, with a defensive and residential function. The city of San Giminiano stands on a hill approximately 350 m high, which dominates the Val d'Elsa in Tuscany and has a limited number of massive stone towers. The towers of Pisa, fewer in number and built between the twelfth and thirteenth centuries, were part of fortified palaces. Finally, we must not confuse the towers of Pisa with the most famous leaning Tower of Pisa—which is a bell tower located in the well-known Piazza dei Miracoli (Square of Miracles)—designed and begun in 1173 and completed in 1350, representing an architectural masterpiece.

2.3.4 "Tower-of-Babel" System

Among the most powerful, widespread and easy to verify combinatory systems are the linguistic systems. For simplicity's sake, we will refer to systems of spoken language, because it is easier to see in such systems variations on a main base that is preserved in the written form. Surely no one will argue that the most deeply rooted tradition in any social system is the learning and maintenance of the *mother tongue* by the inhabitants of an area of varying size. This is a very powerful Control System given that, as reported by SIL International the Ethnologue (2013) site, there are at present 7105 languages, of which 2304 are in Asia and 2146 in Africa (Lewis et al., 2013, on line). Although researchers often present slightly different data, Fig. 2.19 lists the 20 most widely spoken languages in the world.

Why does each of us—except for those lucky enough to be multilingual from birth—speak a certain language, which is considered the mother (natural) tongue, and learn other languages that are considered acquired? Why are there hundreds of different languages that are maintained even though a common language would obviously be useful to facilitate communication? Why do dialects exist and why are they maintained for the same language? Why in the case of two urban settlements, perhaps only a hundred meters apart, are there distinct dialects, so that we can unmistakably determine to which community someone belongs?

The Catholic religion has found in the presumption of the Tower of Babel and the failure of its construction an allegorical explanation to justify the presence of many different languages.

> The Lord said, "If as one people speaking the same language they have begun to do this, then nothing they plan to do will be impossible for them. Come, let us go down and confuse their language so they will not understand each other." (NCB, New Catholic Bible (2019) Genesis 11:1–9)

The theory of combinatory systems provides the justification for their maintenance across generations as well as for their changes. Why is English becoming the "universal" second language? Why hasn't Esperanto spread outside of a small circle of proponents?

We are not interested in knowing when and why a certain language arose among the members of a certain community, but one thing is certain: if "by chance" an individual acquires a certain mother tongue, then "by necessity" he will teach this language to his descendants, and this micro behavior will cause the macro behavior characterized by the perpetuation of the teaching of the language, which can be

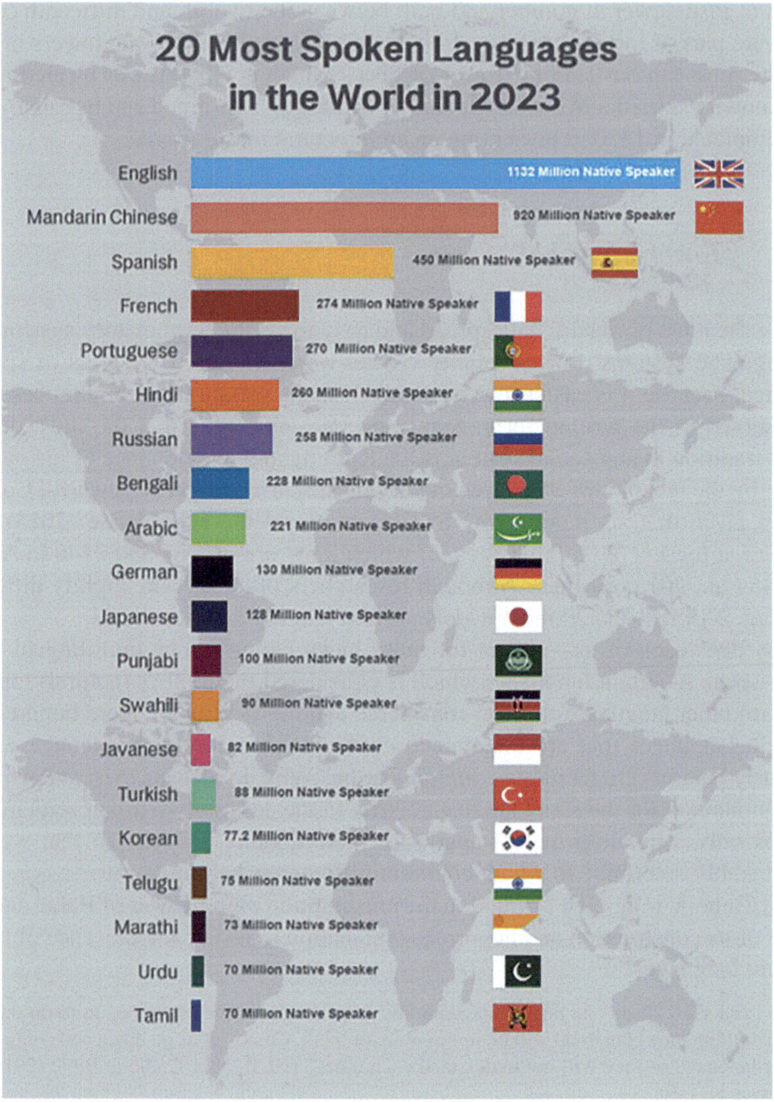

Fig. 2.19 The 20 most widely spoken languages in the world (Source: https://www.ccjk.com/most-spoken-languages-in-the-world/)

viewed as the *macro effect* of the preservation of the language. And this in turn conditions the individual micro behavior of the speakers, as shown in Fig. 2.20.

The spoken language—as opposed to the written one, with its more narrowly codified syntax, which is less affected by practice (how often have scholars despaired over the many errors in syntax and orthography!)—thus appears as one of the most powerful combinatory systems operating in human society.

At times the family of a newborn knows only one language that is different from those of the other members of the community (due to immigration, deportation,

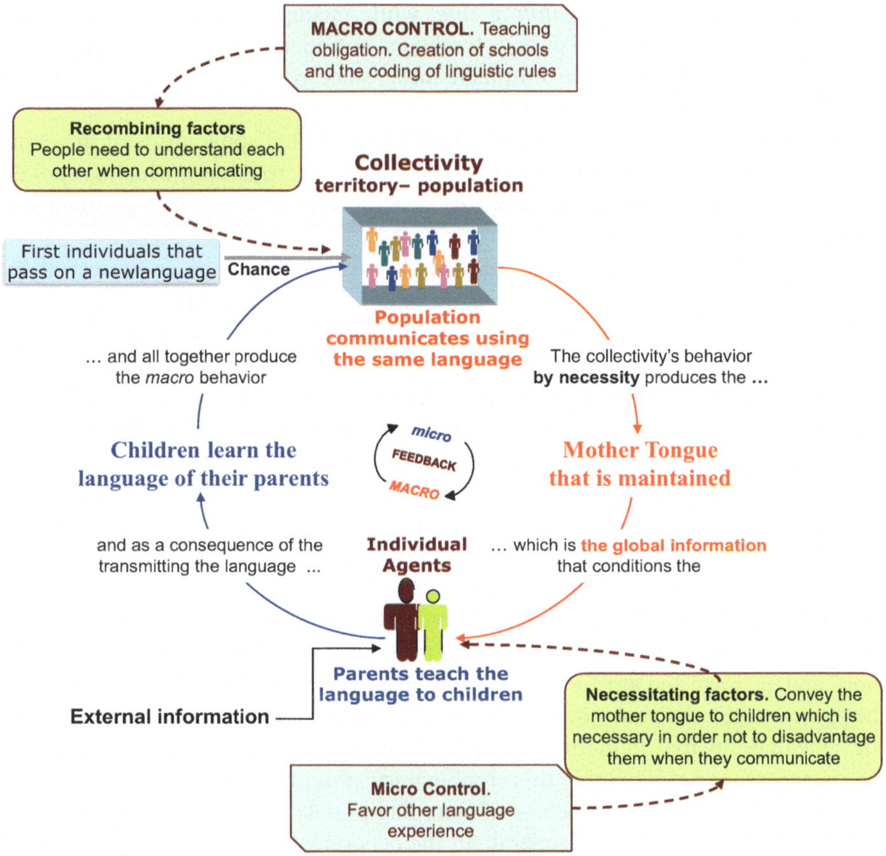

Fig. 2.20 Graphic model of the "maintaining-the-language" system

etc.); in this case, this language is passed on as the mother tongue of the new-born, but the combinatory system is not activated since the newborn is forced to learn the second common social language that is dictated by the educational system of the community. Usually, this second language is held to be rewarding, and is thus passed on to the offspring (dialects are spoken today only by the elderly, for whom they represent the mother tongue, since the youth are taught the national language). At other times, in the same environment, several communities co-exist, each characterized by its own language; in this case, there are as many combinatory systems as there are different languages. If the languages are equally important with regard to interpersonal relations, and if marriages between people with different mother tongues are frequent, then the parents may pass on different mother tongues to their children at the same time (Switzerland, for example). If the number of multilingual newborns exceeds the minimum for activation, then a new combinatory system that "preserves multilinguism" is formed.

The rules forming the heuristic model that describes this system can be represented as follows:

Micro behavior and necessitating rule: while raising and educating your children, teach them the language of your forebears; consider as an error any syntactical or semantic deviation and try to eliminate these, since your children must be able to communicate within the collectivity.

Macro behavior and recombining rule: the educational apparatuses of the community preserve and diffuse the ancestral language (macro effect), which is spoken using the same structure, generation after generation; the language is codified and passed on without changes.

Micro-macro feedback. Chance and necessity: the language is preserved by being taught to the newborns; it is thus the result of the combination of past *micro behaviors*, but it also conditions future ones. A new language, or a variant of the old one, can be introduced in a given territory "by chance"; if this new language, or new variation, comes out the loser in the competition with the pre-existing one, then no combinatory system is activated and the pre-existing language prevails. But once the system is activated, it maintains "by necessity" the linguistic foundation through the transmission to the newborns.

Strengthening, weakening, and control actions: the creation of schools and the cataloguing of linguistic rules in a written form (syntax) considerably strengthen the system; periodic waves of immigration can instead weaken it.

Among the many combinatory systems with a functioning logic similar to the one we have just seen are those that lead to the preservation of the "national religion" and of "local and territorial uses" and customs. It would be difficult to "export" the proverbial English custom of waiting in line in an orderly fashion to Latin countries; the inability to wait in line would not allow the minimum density to be reached that is necessary for activating this combinatory system for the diffusion and preservation of such a custom. The use of Georgian portals, so typical to the houses in Dublin, Scottish kilts and, in general, all the decorative and clothing styles typical of a region that are passed on from generation to generation, are the effect of combinatory systems entirely similar to the one we have just described. The "Tower-of-Babel" System described in the previous section presents a structure and mechanism of action that is useful for also describing the phenomenon of the "persistence of a language or a dialect" in certain communities over long periods, albeit with slow variations over the centuries.

2.3.5 "Cultural Models" System. Persistence and the Slow Pace of Change

In a broad sense, "culture" refers to the complex set of behaviors, beliefs, scientific, practical, and artistic knowledge, common products, customs, traditions, and habits that, along with a collection of value systems and beliefs—typically codified into norms—characterize a community of people within a given geographic context and a specific historical period. These elements are passed from one generation to another through a process of "socialization."

> Culture or Civilization, taken in its wide ethnographic sense, is that complex whole which includes knowledge, belief, art, morals, law, custom, and any other capabilities and habits acquired by man as a member of society. (Tylor, 1871, Vol. 1, p. 1)

The concept of culture can be expanded to encompass all relevant social elements within its meaning: language, religion, traditions, visual arts, practiced sports, literature, music, and even typical cuisine. Most importantly, cultures have shared systems of symbols.

> [Culture] denotes a historically transmitted pattern of meanings embodied in symbols, a system of inherited conceptions expressed in symbolic forms, by means of which men communicate, perpetuate, and develop their knowledge about and attitudes toward life. (Geertz, 1973, p. 89)

The term "culture" is often considered synonymous with "civilization," as Taylor himself stated in the previous quote. However, the concept of civilization is frequently associated with a meaning related to a collective entity—identified as a state, nation, people, or society—to denote the condition of a community with a consolidated culture that has translated into concrete institutions, physical or mental, and is recognized in its modes of production and way of life.

At the state level, culture is maintained by "nationalism." Culture, along with the language that integrates it, represents the "glue" between individuals, as it influences how people think, perceive the world, act within it, and potentially seek to change it. For culture to be learned, it must be shared within the community. Language, therefore, plays a significant role in the structure, maintenance, and dissemination of culture. Linguists Edward Sapir (1968) and Benjamin Lee Whorf (1956) proposed the "theory of linguistic relativity"—(known as the "Sapir-Whorf hypothesis"—which suggests that the language spoken by a community influences how individuals perceive and evaluate the world, since different languages can provide different categorizations of objects and thoughts (Schlenker, 2004; Widhiarso, 2009)).

Around the world, cultures are so numerous that it seems difficult to list and describe their characteristics, especially because within larger cultural frameworks, countless local cultures can be identified. For example, one can distinguish so-called "Eastern" macro-cultures, each divided into specific cultures characterized, as is well known, by distinct languages, religious traditions, and culinary practices.

1. *Chinese culture*: This can be considered one of the oldest and most influential cultures in the world, with a rich history encompassing art, literature, philosophy, cuisine, traditional medicines such as Ayurveda, sciences and technologies, religious traditions like Taoism and Confucianism, and practices such as Tai Chi and the concept of Yin and Yang.
2. *Japanese Culture*: This includes traditional arts such as the tea ceremony, ikebana (flower arranging), calligraphy, martial arts, popular culture like manga and anime, and traditional rituals and festivals.

3. *Korean Culture*: Distinguished by its cuisine, martial arts such as Taekwondo, traditional music, traditional ceremonies, and practices such as traditional Korean medicine.
4. *Indian Culture*: Along with Chinese culture, it is among the oldest; however, it is more complex to describe as it encompasses a wide range of languages, religious traditions such as Hinduism, Buddhism, and Sikhism, culinary traditions, architecture, fashion, dance, music, and art, traditions that characterize specific cultural areas.

Europe, with its numerous states, is home to many cultures that coexist and influence each other, characterized by a remarkable diversity in language, religion, history, education, healthcare, regional traditions, and culinary practices, creating a rich and dynamic social fabric. Each European country has its own unique culture: the English culture is distinct from the French, which in turn differs from the Spanish and Italian cultures, and even Greek culture stands apart from the others. In addition to these national macro-cultures, there are ethnic groups with specific cultural traits and regional micro-cultures that further enhance the cultural richness of the world.

Culture is not an innate trait or a biological inheritance; it is learned by individuals as members of a community and passed down from generation to generation through the action of a powerful Combinatory System that functions similarly to the learning and maintenance of a mother tongue. Although cultures can be transmitted from generation to generation for centuries, they are not static but dynamic, continuously evolving in response to social, political, and economic changes, which can be driven by various factors, such as *historical events* that may lead to the disappearance of the original community (migrations, wars, and domination by other peoples), *social changes* (rapid shifts in living standards, internal upheavals, and even "revolutions"), technological innovation, globalization, and interactions between different cultures facilitated by the development of new forms of communication, such as the advent of the internet and social media. Often, macro-cultural changes may occur gradually over many generations, while others can happen more suddenly and drastically, driven by significant events or forced social and political transformations.

The heuristic model illustrating the "*Cultural Models" System* is represented as follows (see Fig. 2.21):

Micro-behavior and necessitating rule: While raising and educating your children, you transmit to them, through imitation or specific linguistic interactions, the religion, education, customs, legal, and social norms (as well as the language) of the cultural community, so that they respect traditions and institutions and are not disadvantaged in their interactions with others in the community. If children exhibit behavior that deviates from culturally accepted and widespread norms, attempt to correct them through reasoning, persuasion, and, in some cases, coercion, so they can integrate into the community (*micro-effect*).

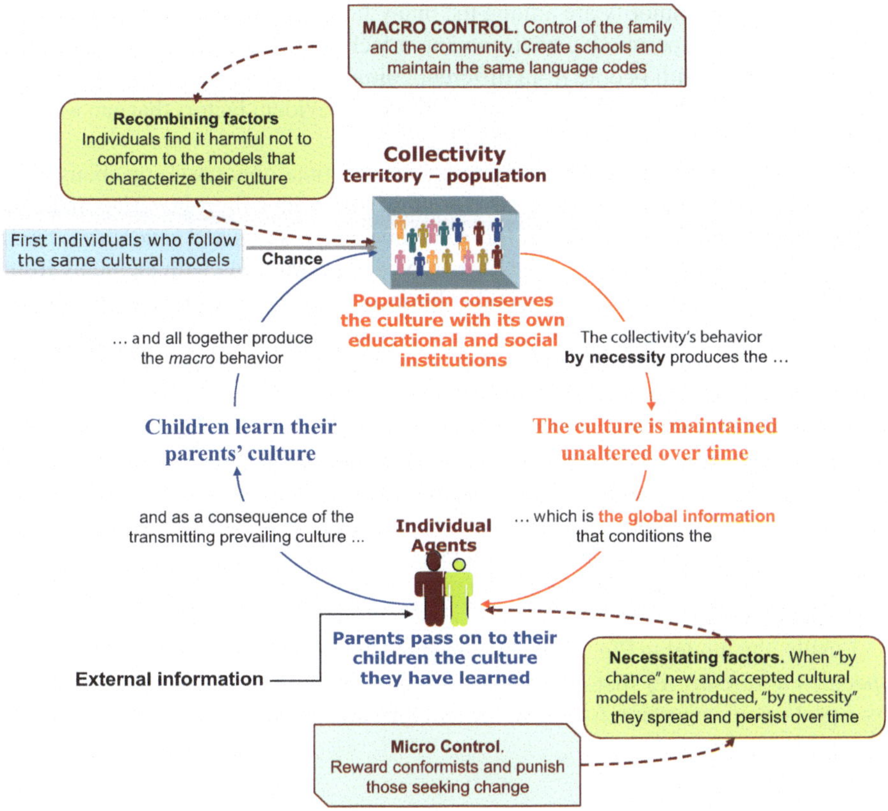

Fig. 2.21 Graphic model of the "Cultural Models" System

Macro-behavior and recombining rule: The educational apparatuses of the family and community preserve and disseminate traditional culture (*macro-effect*), which can be maintained from generation to generation; therefore, the accepted traditional culture is transmitted without changes.

Micro-macro feedback. Chance and necessity: Culture is preserved by being transmitted to newborns and is thus the result of the combination of past *micro-behaviors*; however, it also influences future ones. A new cultural model, or a variant of the old one, can be introduced into a given community living in a territory "by chance." If this new cultural model or new variant loses out in the competition with the pre-existing one, then no new combinatory system is activated, and the pre-existing cultural models persist. However, once the new combinatory system is activated, it introduces "by necessity" the new cultural models, initiating their transmission to newborns.

Strengthening, Weakening, and Control Actions: The creation by the state or local authorities of static educational structures, rigid mental models within the family and community, and the ostracism of individuals who exhibit behavior different from the majority are among the many factors that can *strengthen* the system. Conversely, a permissive environment, a lack of or lax education, freedom to maintain contact with people from different cultures, acceptance of migration, and the spread of new ideas, techniques, and fashions are evident factors that can *weaken* the system, as illustrated in the graph in Fig. 2.21.

There can be cases—such as migrations, invasions, conquests, or changes in work and living standards—where two (or more) different cultures overlap. In these cases, the two cultures may coexist for some time, but as years and generations pass, each community may replace some of its typical cultural traits with elements from the other community. The two original cultures blend and, over time, are forgotten, replaced by a new common culture. This process is directed by a Combinatory System that follows a pattern like that indicated in Fig. 2.21 and develops through the micro and macro-behaviors described above.

When factors that favor cultural change intervene, often in families where the innovative culture coexists with the traditional one, parents prefer to transmit the new culture to their children. When the number of those who have adopted this culture exceeds the minimum activation threshold, a Combinatory System develops that preserves and disseminates the cultural innovations. The new culture eventually prevails over the traditional one and remains in the community until other modifying factors emerge.

Finally, the model in Fig. 2.21 also interprets the maintenance and innovations of *specific micro-cultures* that characterize small communities of individuals, united by a profession, a passion for sports, or by clubs and societies.

2.3.6 "Spread-of-Drugs" System

Let us now look at the devastating and worrying effects from the spread of drugs, a scourge that seems unstoppable. This is a phenomenon with many features that can also be analyzed in terms of combinatory systems of diffusion.

Assume that "by chance" in a certain community (base) in a given territory (environment) someone becomes enslaved to drugs and, to support the habit, must find the money to buy the drugs by selling them in turn. If the number of addicts is not very high—and does not exceed the minimum activation number—the upbringing and moral principles in the community are solid, and the police are efficient in preventing and repressing the use of drugs, then the phenomenon can remain isolated. The few addicts can be rehabilitated or moved to an environment where proselytism is easier. If the weakening factors mentioned above are not in operation, no combinatory system is activated. However, when the number of addicts exceeds the minimum number for activation—which usually is not very high and depends on the socio-cultural conditions of the collectivity and on its political and police organization—then an unstoppable combinatory system develops where several individuals undertake micro behaviors that consist in individual drug-pushing actions that produce as micro effects the initiation of new addicts.

The *macro behavior* is represented by the spread of drugs in the community, with an increase in the average number of users and in the quantity of drugs taken and sold (macro effect). The spread of drugs (macro behavior) has as an effect the incentive for individuals to turn others on to drugs (micro behavior); but this in turn has the effect of spreading drugs, according to a typical *micro-macro feedback.* The effects of this combinatory system are apparent to all of us: the cultivation of drug-producing plants expands, the production of synthetic drugs intensifies, international trafficking from producing countries to those selling the drugs increases through increasingly sophisticated land, sea, or air transport methods. Smuggling flourishes, and there is an expansion in distribution channels, dealing networks, direct sales, or sales through the Dark Web, but above all, through social "word of mouth." Social tolerance can also increase because the perception of risk associated with drug use becomes less negative, and the change in cultural norms can influence the level of consumption and the spread of drugs.

The heuristic model that describes this system can be represented as follows. And is also shown in Fig. 2.22.

Micro behavior and necessitating rule: if you need drugs try to buy them; if you need money for drugs, obtain it by selling them by influencing others to use drugs.

Macro behavior and recombining rule: the acts of individual drug pushing spread the use of drugs at the collective level, but the spread of the drug is the condition that favors drug pushing; the higher the number of drug addicts, the higher the number of pushers, which in turn leads to an increase in the number of addicts.

Micro-macro feedback. Chance and necessity: the spread of drugs (macro behavior) has as an effect the incentive to proselytize (micro behavior), but this in turn leads to the spread of drugs; the first addicts arrive "by chance," and if they are not rehabilitated, then the macro behavior begins and is maintained "by necessity," with an increase in the number of new addicts.

Strengthening, weakening, and control actions: the strengthening of the police force, the destruction of distribution channels, the repression of drug selling organizations, the tendency to denounce drug pushers, and the creation of rehabilitation and detoxification centers are examples of *weakening factors* for the system. The strong presence of the Mafia-like organizations, the absence of social constraints, the proximity of import centers for drugs are examples of *strengthening factors*. The *macro control* must act upon the drug supply lines and the social conditions that favor drug addiction. The *micro control* must act by means of preventive campaigns that provide information about the dangers of drug use, along with rehabilitation facilities and, in general, social support and psychological measures that help drug addicts overcome their addiction.

The above system can activate other combinatory systems that are connected to it. For example, the need to obtain drugs when money cannot be obtained from selling them leads to the need to rob and steal (micro behavior), which results in an increase in petty crimes (macro behavior) in the social system, which, in turn, permits the spread of drugs along with the need for new criminal acts, according to a typical micro-macro feedback, whose effects cannot be predicted.

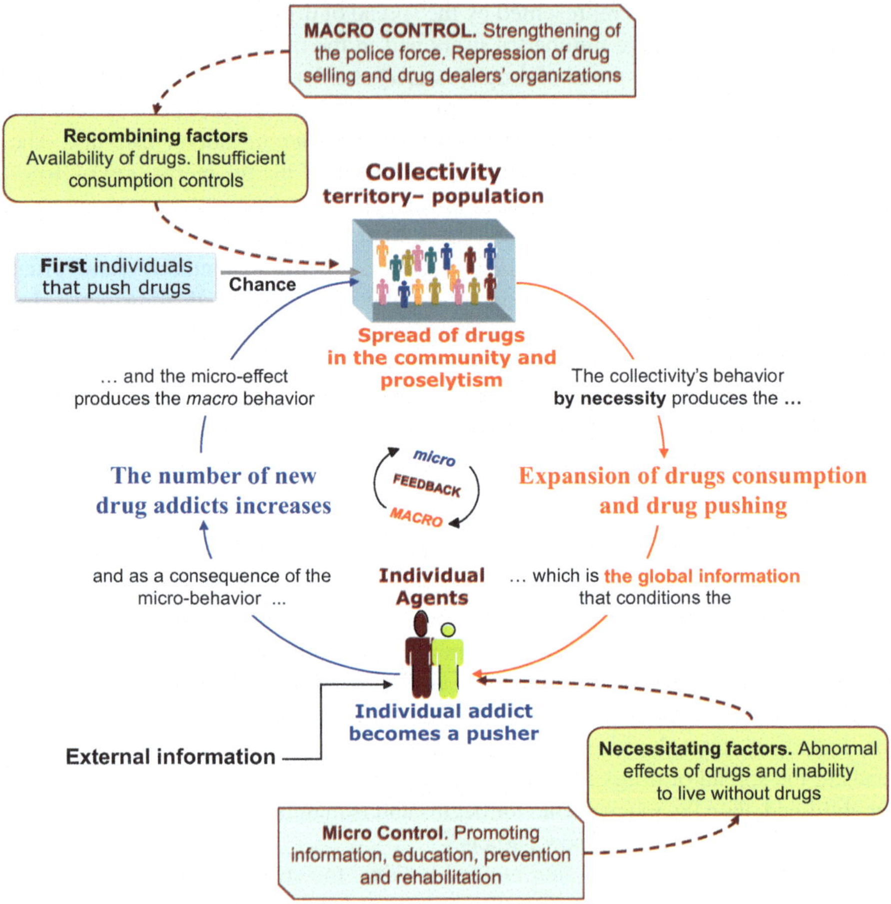

Fig. 2.22 Graphic model of "drug" system

The logic of this combinatory system is entirely similar to that of the spread of corruption, which represents a serious problem in many societies and which can be described using the same heuristic model adopted for drug use.

2.3.7 *"Break-out-of-Applause" System*

Applause, an act of consensus, appreciation, and recognition that is widespread throughout Western culture, is a collective phenomenon that manifests, for example, as a form of gratitude, appreciation, or social recognition for an action worthy of praise (such as a theatrical performance, a speech, or a sports achievement); or as a collective ritual after a performance, a victory, or the conclusion of an event; or, again, to express unity and celebration, to create a sense of oneness or solidarity in a group, or the success of a shared experience.

Applause often follows the rules of combinatory systems. For example, a crowd (base) is listening to a speaker in a room or a square (environment); the speaker interrupts his speech, and there is a prolonged pause...while waiting for applause. How many times have we experienced this! The phenomenon of applause can be concisely described as follows. Suddenly someone "by chance" claps (micro behavior), producing the typical noise (micro effect). If the number of those who applaud does not reach the minimum activation number, the applause does not begin; but if the initial applause does not "die out," others will join in and the applause will rain down. The micro behavior turns into a macro behavior (everyone applauding) of which the applause, understood as a typical noise, represents the macro effect. The clapping individuals form the applause; but this forces everyone "by necessity" to continue clapping to maintain the applause until someone stops and the macro effect diminishes. The feedback inevitably acts in the opposite direction as well, and slowly the applause fades away, as described in Fig. 2.23; the combinatory automaton model simulating the Applause System will be shown in Sect. 3.4.2. The speaker can strengthen or weaken the applause through obvious gestures of satisfaction aimed at quieting the crowd.

This combinatory system is similar to that which occurs when, in a crowd of persons, someone at the rear stands on her toes to see better, forcing everyone else behind her to stand on his or her toes as well. In particular, this combinatory system—which along with many others has as its object collective behavior characterized by a feedback relation with the individual micro behaviors and the collective macro behavior, as described by Alessandro Manzoni (as observed in Sect. 1.3.4): "*And everyone, rising on their toes, turns to look to where the unexpected arrival [of the grand chancellor Antonio Ferrer] was announced. With everyone rising, they could see, some more some less, as if they had remained with their feet on the ground; in any event [pushed by the micro-macro feedback] everyone stood up*" (Manzoni, 1985, Chapter XIII, p. 240).

2.3.8 "Nudging" System: Diffusion of Useful and Virtuous Behavior

The concept of "nudging," originally introduced by economists Richard Thaler and Cass Sunstein in their book *Nudge: Improving Decisions About Health, Wealth, and Happiness* (Thaler & Sunstein, 2008) has gained considerable attention in recent years in the field of behavioral and organizational sciences, as well as in public policy.

Nudging represents a powerful tool to promote useful and virtuous behavior, subtly guiding individuals toward a particular decision or behavior. By exploiting cognitive biases and psychological predispositions, it proposes simple interventions

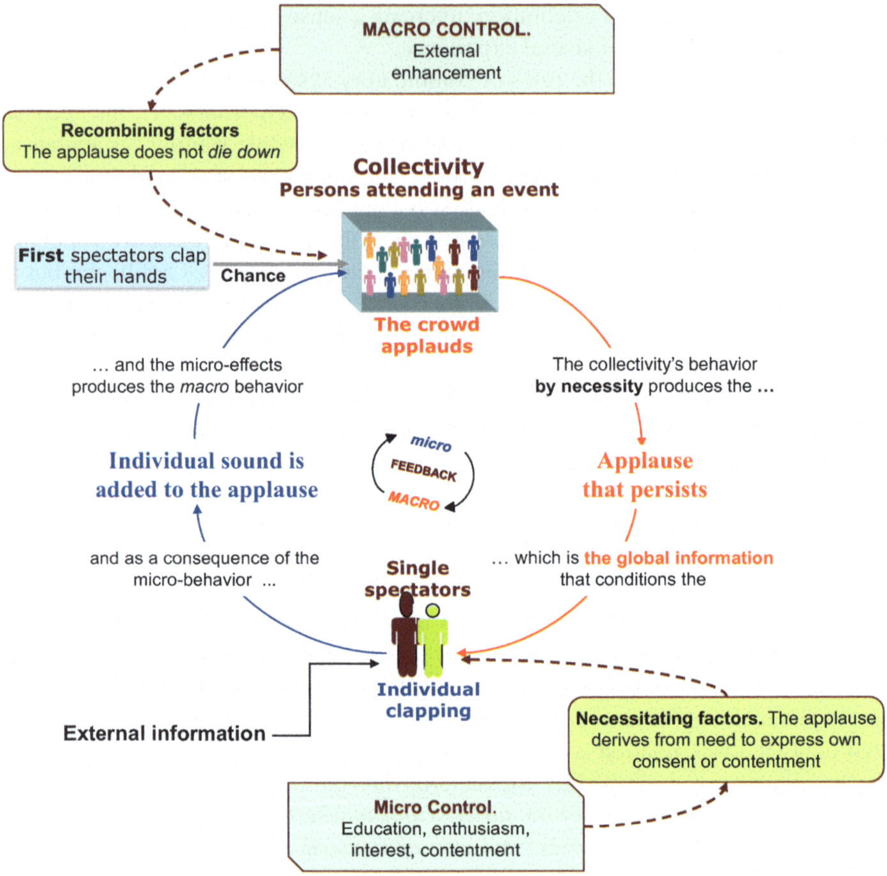

Fig. 2.23 Graphic model of systems of "applause"

that, without imposing obligations or restricting choices, influence people's behavior in predictable ways, often improving *individual* and *collective well-being*.

Thus, "nudge theory" can be considered a behavioral science that encourages social agents to make individual decisions and produce behavior that, when "combined together," produce collective well-being as a macro-effect, which in turn directs and maintains individual behavior, resulting in a useful *Combinatory System*.

Through simple interventions, easy and inexpensive to implement, well-designed and ethically acceptable, which influence *individual* behavior, nudging produces *collective* behavior that improves public health, financial, and personnel management in organizations, as well as environmental sustainability. These interventions can be as simple as they are effective, such as the strategic placement of healthy and appetizing foods in a school cafeteria to encourage better food choices, granting breaks at work to allow for more movement to benefit individual health or the use of motivational messages to increase organ donation response rates.

The fundamental characteristic of nudging is that it preserves freedom of choice: it is not about imposing actions but offering social and organizational "suggestions"—often not explicit or transparent—that make certain individual choices more likely than others. It is useful to present 15 examples of Workplace Nudging presented by Tom Haak (2020, online recalling the three guiding principles on the use of nudging:

(a) All nudges should be transparent and never misleading.
(b) It should be as easy as possible to opt out of the nudge.
(c) There should be good reason to believe that the behavior being encouraged will improve the welfare of those being nudged.

Fifteen examples are presented, which can be summarized as follows:

1. Nudge: *In cafeterias, provide small plates*. Give people smaller plates, and they will eat less.
2. Nudge: *Put workstations on dark carpet* to encourage workers to maintain the desired distance.
3. Nudge: *Remove all wastebaskets except one from the floor*. In offices, leave only one paper bin. This gets individuals accustomed to moving during working hours.
4. Nudge*: (Re)design your office in such a way that the stairs are very dominant, and the elevators hardly visible.* Same as for nudge n. 3.
5. Nudge: *Put something in the urinal that men like to aim at.* Try to encourage men to pee inside the urinal.
6. Nudge: *Choose a lively location for your HQ*. Desired behavior: meet some new people, outside your "bubble."
7. Nudge: *Put pictures of people who are perceived as ethical leaders* on the walls or on the desktop of the computers. This will encourage ethical behavior.
8. Nudge: *Remove all the printers except one from the floor or office*. People will use printing less and walk more.
9. Nudge: *Make fruit and other snacks easily accessible* and put them where they are clearly visible. *Put the unhealthy snacks high or low in the cupboard*. This encourages people to eat more healthily.
10. Nudge: *Provide people with a flexible desk*, at home or in the office. Employees receive the message that it is beneficial for them to be less sedentary.
11. Nudge: *Ask some of your employees to wear their sportswear in the office* or during video calls. Other people will imitate that behavior, especially if the sportswear is worn by opinion leaders.
12. Nudge: *Include a picture of hands that are washed at the beginning of your presentations* or make it your Zoom/Teams background. This spreads the belief that hygiene is important even in the workplace.
13. Nudge: *Paint footsteps on the floor*, and people will have the tendency to follow those steps. Get employees accustomed to following rational paths when moving between offices.

14. Nudge: *When people want to schedule a meeting, ask them what kind of meeting they want (traditional, standing, video, or walking).* Recommend holding a walking meeting to shorten the meeting time and encourage walking.
15. Nudge: *Show a photo of a puppy.* Real puppies are great, but pictures of puppies can also work (postcards, screensavers, posters). People tend to feel better when they see a cute animal.

Nudging, however, is not limited to influencing individual decisions but can also play a crucial role in the *spread* of virtuous behavior within a *community*. When a certain number of individuals are encouraged to adopt a virtuous behavior—such as energy saving, recycling, using eco-friendly packaging, and so on—it can trigger a "chain reaction," prompting others to follow the behavior perceived as virtuous. This phenomenon is a *typical example of a Social Combinatory System* that is activated or enhanced through nudging strategies, though such strategies do not always succeed.

An example of a nudge that operates in urban settings to promote ecological behaviors is the implementation of bike and electric car-sharing initiatives in cities. The increase and strategic placement of bike-sharing stations and electric vehicle charging points, along with subsidized rates for new users, are *reinforcing factors* that can encourage the use of bicycles and eco-friendly cars. As more individuals adopt this behavior, the social perception of bicycles and electric cars as efficient and convenient modes of transportation becomes stronger, representing a *recombining factor* that maintains the virtuous behavior and encourages further adoption.

However, those who use nudges to influence behavior should avoid imposing mandates and must respect individual autonomy, steering clear of paternalistic or manipulative approaches that might affect people's decisions without their full, conscious consent. They should ensure that the proposed interventions are transparent and perceived as aimed at collective well-being.

The Combinatory System that arises from nudging can be described by the following heuristic model and represented by the graph highlighted in Fig. 2.24.

Micro-behavior and necessitating rule: If, in your work, social, or sports environment, you notice that certain messages, advice, or actions make the behavior of your colleagues—whether workers, citizens, or athletes—"better, easier, or more useful," then you should also follow that advice and adjust your current behavior to align with those individuals who have received those messages (*necessity*).

Macro-behavior and recombining rule: When the number of individuals conforming to a nudge becomes large enough, it becomes clear to others that collective behavior is necessary because it improves and makes their own behavior more useful. Therefore, it is advantageous to adopt the behavior that is widespread within the community (*necessity*).

Micro-macro feedback. Chance and necessity: The spread of collective behavior among individuals who follow the nudge (macro-behavior) encourages others to adopt and proselytize this behavior (micro-behavior). This, in turn, leads to the spread of new behavior. The initial individuals adopt the suggested or proposed

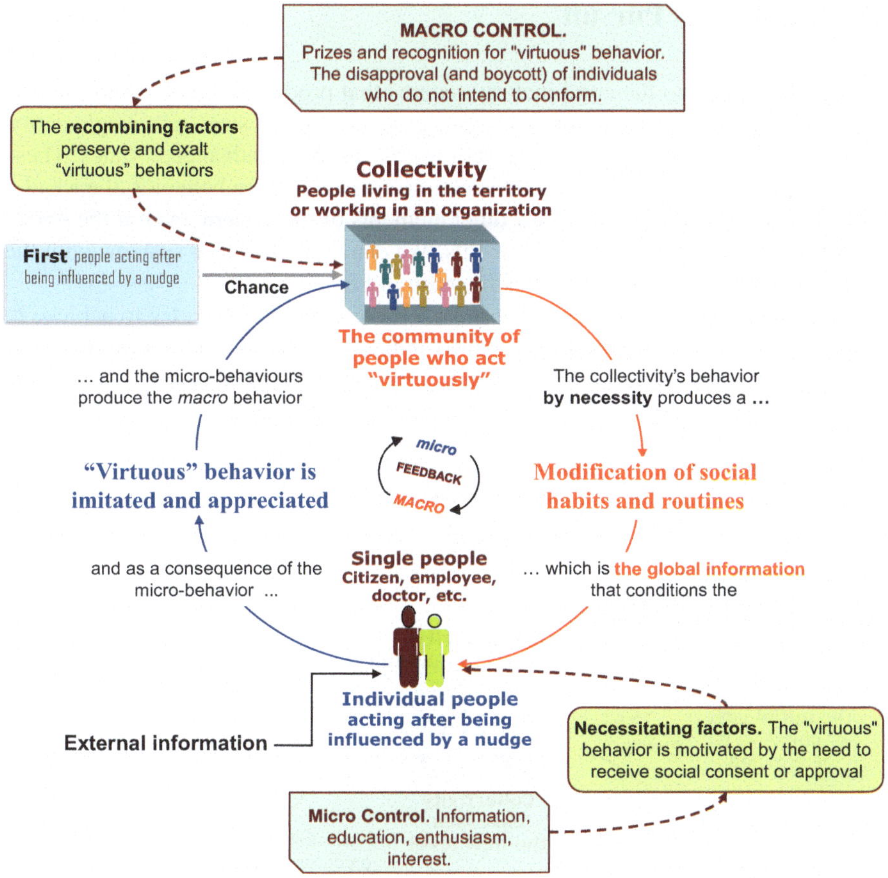

Fig. 2.24 Graphic model of a "nudging" system

behavior "by chance" and, if they are imitated, the macro-behavior begins and is maintained "by necessity," with an increase in the number of new behaviors.

Strengthening, weakening, and control actions: incentives, *rewards*, and *recognition* for those who align with what a nudge proposes are forms of reinforcement for the Combinatory System that disseminates new behaviors. Disapproval and boycotting by individuals who refuse to adapt and continue with traditional behavior serve as a powerful way to weaken the Combinatory System. Additionally, behaviors spread more rapidly in tightly knit communities where individuals have frequent interactions.

2.4 Systems of Pursuit

We give this name to the combinatory systems that produce behavior consisting in a gradual shift of the system toward an objective, as if the system, as a single entity, were pursuing a goal or trying to move toward ever more "advanced" states. These systems typically involve limited information, since the micro behavior of each element depends on that of only a few other elements of the system, even if the objective or limit is known by all the elements. We can describe these systems with the following heuristic model and in the diagram in Fig. 2.25.

Micro behavior and necessitating rule: if there is an objective, try to achieve it; if there is a limit, try to exceed it; if another individual overtakes you (negative gap), regain the lost ground; if you're even with someone, try to go ahead; if you're in the lead, try to maintain or increase your advantage (positive gap).

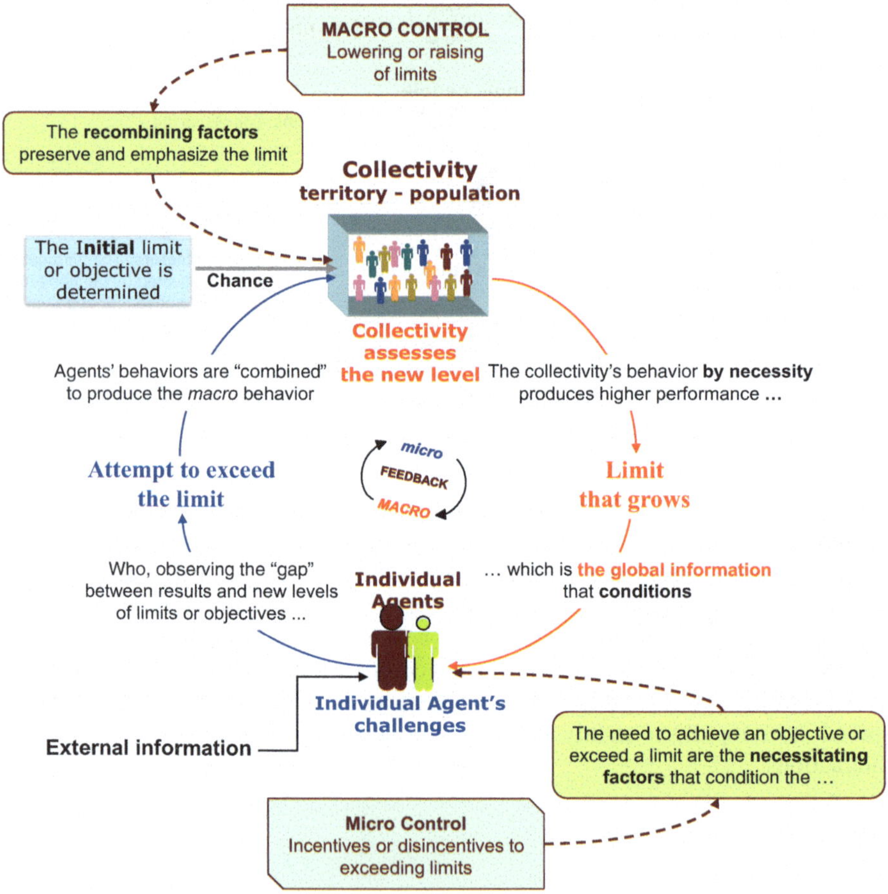

Fig. 2.25 Graphic model of systems of "pursuit"

Macro behavior and recombining rule: the collectivity recognizes the validity of the objective and views limits in a negative way; the more individuals try to exceed the limit, the greater the chance of exceeding it, with a consequent advantage for those who succeed in doing so. This provides the incentive for the pursuit.

Micro-macro feedback. Chance and necessity: if everyone tries to go beyond the limit (macro behavior), then this is raised (macro effect), thereby eliminating the advantage for those who have already reached it (micro effect). This forces the individuals to exceed the limit (micro behavior). The initial attempt to achieve an objective or exceed a limit occurs "by chance" but is maintained "by necessity."

Strengthening, weakening, and control actions—If the individual that exceeds a limit or constraint receives an award, then the system is strengthened due to the increase in gratification. If the achievement of the objective or the setting of the limit causes injuries or the cost of the behavior becomes too high, then there is a weakening.

This model can represent quite a different array of combinatory systems, which will be described in the following sub-sections.

2.4.1 "Voice-Murmur" System

We encounter this system in a room (environment and base) crowded with individuals trying to communicate (micro behavior) with others near them, murmuring or speaking with a certain loudness (micro effect). If the number of individuals undertaking similar micro behavior is sufficient (density of activation), we observe a *macro behavior* in the form of a general communication activity whose macro effect is the typical background buzzing that comes from the "combination" of noises arising from the voice levels (micro effect) of those who are speaking (micro behavior).

The typical feedback relation between voice (micro effect) and murmur (macro effect) is clear: the buzzing (macro effect) stops those engaging in interpersonal communication from making themselves heard; because of this, those present are forced by the system to raise their voices, thereby giving rise to individual micro behavior. If the voices rise (micro effects), then the buzzing increases and the voices must further rise, until a deafening din breaks out. The murmur level represents the information produced by the group of speakers themselves, indicating the minimum voice level needed to be heard. The background murmur conditions the micro behavior and its micro effects, even if we observe the micro behavior as creating the buzz. When the system begins "by chance," it "by necessity" forces the speakers to raise their voices in order to be heard; the combinatory system thus operates according to typical chance-necessity logic. If in a conference room someone, while waiting for the speaker to begin, should "by chance" begin to speak, even if only to say "pshaw!," then the background noise would begin "by necessity" and the combinatory system would probably start to operate. In a movie theatre with acoustic paneling (weakening factor), a buzzing, if it should arise, would either cease or remain stable at an acceptable level. If the environment has acoustic panels, the buzzing would stabilize at an acceptable level (classrooms); if some factor makes the buzzing bothersome, the latter would weaken and even diminish until it disappeared (conference room or theatre).

The buzzing phenomenon also arises if the number of speakers exceeds a minimum, which must be determined by considering both the social ties among the individuals forming the base and the structure of the environment. In a large cathedral, the activation density will presumably be greater than in a small restaurant. The system typically functions with global information: in determining his or her voice level, each speaker only considers the hearing capacity of the person to whom they are speaking, but this is conditioned by the murmur level. If small groups form—as could occur, for example, among people seated together in a restaurant—then the speaker would take into account the hearing capacity of the others in the group; nevertheless, he would ignore the acoustic conditions of those seated at adjacent tables.

The voice-noise system can be represented by the following heuristic model and in Fig. 2.26.

Micro behavior and necessitating rule: if you have to talk and you hear a background murmur or noise, then in order to be heard, you must raise your voice level several decibels above the background noise; the level of your voice will depend on the amount of murmur (information) you perceive and on the number of persons who must be reached by your voice.

Macro behavior and recombining rule: the environment preserves the noise, and the collectivity makes interpersonal communication necessary or favors it. The background murmur results from the voice level of the individual speakers, and its level is determined by the average voice level of each person who is speaking, taking account of the number of speakers and the sound-absorbing coefficient of the environment. Those who are speaking must also take account of the noise generated by external factors: bells ringing, shouting from outside the system's environment, and so on.

Micro-macro feedback. Chance and necessity: the individual, in order to be heard, must speak louder than the level of the background noise that results from the macro behavior. The system produces a background noise which is a function of the micro behavior of those who, in order to be heard, must speak in a loud voice; the macro behavior depends on the noise coefficient of the environment in which the system operates. In a room with bad acoustics, the noise coefficient will be greater than 1, and very soon a noise will arise which is so deafening that the individuals will be forced to stop talking. Silence will intervene! If "by chance" someone should start speaking again, even to say "What silence!," then "by necessity" the background noise will reappear and the combinatory system will start to operate.

Strengthening, weakening, and control actions: the *internal control* is provided by the individual speakers, who vary their voice levels when the buzzing varies. Typical forms of external macro control are the placing of sound absorbing panels on walls to reduce the action of the recombining factor that transforms the voices

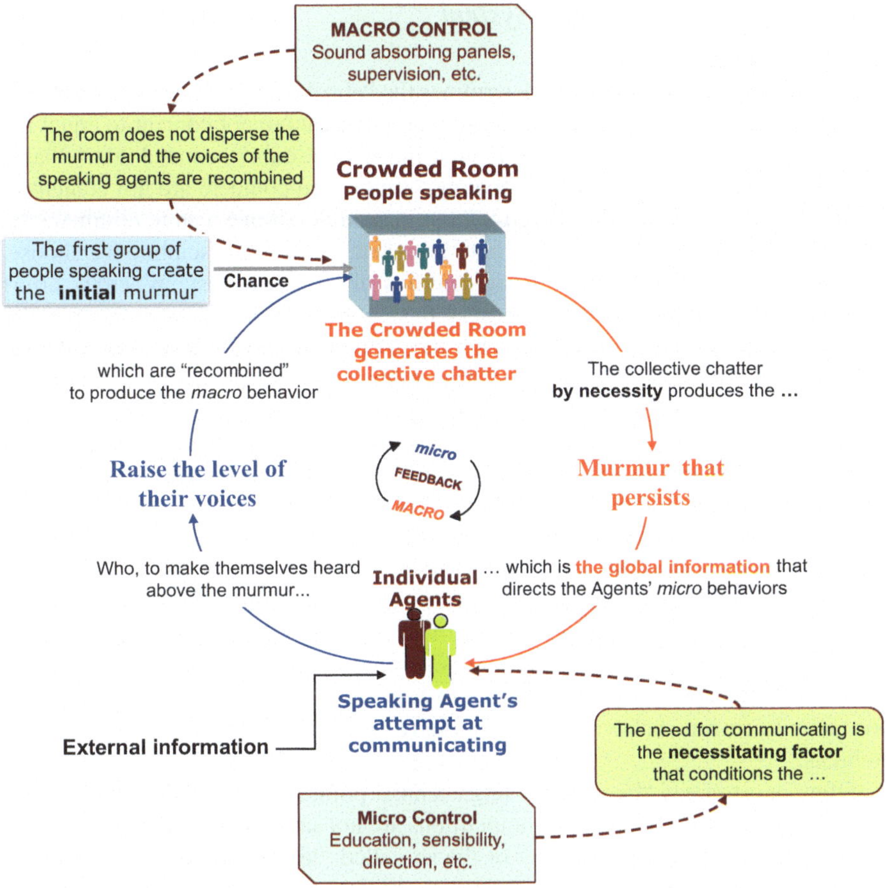

Fig. 2.26 Graphic model of the "Voice-Murmur" System

into a buzzing noise or keeping macro-level external disturbances under control. A form of *micro control* is the education and upbringing of the speakers, which affect all the necessitating factors. Another form is to assign a supervisor to the room (room director, usher) who, at regular intervals, invites those present to moderate their voice levels.

It has been decided to classify the production of a murmur in a crowded room (the murmur can also occur in open spaces with many speakers) as a System of Pursuit. However, from another point of view, it could also be interpreted as a System of Diffusion, since the murmur spreads and is maintained in the group; or even as a System of Improvement and Progress, since the attempt to improves one's communicative performance leads to progress, in the broadest sense, in the murmur level.

2.4.2 *"Beat-the-Record" System*

A record, in any sports category, determines the "absolute best"; because of this, we witness a true race for the record. Those who compete are not content to equal the record but do all they can to beat it. Even if Pierre De Coubertin stated that "what is important is not winning but participating," those who compete are not content to merely equal the record (the objective), but instead do all they can to eliminate the gap and break the record, since the latter, in whatever sports discipline we choose, represents the global information that identifies "the absolute best." When the effort is produced by a multitude of competitors, the record is gradually improved, thereby raising the level of the objective which the individual competitors must achieve. Thus, records are gradually improved. The attempts to improve the records motivate more and more athletes to take part in competitions, which also leads to the continual improvement in the "average performance" of the athletes, so that after a more or less lengthy period, beginning when the first record was set, the average performance of the competitors reaches a very high level.

Today, it is normal for even the last-place finisher in the national 100-m sprint to have a better time than the world record-holder of 30 years ago. The phenomenon of chasing a record and the attempt to break it can also be interpreted with the aid of combinatory systems: the improvement of a record and the raising of the average performance in competitions (macro effect) is caused by the results (micro effect) obtained by the competing athletes (micro behavior), who spread general competitive activity (macro behavior). In turn, the results are conditioned by the records to break.

Often sport arises "by chance" (e.g., tennis, polo, arm-wrestling, windsurfing, snowboarding, etc.). If the local competitions are not satisfying, no one will want to imitate them (consider the destiny of the so-called "lesser" sports such as Greco-Roman wrestling, tug-of-war, etc.). However, if the number of competitions and participants exceeds the minimum number for activation, then the record is recognized, the desire to set the record increases, and the attempts to improve the existing results begin. The record "by necessity" conditions the micro behavior; if the record-holder is praised, if victory leads to considerable monetary prizes, then the system is strengthened, since there is an increase in the level of gratification. If the setting of the record caused accidents, with victims, then there is a weakening. Another weakening factor involves the cost of the competitions: the higher this is, the less interest there is in the record (consider the world land-speed automobile record).

The following heuristic model, which describes the above system, is also represented in Fig. 2.27.

Micro behavior and necessitating rule: if there is a record and you want to stand out, and if you want to show you are "the" best, then you must compete to win and to beat the record; participation in the competitions will bring honor to you and, in the event of victory, you will be awarded a prize.

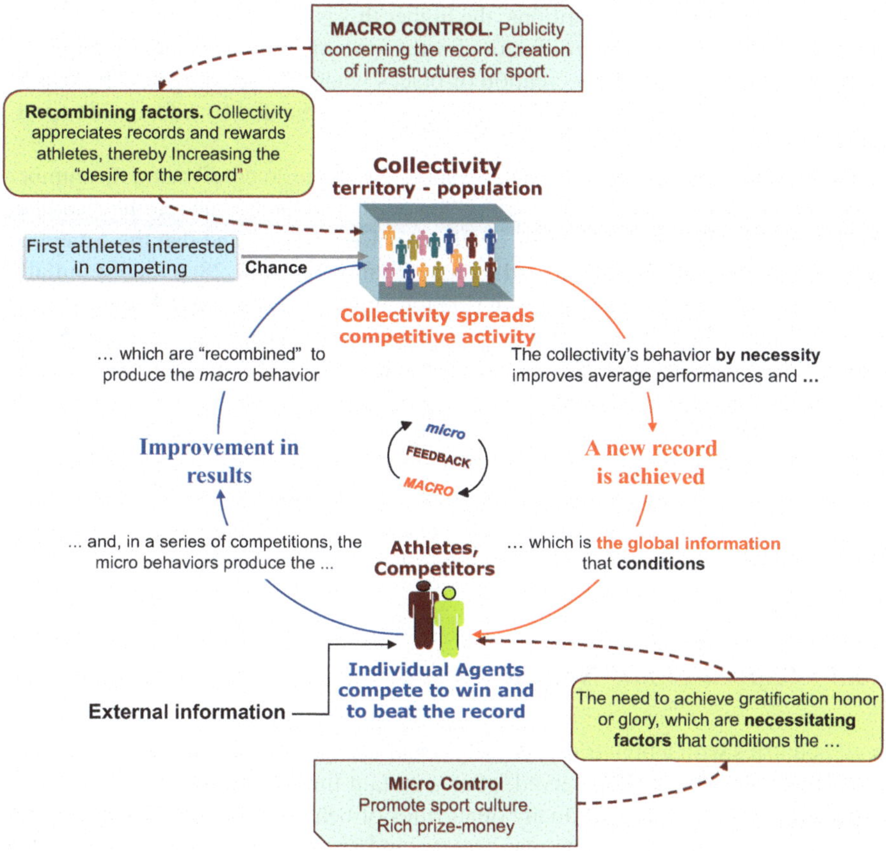

Fig. 2.27 Graphic model of "record" system

Macro behavior and recombining rule: the record is held in esteem by the collectivity; the general attempt to break the record increases the "desire for the record" and improves average performances; thus, in order to compete successfully, more intense training is needed, which leads, sooner or later, to the record being improved as well as to an increase in the average level of quality of the competition.

Micro-macro feedback. Chance and necessity: the improvement in the record and the increase in the average performances in the competitions are caused by the results of the athletes who, in turn, are conditioned by the record to break; the first record arises "by chance" when a minimum number of athletes interested in competing is reached. The record is maintained "by necessity" as long as it is favorably judged by the collectivity.

Strengthening, weakening, and control actions: if honor is heaped on the record-holder and if victory carries with it considerable prize money, then the system is strengthened due to the increase in gratification. If the setting of the record causes accidents, with victims, then there is a weakening. Another weakening factor involves the cost of the competitions: the higher this is, the more difficult the search for the record becomes. Control factors for the system are represented by publicity concerning the record and the creation of places where the sporting activity can be practiced. A typical micro control intervention is represented by a favorable sporting culture.

The model above also applies to these sports, with the "record" being the number of career victories of an individual or team. In Wikipedia, under the heading: "*Athletes with the Greatest Number of Olympic Medals*," we find that the American swimmer Michael Phelps has won the most Olympic medals, 28, a record initially broken at the 2012 Olympic Games in London and then surpassed in Rio de Janeiro in 2016. Skier Marit Bjørgen holds the record for the most medals at the Winter Games. Phelps shares with Soviet gymnast Alexander Ditjatin the record for the most medals in a single Olympics.

Variant: the above system can be broadly generalized, since falling within the logic of this system are all cases where there is a limit—for example, a speed limit—or a rule, and this limit and rule is disregarded, giving rise to micro behavior involving going beyond the limit, which ends up causing the abandonment of the limit itself. One such system is described in the following section.

2.4.3 "Speed-Limit" System

In all countries with a high traffic density, there are speed limits for roads and highways. But, as we have all observed, after a certain time, these limits are no longer respected. This also holds for many other rules of behavior, not only for traffic, for example, no parking rules, one-way streets, and so on. Speed limits in populated areas often appear as a throwback to past times, when the car had no brakes. Today, with more modern cars, speed limits on highways are considered instead as oppressive constraints by many drivers rather than as live-preserving measures. Yet, when it was first introduced, the speed limit was respected, at least for a while. But it is normal to then witness behavior characterized by a disregard for the speed limit, which soon becomes so generalized as to make people forget about the existence of the obligation to obey the limit ("I thought it had been abolished…" is the typical response of the violator who has been stopped by the police).

The mechanism that explains why speed limits are disregarded can be understood if we again consider the action of a combinatory system. The *micro behavior* is represented by people travelling while ignoring the speed limits. Usually such behavior is inspired by the following reasoning: if you are in a hurry and you see that many drivers are exceeding the limit, then this means that the controls are lax, that there is no danger, and that most people hold the limits to be useless; thus you, too, can risk not respecting them. The micro effects are instead represented by the speed that is maintained. The combined effect of the micro behaviors leads to an increase in the number of drivers who exceed the speed limit, and this conditions the individual choice not to respect the constraint, leading to a gradual rise in the average speed. Every driver acts on the basis of limited information, represented by the number of drivers encountered during the trip who are exceeding the speed limit that the driver himself is observing.

The phenomenon begins "by chance" when the number of drivers not obeying the limits exceeds a given minimum; in fact, for a more or less lengthy period of time following the introduction of the speed limit, all drivers respect the maximum speed; those few who exceed it are considered undisciplined and their micro behavior is censured by the majority. But when the activation minimum is reached, the behavior of the undisciplined drivers is no longer censurable and soon becomes imitated, until everyone drives at a speed that is independent of the limit, which ends up being forgotten.

The heuristic model that describes this system can be represented as shown in Fig. 2.28.

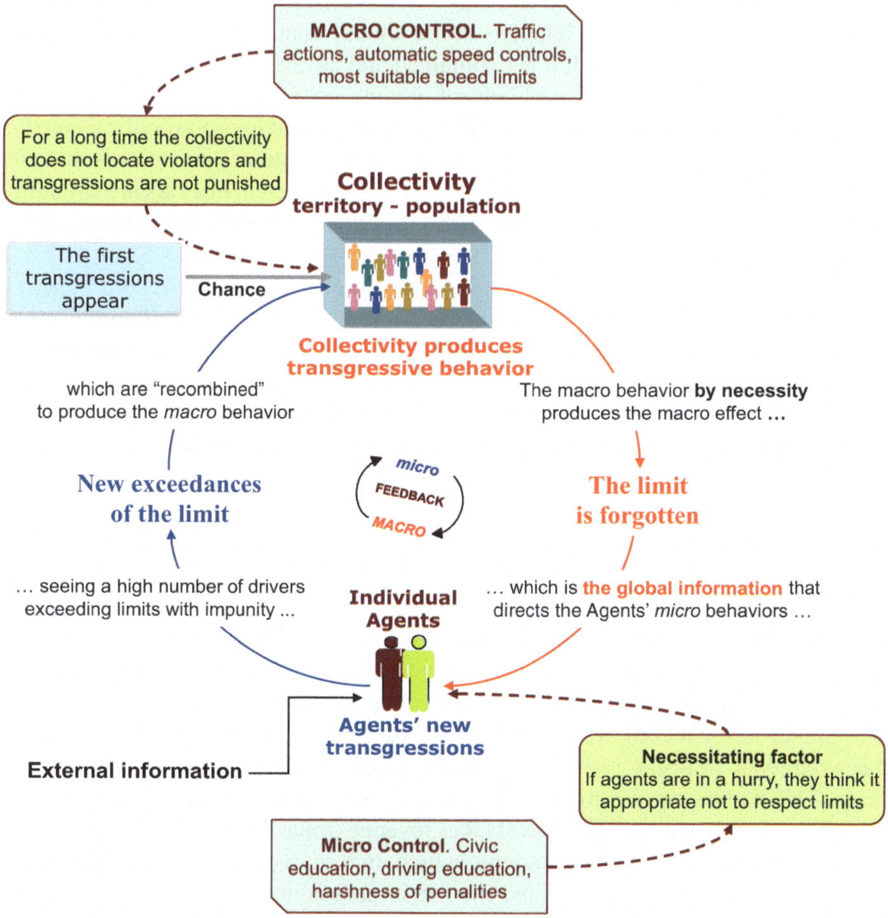

Fig. 2.28 Graphic model of "speed-limit" system

Micro behavior and necessitating rule: if you are in a hurry and feel the limits should no longer be observed, because you see a high number of drivers exceeding them with impunity, then you also think it appropriate not to respect them.

Macro behavior and recombining rule: the transgressors are rewarded, since they arrive first and are not punished; those travelling above the speed limit want to get past those respecting the speed limits, placing the latter in difficulty.

Micro-macro feedback. Chance and necessity: the growing number of transgressions is the result of the micro behavior, but this in turn conditions future micro behavior; the number of drivers going above the speed limit influences the respect the other drivers have for the speed limit, thereby leading to transgressive behavior. The higher the number of transgressors, the more individuals feel justified in transgressing. The limits are exceeded for the first time by drivers who "by chance" are so much in a hurry that they are willing to risk a fine; if the degree of civility and social conscience is high (in Finland, for example, limits are strictly respected), or the system of control, prevention and repression is efficient, so that the initial transgressors are fined (as usually happens in the United States), then the phenomenon would not begin. Once started, it is maintained and advances "by necessity."

Strengthening, weakening, and control actions: an efficient vigilance and prevention service, bad road maintenance, the large number of low-powered cars on the roads all represent obvious weakening factors for the system. Speed limits that are too low on wide, straight roads, the existence of high-powered, fast cars are strengthening elements for the system. The micro control must act on the individual driver's sense of responsibility.

The "speed-limit" system, though similar to the "breaking-the-record" one, is different in an important way: while the record represents a "ceiling," an objective to achieve, that continues to be raised, the speed limit in the "speed-limit" system represents a "floor" from which to flee; the record must be pursued, the limit abandoned. The record system is one *of pursuit*; that of the abandonment of the limit is more properly a system *of flight*, as we will analytically observe in Sect. 3.6.

2.4.4 "Eternal-Feud" System (and Variants)

If Abel had a vindictive son, then the latter would probably have taken revenge on Cain or his son, and the most ancient feud on the face of the earth would have begun. Even feuds can be interpreted as a combinatory system that is easy to describe, since it is similar to the record system. Someone "by chance" is wronged; if he forgives the wrongdoer or seeks justice through the proper authorities, the feud does not begin. This barbarous custom (macro behavior) begins instead when the person who has suffered the wrong (micro effect) takes revenge (micro behavior), thereby creating the occasion for a counter-vendetta, which, in turn, requires a violent reaction.

If "by chance" the feud begins, then it is maintained "by necessity" (macro behavior), since the information involving the wrong that is directly suffered—or the memory of a limited number of previous wrongs in the chain of vendettas—makes the reasons for the start of the feud unimportant; the person who suffers the vendetta of another feels he has been wronged and that he is in the right. It is not uncommon to see feuds between families last for decades.

The feud system is more common than people think, since falling into this category are not only feuds between individual families or factions (Guelphs and Ghibelines, Houses of Lancaster and York, etc.) but also those between town districts, tribes, even peoples and nations, as the centuries-old wars among peoples in the Middle East, North Africa, and the Balkans still testify to today.

The heuristic model that represents these systems can be depicted in Fig. 2.29 and described as follows.

Micro behavior and necessitating rule: if your family has suffered a serious wrong, you must expunge the dishonor by means of a vendetta; if you have no desire to do so and do not want to "wipe out the shame" of the wrong you have suffered, then your relatives and friends will consider you a coward, a traitor, and they will cause problems for you; thus, on the first occasion, give back what you have received.

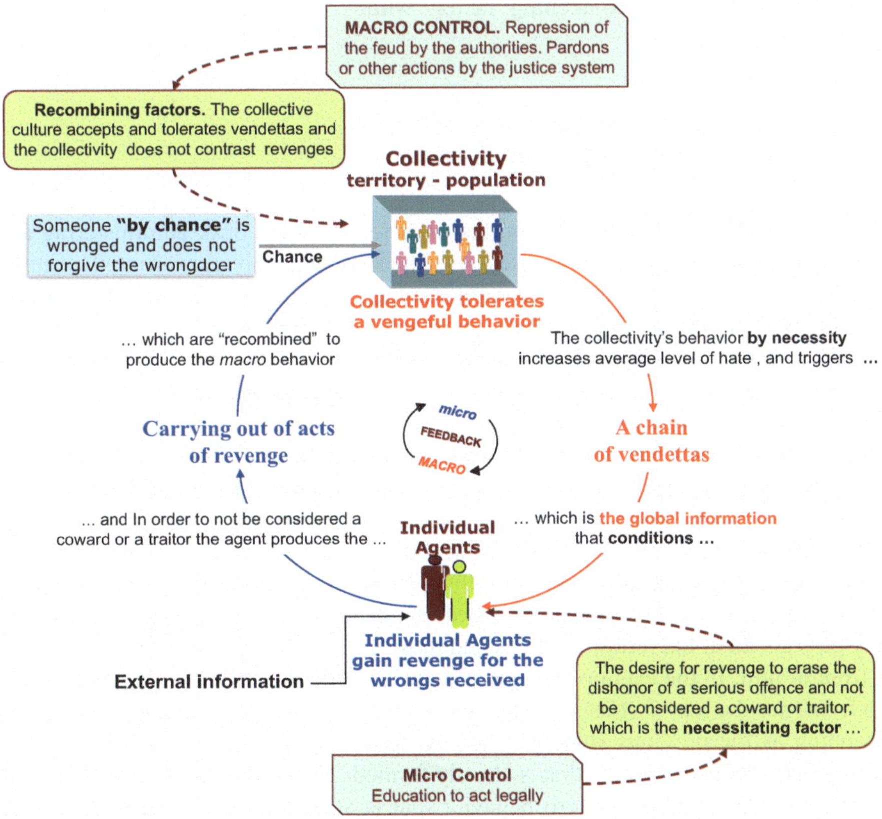

Fig. 2.29 Graphic model of a "feud" system

Macro behavior and recombining rule: the environment does not contest vendettas; instead, the collective culture accepts and tolerates them. Each new vendetta increases the hate among the adversarial families, which becomes proportional to the number of offences received, without any corresponding decrease from the wrongs paid back through the vendettas. The more hate there is, the more it is passed on to succeeding generations, which will seek vendettas as soon as possible.

Micro-macro feedback. Chance and necessity: each vendetta increases the level of hate, which in turn calls for a new vendetta. The first wrong can be considered as a chance event, just as the first vendetta can. After the first exchange in the chain of vendettas, pardon, or the justice system can intervene; in this case, no combinatory system manifests itself. If, however, others follow the first exchange, then the combinatory system arises and is maintained by necessity in order to wipe away the shame suffered and to gain revenge for the wrongs received. The feud ends even when, in addition to a lack of "relatives" on the side of one of the factions, the chain is interrupted by chance for a sufficiently long period of time (which depends on the limited information) so as to cause the feuding sides "to forget."

Strengthening, weakening, and control actions: the repression of the feud by the authorities is a strong weakening factor, while cross-family feuds strengthen the system.

2.4.5 "Assail-the-Professor" System

In an old university classroom, a professor is seated at his desk administering an oral exam along with two other professors. The students, who are waiting their turn or trying to listen to the questions the other students are being asked, occupy all the available seats in the small room. A new student enters and, seeing that all the seats are taken, remains standing at a distance from the professors, thus allowing him to observe and listen to the oral examinations of the other students. Another student comes in and, since he also wants to listen to the questions, he stands next to the other student, or just behind him. A third student enters and, based on the same reasoning, he produces a similar micro behavior. A small line of students begins to form.

If other students come into the room, the line will gradually lengthen until a student joining the line is too far from the desk to either see or hear what is going on. At this point, the last on line may become discouraged and leave the room, or may pass ahead of his fellow students and move next to the first on line, or even cut in front of him. In the event of this micro behavior, the line will lengthen, but from the front, not the rear. Let us suppose that other students enter who adopt the same reasoning as the previous students, thereby producing similar micro behavior by cutting to the front of the line that is already formed. The combination of such micro behavior translates into a macro behavior that is attributable to the entire line of students; in fact, there is a gradual "advancement" of the line of students closer and closer (macro behavior) to the front desk, until it runs up against it and, as students continue to come in, ends up trampling the examining committee and examinee (macro effect).

This situation occurs more often than we think, since it takes place in all types of job competitions, in all conventions during the interval when the speaker is answering questions, during interviews of a famous person by a large group of journalists, and so on. The advancing line is the effect of the macro behavior, but the latter conditions the future micro behavior in a typical *micro-macro feedback* process.

The line of students represents the global information that drives the micro decisions and behavior. However, the combinatory system also functions under limited information; each student that enters decides his own micro behavior by observing the number of students ahead of him in line and by taking account of how well he can hear; the number of students behind him in line is not important. The macro behavior occurs only if the number of students in line exceeds the minimum activation number, which we can calculate as the minimum number after which the student, joining the rear of the line, is too far from the front desk to see or hear efficiently; the macro behavior ends when the maximum saturation number is reached, which forces the professor to suspend the exam due to a lack of space.

The following heuristic model describes the system, which is also illustrated in Fig. 2.30.

Micro behavior and necessitating rule: if you want to follow the exam from a position where you can see and hear something, and if there is already a line of students in front of you, the length of which exceeds the minimum number that would permit you to see and hear, then try to push forward the students ahead of you; if you are not able to do so, get out of line and move to the front of it.

Macro behavior and recombining rule: the exam is important and you need to listen to the questions, or are interested in doing so. The room is set up so that a line will form; for each new student that enters the line lengthens and advances, which forces subsequent new entrants to move to the front and once again lengthen the line.

Micro-macro feedback. Chance and necessity: the line that is getting longer and gradually advancing is a result of the micro behavior of those moving to the front of it rather than to the rear; but the advancement of the line, that is, the producing of the macro behavior, leads in turn to the occurrence of the micro behavior. If "by chance" students do not enter, or if everyone finds a seat, or those at the end of the line leave the room without cutting in front of their fellow students, then the combinatory system does not arise; but if "by chance" the line forms and there is a sufficient number of students to feed it, then "by necessity" the macro behavior is produced.

Strengthening, weakening, and control actions: if the custodian controls access to the room, the professor orders that no one should be standing, or the wait is particularly long, the system will suffer a drastic weakening action; if the room is very large, has a small seating capacity, the desk is far from the seats, there is a large number of students, the exam is important, and the professor is not strict, then the system will be strengthened. The macro control takes the form of barriers or the suspension of the exam; the micro control acts through students being persuaded not to remain standing in the room.

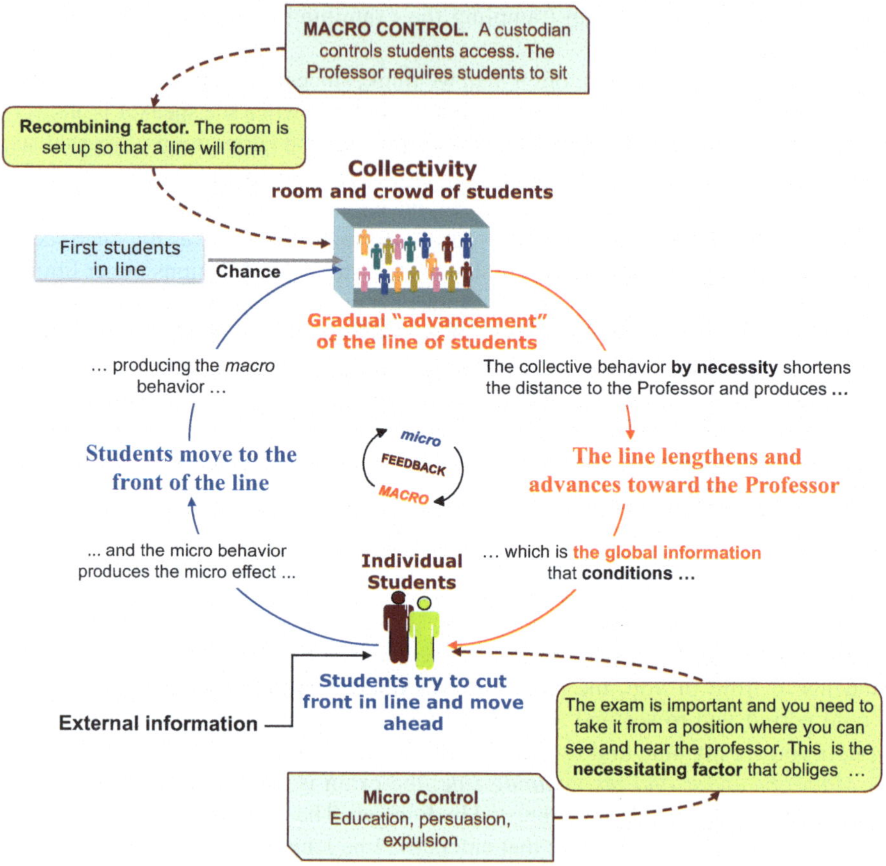

Fig. 2.30 Graphic model of the trample-the-professor system

2.4.6 "Human and Animal Stampede" System

It is very common to observe "collectivities" of *people* or *animals* present at the same time in a given place to perform social or biological functions. Humans collectively participate in many social events, forming crowds or gatherings, such as, for example, attending a religious service, enjoying a game at the stadium or watching it on video screens in a square with many other fans, gathering in some place to listen to a speaker during celebratory or political events, crowding into a classroom for a university lecture, participating in thousands at an outdoor concert or in some indoor venue, and so on. Animals gather to graze or hunt, forming herds, flocks, schools of fish, or even to migrate collectively from one place to another, and so on.

In a 'collectivity' with gatherings or masses of individuals, if some people or animals perceive a danger caused by a sudden and frightening event, such as an explosion, a fire, or the fear of an attack, and they become scared, the individuals *panic* and begin to flee. In the absence of *specific information*, the collective *macro-behavior* acts as *global information* and is almost instantly imitated by other individuals who tend to move quickly (Dupuy, 2005; see also Sect. 4.3.2). Individual *micro-behavior* produces the collective effect that influences subsequent micro-behavior, in a typical *micro-macro feedback loop*.

If individuals (people or animals) panic but are in open spaces without barriers, a disorderly flight usually occurs in all directions, with the effect of increasing the distance between individuals; when the danger subsides, they may regroup. However, when they are inside a confined, enclosed space with natural or artificial barriers, individuals must necessarily move together quickly in the same direction, often in an uncontrolled and chaotic manner, without concern for others, thus forming a *stampede*.

A *stampede* is usually caused by *overcrowding*: when a place is too crowded and individuals are too close to each other, even without the perception of danger, the movement of a few can create a "domino effect," triggering collective behavior. If some individuals begin to move quickly or run, others often follow in a chaotic manner, resulting in a stampede. The collective behavior becomes the *global information* that influences individual behavior (for factors that can increase the risk of a stampede, see Kefan et al., 2019). From the previous considerations, we can agree with the following simple definition. A "*stampede*" is the collective macro-effect resulting from the chaotic and disorderly movement of a crowd of people or animals (macro-behavior) present at a given time in a certain area, running in panic or terror, or even out of a desire to reach a destination (micro-behavior), often crushing and trampling each other, causing injuries or deaths.

The following are some examples from among the hundreds of events where a stampede has occurred:

- Hajj Stampede (2015): During the annual pilgrimage to Mecca in Saudi Arabia, a deadly crowd crush occurred, resulting in the deaths of approximately 2400 people. This event is one of the worst pilgrimage-related disasters in modern history (Wikipedia, 2024e; Mina Disaster).
- Stampede in Baghdad (2005). In Iraq, during a religious celebration, panic spread among the crowd due to a false alarm of a suicide attack. This caused a mass stampede on the al-Aimmah bridge, resulting in the deaths of about 1000 people who were trampled or drowned in the Tigris River (Wikipedia, 2024a; Al-Aimmah Bridge disaster).
- Hillsborough Stampede (1989). To access the stadium where the football match between Liverpool and Nottingham Forest was to be held, a large crowd attempted to enter the stadium through a tunnel that, unfortunately, was blocked. Ninety-six fans died, crushed against the barriers at the stadium entrance (Wikipedia, 2024b; Hillsborough disaster).
- Stampede in Turin, Italy (2017).* To allow the fans without tickets to the match between Juventus and Real Madrid to watch the contest, the city set up a large screen in Piazza San Carlo, which soon filled with a huge crowd. When some

Fig. 2.31 Graphic model of the "stampede" system

pickpockets, trying to escape after a theft, sprayed irritating spray, it caused those affected to flee, triggering panic across the entire square and resulting in a stampede that caused 3 deaths and 1620 injuries (Wikipedia, 2024f; Turin stampede).

Wikipedia, under the entry *List of fatal crowd crushes* (2024c), reports many dozens of stampedes, ranging from Roman times to recent years (see also: Ngai et al., 2009; Illiyas et al., 2013).

It is immediate to recognize that the combinatory system described in Sect. 2.4.5 "Assail-the-Professor" System highlights a phenomenon that could theoretically lead to a stampede.

The graphic model of this system is shown in Fig. 2.31. The heuristic model of this system can be developed as follows.

Micro-behavior and necessitating rule: If you see many people fleeing and you are surrounded or pushed from behind by the flow of those escaping and are in a panic, do not try to stop but adjust to the flow and, out of necessity, run in the same direction as the crowd (*micro-behavior*).

Macro-behavior and recombining rule: Observed as an organism, the crowd (collectivity) produces the macro-behavior of moving in one direction, causing the stampede as a collective macro-effect. If the crowd encounters barriers in the direction of movement, the direction changes, forcing individuals to adapt because they are not "free" to perform autonomous *micro-behaviors* but must follow the direction of movement.

Micro-macro feedback. Chance and necessity: The movement of the crowd (collectivity) in an apparent direction is the collective *macro-behavior* produced by the *micro-behaviors* of individuals; but in turn, this conditions the *micro-behaviors* of the individuals who make up the crowd, in a typical *micro-macro feedback* loop. The macro-behavior starts "by chance" when a group of individuals begins to move quickly and often chaotically in the same direction; once started, the collective movement of the crowd in the assumed direction is maintained "out of necessity," as it becomes impossible for individual members to change it.

Strengthening, weakening, and control actions: The lack of safety measures and crowd management, the narrowing of space and/or escape routes, the placement of barriers to the flow of the crowd, the sound and danger lights that increase panic, and the lack of timely information are all strengthening factors for a stampede. The presence of factors that act in an opposite way facilitates the weakening of the stampede. Limited space, panic, and crowd pressure are the most evident *recombining factors*. The strongest *necessitating factor* is represented by the instinct to save oneself from danger.

2.5 Systems of Order and Conservation

These are combinatory systems which produce a macro behavior or a macro effect capable of interpretation as the attainment and maintenance of an arrangement, an ordered disposition, among the elements that form the system (flocks of birds, schools of fish, herds of mammals, queues of people at the bus stop, etc.). These are typically limited and incomplete information systems and they can be described by the following heuristic model and illustrated as in Fig. 2.32.

Micro behavior and necessitating rule: there are advantages in maintaining a particular order and disadvantages in breaking it; if you want to gain the advantages or avoid the disadvantages, try to adjust your behavior so that you maintain or achieve the order that is indicated by the rules establishing it;

Macro behavior and recombining rule: the more the particular order is maintained, the greater the advantages from adjusting one's behavior to maintain it and the disadvantages from breaking it;

Micro-macro feedback. Chance and necessity: the order (macro effect) creates the convenience for individuals to maintain the arrangement and respect the rules (micro behavior); everyone maintains a coordinated behavior (macro behavior). The first ordered behavior may rise "by chance," but if the order is judged useful then the *micro-macro feedback* will assure the maintenance of the ordered behavior.

Strengthening, weakening, and control actions: the individual being brought up to value order, the social approval of order, punishment for those who do not respect order: all these factors considerably strengthen the system.

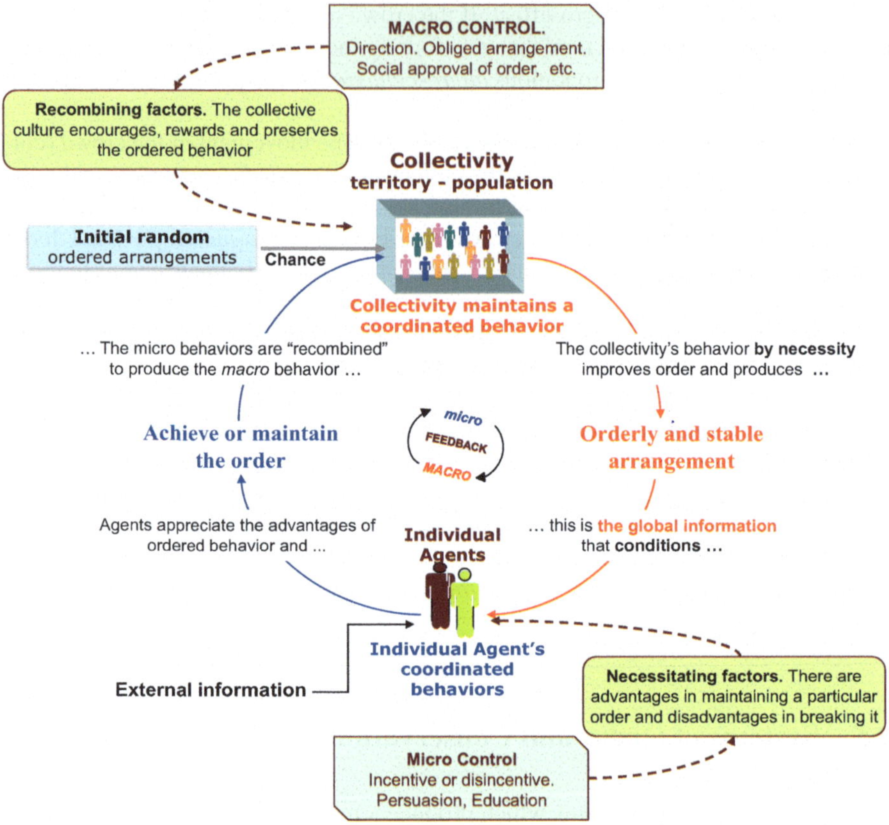

Fig. 2.32 Graphic model of systems of "order"

The systems of order can be used to interpret a large number of phenomena—the spontaneous formation of ordered dynamics in crowded places (dance halls, pools, etc.), the behavior of groups that proceed in a united manner (herds in flight, flocks of birds, crowds, etc.), the creation of paths in fields, of wheel-ruts on paved roads, the appearance of artifacts and of furniture of the same kind and/or quality, the spontaneous formation of the best or the worst practices, the acceptance of the same vision or mission or strategic guidelines of the organization—many of which are shown in the following sub-sections.

2.5.1 "Waltz-Spin" System

Suppose we are in a dance hall (environment) where, after a modern dance, a waltz suddenly is played. People pair off (elements of the system); some dancers leave the dance floor while others go onto it. The floor is very crowded. Since the waltz is a typical rotating dance, it meets a large obstacle due to the fact that no couple can spin as it would like to or as the music requires. Some try turning to the right, others dancing in the opposite direction, but the spectator along the perimeter of the floor only perceives a disorderly movement of couples. This situation can go on for a while, even until the music stops, in which case no combinatory system forms.

Let us instead assume that, several seconds after the music starts, a short empty "corridor" forms among the throng of dancers, into which couple A immediately "enters" (or several couples, if the corridor is large enough) and tries to dance at the left-hand edge (with respect to the observer along the edge of the dance floor). The two dancers cannot believe it is possible to finally launch into a series of spins in the direction of the corridor (to the right of the observer). Behind couple A, which has begun to spin and has thus freed up a corridor behind them, the pair next to them, couple B, enters the corridor, as does another couple behind them.

There are now two possibilities: couple A encounters an impenetrable "wall" of other dancers before it, which forces it to stop its spinning movement; or it manages to continue its rotation, thereby providing an incentive to other couples to do the same. In this case, as we can easily see, after a short while, all the pairs on the floor are following a rotating movement (micro behavior) from left to right (clockwise with respect to the spectator on the perimeter of the floor), so that the group of couples form a combinatory system that produces the rotating dance (macro behavior).

If "by chance" couple B had entered the corridor on the right-hand side and the process had followed its course, then "by necessity" the combinatory system would have produced as macro behavior a rotation in the opposite direction to the preceding one. In fact, the group rotation obliges all the dancing couples to follow the rotating movement, which represents the global information produced by the system itself. Nevertheless, the information the dance pairs use to decide the direction to take is limited; in fact, each couple takes account of only the position of a limited number of nearby pairs, not noticing the position and movement of those on the other side of the floor, who are hidden from their sight.

The heuristic model of the system can be represented as follows (Fig. 2.33).

Micro behavior and necessitating rule: if you want to dance and not be pushed, try to dance by rotating where you see an empty corridor, and do not go in a direction opposite to that of the couples that follow you.

Macro behavior and recombining rule: The rotating dance of the individuals imparts to the mass of couples a similar rotational movement made up of concentric circles or even whirls. The rotation is generated by the individual couples dancing in rotating fashion; but, once started, this conditions the manner in which each couple dances, according to the micro rule.

Micro-macro feedback. Chance and necessity: the presence of a group of dancing couples on a dance floor does not always activate a combinatory system—that is, generate on overall rotating movement; it is necessary to have a pair of dancers actually begin to dance in a rotating fashion, and the number of couples dancing in the same direction must exceed the minimum activation number. This can occur by chance, but when "by chance" the system starts up, it forces the dancers to dance in the same direction "by necessity" in order not to get run into. The pairs are forced

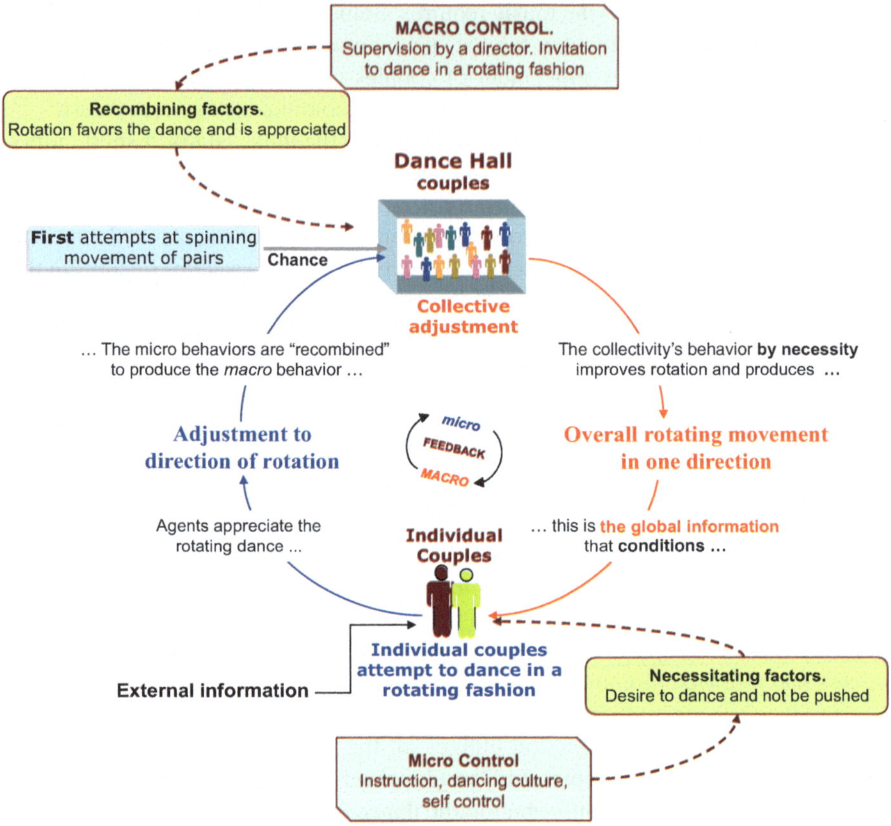

Fig. 2.33 Graphic model of the "waltz-spin" system

to dance in a circle and by doing so they themselves produce the rotation. We must also point out that the sense of direction is also due to chance: if, at the moment when by chance the open "corridor" has formed, the pair to the right of the corridor enters it and begins to circle in that direction, then the final rotating movement of all the dancers would be in an anti-clockwise direction. Chance operates twice here: in starting the rotation of the system and in imparting a direction to this movement.

Strengthening, weakening, and control actions: the strengthening of the system can be represented by the invitation from the band leader to dance in a rotating fashion, the circular shape of the dance floor, or the presence of spectators for whom the dancers want to put on a good show; the high density of people on the floor can cause, on the other hand, a weakening effect. The macro control can be carried out through the supervision of a director who forces the dance pairs to dance in a rotating way. The prior instruction of the pairs represents a form of micro control.

Even the Can-Can dance, the well-known theater show (entertainment) in which a certain number of dancers spin around individually but in a coordinated manner, forming a line that rotates as one, can be interpreted as a combinatory system of this

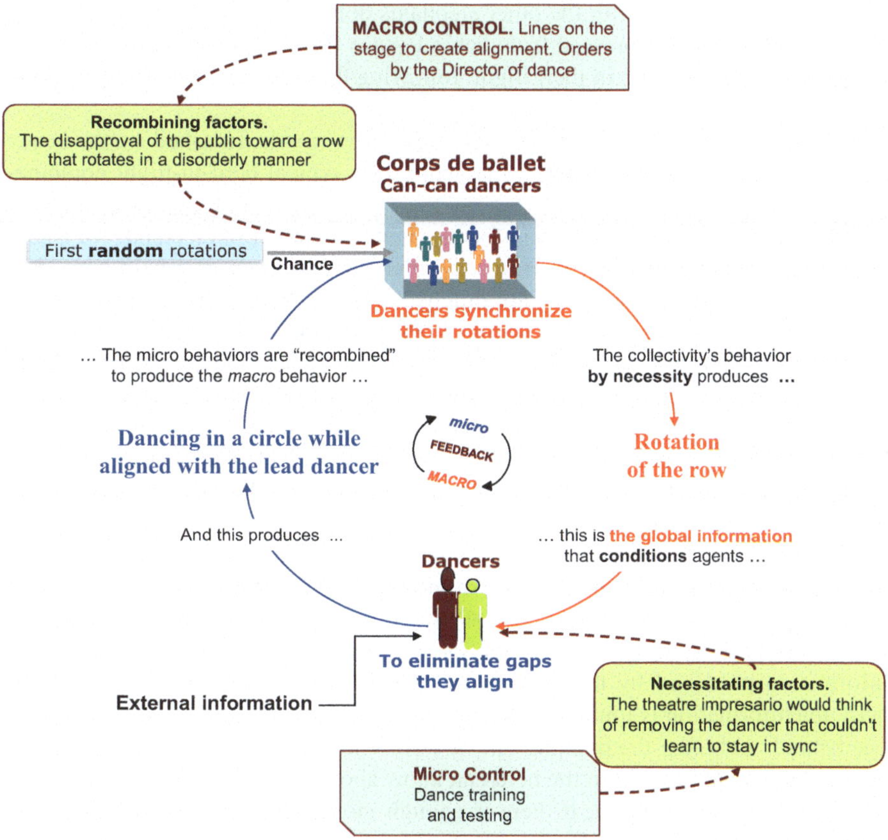

Fig. 2.34 Graphic model of "can-can dance" system

type. As shown in Fig. 2.34, this "phenomenon" derives from the *micro-macro feedback* between the rotation of the individual dancers who, in order to dance in a circle aligned with the lead dancer, must eliminate any gap between their alignment and that of the lead dancer.

2.5.2 "Herd-in-Flight" System

The young African was observing a herd of elephants downwind. Suddenly the wind changed direction and the elephants detected the presence of the young man by his smell. A terrible trumpeting came from the throat of the female head of the herd, who began to run in a certain direction; the other elephants followed her, and soon the herd formed a compact mass moving behind the lead elephant. The galloping herd came upon some woods, at which point the lead elephant slipped between two trees like a wedge, with the other elephants doing the same. The lead elephant wanted to stop but was pressed from behind, since the elephants that followed had no choice but to continue in their flight, following those in front that were marking the way.

The result was a wide path of felled trees in the woods, smashed by the heavy legs of the elephants, testifying to the action of a typical herd-in-flight combinatory system.

This type of combinatory system forms when the number of elephants (or a herd of great mammals from any other species) that start to move as a herd exceeds a given number, so that the others are dragged along in flight. Each elephant has a micro behavior which consists of fleeing while following the elephant that precedes him, except for the lead elephant, who "by chance" sets the direction of flight through an initial behavior (trumpeting and starting to flee). This *micro behavior* causes the movement of the herd, which represents the *macro behavior*; the moving herd conditions the successive micro behavior, since each elephant has to follow the movement suggested by the herd, in this way, maintaining the movement in a manner similar to that of the waltz-spin system.

This combinatory system is very common in nature among animals as well as man. We only need to observe a school of sardines, a flock of birds, a swarm of bees, or an angry crowd to become aware of this. In fact, the direction of the compact heard forces each animal to follow the same direction, which represents the global information produced by the system itself. Nevertheless, for each animal in the herd, the information is limited to the direction taken by the animal in front and the one behind, and often the flanking one as well. The herd can also fall into a ravine, but the individual running in the herd can know about the ravine only when the one in front of him is falling in; if there is enough momentum or if the pushing from behind does not allow the herd to stop, then even the last member ends up in the ravine.

Since collective movement phenomena occur in large spaces and, usually, even with the perception of imminent danger, it is unlikely that a stampede will occur; in this case, the Combinatory System "Stampedes" described in Sect. 2.4.6 would not form.

The heuristic model of this system can be illustrated as in Fig. 2.35.

Micro behavior and necessitating rule: if you must flee, follow the lead animal or those in front of you; if those following you are pressing you, maintain your direction to avoid being trampled.

Macro behavior and recombining rule: the direction of the individual animal conditions that of the herd, which, running in a given direction at a given speed, in turn conditions the direction and speed of the elements it is composed of.

Micro-macro feedback. Chance and necessity: the moving herd is the result of the direction of the individual animals in flight (micro behavior), which in turn conditions the direction and speed of flight of the animals that make up the herd. The herd forms "by chance" when the lead animal gives the agreed upon signal; once started, the flight of the herd in a given direction is maintained "by necessity."

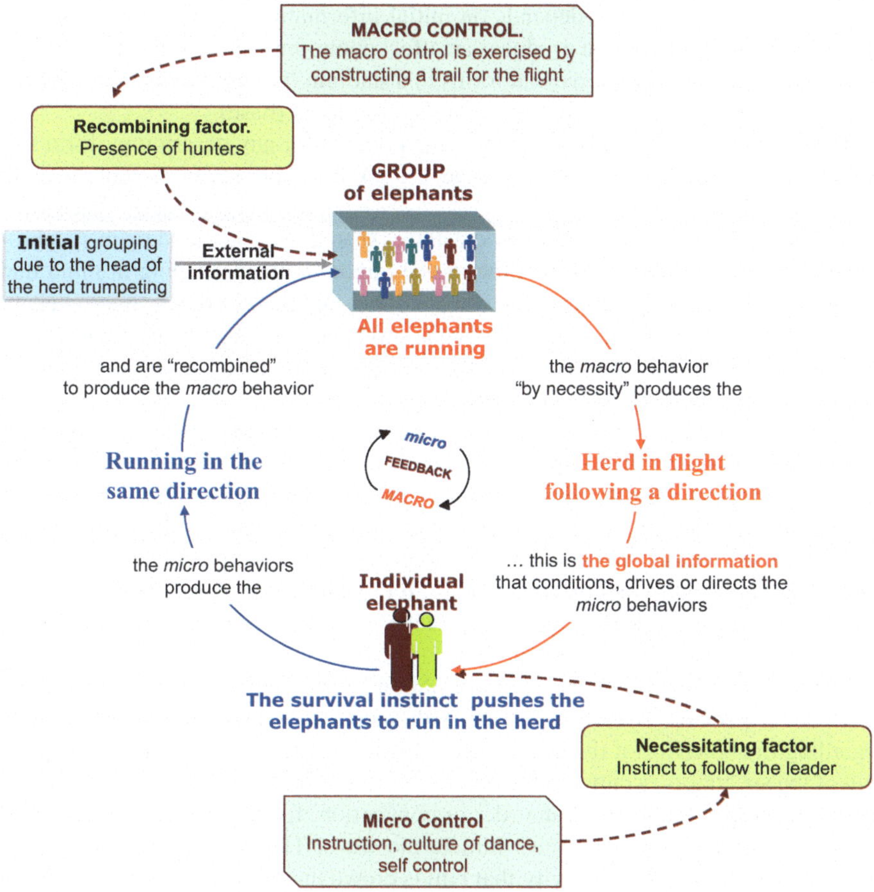

Fig. 2.35 Graphic model of the "herd-in-flight" system

Strengthening, weakening, and control actions: the wind that impedes the herd from hearing the order of the lead animal, or the presence of obstacles along the route represent weakening factors; a fire in the savannah or shots from hunters instead represent strengthening factors. The macro control is exercised by constructing a trail for the flight; the micro control through some form of individual training.

2.5.3 "Stadium-Wave" System

The "herd-in-flight" system differs from the "waltz-spin" one, in that the macro behavior starts subsequent to some initial information (instructions, orders, conventions, etc.) produced by some individual or event. There are many combinatory systems whose functioning depends on initial information. An example of such a system is the unique one that produces an effect commonly known as "the wave," or "ola," which consists in a wave that forms in a stadium (environment) following the rhythmic standing up and sitting down of the spectators (base).

Those who have seen this in person have no need for any description; for a significant example, see YouTube, 2022, https://www.youtube.com/watch?v=hq387m0PRac.

In short, the wave forms when the spectators seated in a row stand up in synchronization, count to three (or some other agreed upon number, keeping in mind that the higher the number the slower the wave will be) and then sit down (micro behavior); the other spectators produce the same micro behavior, the only difference being that they stand up when those to their right or left sit down, according to the direction of the wave that has been decided upon.

From the combination of this (or a similar) micro behavior—when the spectator to your left sits down, stand up, count to three and sit down—there arises a *macro behavior* represented by the wave formed by the spectators' heads. The wave circles the stadium several times until the stop signal intervenes, either explicitly or implicitly. Micro and macro *effects* cannot be observed unless we postulate some effect connected to the micro and macro behavior; for example, the wave that moves a flag could represent the macro effect, the passing of a flag from one spectator to an adjoining one would be the micro effect.

The wave constitutes the global information for the individual spectators; however, each spectator usually operates with incomplete and limited information regarding the behavior of the one seated to his left (or right), not noticing the position of the spectators occupying the other seats in the stadium. Initial information is required, even in the form of an order or convention, for the spectators in the first row (or circular sector) who must initiate the wave. The wave by spectators in a stadium is a fun, unifying activity that builds crowd energy and camaraderie, but it can take on different forms depending on the context.

The heuristic model for this system can be described as follows (Fig. 2.36).

Micro behavior and necessitating rule: if you are in the first row, count until three and then sit down again; if you are in another row, when you see the person on your left (or right) stand up, wait until he sits down, then stand up, count to three, and sit down.

Macro behavior and recombining rule: the spectators must agree to participate in the wave; each spectator that stands up moves the wave along by one row, but when this reaches a row of spectators, it transmits to them information to follow the micro rule.

Micro-macro feedback. Chance and necessity: the vertical rhythmic movement of each spectator (stand/sit micro behavior) causes the horizontal movement of the wave (macro behavior) that circles the stadium. A spectator rises when the wave arrives from his side, but his micro behavior feeds the wave and moves it along in a typical feedback process. The person directing the wave can randomly choose both the row of spectators that will start the wave as well as the direction it will take

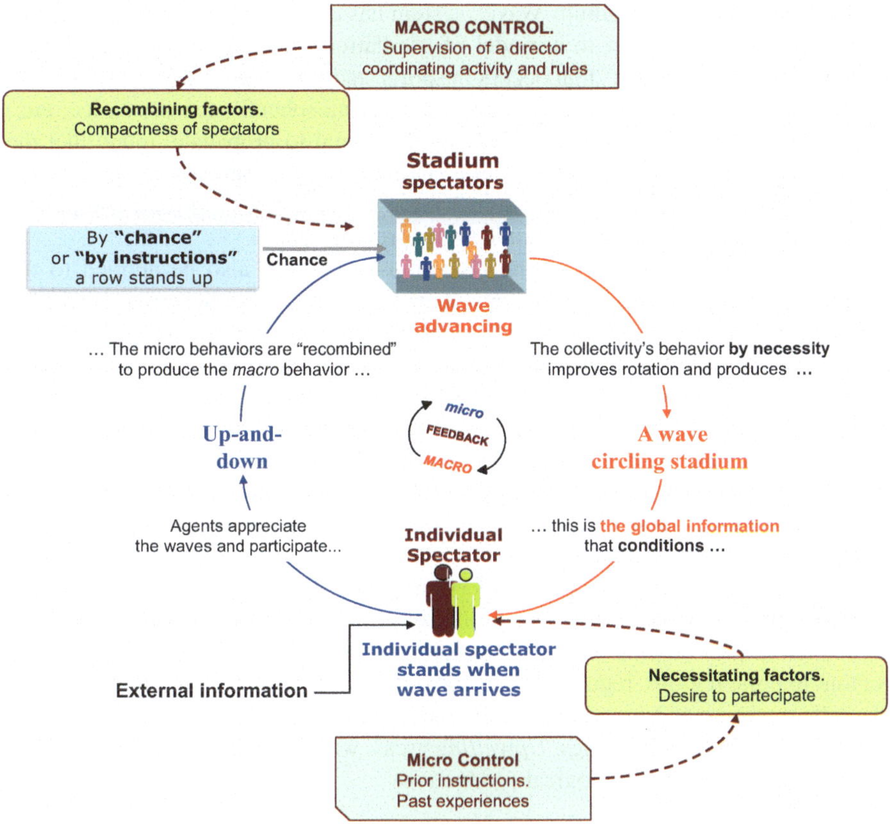

Fig. 2.36 Graphic model of the "stadium-wave" system

(micro rule); if there is a spirit of collaboration, however, then once underway the wave will be maintained "by necessity," until information to stop arrives.

Strengthening, weakening, and control actions: the impossibility of perceiving the order of the director, the presence of obstacles or rows of empty seats, or even just the lack of interest in collaborating, all represent weakening factors; the presence of television, or cheering for the home team instead represent strengthening factors.

Similar to what we have just presented is the combinatory system that describes the human chains that form to put out small-scale fires using water from a well. The water is brought in pales that, after being filled, pass from hand to hand until they are emptied onto the flame and subsequently taken back to the well.

A stadium wave is a good example of a "traveling wave", like a wave in water or another liquid (macro-effect) that moves in a horizontal direction (macro-behavior), formed by "particles" whose motion (micro-behavior) is oriented vertically, just like the spectators in a stadium.

The concept of the "Stadium-Wave" system has also been used in the context of climate variability to refer to a model of oscillation involving the global climate system, similar to a "wave" that occurs in sports stadiums. This model suggests that the various components of the climate system (atmosphere, oceans, sea ice, etc.) interact in a coordinated manner, allowing the signal to propagate, much like the wave that spreads among spectators in a stadium. In this context, the wave represents the propagation of a climate signal through the Earth, with each component responding with a slight delay relative to the previous one.

The Combinatory System of the "Stadium-Wave" can also be applied to the interpretation and study of "Waves in Rotary Seas," complex phenomena that play a crucial role in ocean dynamics and in Earth's climate. Understanding these phenomena is essential for improving climate forecasts, maritime safety, and marine ecosystem management. The concept of "Waves in Rotary Seas" refers to waves and oscillations of various types and orientations that occur on a large scale in "rotating seas," i.e., seas and oceans, significantly influenced by Earth's rotation (Coriolis effect), winds and other dynamic forces that generate waves oriented in certain directions, deep-sea currents, or rotary wave phenomena (Rotary wave, online, 2014), gyres, and eddies.

Such phenomena are crucial for understanding ocean dynamics and their interaction with the atmosphere. Ocean circulation influenced by Earth's rotation has a significant impact on global climate. Ocean currents transport heat from equatorial regions to polar regions, regulating the climate (e.g., the Gulf Stream). Rotary waves also affect marine energy production (Galangashi et al., 2017) and the distribution of nutrients and marine species. *Upwelling* areas, where nutrient-rich water rises to the surface, are vital for biological productivity.

2.5.4 "Trace-a-Path" System

When we observe the side of a hill, a vast stretch of countryside, woods, or simply a grass field it is easy to be aware of a path. At times there is only one, which is well marked; at other times, numerous lanes cross the same area, some clearly marked and others hardly evident. How do paths form? Once again, the Theory of Combinatory Systems immediately makes clear the origin of this phenomenon: the path—suppose we observe it in a grass field—is the macro effect of a combinatory system that is fed by individual micro behavior (the crossings of the observed site), each of which produces a micro effect represented by the trail that is left behind. Naturally the first crossing takes a direction that to an outside observer (the system observer) appears to owe "to chance," even though it is the result of the passers-by individual decisions based on the choice of the route held to be most efficient for the crossing. If no one else takes this route, then the grass has time to straighten out; to a second individual, the field would still appear intact, so that even his crossing would occur based on a route once again owing to chance.

This phenomenon could repeat itself several times, according to the preceding hypothesis, without a path forming; no combinatory system would be activated. But if, again by chance, many individuals should cross the field in the same direction at the same time, in single file, or individually but in rapid succession, the grass trampled by one would have time to straighten out. In this case, we can assume that the next person, observing a path already almost outlined (grass stomped by the previous walker), will decide to take it and save himself the fatigue of opening a path through the grass and the inconvenience of getting his shoes wet with the dew. In this second case, it is clear that the path will be better marked, making its use by a third person having to cross the field even more inviting. The path arises "by chance" from the effect of the individual micro behavior, but it is then maintained "by necessity"; the existence of the path then motivates other individuals to undertake a similar behavior, as part of a typical *micro-macro feedback* process.

The outline of the path represents the global information the individual uses for his decision regarding the direction to take, even if each individual only takes account of the visible tracks and of subjective rationality parameters (objectives, preferences, physical makeup, etc.). At times, the same field, the same slope, the same rock-face are crossed by several intersecting paths; this is reasonable, since those paths result from the action of different combinatory systems that combine their macro effects.

The following heuristic model of the system can also be represented in Fig. 2.37.

Micro behavior and necessitating rule: if you need to cross an area, try to cross the environment (field, slope, rock-face, etc.) using the maximum rationality. If there is a path in the right direction, or a direction not too different from the desired one, take it; otherwise look for a route that optimizes your individual rationality parameters.

Macro behavior and recombining rule: from the successive superimposition of individual tracks the path forms, guiding a flow of persons in the direction it indicates, thereby producing the macro behavior of the combinatory system; the path (macro effect) is created by the individual crossings but in turn conditions the crossings themselves according to the micro rule.

Micro-macro feedback. Chance and necessity: we cannot know from where chance will bring us the individuals and where these will be headed; but once the path is formed in a given direction after the macro effect has occurred, then all the succeeding individuals are forced (rationality assumption) by the system to take this path and thus cross in the predetermined direction, even if it appears as if the micro behavior is marking the path. It should be noted that the direction of the path owes to chance: if the field had been crossed by a large number of individuals who always took a different direction, no path would have formed; at most the field would have become bare without any combinatory system forming.

Strengthening, weakening, and control actions: if the path arises in an open environment and is viewed as an efficient route, then strengthening actions can be carried out: the placing of kerbstones, paving and asphalting (the first Roman paths became roads that still today provide the most efficient route), indications on maps, and so on. If the path crosses private territory, then weakening actions can be carried

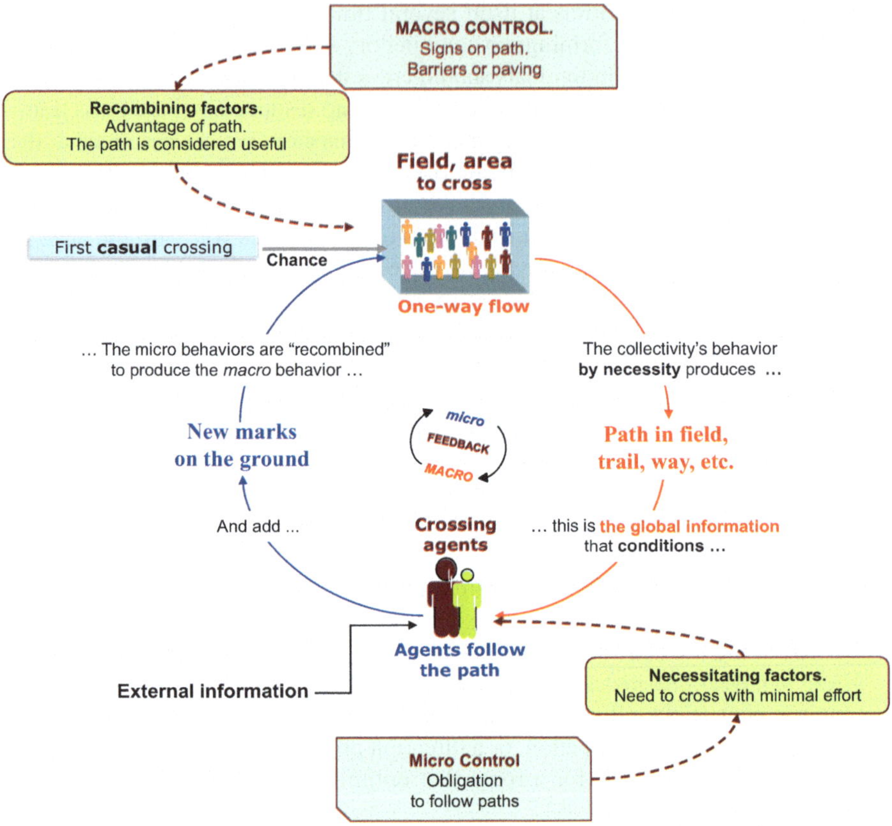

Fig. 2.37 Graphic model of "trace-a-paths" system

out, the most drastic of which is the "Don't Trespass" sign and surveillance. The macro control takes the form of actions on the path (asphalting, directional signs, etc.); the micro control of actions directed at the passers-by (persuasion, threats, etc.).

The explanation of the formation of paths provided by Combinatory System theory is in part similar to that given by Friedrich von Hayek, who, considering only "chance," sees the formation of paths as a *spontaneous order* without recognizing the *micro-macro feedback action* in an explicit form.

> The way in which tracks are formed in a wild broken country is such an instance. At first everyone will seek for himself what seems to him the best path. But the fact that such a path has been used once is likely to make it easier to traverse and therefore more likely to be used again; and thus, gradually more and more clearly defined tracks arise and come to be used to the exclusion of other possible ways. Human movements through the district come to conform to a definite pattern which, although the result of deliberate decisions of many people, has yet not been consciously designed by anyone. This explanation of how this happens is an elementary "theory" applicable to hundreds of particular historical instances; and it is not the observation of the actual growth of any particular track, and still less of many, from which this explanation derives its cogency, but from our general knowledge of how we

> and other people behave in the kind of situation in which the successive people find themselves who have to seek their way and who by the cumulative effect of their action create the path. (von Hayek, 1942, p. 288)

2.5.5 *"File-of-Ants" and "Stigmergy" Effect*

The modus operandi of the previous combinatory system is entirely similar to that which explains the forming of long uninterrupted lines of ants appearing to follow a predetermined trail or path. Nevertheless, this cannot be observed from the outside since the ants leave no visible trace on the ground, only pheromonic traces, so that the ants produce their behavior through a process characterized by spontaneous ordering or self-organization, known as "stigmergy."

The term "stigmergy" (stigmergie) was introduced by the biologist Pierre-Paul Grassé in 1959 to refer to the typical *micro-macro feedback* effect, not explicitly formalized by Grassé, according to which "stigmergy" is defined as the phenomenon that produces the "*Stimulation of workers by the performance they have achieved.*" This term indicates that an agent's actions leave a micro effect, a sign, in the environment which allows other agents to perceive the sign as information, based on which they determine their subsequent actions (Grassé, 1959, 1960, 1967). Stigmergy describes, in fact, a *micro-micro* feedback more than it does a *micro-macro feedback*, since Grassé primarily considers the behavior of individual agents, and not that of groups of agents, that produce *global information* causing "*stimulation.*"

> Stigmergy is more than blind variation and natural selection, though: the visible traces of the work performed previously function as a mediator system (Heylighen, 2007), storing and (indirectly) communicating information for the community. In that way, the mediator coordinates further activity, directing it towards the tasks where it is most likely to be fruitful. This requires a shared workspace accessible to all contributors (similar to what in AI is called a "blackboard system"). This external memory registers which tasks have already been performed and what problems still need to be tackled. The Web has provided a very powerful such workspace, since it enables the storage and public sharing of any "work-in-progress" information product. (Heylighen, 2007, p. 8)

This is the case of populations of insects, typically ants, which act by creating an "aromatic potential field" by spreading *pheromones* or other permanent chemical messages. With their micro behavior, the agents spread pheromone across one site (micro information); the increasing concentration of pheromone (global or macro information), increases the probability that each agent will move in the direction of that site. The *micro-macro feedback* is quite evident (Zollo et al., 2001). This behavior is the evident consequence of "stigmergy," as it derives from the ability of ants to communicate by means of the small aromatic signals able to trigger a chain reaction and from the *necessitating factors*, biological in nature, that force individual ants to follow the aromatic path when this is perceived. Obviously, *recombining factors* need also be present: the environment must not eliminate too quickly the pheromone left by the individual ants, and the latter must use the aromatic path frequently enough to strengthen it. This sequence requires a certain number of ants (Sect. 1.2.2). Only above a *critical activation mass* of ants can the pheromone amplify and become effective, and thus lead to the order effect (Deneubourg & Goss, 1989), as if the ants synchronized their micro behavior to move in an orderly line.

Clearly, the aromatic field can give rise to behavior different from the alignment in an ordered row: we can also observe circular dynamics with ants and other animals that release chemical signals, as shown in Fig. 2.38.

The aromatic field is the macro effect of the population of moving ants, acting as an ordering parameter that guides their micro behavior, which, in turn reinforces the macro effect, in a typical micro macro feedback that interprets the slaving principle as the key to understanding self-organizing systems as shown in the model in Fig. 2.39. Following the logic of synergetics, the global information produced by a combinatory system—the aromatic pheromonic fields—may be considered an ordering parameter that enslaves all the agents of the system and forces them to self-organize and produce collective phenomena. We could also consider the *micro-macro feedback* as the expression of the so-called *circular causality* that emerges when the ants collectively determine the order parameters and the order parameters determine the behavior of the entire collectivity (Sect. 1.1.5).

Note. Grassé introduced the concept of stigmergy not with ants in mind but termites, which, thanks to the accumulation of pheromone, are stimulated to accumulate grains of sand and other material from the ground in order to build nests with pilasters, arches, and other more or less elaborate structures.

> The emergence of pillars, walls and royal chambers in termite nests can be accounted for once the self-organizing properties of this kind of interactions are understood. In particular it has been suggested that termites' behavior follows just a single pattern (i.e. picking up and depositing a soil pellet, if pheromone is present) …| and Ibis building behavior is modulated by several environmental conditions (e.g. when an air stream drives molecules of pheromones in a given direction the pattern is influenced), not last the fact that the environment is changed by work of other termites. It is self-organization that explains the emergence of a structure at the global level from interactions among its low-level components, without any explicit coding of this process in the individual agents … The basic intuition of

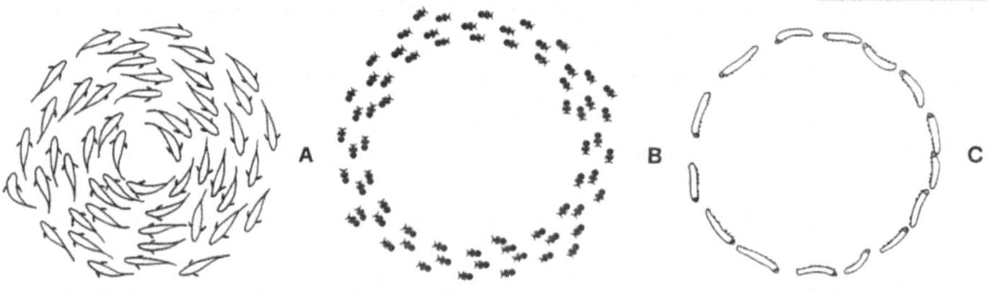

Fig. 2.38 Three examples of circular behavior caused by the release of chemical signals, based on the logic of stigmergy (source: Deneubourg & Goss, 1989, p. 300)

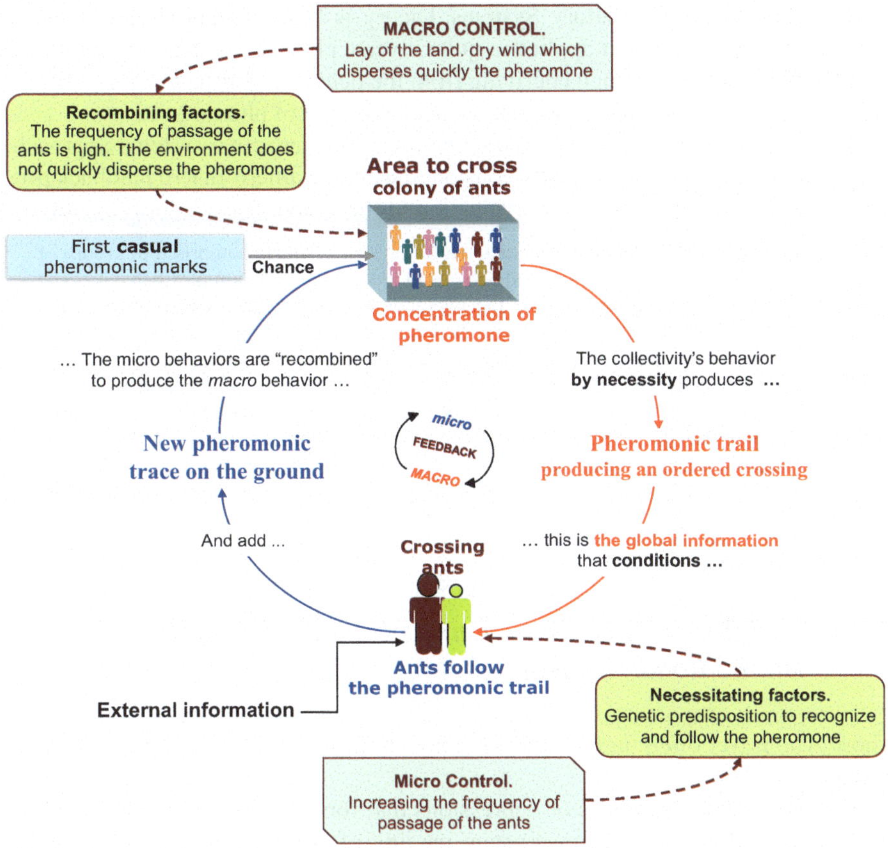

Fig. 2.39 Graphic model of "ant's pheromonic trail"

> Grassé was in the end right: no sophisticated cognitive processing is performed by the termites in order to construct, their nest. However, it is the self-organizing dynamics enabled by the properties of pheromone (i.e., its additivity and its decadence rate) that regulate and modulate termites' behavior. (Uhrmacher & Weyns, 2009, p. 246)

The previous quotes clearly show that in the case of termites—more so than that of ants—stigmergy represents the typical modus operandi of combinatory systems of accumulation (see Sect. 2.2). The first deposits of materials are made by several termites "by chance," and since the accumulated elements are impregnated with pheromone, if there are *recombining factors* that do not disperse the pheromone, the biological necessitating factors will push other agents-termites to deposit new material, thereby increasing the accumulation. As a result of this increased quantity of pheromone that is released, the accumulation represents the global information that modifies the subsequent micro-behaviors of the agents-termites, leading them to undertake new accumulations along the lines of the "Pile-of-Garbage" System (Sect. 2.2.1).

It is clear that Combinatory Systems Theory is broader than Grassé's theory (which it includes), since it applies to quite diverse phenomena that is more varied than the biological phenomena deriving from the behavior of insects living in colonies that are self-coordinated through the accumulation of pheromone.

Stigmergy can also be interpreted using a complex adaptive systems approach, which, however, does not explicitly consider the evident micro-macro feedback process to explain the spontaneous coordination of the agents.

> Complex adaptive systems can be characterized as systems that are comprised of groups of agents following simple rules that, collectively, produce emergent, complex behavior. The key to these emergent properties are the interactions—the exchanges of information—between the agents. Many biological systems can be studied using a complex adaptive systems approach, such as colonies of bees or ants. In some of these biological systems, the communication between individual agents is indirect. This type of communication is termed stimergy: a relatively small amount of information being shared through the environment, rather than directly from agent to agent. This information is nonetheless crucial to the self-organizing properties of the system, and is used by the agents to inform decision making, such as when ants follow a trail of pheromones left by other ants. (Olsen, 2019, p. 143)

2.5.6 "Highway-Ruts" System and "Chains of Potholes on the Road" System

Those who travel often on the highway know how annoying and dangerous wheel ruts are.

These are parallel grooves or depressions that form on the asphalt in the wheel paths of a roadway due to the constant passing of heavy trailer-trucks along the same trajectory and the action of a very evident Combinatory System.

Continual pounding by the wheels on the same section of asphalt, especially during hot days when the surface is softer, represents the macro-behavior of the collectivity of drivers going down that stretch of road. The macro-behavior causes a typical subsidence (macro-effect) that turns into a long track if observed over the entire stretch of highway or even on any unpaved straight road with heavy traffic. Individual drivers (agents), following a straight trajectory on the road (micro-behavior), applying pressure to the asphalt (micro-effect); if the asphalt is not compact enough, repeated compressions create, through recombination, an unwanted macro-effect. This phenomenon is general, however, and also occurs on dirt roads and even on Roman paved roads, where carts passing in long lines left permanent wheel tracks, forcing other carts to follow them to avoid damaging their wheels (Fig. 2.40, top).

Wheel tracks are present especially over the long straightaway sections with narrow lanes, since the drivers of large trucks end up going along the same strip of road surface in order to stay within the lane and not pass over into the adjacent one; they are rarer over sections with many curves, since here it is harder to maintain the same trajectory. Once formed, wheel tracks force drivers of trailer-trucks to drive "inside" the tracks, or else risk dangerous swerves and route corrections, with the risk of invading the adjacent lane. The *micro-macro feedback* continues to act until the authorities intervene with control measures to eliminate the *ruts*. The models that describe this combinatory system are similar to those of the "File-of-Ants" examined in Sect. 2.5.5, though with some variations. Certain *reinforcing factors* can intensify and exacerbate the phenomenon, such as heavy traffic, insufficiently compacted asphalt leading to micro-level collapses, and a narrow, straight road that encourages speed and makes it difficult to avoid already-formed ruts. Several control measures are possible, but the most evident are the frequent replacement of the road surface using compact asphalt and regulating vehicle speed.

A similar line of reasoning, though with different mechanics, applies to the formation of holes and subsidences one next to the other, which make the driver running into a series of these feel his car hopping along. However, the driver is forced to enter it, causing a widening and increase in the number of potholes that form the sequence (Fig. 2.40, bottom). The first hole forms by chance when, on a rainy day, the wheel of a vehicle enters a small crack in the asphalt, causing it to break further; other drivers, forced to drive into the water-filled hole, compress the liquid, which, splashing violently, deepens and widens the hole. As the wheel enters it, it sinks down, only to rise abruptly when it reaches the edge and to fall heavily a few centimeters further on, thus producing a new crack in the road surface. The subsequent holes form by necessity due to the effect of the wheel hitting the asphalt, bouncing off the first hole, coming down again heavily, pushed by the shocks, and then forming a subsidence that, after being hit hard by subsequent wheel landings, becomes a new hole.

The series of holes is the macro-effect from the succession of hard landings by the wheels of cars passing one after another (macro-behavior), each of which

Fig. 2.40 Highway-ruts system and chains of potholes on the road

contributes (micro-behavior) to the sinking of the road (macro-effect) while, in turn, being produced by that sinking, in a typical micro-macro feedback process. The most evident reinforcing factors are high speed, poor visibility, lack of signage indicating the presence of potholes, and inadequate maintenance. To control the phenomenon and weaken the Combinatory System, it is useful to ensure frequent maintenance, use compact asphalt, impose speed limits, apply a seal coat to protect the asphalt from water and UV damage, and promptly fill potholes using durable materials to prevent further deterioration. Lastly, clearly mark the potholes with visible warnings.

2.5.7 *"Macedonian-Phalanx" System*

Phillip II, the father of Alexander the Great, was satisfied with "his" phalanx, the potent war machine that helped him win every battle. How did this idea come about? How do we transform a group of vulnerable soldiers into a compact testudo? It was a stroke of genius! He formed large divisions of hoplites that marched in closed formation, like a single soldier, armed with double-pointed pikes—the sarissas—held horizontally by five soldiers, one behind the other. The problem was to maintain the phalanx during combat even when the hoplites carrying the sarissas were hit, which happened often, despite the formidable shields that were held erect for protection. Phillip had a second winning idea: it was enough to set a simple rule that each soldier had to take up position next to another and hold it. The first rows were occupied by the hoplites carrying the sarissas, behind them a formation of soldiers protected by shields. When a soldier was hit, the one behind him had to move up and cover the open position; at the rear of the phalanx was a large number of reserve soldiers who had to take over when a gap opened up in the tail of the phalanx when the soldiers up front were charging. The phalanx was thus invulnerable and unstoppable: a reserve soldier immediately replaced every loss, and the phalanx could thus present itself as a single combatant. Alexander reinforced the Phalanx structure, making it an almost invincible combat unit.

> It is certainly true that Alexander increased the depth of the phalanx to sixteen men and that the Successor armies expanded its size and depth to where the phalanx could hardly maneuver at all. Thus, the phalanx's immobility dates after Philip's reign. As Adcock cogently observes, Philip's phalanx "had legs" from the beginning and was capable of much greater mobility than the hoplitic phalanx was." Within the long Macedonian phalanx infantry line, any given battalion or regiment could be ordered to take several formations … It could, for example, form itself into a solid triangular wedge to penetrate the enemy's line. Drawn up in tight compact order, the battalion could also form into a hollow wedge. This formation attacked the enemy with double the number of spear points of the solid formation, not only head-on but laterally as well, so that the individual hoplite could not rescue the comrade to his side. At Gaugamela, Alexander formed his phalanx into a hollow outward-facing rectangle with orders that if it broke under attack, it was to reform itself into two similar rectangles. (Gabriel, 2010, p. 67)

As a *combinatory system of order*, the phalanx can be observed under two aspects: the direction of march and the maintenance of the structure. Regarding the former, the direction of march of the phalanx can be interpreted as the effect of a combinatory system of order entirely similar to the "*Herd-in-Flight" system* described in the preceding sections: the movement of the phalanx is caused by that of the soldiers forming it, but this is conditioned by the position and direction of the phalanx itself. The direction of the phalanx depends on both the instructions received as well as on where the enemy is coming from; "chance" can direct the phalanx in a given direction, but "by necessity," this conditions the direction of the soldiers it is composed of. We are not interested in the *direction of march* as much as the preservation logic of the phalanx, which can be interpreted in terms of combinatory systems with the following heuristic model (Fig. 2.41).

Micro behavior and necessitating rule: soldiers in the first row: follow the commander or the officers that give the direction of march of the phalanx. Soldiers further back in the formation: follow the soldier in front of you; if he is hit and the

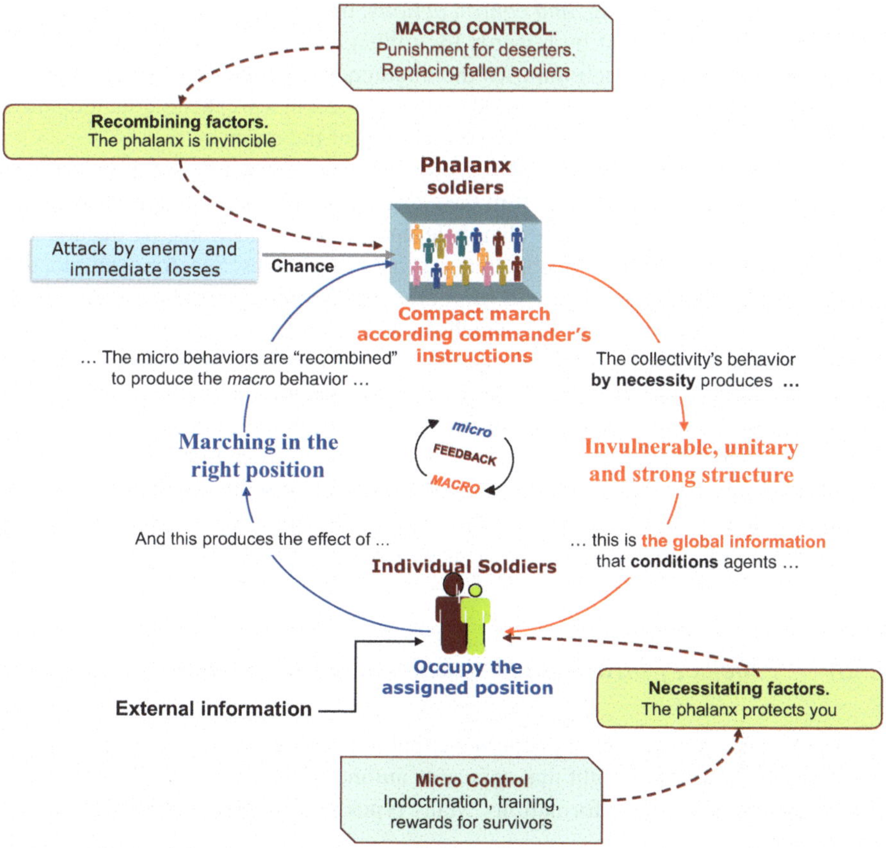

Fig. 2.41 Graphic model of the "Macedonian phalanx" system

position is freed up, move forward to occupy it. If the phalanx does not reform, then run the risk of being killed by the enemy; if you flee, you will be hit by your own soldiers behind you. Each soldier acts under limited information, since he knows the position of the companions near him and the one that precedes him in the phalanx.

Macro behavior and recombining rule: by filling the gap, the phalanx remains compact and strong, and it can march toward the enemy according to the commander's tactical instructions; if the phalanx is compact, the invulnerability of its members is guaranteed.

Micro-macro feedback. Chance and necessity: the consistency of the phalanx is the effect of the micro behavior that substitutes the fallen comrades, but this is conditioned by the consistency of the phalanx itself. This consistency depends on both the instructions received as well as the strength of the enemy; "chance" can cause the phalanx to be attacked by a given number of enemies that inflict losses, but "by necessity" the maintenance of this consistency conditions the behavior of the component soldiers. If the number of fallen soldiers exceeds the number of reserve soldiers, the phalanx is diminished or breaks up.

Strengthening, weakening, and control actions: the maintenance of the phalanx's consistency is strengthened by rewards for soldiers who gain a position or by the strength of the enemy, which imposes the maintenance of the order so as to avoid defeat. A clearly superior enemy would instead represent a weakening factor, since it would cause withdrawal leading to the break-up of the order.

The model of the Macedonian phalanx system can be considered a prototype of all biological systems that derive from the aggregation of mono-cellular individuals into multi-cellular ones that can also differentiate their functions, as occurs, for example, in the cycle of the amoeba, in corals, and in sponges. See, for example, the description of the amoeba's cycle in Prigogine and Nicolis, Exploring Complexity (Prigogine & Nicolis, 1989, p. 35), which shows the biological self-organization of single individuals. The individual amoebas are attracted to each other—chemotaxis—in order to form a multi-cellular body made up of mono-cellular individuals that, by acting in a coordinated manner, allow the multi-cellular body to act as a single individual. Chemotaxis initiates by chance, due to some individual that emits chemical signals, but then it comes to involve by necessity other individuals, since the greater density of chemical signals attracts other individuals, who produce a larger quantity of signals as part of a typical *micro-macro feedback*.

2.5.8 "Flocks of birds" and "School-of-Fish" Systems

Flocks of birds are fascinating phenomena that we can frequently observe, consisting of groups of birds in flight that appear as autonomous entities and exhibit complex behavior and intricate formations. Many types of birds frequently fly in flocks, among which:

- Swallows, which fly in large, agile flocks during migration.
- Pigeons and sparrows, which often fly in large groups, especially in urban areas.

- Starlings, often observed in spectacular synchronized flights called "murmurations," which often occur near wetlands, open fields, or urban outskirts where starlings find abundant food and safe shelters.
- Ducks and geese, which fly in V formations during migration.
- Herons, cranes, and pelicans: although they belong to different families and orders, they often migrate in large flocks.
- Seagulls, which can form flocks near bodies of water.

Flying in flocks offers some existential advantages. First, flying in flocks protects birds from predators, as it is more difficult for a predator to single out one bird among many and, above all, to prey on a bird flying within the formation.

This natural biological phenomenon is probably a survival strategy that offers several benefits, such as protection of individual birds that move toward the center of the flock by reducing the probability of being eaten by predators that prey from outside the surface and increasing foraging efficiency. Furthermore, the likelihood of detecting predators is greater among the numerous members flying along the outer edges of the flock. The "Herd-in-Flight" System (Sect. 2.5.2) and the "Macedonian-Phalanx" System (Sect. 2.5.7) are appropriate for explaining the formation of a "Flock of Birds."

The heuristic model that interprets the "Flock of Birds" system, with simple adaptations, can also describe the phenomena of the formation of a "School of Fish," a very frequent Combinatory System which can be easily observed in every sea, but above all in those rich in small-sized fish. The "School of Fish" phenomenon refers to a large group of fish swimming together in a coordinated manner, often moving in the same direction and exhibiting synchronized behavior. Since the smaller fish can be food for larger fish, when there is danger, they gather together in large schools. The school is formed by the *micro behavior* of the single individuals that, reaching the boundary of the school, dart and follow their instinct (innate operative program) to slip into the school once again, in order to hide from probable predators, leaving on the outskirts, other fish that will behave in the same way, as described in model in Fig. 2.42.

Among the many fish species, the most notable that exhibit this dynamic behavior are:

- Herring, Anchovies, and Sardines, known for moving in large schools that can include even millions of individuals.
- Tuna, which form schools, especially during migration. They are known for their speed and agility. Tuna schools are easily intercepted by the "tonnare," specialized fishing structures designed to channel the tuna thanks to a traditional fishing technique, the "mattanza," used mainly in the Mediterranean, particularly in Italy, Spain, and North Africa.
- Cod, often found in large schools in the cold waters of the Atlantic and Pacific Oceans.

This natural biological phenomenon is probably a survival strategy that offers several benefits, such as the protection of individual fish that move toward the center

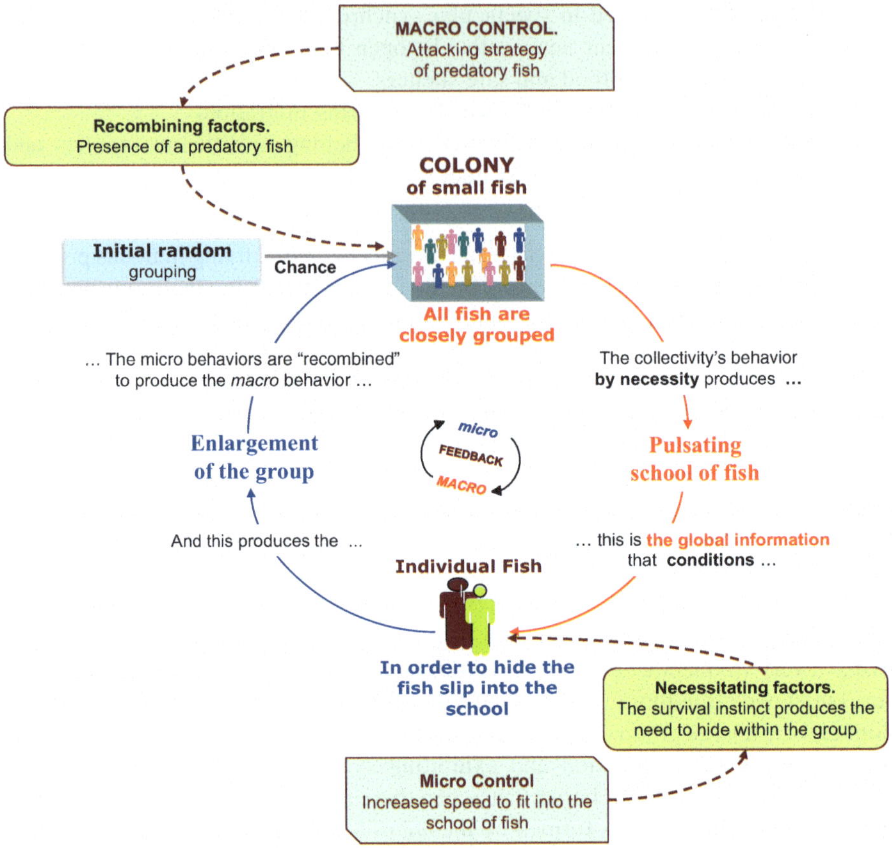

Fig. 2.42 Graphic model of the "School-of-Fish" system

of the "cloud," reducing the likelihood of being eaten by predators below the cloud's surface and increasing foraging efficiency (especially for dolphins), and in some cases, greater hydrodynamic efficiency. Additionally, there is a greater chance of detecting the presence of predators by the numerous members swimming at the surface of the cloud.

The school thus appears to be an autonomous, pulsating entity that offers a fascinating spectacle for underwater photography lovers. Nevertheless, this effect represents the macro effect of a combinatory system that not only is produced by the *micro behavior* but is also able to condition such behavior. The micro behavior is motivated by a voluntary necessitating factor: to remain hidden in the school in order not to be attacked by predators, rather than flee in disorder in every direction.

The school produces the *recombining factor,* since it hides from view the fish inside it. It is an autonomous entity with its own macro behavior: it can navigate in a particular direction or remain in the same place (like the herd of elephants); it can become oblong, reform into a sphere, or break up into two smaller schools, according to how the single individuals develop their micro behavior as they re-enter the school.

2.6 Systems of Improvement and Progress

A very special and important combinatory system is the one that shall be referred to as the Improvement & Progress Combinatory System, since its particular effect is to produce *progress*, understood as an improvement in the overall state of a collectivity that is attained through individual *improvement*. These systems can be classified among those belonging to the classes mentioned above. In particular, they are systems of pursuit and produce accumulation or diffusion; they will be described as an independent class only because of their particular relevance in social collectivities. Individual improvements raise the parameter that measures collective progress; this constitutes the *global information* that leads to the perception of positive and negative gaps that push the individuals to improve in order to increase the gaps (if positive) or eliminate them (if negative). These systems can be represented by the following *heuristic* model and illustrated in Fig. 2.43.

Micro behavior and necessitating rule: if you perceive that the level of your *improvement* parameter is below the level of the system's *progress* parameter—that is, there is a negative gap between your state and that of the others—try to *improve* in order to reduce the gap and, if possible, attain a positive gap; if you perceive there is a positive gap, do nothing or try to improve further in order to increase the favorable gap;

Macro behavior and recombining rule: the system must be able to notice the individual improvement and adjust the progress parameter to the average (or, more generally, to the combination) of the individual improvement measures;

Micro-macro feedback. Chance and necessity: individual improvement (micro effect) raises the parameter that measures collective progress (macro effect); this leads to the formation of positive and negative gaps that push the individuals to improve in order to increase the gaps (if positive) or eliminate them (if negative).

Many relevant phenomena that can be explained using the systems of improvement and progress are described in the following sub-sections. Given the great variety of these systems, several specific Combinatory Automata will be described in Sect. 3.6.

2.6.1 "Increasing-Productivity" System

Productivity is the efficiency of man in producing, and it can be observed at the level of the individual producer (usually a firm or division) or with reference to a sector or the overall economic system (Mella, 2021). The degree of average productivity is

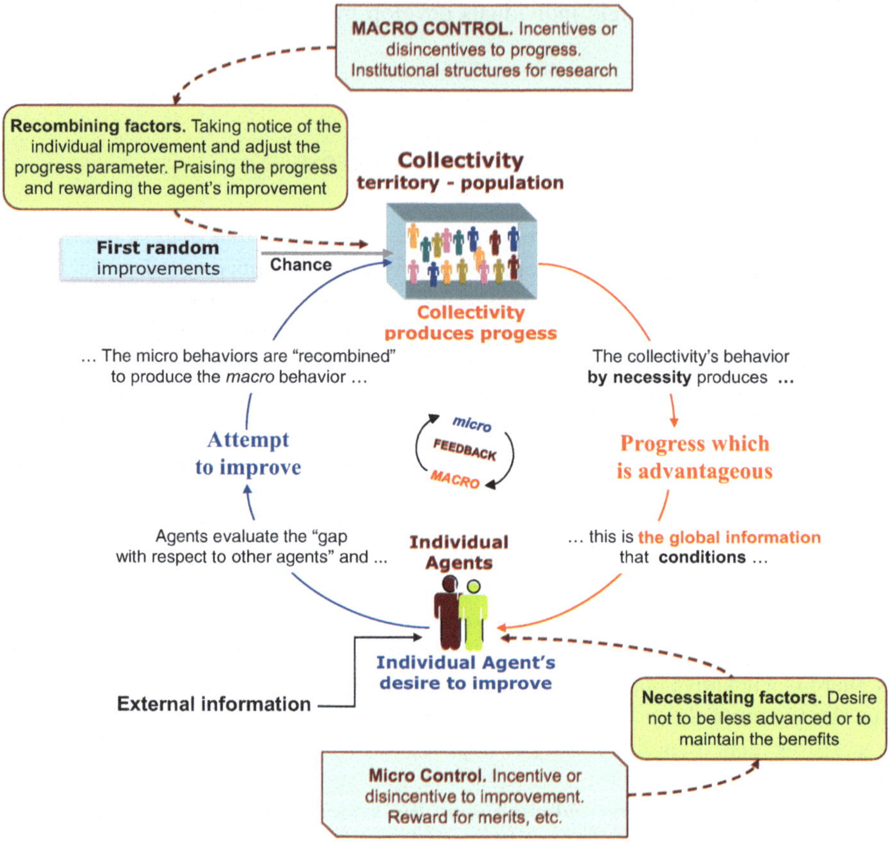

Fig. 2.43 Graphic model of systems of "improvement & progress"

measured by the ratio between the quantity of production obtained from a certain productive process and the quantity of labor used in that process over a given period of observation; or even by the inverse ratio, which indicates the average labor requirement per unit of product.

We all know that some economic systems—such as Japan's, Germany's, China's and USA's—enjoy a very high level of productivity as opposed to others where average productivity is quite low, such as in East European countries. But one thing is certain: when in an economic system competition is possible among producers, the average productivity of the system continually increases and becomes unstoppable. Even the average productivity of the other economic systems shows an increasing tendency for productivity to expand, even if at non-uniform rates of growth. Though varied, the "drivers of productivity" can be grouped into a few categories which, though they interact, must be kept separate in order to facilitate our observation, which is derived from Adam Smith's famous example of the manufacture of pins to illustrate the advantages of the division of labor.

> This great increase in the quantity of work, which, in consequence of the division of labor, the same number of people are capable of performing, is owing to three different circumstances; first, to the increase of dexterity in every particular workman; secondly, to the saving of the time which is commonly lost in passing from one species of work to another; and lastly, to the invention of a great number of machines which facilitate and abridge labour, and enable one man to do the work of many. (Smith, 1776, p. 5)

Generalizing Adam Smith, the following typology is proposed:

A. *Passive drivers*: these increase the production obtained per unit of labor applied to production. These drivers are connected to fertility in all its forms: the fertility of land, water, subsoil; natural (the banks of the Nile) or artificial fertility (irrigated and fertilized land);
B. *Active drivers*: these reduce the of quantity of labor needed to produce with fertility held constant; there are three types of active drivers of productivity: *Skill, Equipment,* and *Specialization*.

   1. *Skill*: it is easy to imagine that a skilled fisherman or hunter can obtain the same amount of goods in less time than other, less-skilled producers since, by applying the same amount of labor as the others, they can obtain a larger quantity of products.
   2. *Equipment*: from the first chipped rock to modern machines and digitally controlled plants, equipment extends the capacities of "human hardware" (represented by our body, its limbs, and our brain), reducing the effort and danger associated with production and rewarding skill. Holding the amount of labor employed constant, we can greatly increase the amount of clothes produced by using a loom rather than a crocheting needle. With labor hours held constant, we can increase beyond measure the amount of arable land by using a modern-day tractor, which uses a six-bladed plough, compared to what we could obtain with a single-bladed plough pulled by a pair of oxen. It is easy to understand how few cars would be produced without the modern robotized assembly line; and the construction of three pyramids with the limited equipment at the time required more human labor that the construction of all of Manhattan's skyscrapers put together. We can easily recognize the progress computers and robots have produced until now and will produce looking forward.
   3. *Specialization*: a characteristic of the organization is the functional division of labor: each worker in the organization provides his skilled and equipment-aided labor to undertake a specific activity, and this specialization, together with the equipment available, further increases productivity (Smith, 1776, p. 3–5).

C. *Endogenous or psychological drivers*: these are the psychological conditions that lead man to supply his labor to a given organization; these drivers can be divided into *Motivation* and *Satisfaction*.

   1. *Motivation*: man is willing to supply his labor only if adequately motivated and if there are expectations this will satisfy his needs or motivations. Once

the main motivation was pay, either in the form of wages, salary, or profit. Today, other motivations, intellectual in nature, have been added to those regarding monetary payments: a satisfying job is often preferable to a better paid but boring one. Motivation drives man not only to be more efficient but also to continually learn to improve his preparation. Motivation not only is linked to a worker's personal and family situation but also depends on individual "vision" (improving one's status) and the collective culture (advance the group, the homeland, etc.). This driver also includes the worker's ethical attitude toward his role in production and toward time-wasting (Ohno, 1988).

2. *Satisfaction*: motivation provides work incentives to man. The initial *motivations must be followed by satisfaction; that is, the satisfactory achievement of* motivations.

D. *Extrinsic drivers*, so called because they involve the organization of productive systems, the environment within which the work is carried out, and, in the final analysis, the firm's policies regarding increased productivity. Among the most important of these are the following (Schmitz, 2005):

1. The ongoing *trend of mechanization*, including automated process control systems and continuous mechanization, has been a fundamental aspect of human development. Initially, mechanization stemmed from the worker's innate creativity to craft tools, a skill later integrated into organized systems. In recent decades, this progression has accelerated, driven by advances in electronic and computing technologies.
2. On-line *automatic control systems* for processes: factors influencing the work environment, such as ergonomics, sophisticated information systems, the shift toward dematerialization, and the use of standardized materials, have all contributed to process acceleration (Demartini & Mella, 2011).
3. The *optimization of materials handling and logistics* has also played a key role in enhancing efficiency (Bowersox et al., 2002).
4. The adoption of *standardized designs and new materials* has enabled the creation of products and processes based on consistent criteria. These materials and components, designed with predefined, interchangeable standards aimed at making the processes more efficient, contribute to faster, more accurate, and more precise assembly while reducing the need for quality corrections.
5. Advances in *scientific and technological research*, above all in energy supply systems and the productive application of energy, have significantly impacted progress, as Carlo Cipolla highlighted in his work, *The Economic History of World* Population (Cipolla, 1962).
6. Advances in the productive use of *Artificial Intelligence systems* in the design, execution, and control of production processes (see: Sect. 2.6.9).

The primary challenge to developed economies will be to create the endogenous drivers of productivity; that is, to motivate and satisfy workers (Baumol et al., 1989). Certainly, the problem for the future (which already today has become a crucial point) will be to improve productivity through the quality of both products and working conditions.

The improvement in productivity increases *welfare* (Coyle, 1999), which we can see not only in the spread and accumulation of wealth among consumers but also, and perhaps above all, in the gradual reduction in labor times and the improvement in working conditions (Blank & Shapiro, 2001).

The following hypothesis is proposed: the *hypothesis of increasing productivity*: the network of production firms tends to achieve ever greater productivity, but it is itself governed by the continued increase in productivity and quality, following a typical *micro-macro feedback* of a particular combinatory system (Mella, 2018).

Suppose that in a given environment, different producers operate who manufacture substitute goods; if "by chance" one of these should increase his average productivity—by reducing average unit costs of production—it would create advantages for him and a threat for the others, since he can attract customers by reducing the sales price. In order to survive on the economic scene, the other producers will "by necessity" be forced to increase their productivity in a typical feedback of the combinatory system. The result of this micro behavior is the increase in the average productivity of the system, an increase that becomes continuous due to the incessant competition among producers and the need for each producer to maintain sufficient profit margins to remain in business.

If "by chance" a producer introduces a new technological invention—for example, robots for production—then the combinatory system guarantees that "by necessity" the robots will spread to all the sectors. If a producer introduces optical fibers, then sooner or later everyone will adopt them, unless alternative inventions are found that, if efficient, could supersede the previous ones in terms of productivity. The increase in average productivity at the level of the productive system is the result of past micro behavior, but this also conditions the search by individual producers for new factors of productivity. The history of mankind is the history of the increase in productivity, and events at any time confirm the action of the combinatory system we are looking at.

The heuristic model of the system can be described as follows (Fig. 2.44):

Micro behavior and necessitating rule: if your unitary profit falls and you want to remain in the economic system as a producer—and you cannot influence the sales price—then you must reduce your average unit cost of production by increasing productivity to a level equal to or greater than the average level of the other producers you are competing with, by searching for some factor of productivity.

Macro behavior and recombining rule: the introduction of a factor of productivity improves the system's average level of productivity by cancelling the advantages of the producer; the producers try to equal—or preferably exceed—the average productivity level of the system.

Micro-macro feedback. Chance and necessity: the increase in the average level of productivity of the productive system is the result of past micro behavior, but this also conditions the search for new factors of productivity by the individual producers. A productive innovation—or more generally a new factor of productivity—is introduced or utilized "by chance" (was not the discovery of America due to chance,

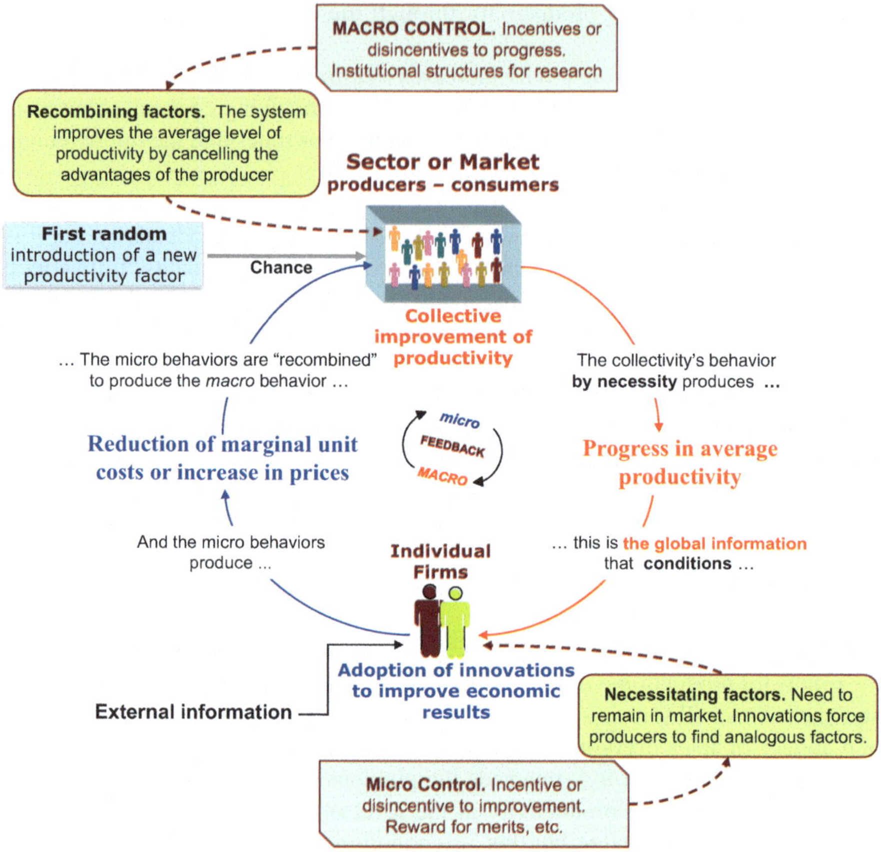

Fig. 2.44 Graphic model of the increasing-productivity system

in the attempt to find the fertility that was missing in Europe? Wasn't nuclear radiation discovered "by chance"?), but the new factor of productivity the producer turns to has an effect on the costs of production (and often also on price, if the innovation improves the quality of the product) by forcing the other producers to find "by necessity" similar factors (often by simply imitating the innovation—and we all know about industrial spying; at other times, with a substitute innovation).

Strengthening, weakening, and control actions: incentives to innovation, the presence of specialized research centers, high profit motivation, are all examples of strengthening factors for the system; monopoly conditions and scarce consumer awareness are examples of weakening factors for the system. The macro control can be exercised through laws that favor or inhibit the introduction of innovations, or through measures affecting the educational system of the country the firms get their human resources from; the micro control instead acts on the individual producers through various kinds of incentives.

2.6.2 *"Increasing-Quality" System*

A truly important combinatory system, which operates in a similar way to that of the increasing productivity system and leads to progress in the way of producing and utilizing the goods obtained, is the one that operates on the *quality* of the goods produced. The concept of quality is not easy to define. It is "elusive," and understanding it is normally left to intuition. Several scholars have proposed giving up the attempt to precisely define the concept, viewing it as an intuitive, non-definable, almost primitive or philosophic notion derived from the concepts of differentiation, excellence, perfection, and consistency. Robert Pirsig (1974), in his book *Zen and the Art of Motorcycle Maintenance: An Inquiry into Values*, introduces the idea of the "Metaphysics of Quality," where quality is a direct experience independent of and prior to intellectual abstractions.

> Quality is the continuing stimulus which our environment puts upon us to create the world in which we live. All of it. Every last bit of it. Now, to take that which has caused us to create the world, and include it within the world we have created, is clearly impossible. That is why Quality cannot be defined. If we do define it we are defining something less than Quality itself. (Pirsig, 1974, p. 114)

I propose identifying three distinct, though connected, notions that sum up the majority of definitions found in the literature and allow us to focus on quality from a threefold perspective: the customer, the product, and the environment: *functional*, *design*, and *environmental* quality (Mella, 2011).

Functional (or market) quality defines the set of characteristics which, from the customer's point of view, make the *product* appropriate for *use*, that is, capable of satisfying a specific *use or utility function* of the good or service (Band, 1991), taking into account a desired standard of *reliability* (the product must provide use that is not interrupted due to imperfections) and *safety* (the product must be capable of use without damage or risks).

Design (or *intrinsic, productive*) *quality,* or conformance to specifications, is the set of characteristics that, from an internal point of view (in terms of production processes), make all the *product units* conform to a *standard* of reference (prototype, sample, model, design). To Philip Crosby (Crosby (1979), any product that falls within its design requirements is a quality product (Garvin & March, 1986), and for this reason, the message to managers is "*do it right the first time*" (DIRFT). These forms of quality refer not so much to the product as to the *production*; that is, to the flows of production units obtained during a process. Normally, the standard of reference is the initial project and the specifications provided therein. Each company, whether large or small, must continually verify that the intrinsic quality of its products always conforms to set qualitative standards or improves over time. The need to at least distinguish between functional and design quality is explicitly recognized by Nelsen and Daniels (2007), according to whom: "*quality has two meanings: 1. the characteristics of a product or service that bear on its ability to satisfy stated or implied needs; and 2. a product or service free of deficiencies*" (p. 54).

Environmental (or *context*) quality is the set of characteristics which, from the point of view of external impact, make the product compatible with the environment, both in terms of pollution, waste disposal, environmental risks, or suitability for introduction into the context in question.

In 1983, the American Society for Quality Control (ASQC) proposed a specific definition which confirms our previous tripartition.

> [Quality is a] subjective term for which each person or sector has its own definition. In technical usage, quality can have two meanings: (1) the characteristics of a product or service that bear on its ability to satisfy stated or implied needs; (2) a product or service free of deficiencies. According to Joseph Juran, quality means "fitness for use"; according to Philip Crosby, it means "conformance to requirements." (ASQ, 1991, online)

Moreover, these notions of quality consider the products as *fit for a purpose*: "*One of the possible criteria for establishing whether or not a unit meets quality, measured against what is seen to be the goal of the unit*" (Campbell & Rozsnyai, 2002, p. 132). "*Fitness for purpose sees quality as fulfilling a customer's requirements, needs or desires*" (Harvey & Green, 1993). For Feigenbaum (1951, p. 10): "*Quality is best for certain customer 'conditions.'*" In a similar vein, Flood (1993, p. 48) states that: "*Quality means meeting customer (agreed) requirements, formal and informal, at the lowest cost, first time every time.*"

> After-sales service quality affects satisfaction, which in turn affects behavioural intentions. Hence, after sales services affect the overall offering, and thus the quality of the relationship with customers. (Rigopoulou et al., 2008, p. 522)

These definitions reveal how quality, together with productivity, is an essential element in judging the efficiency and effectiveness of a productive process; it is thus a fundamental element in the production and sales strategies of every firm. Every improvement in the quality of its product brings a temporary competitive advantage to the producer, which lasts until the qualitative improvement is cancelled by the imitating effects of the other producers. The history of mankind—as we can see from a simple reading of the economic and technological environment in which we live—is thus not only the history of the increase in productivity but also that of the continual improvement in the quality of the goods and services produced. We need to only consider the recent history of the automobile, of arms, computers, or smartphones to appreciate the rapidity of the qualitative progress of their design and performance. In increasing-quality combinatory systems, when a producer introduces a qualitative improvement in the product, which is appreciated by consumers and gives him a competitive advantage, this variation is sooner or later introduced by the competitors as well. This causes a continual rise in the average quality of the goods that are produced and sold. The improvement in the average quality of the productive system is the result of past micro behavior, but this itself conditions the search by individual producers for new qualitative improvements. An innovation that produces an increase in quality is introduced "by chance," but the innovation to which the producer turns gives him an advantage in sales, and thus economic benefits, over the other producers, forcing the latter to find "by necessity" the means and forms to in turn improve the quality of their products.

The heuristic model of this system can be represented as follows (Fig. 2.45).

Micro behavior and necessitating rule: if your sales fall with respect to those of other producers who have introduced quality improvements in their products or processes, and you want to remain in the economic system as a producer—and if you cannot influence your productivity by reducing your marginal unit costs—you must in turn try to improve the quality of your products.

Macro behavior and recombining rule: an improvement in the quality of a product raises the average level of quality of similar products in the productive system; the individual quality of each producer must at least equal—or preferably exceed—the average level of quality in the system.

Micro-macro feedback. Chance and necessity: the improvement in the average quality in a productive system is the result of past micro behavior, but this itself conditions the search by individual producers for new qualitative improvements. An innovation that produces an increase in quality is introduced "by chance," but the innovation to which the producer turns has a negative effect on sales, and thus on the

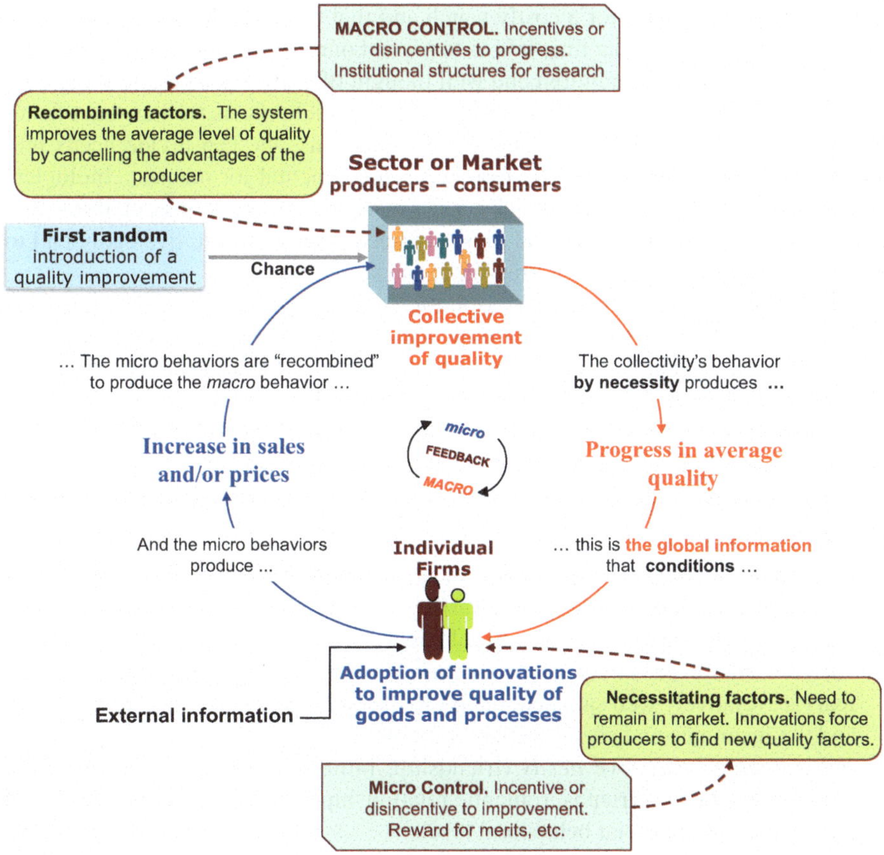

Fig. 2.45 Graphic model of the "increasing-quality" system

economic benefits, of the other producers, forcing the latter to find "by necessity" the means and forms to improve the quality of their products.

Strengthening, weakening, and control actions: these are similar to those for the increasing-productivity system.

2.6.3 "Needs-and-Aspirations" System of "Increasing Well-Being"

Man's life is characterized by the continual appearance of needs and aspirations and by the attempt to satisfy these. Needs are *unpleasant* sensations—connected to psychic-physical states of disequilibrium—felt (or imagined) in a given period, which man wishes to eliminate, attenuate, or avoid. Our experience shows us that needs are large in number and in most cases appear with regularity. Aspirations differ from needs, in that they are *pleasant* states men wish to acquire, maintain, or increase. Both the desire for a costly watch and that for a quality fur do not derive from the need to know what time it is or to protect ourselves from winter's chill, but from the aspiration to be admired and well thought of in the social circle we live in. Both the desire to improve our knowledge of a specific topic as well as that of succeeding in a sport are further examples of aspirations. We can divide needs into primary and secondary needs. The former are fundamental for man as a biological system, without which "system-man" could not survive: food, refuge, clothing, and so on. Secondary needs are instead those we need in order not to live badly and to have a better existence: hygiene, free-time, and so on.

The distinction between need and aspiration—and in general secondary need and aspiration—is often difficult to determine clearly, but, at least in theory, one criterion appears to be reliable: the intensity of the need—whether primary or secondary—decreases with its satisfaction, except when it reappears at a subsequent period. On the other hand, the intensity of the aspiration increases with its satisfaction (at times only to then suddenly cease when boredom sets in).

The psychologist Abraham Maslow (1943, 1954) suggested a Hierarchy or pyramid of needs:

1. *Biological and physiological needs*: in advanced industrial societies physiological needs (air, food, drink, shelter, warmth, sex, sleep, etc.) are not very motivating, since they are reasonably well satisfied.
2. *Safety needs*: even safety needs (security, order, law, stability, freedom from fear) are relatively well satisfied, thanks to the continuity in the level of resources and the insurance and social welfare systems.
3. *Belongingness and love needs* (friendship, intimacy, affection and love from work group, family, friends, romantic relationships): these cannot be renounced and operate at the social behavioral level.
4. *Esteem needs:* these correspond to the aspirations for self-esteem and social status (achievement, mastery, independence, status, dominance, prestige, self-

respect, respect from others); they join together to express the individual's need for his own identity, to stand out in his environment; a favorable sense of self-identity is strengthened by the recognition and approval of others, by the prestige and status one enjoys.
5. *Need for self-actualization:* at the top of the hierarchy of human needs is self-actualization (realizing personal potential, self-fulfillment, seeking personal growth, and peak experiences). This is the need to realize one's own potential capacities, to develop and grow in an autonomous and constant manner, to be creative and full of ingenuity.

> It is quite true that man lives by bread alone—when there is no bread. But what happens to man's desires when there is plenty of bread and when his belly is chronically filled?
>
> At once other (and "higher") needs emerge and these, rather than physiological hungers, dominate the organism. And when these in turn are satisfied, again new (and still "higher") needs emerge and so on. This is what we mean by saying that the basic human needs are organized into a hierarchy of relative prepotency'. (Maslow, 1943, p. 375)

The probability of a motivating appearance of a need is a function of the degree of continuity in the satisfaction of the need itself: that is, a need that is regularly satisfied ceases to be motivating. A need is not motivating if lower-level needs have not been satisfied in the hierarchy at whose base are the physiological needs for survival (see also the discussion in Weinshall and Raveh (1983, p. 148)).

> The late Abraham H. Maslow, the father of humanist psychology, showed that human wants form a hierarchy. As a want of a lower order is being satisfied, it becomes less and less important, with a want of the next-highest order becoming more and more important. Maslow applied to human wants what might be called "marginal utility"—and his was a profound and lasting insight. Maslow put economic want at the bottom and the need for self-fulfillment at the top. But the order is not of first importance. What matters is the insight that wants are not absolute; the more one want is being satisfied, the less its satisfaction matters. [...] This is, we now know, to be true of every one of Maslow's wants. As a want approaches satiety, its capacity to reward and with it its power as an incentive diminishes fast. But its capacity to deter, to create dissatisfaction, and to act as a disincentive rapidly increases. (Drucker, 1977, p. 246)

During the decade 1960s and 1970s, Maslow enlarged the original five-level Hierarchy proposing an eight-stage model (McLeod, 2007):

1. *Biological and physiological needs.*
2. *Safety needs.*
3. *Love and belongingness needs.*
4. *Esteem needs.*
5. *Cognitive needs* (knowledge, meaning, culture, etc.)
6. *Aesthetic needs* (search for beauty, balance, form, etc.)
7. *Self-actualization needs.*
8. *Transcendence needs* (helping others to achieve self-actualization).

We are easily aware of the fact that the quality of our lives depends on the number and types of needs and aspirations we can satisfy, and above all on the way in which we achieve this satisfaction. Until a few decades ago, man could satisfy only his primary needs for food, clothing, and protection from disease, and only a few people managed to satisfy their aspirations for culture, art, and entertainment, which today are available to everyone.

This is a general phenomenon: in all ages, all civilizations, and in every area of the world, we observe a twofold transition: from low-level needs to high-level ones, and from needs to aspirations. And this leads to an increase in welfare and to the search for wealth as an opportunity and means of welfare.

We can explain this phenomenon again in terms of combinatory systems, by referring to the increasing-productivity and the increasing-quality system.

The logic of this system is simple: the micro behavior is represented by the search for individual improvement, understood as an increase in the degree of satisfaction of needs and aspirations, by the desire for greater wealth and better working conditions, and by the demand for ever greater quantities of goods of ever higher quality and for new goods capable of satisfying motivations; the micro effects represent a greater satisfaction regarding our existence. The state of the system is a given perceived average level of the quality of life, understood as the number and types of needs and aspirations satisfied in the context of a given community. The individual elements of the base—that is, the families and consumers—compare their quality of life with the average level; if they see theirs as inferior to the latter, they tend to reduce the gap, and if they perceive it as superior, to increase the gap (micro behavior).

The system thus tends to modify its own state by producing a macro behavior that in turn will guide the subsequent micro behavior: the improvement in the average quality of life in the environment is the result of past micro behavior, but this itself conditions the search by individual consumers for new improvements in their quality of life.

We can also say that the combinatory systems we have previously examined—increasing productivity and quality—derive their operation from the needs-aspirations system. This system can be represented by the following heuristic model (Fig. 2.46).

Micro behavior and necessitating rule: since you desire to improve your and your family's quality of life, try to improve it, if it is below that of the average for those families you can directly observe; in any event, try to increase the level of satisfaction of the needs and aspirations for you and your family.

Macro behavior and recombining rule: the collectivity appreciates those who have a high standard of living; when individuals raise their standard of living, there follows a rise in the general standard; the search by individuals for more and better-quality goods leads to a quantitative and qualitative improvement in production.

Micro-macro feedback. Chance and necessity: the improvement in the average standard of living in the environment results from past micro behavior, but this itself conditions the search by individual consumers for new improvements in the standard of living; in some populations, the standard of living does not change for centuries: until a few years ago, some tribes did not know about the wheel or metalworking. However, almost everywhere, the average level in the standard of living has risen well above the minimum vital limits for survival. "Chance" of

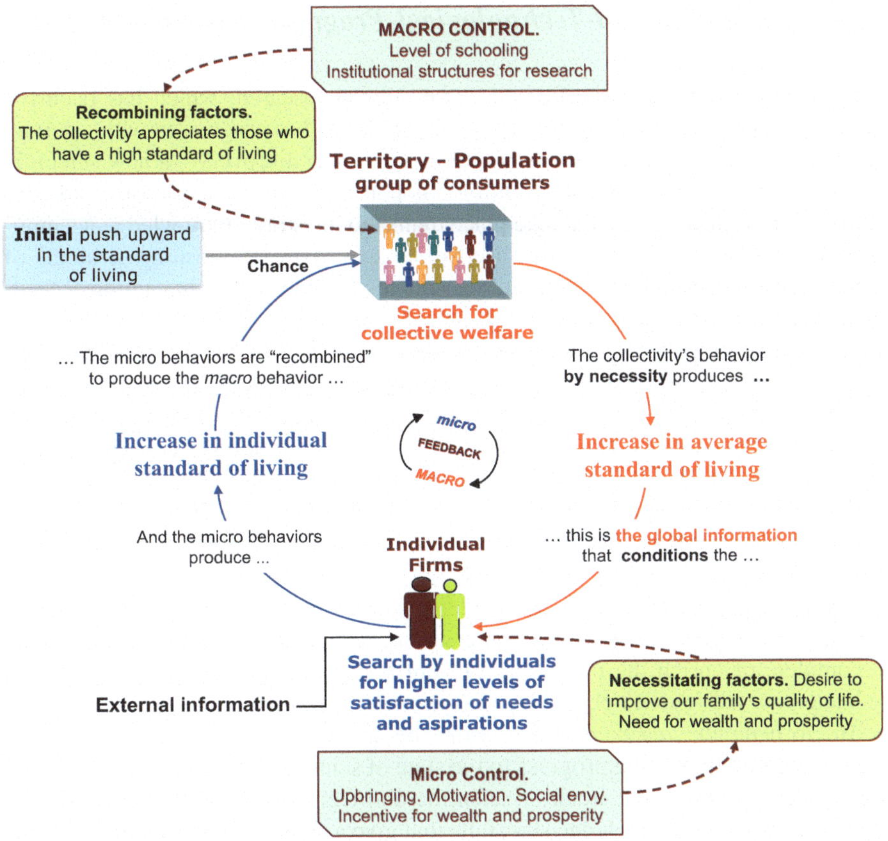

Fig. 2.46 Graphic model of the "needs-aspirations" system

course plays a fundamental role in causing the initial push upward in the standard of living, but the innovations that by chance have permitted the increase in productivity (consider, for example, the influence of the piston engine and the nuclear reactor) and in the quality of products (e.g., the temper of metals, biological crosses, the attempts to attain "zero tolerance") are too numerous to remember. We can show that chance has played a role in triggering the macro behavior by examining the differing distributions of wealth and the differing trends in the standards of living on our planet with regard to the various combinatory systems of the types we are presently discussing. Once set off, the macro behavior "by necessity" is maintained. Today, there is no apparent limit to the satisfaction of needs, so much so that 40% of food resources in developed countries is transformed into garbage.

Strengthening, weakening, and control actions: we shall let the reader imagine the multiple strengthenings and weakenings which in various ways society places on the system under study. The macro control for the system must operate on the basic search, on the level of schooling, and on production trends. The micro control operates on the individual's upbringing and values.

2.6.4 "Scientific-and-Technological-Progress" System

The process involving scientific and technological research, which has occurred since the dawn of civilization, but which only recently has progressed with a certain consistency, seems to be a "one-way" trend. Each new discovery—a new scientific law or theory or technological invention—is part of a cultural patrimony and contributes to new discoveries that take place thanks to the store of knowledge accumulated until that point. The new discoveries contribute to forming a stock of knowledge, but this in turn contributes to new discoveries, following a typical feedback process of combinatory systems.

A scientific or technological discovery usually arises "by chance," even if it often results from teamwork and concludes a research program undertaken specifically to reduce the scientific or technological knowledge gap; but if the discovery is useful, it increases the stock of knowledge and makes other research necessary, thereby setting off feedback between the micro and macro behavior. We can thus try to explain the mechanism involving the growth of scientific and technological knowledge in terms of combinatory systems according to the following heuristic model (Fig. 2.47).

Micro behavior and necessitating rule: if you are motivated to undertake scientific or technological research and you discover lacunae in your knowledge or need to complete or enrich the patrimony of your scientific and technological culture, dedicate yourself to research and try to make new discoveries.

Macro behavior and recombining rule: the new discoveries are appreciated by society because they bring progress to the state of science and technology; if, on the one hand, they enrich the stock of scientific and technological knowledge, on the other, they bring out knowledge shortfalls that give a push to new research, generating the *micro-macro feedback* mechanism.

Micro-macro feedback. Chance and necessity: the results of scientific and technological research form a patrimony of knowledge which, in turn, creates a need for new research; a new scientific or technological discovery usually arises "by chance" but, if held to be useful, increases the stock of knowledge and "by necessity" leads to new research, thereby triggering feedback between the micro and macro behavior.

Strengthening, weakening, and control actions: innovations in areas judged to be useful (computers, lasers, energy, Artificial Intelligence, space missions, medicine, etc.) are stimulated, which represents a powerful strengthening force. Research undertaken in areas held to be ethically suspect (human genetic engineering) undergo powerful weakening pressures from public opinion.

Thomas Kuhn's paradigmatic hypothesis (Kuhn, 1962) is an "operational" variant of this combinatory system. According to Kuhn, scientific progress is guided by *paradigms*—that is, by structures of conjectures, postulates, and laws which are accepted by all those scientists who want to be part of a given scientific community—which, after a pre-science phase, are accepted by scientists, thereby giving rise to a period of normal science.

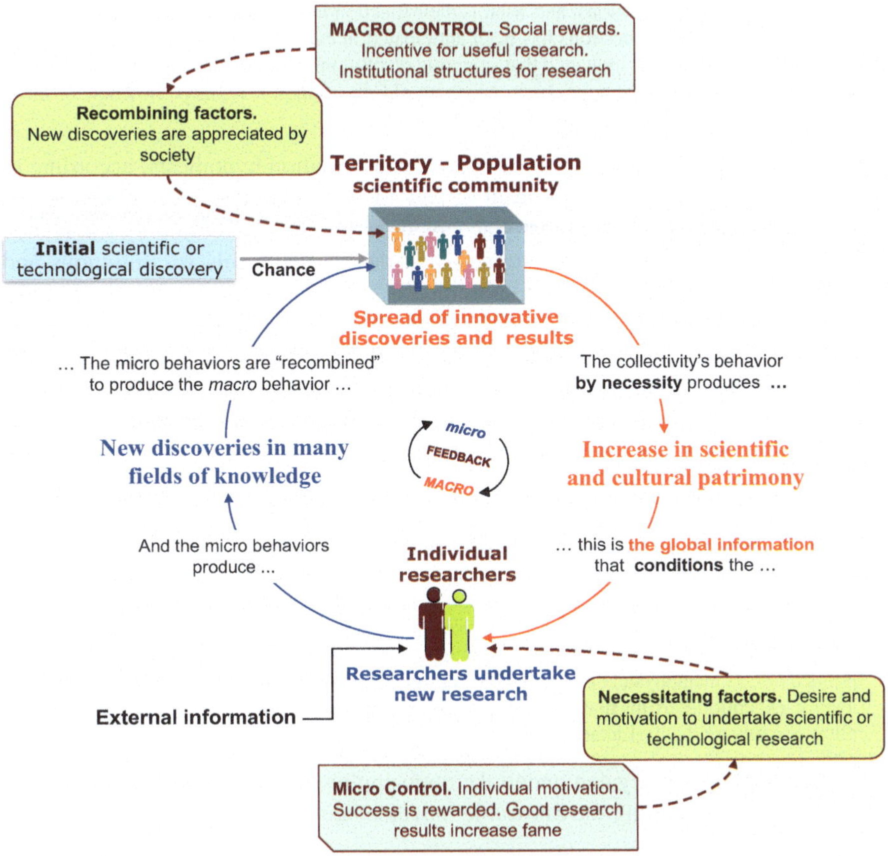

Fig. 2.47 Graphic model of the "scientific and technological progress" system

> Close historical investigation of a given specialty at a given time discloses a set of recurrent and quasi-standard illustrations of various theories in their conceptual, observational, and instrumental applications. These are the community's paradigms, revealed in its textbooks, lectures, and laboratory exercises. By studying them and by practicing with them, the members of the corresponding community learn their trade. The historian, of course, will discover in addition a penumbral area occupied by achievements whose status is still in doubt, but the core of solved problems and techniques will usually be clear. Despite occasional ambiguities, the paradigms of a mature scientific community can be determined with relative ease. (Kuhn, 1962, p. 43)

A scientific revolution occurs when a paradigm is no longer able to solve scientific problems and is replaced by another, with the formation of a new scientific community.

> What are scientific revolutions, and what is their function in scientific development? Much of the answer to these questions has been anticipated in earlier sections. In particular, the preceding discussion has indicated that scientific revolutions are here taken to be those non-cumulative developmental episodes in which an older paradigm is replaced in whole or in part by an incompatible new one. (Kuhn, 1962, p. 92)

Imre Lakatos (1978)—starting from Popper's falsification hypothesis, according to which science is a continual proposal of new theories that have undergone the test of experience and been temporarily accepted or definitively refuted by the results achieved, so that the scientific progress that derives from this is continuous and guided by logic, leading science to continually evolve—states that it is never individual suppositions which are falsified but vast theories. Theories are considered as *research programs*, open structures able to offer a program by which scientific research can be carried out. In the research program, we can identify a negative heuristic—formed by the key theories at the base of the theory—and a positive heuristic represented by the open lines of research.

> The most important such series in the growth of science are characterized by a certain continuity which connects their members. This continuity evolves from a genuine research programme adumbrated at the start. The programme consists of methodological rules: some tell us what paths of research to avoid (negative heuristic), and others what paths to pursue (positive heuristic). Even science as a whole can be regarded as a huge research programme with Popper's supreme heuristic rule: 'devise conjectures which have more empirical content than their predecessors.' Such methodological rules may be formulated, as Popper pointed out, as metaphysical principles. For instance, the universal anti-conventionalist rule against exception-barring may be stated as the metaphysical principle: 'Nature does not allow exceptions' [...]. (Lakatos, 1978, p. 47)

The scientific community can undertake opposing research programs, with the program capable of widening the knowable universe being maintained.

> According to my methodology the great scientific achievements are research programmes which can be evaluated in terms of progressive and degenerating problem shifts; and scientific revolutions consist of one research programme superseding (overtaking in progress) another. This methodology offers a new rational reconstruction of science. It is best presented by contrasting it with falsificationism and conventionalism, from both of which it borrows essential elements. (Lakatos, 1978, p. 110)

This hypothesis can also be interpreted in light of the combinatory system presented in the text. A paradigm represents the knowledge base from which new research develops, which leads to new results that are often in contrast with those obtained by the normal science of that particular period. This leads more and more researchers to try to produce new results, until a new paradigm to guide research is outlined. If this new research program is able to better explain the results obtained, then it substitutes the preceding one. The paradigm is the macro effect of the micro behavior represented by the research by scientists belonging to the same scientific collectivity (community); however, this macro effect guides the subsequent micro behavior, following a typical *micro-macro feedback*. The search for individual improvement leads to new micro effects (new discoveries), which inevitably lead scientists to produce a continual progress in the collective macro effect.

2.6.5 *"Survival and Expansion" System. One population*

Collective phenomena are fundamental to life at all levels, involving humans and other living organisms (animals, plants, fungi, microorganisms, etc.), which function in groups, populations, societies, and various collective forms (nations, tribes, crowds, herds, flocks, schools, etc.). Using the endogenous viewpoint introduced by von Bertalanffy, it becomes evident that any biological collective is made up of individual entities that are distinct from the overall groups they belong to (von Bertalanffy, 1968). Collectives and individuals follow unique dynamics and can be managed differently (Mella, 2017). In this context, a population refers to a collective of similar organisms, typically from the same species, inhabiting a specific geographical region, capable of reproducing through crossbreeding with other members of the same group. When multiple populations, even of different species, coexist and interact in a shared environment for living space and food, they form a *biological community*, creating an *ecosystem*.

As Fritjof Capra explains, early ecologists and those studying organismic biology often used similar terminology, which led to biological communities being likened to organisms. In 1935, plant ecologist Arthur George Tansley coined the term "ecosystem" to describe both animal and plant communities (Capra, 1996; Willis, 1997).

> Tansley (1935) coined the term "ecosystem" to recognize the integration of the biotic community and its physical environment as a fundamental unit of ecology, within a hierarchy of physical systems that span the range from atom to universe. (Schowalter, 2016, p. 368)

Today, the idea of an ecosystem refers to "*a community of organisms and their physical environment interacting as an ecological unit*" (Capra, 1996, p. 33). It is evident that when species are introduced into a specific ecosystem, they typically *survive* and *thrive*, benefiting from the available resources and the *strengthening* and *weakening* pressures of the environmental system (Mella, 2021).

A species can be viewed as a Combinatory System of biological individuals, reproducing once they begin to collectively procreate. The continuity of "autopoiesis" (Maturana & Varela, 1980) compels individual organisms to reproduce, and this micro-behavior maintains the species (the combinatory system). At the same time, the overall dynamics of the species influence the reproductive patterns of individuals, reflecting a typical feedback process within the combinatory system. We can distinguish between:

A. Combinatory Systems that ensure the maintenance and expansion of a species, which we will refer to as Combinatory Systems of quantitative dynamics; these will be the focus of this section and concern the quantitative dimensions of populations.
B. Combinatory Systems that drive the evolutionary dynamics of species when some mutation occurs. These concern the qualitative dimensions of populations. The Combinatory Automata used to simulate the dynamics of these systems will be presented in Sects. 3.6.5 and 3.6.6.

In many instances, *quantitative* and *evolutionary* dynamics are interrelated (Buchanan, 2000). A population in numerical decline can change its trajectory when a significant "mutation" spreads rapidly among individuals, enhancing either defensive capabilities of the population if it is *prey*, or its offensive traits if it is a *predator*. As this altered population interacts with others, its evolutionary changes inevitably impact the dynamics of the populations it interacts with. The predator–prey system is a key example of how qualitative traits influence population dynamics, where any interaction involving these traits results in quantitative fluctuations (Glass, 1975; Kareiva et al., 1990).

In many situations, the "natural" self-sustaining *quantitative* dynamics—those generated by the unconscious actions of individuals within the population—can be managed by *external interventions*. These interventions may support or suppress growth, or reverse population declines, as seen with invasive species (suppressing growth) or useful but endangered species (halting declines). Evidence of this is found in the numerous methods of cultivation, eradication, breeding, and hunting practices that exist. In other scenarios, human activity significantly alters population dynamics, either through protection and tolerance that leads to overpopulation (deer, rats), or through actions that endanger species (e.g., sea turtles losing their eggs or whales being hunted globally).

The Combinatory System that maintains and increases a population is described in the following *heuristic model*.

Micro-behavior and necessitating rule: if you have no intellect (in the common meaning of the term), follow your instinct and reproduce with a partner who might guarantee the greatest reproductive success; if you possess an intellect, choose a partner who might guarantee the greatest reproductive success.

Macro-behavior and recombining rule: the presence of many individuals of the same species, or with the same features, makes reproduction easier; for man the presence of many individuals as a rule increases the stimulation to reproduce.

Micro-macro feedback. Chance and necessity: the spread of the species is the result of the reproductive micro-behavior of individuals, but this conditions future reproductive micro-behavior. The species, or its variant, is born "by chance" when "triggered" by a *minimum activation density* of individuals; however, once the expansive phenomenon is initiated, it is then maintained "by necessity" until a *maximum saturation density* is reached that is compatible with the co-existence of other species of *predators* and the availability of resources.

Strengthening, weakening, and control actions: species enjoy varying degrees of survival capacity (teleonomy). In many ecosystems, species deemed to be dangerous are fought against, and their existence is weakened (horse flies and Anopheles mosquitos are considered damaging species; thus, considerable weakening actions arise in the system). The useful species are preserved (ants raise aphids) or increased through appropriate reinforcement actions (agriculture and farming are the most evident examples). A significant strengthening element for the maintenance and increase of species is the facility with which permanent habitations arise (clay nests, skyscrapers, dens). Humans have the highest survival rate (nowadays, heart transplants are even performed on fetuses just a few months old, with the aim of preserving life in the human species. In certain situations, this results in what can be described as futile care or unnecessary therapeutic persistence).

When there are no limits to growth and no interaction with other populations, a population of size A_n, at "time" "*n*," with an *increasing* (if positive) *rate*, "ri = rb – rd," in the difference between the *birth rate*, "br," and *death rate*, "dr," tends to grow infinitely. This is indicated in the model of a simple *reinforcing loop* in Fig. 2.48, which assumes that the periodic increase in population is added to the pre-existing population based on the *rate of accumulation*, "racc," which is usually set at 1. If a part of the population increase dies from various environmental causes, than racc <1. If there were no *limits to growth*, a population would result that could populate a very vast territory.

It is easy to derive the law of growth, which takes the following form:

$$A_{n+1} = A_n + (\mathrm{ir}\ A_n)\,\mathrm{racc}$$

Setting racc = 1 and generalizing, after specifying the number of initial individuals, A_0, we obtain the well-known *law of accumulation* for each "*n*":

$$A_n = A_0\,(1+\mathrm{ri})^n$$

Unrestricted growth is uncommon due to limiting factors such as territorial expansion, food availability, and the presence of competing populations. Cases of near-unlimited, rapid growth are found only in large areas with abundant food and few predators. A notable example is the rapid expansion of rabbits in Australia, which has become a significant ecological issue. Rabbits were introduced to Australia at the end of the eighteenth century (likely around 1788) to provide meat for the settlers. By the mid-nineteenth century, 24 rabbits were imported from England, together with other species, primarily for sport hunting. These rabbits, being particularly fast and strong, bred with the existing local rabbits, resulting in a rapid population boom. This triggered a classic rabbit *population explosion*. Efforts to control them through hunting, warren destruction, and introducing the Myxoma virus, among other methods, proved unsuccessful (PestSmart Connect, online, n.d.; Dettre, 2022).

This population explosion for rabbits follows the typical reinforcing *Loop* in Fig. 2.48, which produces an exponential increase, clearly shown in the deterministic automaton in Fig. 2.49.

For the construction of the automaton, I have introduced the (didactic) assumptions a–f (note that the simulation model allows for these assumptions to be modified):

(a) For simplicity's sake, the population of rabbits is indicated by the variable X_n and its increase by Y_n; the growth rate of Y is indicated as a function of X by $g(Y/X)$.
(b) A period of 150 years; an initial population of $X_0 = 24$ pairs; each of the newly imported specimen pairs with one already present.

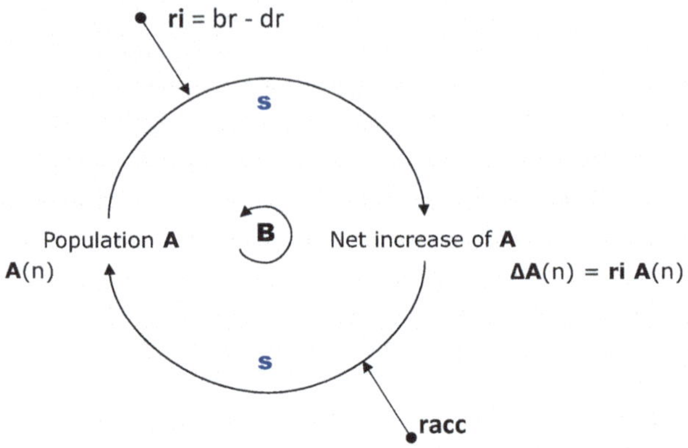

Fig. 2.48 Reinforcing loop producing an increase in a population without limits to growth: "br" = birth rate; "dr" = death rate; "ir" = increasing rate (if positive). "racc" = rate of accumulation

(c) A rate of population increase (fecundity rate-mortality rate) at the start of the process of $g(Y/X) = 25\%$; it is assumed that, partly as a result of the attempts by authorities to limit the phenomenon, the rate "g" diminishes with the increase in population: $g_{99} = 10\%$; $g_{113} = 2{,}5\%$.

(d) It is assumed that for each generation, only a percentage $h = 75\%$ of the new individuals contributes to increasing the population; the remaining ones are prey for the predators.

(e) We assume there are 5 pairings per year; this datum is inserted into the model by simply setting $r = 1/5 = 0.20$ to indicate that in one year the population undergoes a six-fold increase;

(f) Several catastrophic events have been hypothesized for the rabbit population; at $n = 130$, a cataclysm reduces the population by 250 million pairs in a single year. In the three-year period from $n = 134$ to $n = 136$, the *Rabbit haemorrhagic disease* epidemic reduces the rabbit population by 100 million pairs in each year.

The model shows that after 150 years—the equivalent of 750 generations—the rabbit population reaches about $X_{150} = 2.535$ billion pairs. Australia, with a total surface area of 7,617,930 km^2, is estimated to reach an average population density of 322 pairs of rabbits per square kilometer after 150 iterations of the model. If we assume that authorities manage to introduce new rabbit predators, reducing the number of newborn rabbits not yet of reproductive age to 50% on average, even without any catastrophic events, the population density would decrease to just one and a half pairs of rabbits per square kilometer. Of course, the model is presented for the sole purpose of illustrating the action of the *reinforcing loop* in Fig. 2.48, representing a typical single population growth phenomenon. With knowledge of the true birth and death rates of Australian rabbits, we could refine the model's results. The validity of the population explosion model applies to all "populations" with exponential growth rates.

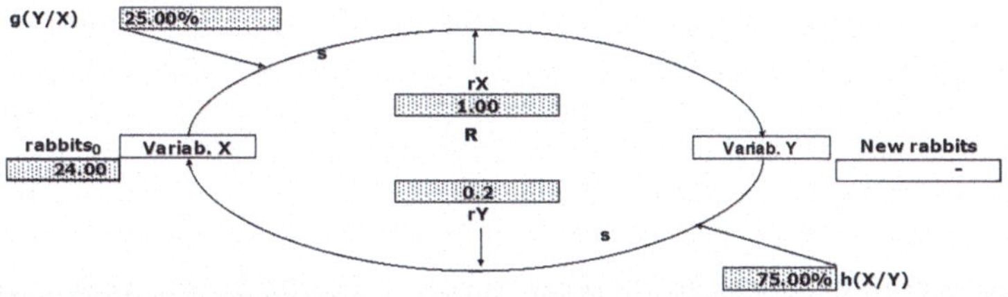

time	Rabbits	g(Y/X)	ΔY	New rabbits	h(X/Y)	ΔX	External events
0	24	25.0%	6.0	6.0	75.0%	4.5	
1	29	25.0%	7.1	7.1	75.0%	5.3	
2	34	25.0%	8.5	8.5	75.0%	6.3	
3	40	25.0%	10.0	10.0	75.0%	7.5	
4	48	25.0%	11.9	11.9	75.0%	8.9	
5	57	25.0%	14.2	14.2	75.0%	10.6	
6	67	25.0%	16.8	16.8	75.0%	12.6	
7	80	25.0%	20.0	20.0	75.0%	15.0	
8	95	25.0%	23.7	23.7	75.0%	17.8	
9	113	25.0%	28.2	28.2	75.0%	21.1	
10	134	25.0%	33.5	33.5	75.0%	25.1	
OMISSIS ...							
126	2,058,434,420	2.5%	51,460,860.5	51,460,860.5	75.0%	38,595,645.4	
127	2,097,030,066	2.5%	52,425,751.6	52,425,751.6	75.0%	39,319,313.7	
128	2,136,349,379	2.5%	53,408,734.5	53,408,734.5	75.0%	40,056,550.9	
129	2,176,405,930	2.5%	54,410,148.3	54,410,148.3	75.0%	40,807,611.2	
130	1,967,213,541	2.5%	49,180,338.5	49,180,338.5	75.0%	36,885,253.9	-250,000,000.0
131	2,004,098,795	2.5%	50,102,469.9	50,102,469.9	75.0%	37,576,852.4	
132	2,041,675,648	2.5%	51,041,891.2	51,041,891.2	75.0%	38,281,418.4	
133	2,079,957,066	2.5%	51,998,926.7	51,998,926.7	75.0%	38,999,195.0	
134	2,018,956,261	2.5%	50,473,906.5	50,473,906.5	75.0%	37,855,429.9	-100,000,000.0
135	1,956,811,691	2.5%	48,920,292.3	48,920,292.3	75.0%	36,690,219.2	-100,000,000.0
136	1,893,501,910	2.5%	47,337,547.8	47,337,547.8	75.0%	35,503,160.8	-100,000,000.0
137	1,929,005,071	2.5%	48,225,126.8	48,225,126.8	75.0%	36,168,845.1	
138	1,965,173,916	2.5%	49,129,347.9	49,129,347.9	75.0%	36,847,010.9	
139	2,002,020,927	2.5%	50,050,523.2	50,050,523.2	75.0%	37,537,892.4	
140	2,039,558,819	2.5%	50,988,970.5	50,988,970.5	75.0%	38,241,727.9	
141	2,077,800,547	2.5%	51,945,013.7	51,945,013.7	75.0%	38,958,760.3	
142	2,116,759,308	2.5%	52,918,982.7	52,918,982.7	75.0%	39,689,237.0	
143	2,156,448,545	2.5%	53,911,213.6	53,911,213.6	75.0%	40,433,410.2	
144	2,196,881,955	2.5%	54,922,048.9	54,922,048.9	75.0%	41,191,536.7	
145	2,238,073,491	2.5%	55,951,837.3	55,951,837.3	75.0%	41,963,878.0	
146	2,280,037,369	2.5%	57,000,934.2	57,000,934.2	75.0%	42,750,700.7	
147	2,322,788,070	2.5%	58,069,701.8	58,069,701.8	75.0%	43,552,276.3	
148	2,366,340,346	2.5%	59,158,508.7	59,158,508.7	75.0%	44,368,881.5	
149	2,410,709,228	2.5%	60,267,730.7	60,267,730.7	75.0%	45,200,798.0	
150	2,455,910,026	2.5%	61,397,750.6	61,397,750.6	75.0%	46,048,313.0	

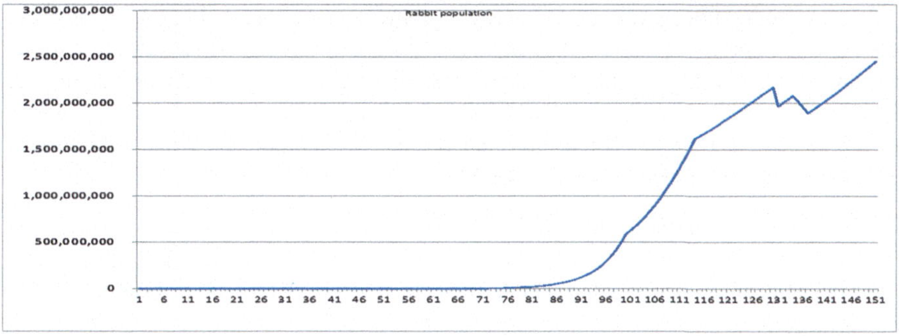

Fig. 2.49 Rabbit explosion

The *reinforcing loop* in Fig. 2.49, which regulates this phenomenon, can also represent an annihilation process if the action rate "*g*" is negative, representing a rate of decrease that is self-reinforcing to the point that *X* is eliminated.

The case of the growth of rabbits in Australia can be extended to most living beings and to the human population of some nations when it has increased too much, either due to the natural fertility of individuals or to boost the agricultural, industrial, or military workforce. In these cases, the authorities have undertaken *weakening actions* to curb population growth through legally sanctioned restrictive actions. For example, in the People's Republic of China, the one-child policy was introduced to control demographic expansion (laws enacted in 2001 prohibited families from having more than one child), coupled with a cultural preference for male children, especially in rural regions (Von der Pütten, 2008). This policy was revised in December 2013, and by 2014, Chinese couples were permitted to have two children.

In situations of a *diminishing* population, *strengthening actions* were introduced to stimulate population growth by taking actions to recover human losses suffered during wars and promote the economic and social development of the country, as occurred in Italy from the end of 1925 to the end of the First World War, even continuing in recent years. Among the many incentives, the following have been particularly effective:

- *Birth incentives*: subsidies and tax exemptions for families with many children; birth bonuses given upon the birth of a new child; housing policies that favor access to public housing;
- *Labor and social policies*: opening schools in every town, even small ones; free medical support for mothers with many children and for children; regulations to promote work that is compatible with family life; incentives for companies that hired workers with large families;
- *Healthcare and health*: creation or strengthening of healthcare facilities for maternal and child care; free or low-cost public healthcare for mothers and children; penalties for illegal abortions.

With simple modifications, the deterministic automaton in Fig. 2.49 can also represent the spread of pests—parasites, bacteria, and viruses—in a human, animal, or plant population, as indicated in the model in Fig. 2.50, which highlights the simulations of the *dynamics of individuals in a population infected by a pathogen* for a period $T = 20$ weeks (or some other period). The model is "extremely simplified," in that it considers only two variables:

(a) Variable X = number of infected individuals. Initial value of Variable X is $X_0 = 1$. The epidemic starts with a single infected individual named "individual zero." The Maximum value of Variable X is limited to X_{max} = 1 billion units (for simplicity).
(b) Variable Y = number of infected individuals.

In the *simulation*, the following rates are assumed:

(c) $g(Y/X)$, *contagion rate* (*reproduction rate* or *transmission rate*) is indicated as a percentage: $g(Y/X)$ = 200%; this means that each element of the population infects two elements at each step (delays are not introduced in the simulation).
(d) $h(X/Y)$, *rate of increase of infected individuals* in the population, $h(X/Y)$ = 100%: this rate indicates how many of the newly infected individuals increase the infected population; if $h(X/Y)$ < 100%, many of the infected recover before infecting other individuals.
(e) The reaction times of the variables are $rX = rY = 1$, which means no delays.

The simulation demonstrates that, given such extremely high rates (which are unrealistic for human populations), the entire population would be infected by the end of Week 19. To make the model more realistic, we can incorporate the response of authorities to the spread of infection. For human populations, we can first assume the implementation of travel restrictions, mandatory use of containment measures (such as masks), and social distancing protocols, all of which help lower the transmission rate. Additionally, we can assume that hospitals are adequately prepared to treat the infected, which would lead to a significant reduction in the infection rate as well.

The model in Fig. 2.50, depicting the explosive dynamics of individuals infected by a virus spread seems unrealistic since it fails to account for virus containment measures and the decline in the number of infected individuals brought about by public and private entities (Burgiel, 2020). In reality, the "General Law of Dynamic Instability" asserts that any model built using Systems Thinking cannot solely be based on reinforcing (or balancing) loops without also incorporating balancing (or reinforcing) loops that offset the primary loop's dynamics (Mella, 2012, p. 73).

Figure 2.51 indicates that to counteract expansive dynamics, some external intervention may be sufficient to produce a "braking action" that:

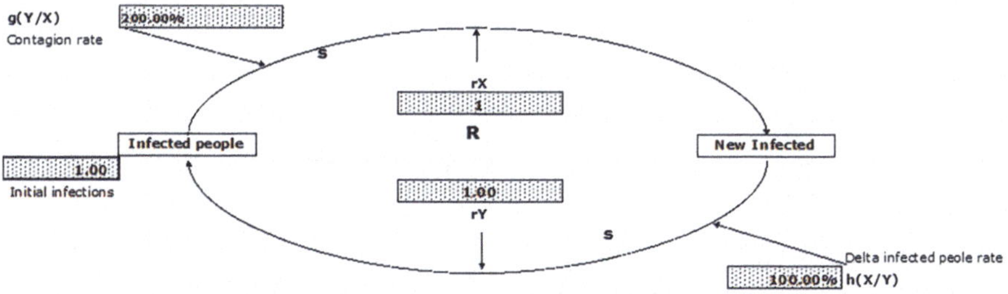

	Variab. X			Variab. Y		
Weeks	Infected people	g(Y/X)	ΔY	New Infected	h(X/Y)	ΔX
0	1.0	200.0%		-	100.0%	-
1	1.0	200.0%	2.0	2.0	100.0%	2.0
2	3.0	200.0%	6.0	6.0	100.0%	6.0
3	9.0	200.0%	18.0	18.0	100.0%	18.0
4	27.0	200.0%	54.0	54.0	100.0%	54.0
5	81.0	200.0%	162.0	162.0	100.0%	162.0
6	243.0	200.0%	486.0	486.0	100.0%	486.0
7	729.0	200.0%	1,458.0	1,458.0	100.0%	1,458.0
8	2,187.0	200.0%	4,374.0	4,374.0	100.0%	4,374.0
9	6,561.0					13,122.0
10	19,683.0					39,366.0
11	59,049.0					118,098.0
12	177,147.0					354,294.0
13	531,441.0					1,062,882.0
14	1,594,323.0					3,188,646.0
15	4,782,969.0					9,565,938.0
16	14,348,907.0					28,697,814.0
17	43,046,721.0					86,093,442.0
18	129,140,013.0					258,280,026.0
19	387,420,039.0					774,840,078.0
20	1,162,260,117.0					2,324,520,234.0

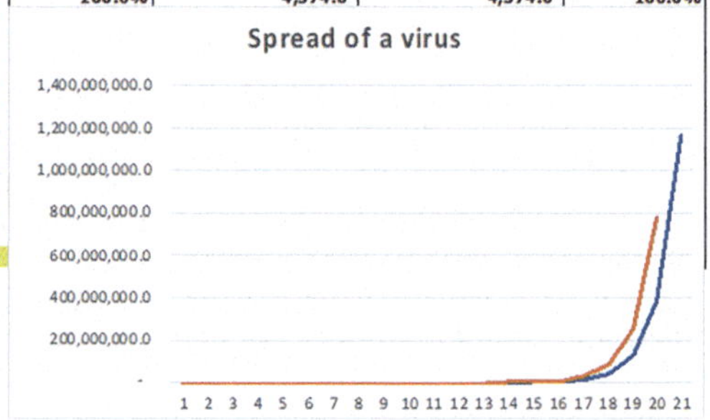

Fig. 2.50 Spread of a virus simulated by a very simplified linear automaton

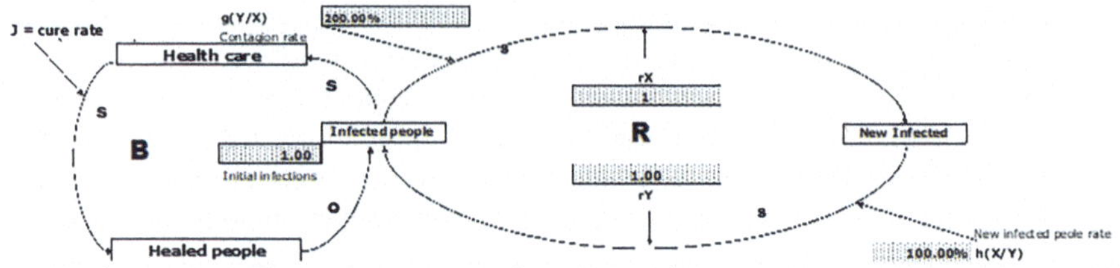

Weeks	Variab. X Infected people	g(Y/X)	ΔY	Variab. Y New Infected	h(X/Y)	ΔX	cure rate	Healed
0	1.0	200.0%		-	100.0%	-	0%	
1	1.0	200.0%	2.0	2.0	100.0%	2.0	0%	0.0
2	3.0	200.0%	6.0	6.0	100.0%	6.0	0%	0.0
3	9.0	200.0%	18.0	18.0	100.0%	18.0	40%	3.6
4	23.4	200.0%	46.8	46.8	100.0%	46.8	40%	9.4
5	60.8	200.0%	121.7	121.7	100.0%	121.7	40%	24.3
6	158.2	200.0%	316.4	316.4	100.0%	316.4	40%	63.3
7	411.3	200.0%	822.6	822.6	100.0%	822.6	50%	205.6
8	1,028.2	200.0%	2,056.4	2,056.4	100.0%	2,056.4	50%	514.1
9	2,570.5	200.0%	5,141.0	5,141.0	100.0%	5,141.0	50%	1285.2
10	6,426.2	200.0%	12,852.5	12,852.5	100.0%	12,852.5	50%	3213.1
11	16,065.6	200.0%	32,131.1	32,131.1	100.0%	32,131.1	50%	8032.8
12	40,162.9	150.0%	60,245.9	60,245.9	100.0%	60,245.9	50%	20082.0
13	80,327.8	150.0%	120,491.7	120,491.7	50.0%	60,245.9	50%	40163.9
14	100,409.8	150.0%	150,614.6	150,614.6	50.0%	75,307.3	70%	70286.8
15	105,420.3	150.0%	158,145.4	158,145.4	50.0%	79,072.7	70%	73801.2
16	110,701.8	80.0%	88,561.4	88,561.4	50.0%	44,280.7	70%	77491.2
17	77,491.2	80.0%	61,993.0	61,993.0	50.0%	30,996.5	75%	58118.4
18	50,369.3	80.0%	40,295.4	40,295.4	50.0%	20,147.7	75%	37777.0
19	32,740.0	80.0%	26,192.0	26,192.0	50.0%	13,096.0	75%	24555.0
20	21,281.0	80.0%	17,024.8	17,024.8	50.0%	8,512.4	75%	15960.8
21	13,832.7	80.0%	11,066.1	11,066.1	50.0%	5,533.1	90%	12449.4
22	6,916.3	80.0%	5,533.1	5,533.1	50.0%	2,766.5	90%	6224.7
23	3,458.2	80.0%	2,766.5	2,766.5	50.0%	1,383.3	90%	3112.4
24	1,729.1	80.0%	1,383.3	1,383.3	50.0%	691.6	90%	1556.2
25	864.5							778.1
26	432.3							389.0
27	216.1							194.5
28	108.1							97.3
29	54.0							48.6
30	27.0							24.3
31	13.5							12.2
32	6.8							6.1
33	3.4							3.0
34	1.7							1.5
35	0.8							0.8
36	0.4							0.4
37	0.2							0.2
38	0.1							0.1
39	0.1							0.0
40	0.0							0.0
41	0.0							0.0
42	0.0							0.0

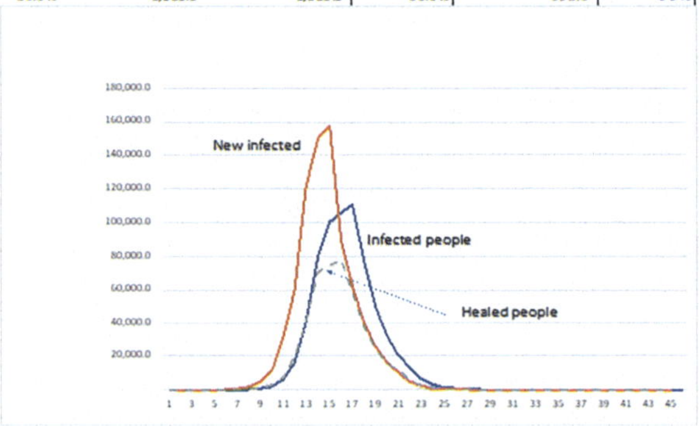

Fig. 2.51 Counteracting the spread of a virus (Source: Mella, 2021, p. 45)

1. Reduces the *contagion* rate, *g*(*Y*/*X*), or Rt (red zones, social distancing, immunizing equipment, masks, tampons, etc.)
2. Reduces the *infected* rate, *h*(*X*/*Y*) (enhanced immune defenses, a spontaneous cure, etc.)
3. Introduces the *cure rate*, "*j*," which expresses the effort to enhance and improve *health care* (s), which leads to infected individuals being cured (s), so that the increase in *cured individuals* reduces the number of the *infected*. This important "braking factor" has been accounted for by adding loop B to the model.

I have modified the simulation in Fig. 2.50 by introducing the following hypotheses:

- The *contagion*, or infection, rate, $g(Y/X)$, which initially had a value of 2 (each infected person in turn infected two healthy individuals), has been gradually reduced to the values that appear in the third column.
- The *infection* rate, which had a value of 100% (all the newly infected increased the number of infected persons), has gradually been reduced to the values seen in the sixth column.
- The *cure* rate has the values indicated in the penultimate column.

As shown in Fig. 2.51, these assumptions for the simulation lead to the infected population dropping to zero by Week 40. *Naturally*, a simulation of real-world scenarios would need to consider additional factors that influence infection dynamics. Most importantly, it should incorporate growth rates for contagion, infection, and recovery, which should be determined by means of thorough epidemiological and health studies. Despite its simplicity, the model can also be applied to various other pandemics, including non-viral ones, such as, to name but a few, epidemics of the plague, cholera, measles, lice outbreaks in schools, and Xylella Fastidiosa, which has caused devastation to ancient olive trees in southern Italy's Apulia region and currently has no known cure.

2.6.6 "Survival and Expansion" with Limits to Available Resources

The model in Fig. 2.49 is unrealistic: no population grows infinitely, since growth is always balanced by at least *three* factors operating individually or jointly: (1) a maximum limit of available resources, (2) the presence of a prey–predator population, (3) human external controls.

Let us consider the first limiting factor to the growth of a population: a maximum limit of resources required for consumption. When we consider that individuals need resources to survive, and these resources are finite, population growth cannot continue indefinitely. Instead, it is restricted by the number of resources available, which sets a boundary to growth, as illustrated in the model in Fig. 2.52. Referring back to Malthus's theories, population growth is predicted to reach a point of stabilization, determined by resource limitations (Malthus, 1798). Despite numerous revisions over time, his core idea about the limits of population growth and its impact on quality of life remains relevant.

Figure 2.53 illustrates a standard example of *population dynamics*, where an initial population size (set at 1000 units) is given, and this population changes annually due to births and deaths. Since the birth rate is assumed to exceed the death rate, the population experiences a steady increase over time. However, there is also a limitation on the availability of food, which grows annually by a fixed amount, suggesting that food is a *renewable resource* but cannot exceed a maximum threshold. We assume at first that the total amount of available resources (10,000 units) surpasses what is needed for consumption (5000 units). This allows the population to

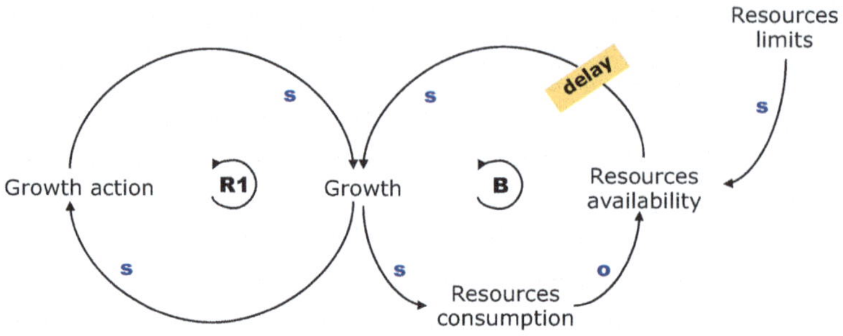

Fig. 2.52 Available resources limit growth

grow until it reaches the limit imposed by the finite resources, which are entirely depleted by time $t = 16$.

The simulation program also considers the scenario where the population, after reaching the saturation point of available resources, initiates a *production process* to increase those resources and continue growing. However, this growth is short-lived as it is assumed that there is a cap (10,000 units) on resource expansion, which could be due to limitations such as the need for other resources (land, water, fertilizer, etc.) for production. Consequently, population growth ends at time $t = 16$. In summary, based on the data input into the control panel, the population will grow until $t = 9$ using the available food, after which a production process will boost resources to 10,000 units. Following this, from $t = 19$, the population will stabilize at 2062 units.

This straightforward simulation program also allows for other assumptions, such as the possibility that once resource saturation occurs, there could be a decrease in the birth rate or an increase in the death rate. Another assumption could involve unit consumption through rationing. In summary, based on the data entered in the control panel, the population grows until time $t = 9$ using the available natural resources, at which point a resource production process begins, raising the available amount to 10,000 units. After that, starting at $t = 17$, the population stabilizes at around 2062 units If there were no restrictions on food production, the population would continue to grow indefinitely, constantly producing more food in an endless cycle, as depicted in Fig. 2.54.

However, when a cap is placed on the renewable resource needed for food production (set at 20,000), the system inevitably moves toward this limit. Production begins at $t = 9$, at which point the available food in the system, which can be replenished annually (in quantities of 7500 units), is fully consumed. Food production then increases gradually beyond 7500 units until $t = 22$, when the scarce resource is completely exhausted (as shown in the last three columns of the table in Fig. 2.54). After this point, food production halts, and the population rapidly decreases to 1579 units by $t = 26$.

		reaction time	1	
initial population	1,000	production	5.0	
birth rate	10%	if famine	9%	negative percentage
mortality rate	5%	if famine	6%	negative percentage
per capita food cons.	5.0	if famine	5.0	negative percentage
food as resource	7,500	maximum food	10,000	(about)

years	old populat. + births - deaths	birth rate	births	mortality rate	deaths	unit food consumption	food requirement	food available	humanitarian aid	productivity	production	food (surplus /deficit)	years
0	**1,000.0**	10.0%	100	5.0%	50	5.0	5,000	7,500		5.0	-	-2500	0
1	**1,050.0**	10.0%	105	5.0%	53	5.0	5,250	7,500		5.0	-	-2250	1
2	**1,102.5**	10.0%	110	5.0%	55	5.0	5,513	7,500		5.0	-	-1988	2
3	**1,157.6**	10.0%	116	5.0%	58	5.0	5,788	7,500		5.0	-	-1712	3
4	**1,215.5**	10.0%	122	5.0%	61	5.0	6,078	7,500		5.0	-	-1422	4
5	**1,276.3**	10.0%	128	5.0%	64	5.0	6,381	7,500		5.0	-	-1119	5
6	**1,340.1**	10.0%	134	5.0%	67	5.0	6,700	7,500		5.0	-	-800	6
7	**1,407.1**	10.0%	141	5.0%	70	5.0	7,036	7,500		5.0	-	-464	7
8	**1,477.5**	10.0%	148	5.0%	74	5.0	7,387	7,500		5.0	-	-113	8
9	**1,551.3**	10.0%	155	5.0%	78	5.0	7,757	7,757		5.0	256.6	0	9
10	**1,628.9**	9.0%	147	5.0%	81	5.0	8,144	8,144		5.0	644.5	0	10
11	**1,694.1**	9.0%	152	5.0%	85	5.0	8,470	8,470		5.0	970.3	0	11
12	**1,761.8**	9.0%	159	5.0%	88	5.0	8,809	8,809		5.0	1,309.1	0	12
13	**1,832.3**	9.0%	165	5.0%	92	5.0	9,161	9,161		5.0	1,661.4	0	13
14	**1,905.6**	9.0%	172	5.0%	95	5.0	9,528	9,528		5.0	2,027.9	0	14
15	**1,981.8**	9.0%	178	5.0%	99	5.0	9,909	9,909		5.0	2,409.0	0	15
16	**2,061.1**	9.0%	185	5.0%	164	5.0	10,305	10,000		5.0	2,500.0	305	16
17	**2,082.4**	9.0%	187	6.0%	207	5.0	10,412	10,000		5.0	2,500.0	412	17
18	**2,062.5**	9.0%	186	6.0%	186	5.0	10,312	10,000		5.0	2,500.0	312	18
19	**2,061.9**	9.0%	186	6.0%	186	5.0	10,309	10,000		5.0	2,500.0	309	19
20	**2,061.9**	9.0%	186	6.0%	186	5.0	10,309	10,000		5.0	2,500.0	309	20
21	**2,061.9**	9.0%	186	6.0%	186	5.0	10,309	10,000		5.0	2,500.0	309	21

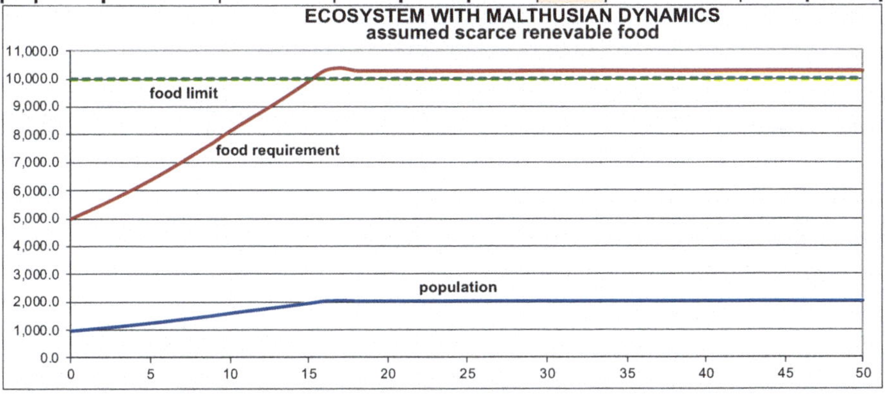

Fig. 2.53 Ecosystem with Malthusian dynamics and scarce nonrenewable resources (for a period of 21k years)

Is there a possibility for humanitarian aid? Certainly! However, the simulation indicates that such aid—starting at $t = 35$ and lasting a decade—would only have a temporary impact. Once this aid stops, at $t = 48$, the *population* reverts to the *maximum limit* of 1579 units. This second scenario, though simplified, underscores the fact that population growth and the resulting increase in production must account for the limitations imposed by the sustainability of the ecosystem.

		reaction time	1	
initial population	1,000	production	5.0	
births rate	10%	if famine	10%	negative percentage
deaths rate	5%	if famine	5%	negative percentage
per capita food cons.	5.0	if famine	5.0	negative percentage
food as resource	7,500			
cons. of resource	1	maximum depletable resource	20000	non-replenishable

years	old population + births – deaths	birth rate	births	mortality rate;	deaths	unit food consumption	food requirement	food available	humanitarian aid	productivity	production	(surplus) or deficit	consumption of resource	resource available	maximum production
0	**1,000**	10%	100	5.0%	50	5.0	5,000	7,500		5.0	0	-2500	-	20,000	20,000
1	**1,050**	10%	105	5.0%	53	5.0	5,250	7,500		5.0	0	-2250	-	20,000	20,000
2	**1,103**	10%	110	5.0%	55	5.0	5,513	7,500		5.0	0	-1988	-	20,000	20,000
3	**1,158**	10%	116	5.0%	58	5.0	5,788	7,500		5.0	0	-1712	-	20,000	20,000
4	**1,216**	10%	122	5.0%	61	5.0	6,078	7,500		5.0	0	-1422	-	20,000	20,000
5	**1,276**	10%	128	5.0%	64	5.0	6,381	7,500		5.0	0	-1119	-	20,000	20,000
6	**1,340**	10%	134	5.0%	67	5.0	6,700	7,500		5.0	0	-800	-	20,000	20,000
7	**1,407**	10%	141	5.0%	70	5.0	7,036	7,500		5.0	0	-464	-	20,000	20,000
8	**1,477**	10%	148	5.0%	74	5.0	7,387	7,500		5.0	0	-113	-	20,000	20,000
9	**1,551**	10%	155	5.0%	129	5.0	7,757	7,500		5.0	257	257	257	19,743	19,743
10	**1,578**	10%	158	5.0%	105	5.0	7,888	7,757		5.0	388	131	388	19,356	19,356
11	**1,630**	10%	163	5.0%	134	5.0	8,151	7,888		5.0	651	263	651	18,704	18,704
12	**1,659**	10%	166	5.0%	112	5.0	8,295	8,151		5.0	795	144	795	17,909	17,909
13	**1,713**	10%	171	5.0%	140	5.0	8,566	8,295		5.0	1066	270	1,066	16,843	16,843
14	**1,745**	10%	174	5.0%	119	5.0	8,724	8,566		5.0	1224	158	1,224	15,620	15,620

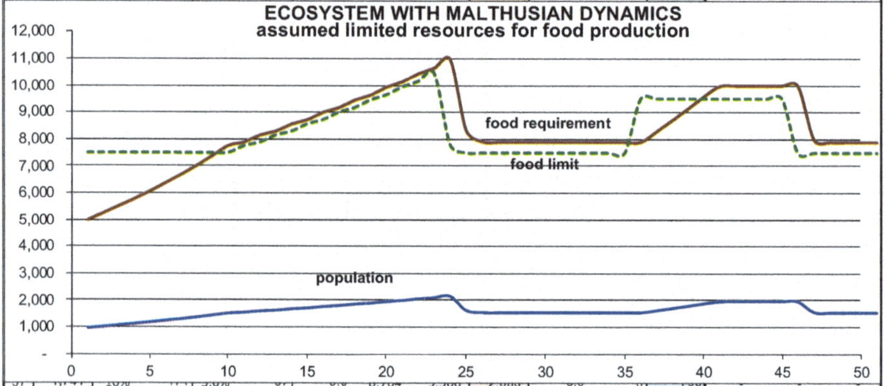

years	old population + births – deaths	birth rate	births	mortality rate;	deaths	unit food consumption	food requirement	food available	humanitarian aid	productivity	production	(surplus) or deficit	consumption of resource	resource available	maximum production
38	**1,828**	10%	183	5.0%	91	5.0	9,139	9,500	2,000	5.0	0	-361	-	-	-
39	**1,919**	10%	192	5.0%	115	5.0	9,596	9,500	2,000	5.0	0	96	-	-	-
40	**1,996**	10%	200	5.0%	196	5.0	9,980	9,500	2,000	5.0	0	480	-	-	-
41	**2,000**	10%	200	5.0%	200	5.0	9,999	9,500	2,000	5.0	0	499	-	-	-
42	**2,000**	10%	200	5.0%	200	5.0	10,000	9,500	2,000	5.0	0	500	-	-	-
43	**2,000**	10%	200	5.0%	200	5.0	10,000	9,500	2,000	5.0	0	500	-	-	-
44	**2,000**	10%	200	5.0%	200	5.0	10,000	9,500	2,000	5.0	0	500	-	-	-
45	**2,000**	10%	200	5.0%	600	5.0	10,000	7,500		5.0	0	2500	-	-	-
46	**1,600**	10%	160	5.0%	180	5.0	8,000	7,500		5.0	0	500	-	-	-
47	**1,580**	10%	158	5.0%	159	5.0	7,900	7,500		5.0	0	400	-	-	-
48	**1,579**	10%	158	5.0%	158	5.0	7,895	7,500		5.0	0	395	-	-	-
49	**1,579**	10%	158	5.0%	158	5.0	7,895	7,500		5.0	0	395	-	-	-
50	**1,579**	10%	158	5.0%	158	5.0	7,895	7,500		5.0	0	395	-	-	-

Fig. 2.54 Ecosystem with Malthusian dynamics and scarce but renewable resources (for a period of 50k years)

2.6.7 *"Survival and Expansion" for Co-existing Populations*

The second factor conditioning population growth is the co-existence with another population, where both populations occupy the same territory. According to Nguyen and Yin (2016), interactions among species may occur by means of cooperation, a predator–prey relationship, or competition for food (Nguyen & Yin, 2016; Krizna et al., 2018).

If two populations interact to produce a *trophic food-chain* (Post, 2002; Schoener, 2011), meaning they are connected by a prey–predator relationship, we can easily see that it is impossible for their population dynamics to be independent of one another. Focusing on the predator–prey relationship, the size of the prey population depends on that of the predators, and vice versa. Note, however, that the prey–predator populations, and species, in general, typically maintain a *natural* self-balance through some form of control. This balance is often the result of conflicts with one or more other populations that prevent unchecked growth. It is important to note that oceans are not solely inhabited by sharks, nor savannahs exclusively by lions, even though these are among the most effective hunters in their habitats. The seas are rich with sardines, and the savannahs are home to antelopes, despite the fact both species serve as prey for many predator populations. To interpret and potentially predict the dynamics of two prey–predator populations, it is useful to refer to the general model developed by Vito Volterra (1926, 1931) and Alfred Lotka (1925). According to this model, the dynamics of the sizes of two populations—prey and predators—denoted by An(192 Bn) in year "*n*," depend on the birth and mortality rates of both populations.

Let us assume the observations originate in year $n = 0$, when the number of prey and predators is determined (we could also have indicated the pairs). Subsequently, the populations increase and decrease year after year at a fixed rate. The *prey* (***A***) increase by a certain yearly number according to the rate "*a*" (which takes account of the difference between the natural birth rate and the natural death rate), which is applied to the population of the previous year, $\boldsymbol{A}_n$, and decrease at a rate "*b*," which expresses the average voracity of the predators (***B***), taking into account the number of predators from the previous year, $\boldsymbol{B}_n$. The *predator* population ***B*** dies out from natural causes at a death rate "*c*" and increases by feeding on the prey at a reproduction rate "*d*," which is tied to the food supply.

To make the dynamics more realistic, Volterra and Lotka proposed a variant to take into account the fact that the rate of decrease of ***A*** and of increase of ***B*** are also a function of the number of individuals in *both populations*. In fact, it is intuitively clear that the chance of being the victim of a predator depends not only on the number of predators but also on the number of preys, since a larger population means more likelihood of escaping from the hunt. Moreover, for a predator, the possibility of hunting a prey does not depend only on the number of prey but also on the size of the predator population, since the higher the number of predators, the more difficult it is to contend for the prey. In prey–predator interaction, the rate of prey capture and prey density are connected by a relationship which represents the functional response of a predator–prey system (Staddon, 2016). In fact, as Staddon proposed in 2016, the risk to which the prey is exposed and the size of its population are strictly connected and depend on the predation rate (Staddon, 2016).

The variation in the model consists precisely in setting the number of deaths in population ***A*** and births in ***B*** at a critical mass: Critical Mass = $\boldsymbol{A}_n\,\boldsymbol{B}_n$, which is given by the product of the number of prey and predator populations:

$$\begin{cases} A_{n+1} = A_n + a\,A_n - b\,A_n B_n \\ B_{n+1} = B_n - c\,B_n + d\,A_n B_n \end{cases}$$

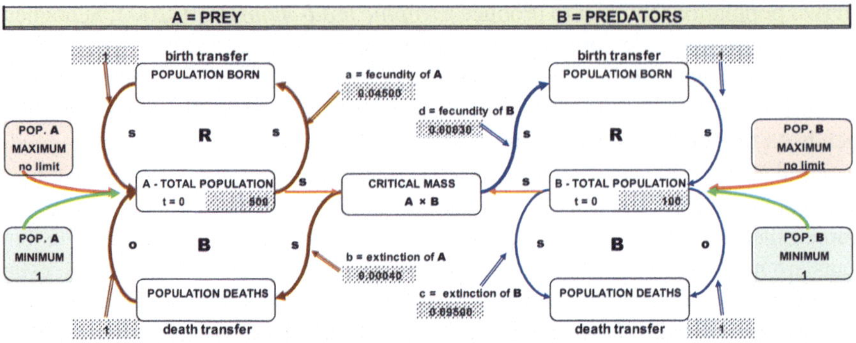

POPULATIONS			PREY		CRITICAL MASS	PREDATORS	
year (n)	PREY A	PREDATORS B	births a A	deaths (-) b A B	A B	births d A B	deaths (-) c B
0	500.00	100.00			50,000.00		
1	502.50	105.50	22.50	20.00	53,013.75	15.0000	9.5000
2	503.91	111.38	22.61	21.21	56,125.98	15.9041	10.0225
3	504.13	117.64	22.68	22.45	59,305.21	16.8378	10.5813
4	503.10	124.25	22.69	23.72	62,511.78	17.7916	11.1756
5	500.73	131.20	22.64	25.00	65,697.65	18.7535	11.8041
6	496.98	138.45	22.53	26.28	68,806.77	19.7093	12.4643
7	491.83	145.94	22.36	27.52	71,776.10	20.6420	13.1526

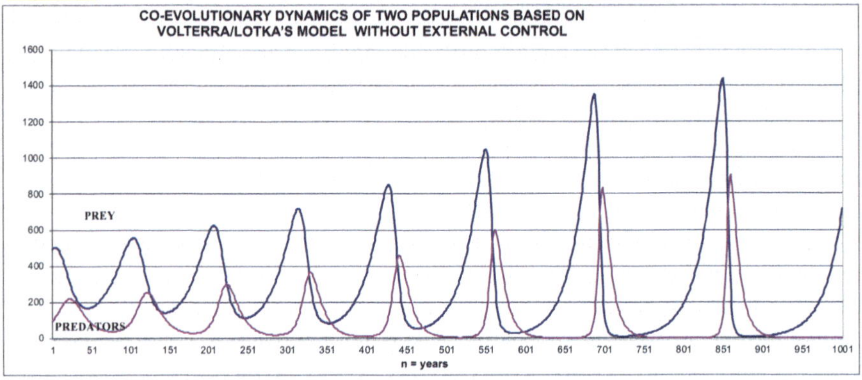

Fig. 2.55 Volterra-Lotka. Model simulating the dynamics of two populations based on Volterra-Lotka equations without external control

To simulate the dynamics of the two populations, the simple deterministic automaton in Fig. 2.55 was constructed, where both the initial number of the two populations and the rates of increase "*a*" and "*d*" and decrease "*b*" and "*c*" were chosen arbitrarily. In constructing the simulation models referring to real populations, these rates must be quantified by studying the historical dynamics registered in the archives.

If we exclude the case of initial values of zero, the populations have a mutual equilibrium when the initial values of each are proportionate to the rates of extinction and reproduction of the connected population. With the data from Fig. 2.55, the equilibrium values are determined by the following relations: $A = c/d = 316{,}7$ and $B = a/b = 112{,}5$ (Takeuchi, 1995).

The third factor that limits the growth of a population is represented by human interventions that develop external controls to set minimum and/or maximum limits on the sizes of one or both populations, carrying out, when necessary, actions of repopulation, or culling of individuals from the populations that are to be controlled. These interventions seek to balance the reciprocal numbers of the populations, resulting in a regular trend over time in their dynamics.

An example of an automaton where the prey–predator dynamics are subjected to external alterations is shown in Fig. 3.69. In Sect. 3.6.6, Fig. 3.70 presents a simulation model of a trophic food chain composed of five populations (possibly of different species).

2.6.8 "Survival and Evolution" System

In summary, the Darwinian theory of evolution, as articulated in the Huxley's "modern theory" of evolution (Huxley, 1942), is grounded on two main principles: the inheritance of mutated traits, known as "descent with modification," and "natural selection," which are key drivers of species evolution. The process can be outlined by the following postulates:

- Individuals within a species exhibit a wide range of variations and mutations as a result of chance and environmental factors.
- The variations or mutations that impact an organism's physical characteristics (the phenotype) can be transmitted to future generations.
- Individuals whose traits are "best suited" to their environment have a better chance of survival and reproduction.
- Favorable mutations are preserved and passed down to future generations.
- Individuals that are not well-adapted to their environment have a lesser chance of survival and reproduction, making it less likely that their mutations will be inherited by the next generation.

As a result, when a variation or mutation in the phenotype proves advantageous for survival, it is passed on to offspring, and over time, this process can alter the entire population, potentially leading to the emergence of a new species. On the other hand, unfavorable mutations tend to disappear. The essence of Darwin's natural selection lies in the concept of the "survival of the fittest" within the "struggle for existence," where a well-adapted phenotype in a competitive environment has a higher chance of surviving, whether as prey or predator.

In the *competition for survival*, which includes the search for food and defense against predators (Darwin, 1859, p. 43), as well as in the *competition for reproduction*—whether it be asexual, through forceful dominance over rivals, or by attracting mates through sexual selection (see Sects. 3.6.5 and 3.6.6 for further details)—reproduction (micro-behavior) and the passing on of hereditary traits take place, sometimes accompanied by genetic mutations (micro-effect). These mutations can help the species thrive and propagate (macro-behavior), while preserving the advantageous traits of the phenotype (macro-effect). However, this process is influenced by the species' overall dynamics, following a natural *feedback* loop within the combinatory system.

If *by chance* an individual is born with a random mutation that differs from the species' usual traits, and if this new characteristic provides an advantage in survival, those with the mutation will have better success in hunting and reproducing. This advantage will allow the trait to spread through future generations, eventually replacing the original trait and potentially leading to the formation and evolution (macro-effect) of a new species or a variation of the original one. This occurs through reproduction (micro-behavior) among the mutated individuals, who are likely to pass on the mutation to their offspring. The reproduction takes place because the Combinatory System "Species" exists and because of the typical *micro-macro feedback.*

Although the original and new species may coexist for a time, the new species might eventually replace the old one if they are in direct competition for resources or reproductive success (Darwin, 1859; Huxley, 1942). Evolution favors the mutated individuals of a species by allowing them to compete more effectively in the "struggle for life" and reproduction, as Darwin keenly observed.

> Owing to this struggle for life, any variation, however slight and from whatever cause proceeding, if it be in any degree profitable to an individual of any species, in its infinitely complex relations to other organic beings and to external nature, will tend to the preservation of that individual, and will generally be inherited by its offspring. [...]. I have called this principle, by which each slight variation, if useful, is preserved, by the term of Natural Selection, in order to mark its relation to man's power of selection (Darwin, 1859, p. 61). What natural selection cannot do, is to modify the structure of one species, without giving it any advantage, for the good of another species; and though statements to this effect may be found in works of natural history, I cannot find one case which will bear investigation. [...] And this leads me to say a few words on what I call Sexual Selection. This depends, not on a struggle for existence, but on a struggle between the males for possession of the females; the result is not death to the unsuccessful competitor, but few or no offspring. Sexual selection is, therefore, less rigorous than natural selection. Generally, the most vigorous males, those which are best fitted for their places in nature, will leave the most progeny [...] A hornless stag or spurless cock would have a poor chance of leaving offspring... (Darwin, 1859, p. 86–87)

Darwin was unaware that mutations in *phenotypes* were caused by changes in genotypes. It was not until 1953, when James Watson and Francis Crick discovered the structure of DNA (Pray, 2008), that the genetic basis of heredity was understood. Jacques Monod, in his renowned work "Chance and Necessity" (1970), explained that evolution is driven by "random" genetic mutations in the genotype, which "*by necessity*" are passed on through the consistent reproductive process of cells, provided the mutations offer a survival advantage; otherwise, they tend to be eliminated. This highlights the role of a typical combinatory system governed by both "*chance*" and "*necessity.*"

> A mutation represents in itself a microscopic, quantum event, to which, as a result, we apply the principle of indetermination: an event which thus is unpredictable by nature (Monod, 1970, p. 97). ... We call these events [mutations] accidental; we say they are random occurrences. And since they constitute the only possible source of modification in the genetic text, itself the sole repository of the organism's hereditary structures, it necessarily follows that chance alone is at the source of every innovation, of all creation in the biosphere. Pure chance, absolutely free but blind, at the very root of the stupendous edifice of evolution: this central concept of modern biology ... is today the sole conceivable hypothesis ... (Monod, 1970, p. 112)

Monod's thinking makes it clear how the Combinatory Systems of biological evolution evolve as a result of the combined action of "chance" and "necessity."

It is also useful to keep in mind Dawkins' view (1989, 2004) that considers the *gene* as the main unit of natural selection. He places the search for stability in genes at the center of evolution, as these genes transform the phenotypes of organisms (their "survival machines") to guarantee a greater "chance of survival" for the genes in future offspring (Dawkins, 1989, 2004).

> What I am doing is emphasizing the potential near-immortality of a gene, in the form of copies, as its defining property. (Dawkins, 1989, p. 35)

> Genes are competing directly with their alleles for survival ... Any gene that behaves in such a way as to increase its own survival chances will, by definition, tautologously tend to survive. The gene is the basic unit of selfishness. (*ibidem*, p. 36)

> An entity is said to be altruistic if it behaves in such a way as to increase another entity's welfare at the expense of its own. Selfish behaviour has exactly the opposite effect. (*ibidem*, p. 4)

> Replicators began not merely to exist but to construct containers for themselves, vehicles for their continued existence. The replicators that survived were the ones that built survival machines for themselves to live in. (*ibidem*, p. 19)

The evolution is possible because biologic individuals are "survival machines" that operate to propagate the selfish genes, with such selection being necessary for the stability and immortality of genes.

> We are survival machines—robot vehicles blindly programmed to preserve the selfish molecules known as genes. This is a truth which still fills me with astonishment. (Dawkins, 1989, p. xxi)

Two species may coexist for some time, but if they compete for resources or reproduction, the new species might eventually replace the older one. As a result, evolution can either modify the traits of a species or lead to one species being replaced by another. From the perspective of autopoietic logic, evolution results in the improvement of a species, driven by the potential for self-creation (autopoiesis) and based on specific parameters that vary over time.

> Evolutionary processes can be depicted by a dynamic activity in a redundant network of interacting regulators (each of which has to secure its own closure) involving the selective formation of linkages between initially independent loci of control. The overall context of stability will shift and actual realizations of stable local configurations in such a network continuously change as linkages are formed and reformed. While such changes depend on various "chance" events, they are not entirely random. They are "directionally" based on

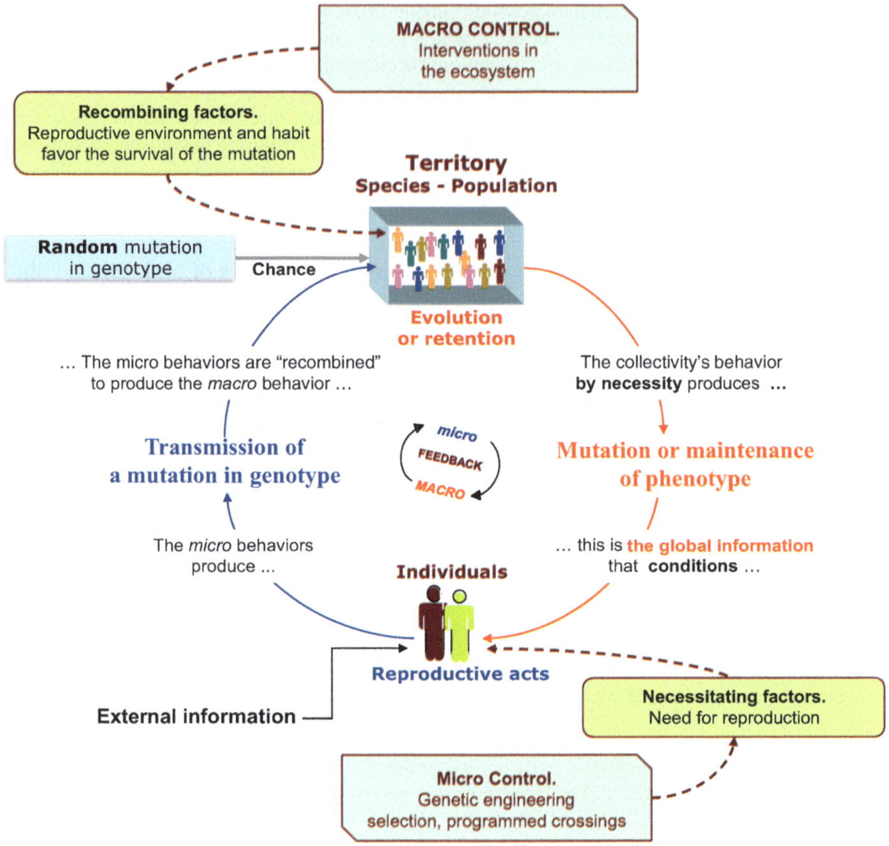

Fig. 2.56 Graphic model of the "survival-and-evolution" system

> selection processes that reinforce particularly "survival-worthy" patterns, thus altering the condition probabilities of their own further realization. As a result, evolution appears to an observer as a consistent trend involving a directed increase in organization and suggesting a built-in-drive for consistent self-improvement. Both features are fundamental to a perception of life as a process that not only maintains but also improves itself [...] Above all, evolution is regenerative and unlimited, for as long as sufficient diversity is generated locally, and sufficient distinction is maintained, among interacting viable organizations. (Ben-Eli, 1979, p. 294–95)

Combinatory Systems that produce and maintain evolution operate as described in the *heuristic model* below (which uses "technical terminology" bereft of emotional significance) and in the associated model shown in Fig. 2.56.

Micro behavior and necessitating rule: if you have no intellect (in the common meaning of the term), follow your instinct and reproduce with a partner who guarantees the greatest reproductive success; if you possess an intellect, in choosing a partner follow the rites (strengthening) that lead to social gratification (courting, getting engaged, matrimony, pregnancy, birth celebrations, etc.).

Macro behavior and recombining rule: the presence of many individuals of the same species, or with the same features, makes reproduction easier; for man, the presence of many individuals as a rule makes the stimulation to reproduce stronger.

Micro-macro feedback. Chance and necessity: the spread of the species is the result of the reproductive micro behavior of individuals, but this conditions future reproductive micro behavior. The species, or its variant, is born "by chance" when it is started by a minimum number of individuals; but once the expansive phenomenon is initiated, it is maintained "by necessity" until the saturation density is reached (compatible with the available resources and the coexistence with other species).

Strengthening, weakening, and control actions: species enjoy varying degrees of exogenous teleonomy. In man's ecosystem, horse flies and Anopheles mosquitos are considered damaging species; thus, considerable weakening actions arise in the system. The human species has the highest degree of teleonomy (today heart transplants are attempted even on fetuses only a few months old, in order to keep alive a member of the human species, and in some cases, forms are obtained that can be defined as futile care or uselessly therapeutic persistence). The ease with which permanent habitations are formed (dens, clay nests, skyscrapers) represents a considerable strengthening factor for the maintenance of the species; the presence of species with rival features (regarding the division of resources) or superordinate species (that feed on the subordinate species or use it for various purposes) represent weakening factors (wars for food, the struggle for survival).

Bibliographic note. Although in describing the Combinatory Systems that guide evolution we have looked to Charles Darwin's theory, it is useful to recall that there is no "unique" theory of evolution. We shall only mention the main theories of evolution that provide interpretations of evolutionary phenomena in the domain of life.

(I) Lamarckism, or the Theory of the Inheritance of Acquired Characters (Lamarck, 1809), and Packard's Neo-Lamarckism (Packard, 1901).
(II) Darwinism, or the Theory of Natural Selection (Darwin, 1859).
(III) De Vries' Mutation Theory (Vries, 1905).
(IV) Huxley's Neo-Darwinism (Huxley, 1942), or the "modern theory" of evolution, supported by studies by Watson and Crick (1953) and by Monod (1970).
(V) Stephen Gould's Punctuated Equilibria Theory (Gould & Eldredge, 1977).
(VI) Selfish gene theory of evolution, popularized by Richard Dawkins (Dawkins, 1989, 2004).
(VII) Humberto R. Maturana and Francisco J. Varela's Autopoietic theory of evolutionary processes (Maturana & Varela, 1987).
(VIII) Conrad Waddington's Epigenetics view of evolution (Waddington, 1957).
(IX) Charles Thaxton's Intelligent Design as a challenge to Neo-Darwinism (Thaxton et al., 1984).

2.6.9 *The Diffusion and Evolution of Non-Biological "Species"*

Some brief considerations are warranted here about the "future" evolution of a new, non-biological "species" that has recently appeared, the "robot species," in all its forms, whose *quantitative* dynamics and *qualitative* evolution are for the moment controlled externally by man. The entire world robot population (industrial and service robots) reached 8.6 million at the end of 2010 (Guizzo, 2010). As of 2024, the total number of industrial robots of all types (in manufacturing, agriculture, logistics, installation, etc.) is estimated at around 3.9 million operational units worldwide (Castigli, 2022, online).

The most common uses are:

- Handling (loading/unloading machine tools, complex manipulations).
- Welding (in all its forms).
- Assembly.
- Applications in "clean rooms" to maintain sterile environments, ensuring compliance with safety and hygiene regulations.

If we also consider service robots (used in tourism, restaurants, hospitals, professional services, etc.) and domestic robots, the total number exceeds 24 million. Compared to 2010, the number of robots has grown at a Compound Annual Growth Rate (CAGR) of about 8.2% per year.

This figure reflects a significant increase driven by advancements in automation, artificial intelligence, and the application of robotics across various sectors, including healthcare, logistics, and household tasks. Many industrial robots are evolving by connecting and collaborating to improve their performance. Robots that perform the same task while operating in different locations have become connected through the cloud and collaborative programs.

> Robots doing the same task connect across all global locations so performance can be compared and improved at the click of a button. Robots automatically download what they need to get started from a cloud library and then start to optimise through "self-learning." (IFR, 2019, online)

In the population of robots (and the various "species" and "families" it is composed of), there is not only a quantitative growth but also a continuous improvement in hardware and software, and consequently in productive and economic performance.

Several studies already exist on "evolutionary robotics" (e.g., Holland, 1975; Koza, 1992; Floreano & Keller, 2010), which highlight how the population of robots evolves according to Darwin's principle of natural selection applied to the robot population.

> The general idea of evolutionary robotics [...] is to create a population with different genomes, each defining parameters of the control system of a robot or of its morphology. The genome is a sequence of characters whose translation into a phenotype can assume various degrees of biological realism. For example, an artificial genome can describe the

> strength of synaptic connections of an artificial neural network that determines the behaviour of the robot. [...] At the beginning, robots have random values for their genes, leading to completely random behaviours. The process of Darwinian selection is then imitated by selectively choosing the genomes of robots with the highest fitness to produce a new generation of robots. (Floreano & Keller, 2010, online)

For example, suppose that in a *"family" of robots* (industrial or service robots) a *mutation* is introduced "by chance" by a given initiator (mutant) to try to improve performance. If the mutation is not advantageous, in term of cost, complexity, and performance, the mutation would remain isolated and no diffusion would develop. If the number of robots that implement the innovation (mutant) reaches the necessary activation density, then that mutation will spread and give rise to a new family of robots. The evolution within the population and families of robots can be gradual and slow, or it can occur suddenly, in leaps and bounds, with a rapid transition from the original robots to the mutated ones.

From these considerations, it becomes clear that the simulation models used to understand the evolution of living species can also be applied to comprehend the evolutionary dynamics of robots.

The heuristic model of the Combinatory System that describes the evolution of robots can be expressed as follows.

Micro-behavior and necessitating rule: if you haven't installed productive robots while your competitors have increased their efficiency by integrating robots into their production, storage, and/or delivery lines, and your profit is decreasing, then to remain competitive, you need to introduce robots into your company. If you have already robotized your processes but your competitors have adopted more modern and efficient robots, gaining a competitive advantage, then you need to replace the obsolete robots.

Macro-behavior and recombining rule: the necessary adoption of robots by individual companies in the same sector has the recombining effect of spreading high-robot-intensity production. This recombining effect improves the system's average level of productivity and quality by canceling or reducing the advantages of the individual producer. Producers try to equal—or preferably exceed—the average productivity level of the system, which compels them to resort to more efficient and productive robots, thus generating the *micro-macro feedback* mechanism.

Micro-macro feedback. Chance and necessity: the general increase in the population of increasingly modern robots that raises the average level of productivity and quality of the productive system is the recombining effect resulting from the micro behaviors of individual entrepreneurs striving to remain competitive. But this also influences the search for new, more efficient robots. If, by chance, an improvement is introduced that makes some function of a class of robots more efficient, then, by necessity, the micro-behavior of the collective entrepreneurs is activated, as they must adopt the improved robots to remain competitive.

Strengthening, weakening, and control actions: innovations in robots to improve production judged useful by consumers, the presence of specialized research centers, the rapid obsolescence of robots in service, incentives for productive innovation and quality improvement, and high profit motivation are all examples of *strengthening factors* for the system. Monopoly conditions, barriers to entry in the sector, and the growing opinion that robots create unemployment are examples of *weakening factors* for the system. *Macro control* can be exercised through laws that favor or inhibit the introduction of robots or through measures affecting the educational system of the country from which firms obtain their human resources; *micro control*, on the other hand, acts on individual producers through various kinds of incentives or disincentives.

The graphic model is indicated in Fig. 2.57.

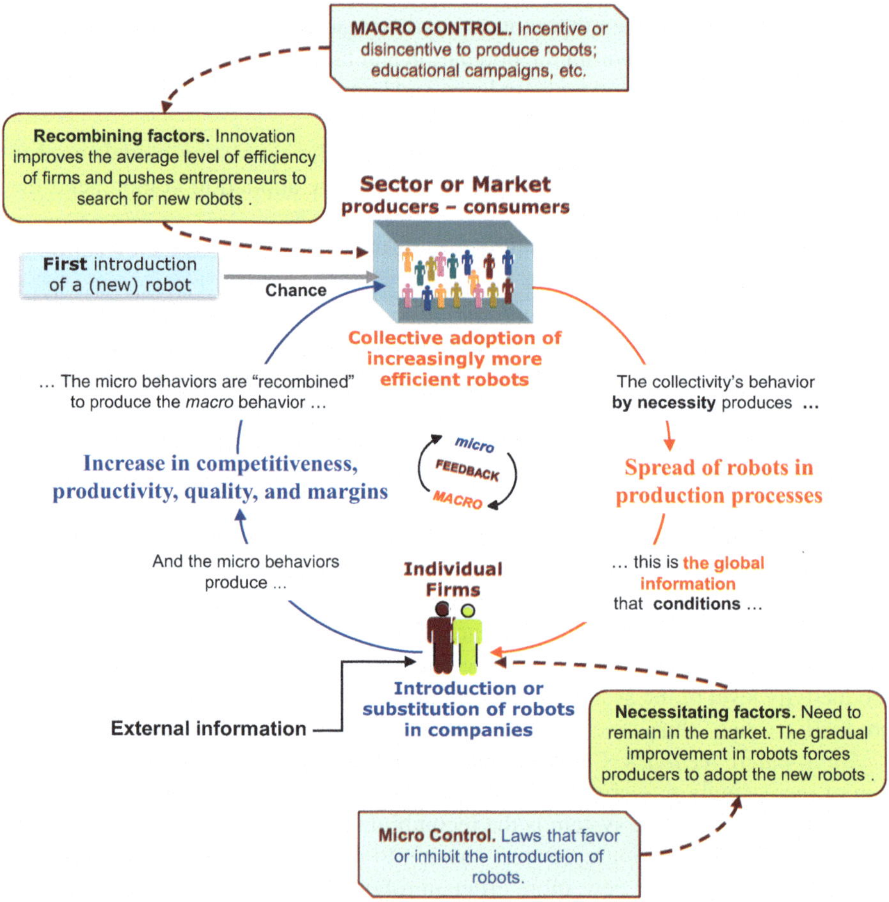

Fig. 2.57 Graphic model of the "diffusion and evolution" of Robots (and other non-biological populations)

2.7 Some Phenomena Produced by the Joint Action of Multiple Combinatory Systems

The previous sections presented a sample of phenomena that can be interpreted as *macro-effects* produced by individual Combinatory Systems. However, some phenomena are often the result of the actions of two or more Combinatory Systems operating together. To guide the reader in exploring this interpretation, I will present some of these phenomena in the following sections.

2.7.1 *Enterprises and Entrepreneurial Dynamics: Co-localization of Firms and the Formation of Productive Clusters, Business and Industrial Districts, and Colonies*

In Sect. 2.2.5, the phenomena of the formation of industrial districts and the genesis of colonies of living beings of every kind were introduced as *macro-effects* of the actions of individual Combinatory Systems of *accumulation*. However, the Combinatory Systems that generate such phenomena were not described, as they represent the macro-effect of the combined action of an *accumulation* system and a *diffusion* system that operate together in all phenomena resulting from the micro-behaviors of agents who, over time, "die" and are replaced by other agents, resulting in the maintenance of the macro-behavior of the community over time.

We will first consider, more generally, the complex topic of business dynamics and, in particular, the phenomenon of the co-localization of firms in a given area. There are two aspects to such dynamics:

1. the *spatial dynamics* of firms and entrepreneurs; that is, the aspect of the *localization* of the firms in a given area or territory, or even—in equivalent though general terms—the dynamics regarding the formation of *clusters* of firms or productive activities; I shall call this phenomenon *exogenous co-localization* (Mella, 2006);
2. the dynamics regarding the *density of firms* and entrepreneurs over a given territory; that is, the aspect regarding the *genesis* of new entrepreneurial initiatives, entrepreneurs, or firms in an area where there is already a cluster of entrepreneurs and productive activities; this phenomenon will be called *endogenous co-localization*.

For this analysis, it is necessary to introduce the hypothesis that entrepreneurs—and by extension, firms and organizations—are "rational agents" who base their production and localization choices on a vector of predictable measures of economic and financial performance, such as the presence of markets and potential customers, the range of variation in selling prices, the availability of credit institutions and the cost of loans, expected profitability, among others.

These basic measures of financial and economic performance not only reveal how internal factors of strength and weakness can heighten or depress the reaction of the firm to opportunities, threats, and risks but can also be viewed as *spatial performance indicators* (or probability) revealing how physical and social characteristics of a territory can support or inhibit the *fitness* of the firm or of a business operating in that territory.

After having mentioned the most significant *performance indicators*, it is possible to introduce the concept of the *firm's fitness landscape*, by which the entrepreneur can derive a *function of attractiveness* for the territory or a *fitness attractiveness landscape*, which will likely present "valleys" of moderate attractiveness, "peaks" of high attractiveness, or "pits" of repulsion (negative attractiveness) to be avoided, if possible. For example, by choosing as indicators of *fitness* the *Return on Equity* (*roe*) and the *Return on Invested Capital* (*roi*), it is plausible to assume that an area full of potential consumers and lacking in competitors is highly attractive, since it has potentially high revenue prospects both in terms of quantity and price, and thus a high *roi*. On the other hand, an area with a reduced tax burden would have, with all other conditions equal, a higher *roe* than others with a higher tax burden. An area with a high amount of pedestrian traffic could favor sales for a small retailer, while one with a large parking area could increase the fitness of a large retailer.

The assumption the firm behaves as a *rational agent* implies that the entrepreneur (hence the firm) must be viewed as an *explorative agent*, continually seeking improvement in the conditions of fitness in all possible areas while following these elements of *optimal behavior*:

(a) Exploring all accessible territories and areas that can be reached by his productive, economic, and financial processes.
(b) Shaping, updating, and continually exploring the *attractiveness landscape* while looking to the future.
(c) Choosing the area(s) with the greatest attractiveness; that is, with favorable conditions for increasing *roe* and *roi* (e.g., the ease with which new businesses can arise, greater sales volumes, expectations for better prices and more favorable supply costs, greater productivity and public subsidies, high levels of social protection, a stimulating environment, an abundance of infrastructures, a lower tax burden, etc.)
(d) Avoiding the "pits" and attaining the highest "peaks".

Since, as observed above (Sect. 2.2.4), a collectivity of agents/businesses can constitute a Combinatory System, to understand business dynamics and the *co-localization* of firms, we present the following thesis:

A. If we refer not to the dynamics of the single firm but to a *collectivity of firms*, which act as *explorative rational agents* continuously attempting to improve their *performance* measures and their *fitness*, we can consider such a collectivity as a *combinatory system* capable both of *co-localizing* in a given territory by forming a *cluster* and of *generating* new enterprises.

B. The *co-localization* of firms and the formation of clusters is basically generated from the action of two *Combinatory Systems*:

B.1. *Accumulation systems*, which promote the *exogenous genesis* of clusters involving units from different territories establishing themselves in the relevant area.

B.2. *Diffusion systems*, which, conversely, encourage the *endogenous genesis* and expansion of a cluster in a specific area, whereby units within the area where a cluster has already formed establish themselves in the same location (typical of districts).

Before attempting to demonstrate this thesis, let us first observe that clusters of firms situated in limited areas are widespread in all contexts and in various forms (Albu, 1997), among which:

1. *Conglomerate clusters*, typical of industrial and commercial areas (Porter & Sölvell, 1998), which we normally observe at the periphery of cities, along the main streets, near a tollbooth, or around universities (Chandler et al., 1999).
2. *Industry cluster,* composed of a group of business enterprises bound together by "*buyer–supplier relationships, or common technologies, common buyers or distribution channels, or common labour pools*" (Bergman & Feser, 1999);
3. *Specialist clusters*, or *districts*, typical of single-business or mainly business and *industrial* areas (Lazerson & Lorenzoni, 1999).
4. *Vertically integrated clusters*, typical of a "filière" or pipeline, composed of independent firms which carry on different phases of a single process along the value-added chain, and are connected "up the line" and/or "down the line" with other firms in the same area (Brusco, 1982, 1986).
5. *Vertically-* and *horizontally integrated clusters* of firms, typical of *networks*, closely linked by inter-company ties in terms of supplies, manufacturing, and process (Hakansson & Snehota, 1994; Thorelli, 1986); the network represents an organized system forming a single productive entity that does not depend on joint location but on the activities of all the firms in the social network, wherever they are located;
6. *Hub* or *constellation* cluster (neck or spider-web) joint location, arriving or departing, which we observe when there are common facilities, a common supplier, or a common client in the center of the web (Lorenzoni & Lipparini, 1999).

Even if clustering represents a general phenomenon that is not limited to the firm, we can easily understand that the clustering of "productive units" in the same area is quite general. In plain words, we must, therefore, assume that a "mass" of firms appears to gain the "force of gravity" and generates *attraction*, representing *global information,* so that rational entrepreneurs are inevitably drawn in to increase the mass of the cluster as part of a positive feedback or reinforcing loop, in the typical effect produced by two Combinatory Systems acting jointly.

B.1. *Exogenous Joint Location Explained by Accumulation Systems.*

When a particular area exhibits advantages for the fitness of a firm, offering a *positive differential* in financial and economic *performance measures* (such as tax reductions, infrastructure, facilities, aid, and subsidies) compared to other areas

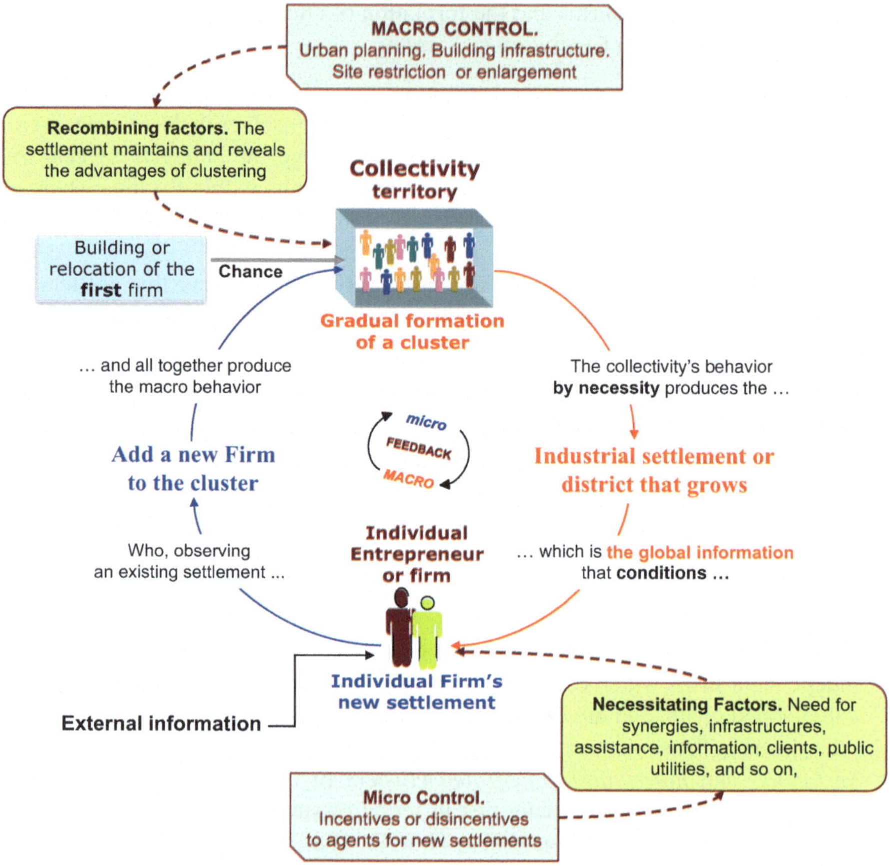

Fig. 2.58 Model of the system of accumulation for exogenous joint-location

[necessitating factor], the *attractiveness landscape* features a peak. Consequently, the probability that a certain number of entrepreneurs will decide to locate [micro-behavior] their productive or commercial units [micro effect] in that area increases rapidly. The combinatory system can then initiate the collective phenomenon of joint location [macro-behavior], leading to the development of distinctive industrial and commercial clusters [macro-effect].

The heuristic model in this case can take on the following structure also shown in Fig. 2.58.

Micro-behavior and necessitating rule: if you need to establish a productive or commercial unit (micro-behavior), seek locations with peaks in the attractiveness landscape, as they provide positive economic differentials.

Macro-behavior and recombining rule: by maintaining economically measurable advantages, the site sustains attractiveness and encourages the influx of new enterprises; many establishments gather there (macro-behavior), leading to the formation of an increasingly larger settlement (macro-effect). By maintaining economically measurable advantages, the site sustains attractiveness and encourages the influx of new enterprises; many establishments gather there (macro-behavior), leading to the formation of an increasingly larger settlement (macro-effect).

Micro-macro feedback. Chance and necessity: the growth of the area with enterprises (macro-effect) creates more advantages and incentives for the establishment (micro-behavior) of new productive units (micro-effects). Locations that initially located in an area by chance result in progressively larger settlements (macro-behavior). *Chance* plays a role in the initial firms together choosing a favorable site. Once the minimum activation density (critical mass) is achieved, necessitating factors set the system into motion, leading to a typical path dependence (Antonelli, 1997). When a *critical activation mass* is attained, the endogenous joint-location system is triggered, provided that the system can effectively "recombine" micro-behaviors within an "*innovative milieu*" that preserves and amplifies the entrepreneurial mindset (Maillat, 1998). This environment enables the selection of optimal routines and procedures (Nelson & Winter, 1982; Nonaka, 1994) "*by imitating the observed behavior of one or more 'masters' within a community of practice*" (Nooteboom, 1999).

Strengthening, weakening, and control actions: *government policies*, such as incentives, constraints, and measures to create logistical infrastructures (roads, harbors, equipped building lots, etc.), research, educational centers (Jaffe et al., 1993; Nelson, 1995), and the recognition of cost advantages (reduced cost of labor, tax, and financial incentives), can play a role. Policymakers may also stimulate local entrepreneurial activity through initiatives like providing venture capital, preferential loan finance, tax incentives, support for business start-ups, and management training and business advice. The strategy of attracting *foreign direct investments* is another crucial economic development approach in many city regions (Gordon, 1999).

B.2. *Endogenous Joint Location Explained by Diffusion Systems.*

The creation of industrial, commercial, and professional areas can result from an endogenous process in an area. This would require the assumption that there is an existing cluster that generates new agents internally that view that area as the best environment for their *economic fitness*. The cluster will then increase through the appearance of its own agents based on a generative rate which is a function of the size of the cluster. The existence of firms in a certain area fosters the development of skills, encourages a willingness to take risks, and stimulates entrepreneurial initiatives to establish new enterprises in the same region. A plausible explanation within a system-procedural framework is provided by the logic of *Diffusion Systems*.

If successful firms locate in an area by *chance* (original nucleus) and can develop their internal talent pool (employees, managers, professionals), some personnel, having *by chance* acquired the necessary skills and assessed the fitness landscape of potential business ventures, may decide to initiate an activity [micro-behavior] to leverage their acquired capabilities for personal gain. New enterprises emerge [micro-effect] and, if successful, a combinatory system can be set in motion, leading to more firms establishing themselves in the area [macro-effect] through endogenous growth (Rabellotti, 1997).

This clustering encourages other individuals with similar capacities to take the personal initiative of starting new enterprises. The process spreads [macro-behavior], and the group of workers gradually transforms into a community of entrepreneurs (Antonelli, 1996). The number of firms continues to grow [macro-effect], increasing the likelihood that individuals will embark on new entrepreneurial ventures [micro-behavior], illustrating a typical micro-macro feedback feature of diffusion systems. An area initially comprised of workers gradually evolves into one dominated by entrepreneurs, who will eventually need to bring in additional workers from other regions.

The heuristic model of the Combinatory System is founded on the following rules derived from Fig. 2.59:

Micro-behavior and necessitating rule: if you observe that numerous individuals are succeeding in business ventures and you desire to increase your wealth and not

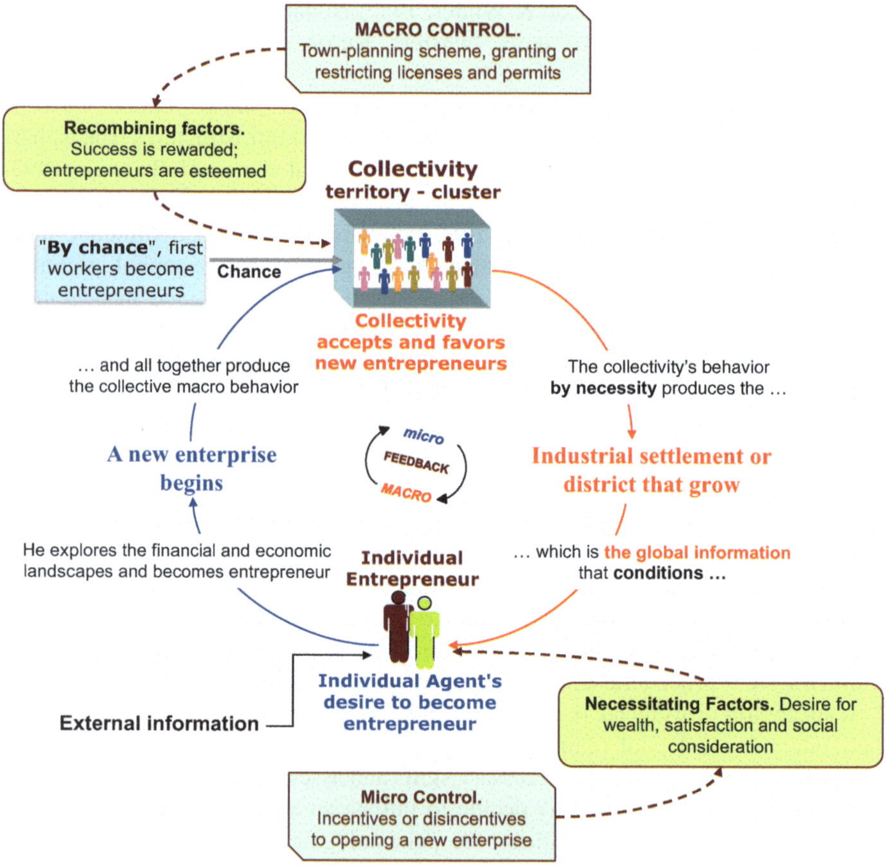

Fig. 2.59 Model of a system of diffusion for endogenous joint-location

be left out (necessitating factor), and you see your *abilities* representing a *fitness factor*, then it becomes imperative for you to "go it alone" by establishing your own business;

Macro-behavior and recombining rule: the environment and the collectivity highly esteem persons who gain wealth by taking on business risks (recombining factor), considering the firms useful and the entrepreneurs successful. Many families hope their children will one day start a business, leading to the formation of exclusive business clubs. Personal wealth and business growth signify social success (recombining factor); the view that success and wealth can be acquired by becoming an entrepreneur spreads throughout the community.

Micro-macro feedback. Chance and necessity: the greater the number of successful businesses in an area (macro-effect), the greater the entrepreneurial initiative (micro-effect). The appearance of a new enterprise (micro-behavior) increases the spread of business activity (macro-behavior) and, even more so, the density of enterprises in the area (macro-effect).

Strengthening, weakening, and control actions: specific strengthening measures can favor the appearance of Combinatory Systems of endogenous joint-location, among which: (a) the establishment of professional schools ensuring initial employment in specific careers; (b) the obtainability of risk and loan capital; (c) encouraging the formation of new enterprises through facilitating measures (e.g., young businessmen); (d) protective measures against unsuccessful ventures; (e) incentivizing the exogenous joint location of small businesses; (f) the creation of *knowledge exchange spaces*. The cluster is viewed as a "ba," "a shared place for emerging relationships" (Nonaka & Konno, 1998). In this context, the success of the cluster can be seen in its potential to represent a natural form of "ba." The start of the endogenous joint location process could benefit from the presence of publicly managed productive units, provided these units engage in activities characterized by discrete processes, allowing them to be managed by new enterprises, or that they necessitate the integration of processes both "up the line" and "down the line," which can be handled by new enterprises and, most importantly, generate the required know-how. "Cathedrals in the desert" have never facilitated the emergence of local entrepreneurs, primarily because they have failed to initiate any Combinatory System, lacking the prerequisites mentioned earlier.

Observation: the two forms of enterprise development in a specific area are not mutually exclusive but typically complementary. On the one hand, an initial external establishment can initiate the diffusion system that results in endogenous joint location, and, on the other, the endogenous formation of entrepreneurs, occurring *by chance*, not only has the potential to trigger the process of endogenous joint location but, with the right incentives from local public authorities, can also initiate the accumulation system of enterprises, attracting the migration of additional productive units to the area.

2.7.2 Joint Action of Combinatory Systems Producing Evolution in IA and Robotics Interaction. A Look at the Future

In Sect. 2.6.9, the Combinatory System that produces improvement and progress in robotics was described as a clear case of the evolution of non-biological "species." This section considers the interaction of two Combinatory Systems operating together: one producing the improvement and progress of AI and the other the improvement and progress of robots that are "intelligent," offering very interesting prospects, though their realization in terms of timing and methods is unpredictable.

Artificial Intelligence (AI) can be defined as a branch of knowledge that deals with the creation of systems (logical or mechanical) capable of performing tasks like those carried out by human intelligence (Russell & Norvig, 2016). These tasks include, among others, image recognition (Szeliski, 2022), natural language understanding, the solving of complex problems, planning, and machine learning. AI can be considered a non-biological species that is rapidly spreading and evolving to continuously improve and progressively enhance itself.

The first Combinatory System influences the diffusion and evolution (improvement) of AI due to the *necessary factors* of increasing power and precision in its processes, which have now made it an essential component in many applications, both in the scientific world, in production, and, more generally, in the *supply chain*. For example, AI serves as the foundation for the creation of *Expert Systems*, programs that—using a "knowledge base" and "inference rules" to solve complex problems—imitate or replace the decision-making process of a human expert in various fields requiring complex reasoning performed autonomously, such as in explorations carried out by rovers where no human intervention is possible. Additionally, AI is employed in Natural Language Processing with applications that allow interaction between computers and human language, enabling machines to understand, interpret, and respond to natural language.

The greatest *necessitating factor* for the spread of AI is likely represented by the numerous and increasingly sophisticated applications in the fields of industrial production and personal services, enabling the construction of robots that, guided by AI, perform tasks autonomously. In this area, the diffusion of AI includes the construction of "autonomous machinery" which, guided precisely by AI, designs work programs according to customer requests, feeds production lines with the necessary resources, sends semi-finished and finished products to warehouses, and ensures their delivery. More generally, AI allows for the development of strategies and the discovery of optimal solutions to complex problems, such as the management of the *supply chain, logistics*, and even *urban planning*.

The *necessitating factors* are joined by the *recombining factors*, represented by the tendency of the research and user communities to develop studies and applications of all kinds with some form of AI incorporated. Think of the vast number of advertisements that emphasize the use of AI in the production or consumption of some product; the information about AI being spread in schools of all kinds, making it attractive to students; the creation of laboratories to experiment with the results achieved; and the widespread collaboration between different users to produce increasingly advanced applications. Finally, consider the high regard that universities place on theses and papers dealing with some aspect of AI.

Driven by powerful *necessitating* and *recombining factors*, the *Combinatory System of AI dissemination and improvement* operates increasingly more widely and quickly. However, this system interacts with a similar system: that of the *dissemination and improvement of robots*, in all their forms and applications (Sect. 2.6.9). In fact, *robotics* combines with AI to create *intelligent machines* capable of performing physical tasks, analyzing their own behavior, learning autonomously from environmental dynamics, and adapting to new situations. Thanks to AI, robots become *intelligent entities*, in the broadest sense of the term, that can interpret verbal and non-verbal language, develop models of understanding useful for making decisions and developing efficient behavior, conduct self-diagnoses, self-repair, reproduce, and even improve themselves by reconfiguring and acquiring new components. Many robots are equipped with vision systems, motion recognition and analysis, sound interpretation (including human speech and music), and can acquire a form of "autonomous existence," increasingly difficult to control externally. Robots will control themselves and even other robots, sharing information and learning from one another through communication networks they design themselves. This exchange of data enables them to collectively improve their "species," accelerating the dissemination and evolution of such systems and their "autonomy."

Some robots can use evolutionary algorithms to improve their capabilities. These algorithms simulate the process of natural evolution to optimize the structures and behavior of robots through successive iterations, selecting the most efficient configurations. Robots can perform tasks of self-repair, maintenance, and reprogramming, introducing forms of modularity that promote "reconfiguration" and ultimately even "reproduction."

Even before the advent of AI, John von Neumann developed the Theory of Self-Reproducing Automata (Von Neumann & Burks, 1966) theorizing that a machine, therefore a robot, could contain the instructions to build a *copy of itself* and execute these instructions autonomously, in a process similar to what happens in biological systems, analogous to DNA replication. Von Neumann introduced the concept of the "Universal Constructor," capable of programming a machine (therefore an automaton) to read a description (or a "tape") of an object and build it, including a *copy of itself*. This universal constructor had to be capable of:

(a) *Copying the tape*: reproducing the instructions needed to build a copy of itself.
(b) *Building the machine*: using the tape to assemble a copy of the universal constructor.
(c) *Reproducing the tape*: providing the new machine with a copy of the original tape.

We can then argue that autonomous robots equipped with AI will not only be able to improve their performance through continuous vision, learning, and adaptation, they will also have the ability to modify, optimize, and replicate their physical structure, becoming an autonomous, albeit non-biological, "species."

The Combinatory System of AI *dissemination and evolution* and that of the *dissemination and evolution* of robots—and, more generally, of "machines"—continuously interact in increasingly sophisticated ways and are expected to revolutionize numerous sectors, improving the efficiency, safety, and capabilities of autonomous machines. This will transform not only industry and society but also our daily lives.

One particular development in robotics and AI that is advancing rapidly is that of "androids," a type of robot specifically designed to have human-like characteristics—such as a face, realistic imitations of skin, hair, upper and lower limbs, and facial expressions—that resemble human beings both in physical appearance and behavior. Androids are *advanced robots* that integrate AI functionalities to enhance their interaction and adaptation capabilities and handle complex and realistic interactions. While still in an early stage, they are progressing rapidly and can learn from experience, understand complex commands, and respond appropriately to environmental stimuli, delivering satisfactory performance. While "robots," in their various forms, are employed in many sectors to perform tasks with precision, speed, and autonomy, "androids" are currently used in contexts that require social interaction, such as elderly care, education, hospitality, entertainment, domestic services, visitor and customer assistance, entertainment, and protection.

AI-powered robots, although they are becoming "autonomous machines," are primarily employed to perform services in the realm of production processes and family communities, broadly defined. Androids, on the other hand, are mainly designed with human features for the specific purpose of interacting with humans. In science fiction writings, in the worlds and settings masterfully described by Isaac Asimov—a well-known science fiction writer, with over 500 books, including the *Foundation* series, the *Galactic Empire* series, and the *Robot* series—androids were omnipresent and always at the service of humans (Wikipedia, 2024g, online, entry Isaac Asimov). However, Asimov also highlighted the aspect of *safety* since androids, through their decisions and behavior, could potentially come into conflict with humans, a theme that today raises concerns and debates among technology experts and the general public alike.

This scenario, often explored in science fiction, involves several considerations, the first of which concerns responsibility and control to prevent harmful actions, even if these stem from the unforeseen or "unintentional" behavior of the androids. Post-event control to correct malfunctions is important, but as Asimov himself masterfully emphasized, the most effective control is preventive control, which is exercised through the careful *programming* of the android to prevent any harm. With great foresight and insight, the Asimov proposed the following "four rules," now known as "Asimov's Laws of Robotics," to ensure the ethical and safe behavior of robot-androids toward human beings:

0. A robot may not harm humanity or, by inaction, allow humanity to come to harm.

1. A robot may not injure a human being or, through inaction, allow a human being to come to harm.

2. A robot must obey the orders given it by human beings, except where such orders would conflict with the First Law.

3. A robot must protect its own existence as long as such protection does not conflict with the First or Second Law.

The "0" rule was added by Asimov to highlight the necessity for a robot-android to be programmed to prioritize the well-being of humanity as a whole over the well-being of individual humans (see the short comment of Singer, 2009).

Alan Turing, who is well-known in the field, believes that *intelligence* is inherently tied to human interactions. Therefore, if we focus solely on linguistic interactions, we can define *intelligence* as the capacity to communicate in a way that *resembles* human conversation. More precisely, it refers to the inability to discern, solely from linguistic interaction, whether an "agent" communicating with a human is actually "non-human"—be it an animal, a machine, artificial intelligence, or an extraterrestrial. Turing proposed a simulation test to evaluate this, famously known as the "Turing Test" or "imitation game." In his original 1950 essay "Computing Machinery and Intelligence," Turing described the test in a specific way:

> I propose to consider the question, "Can machines think?" [...] The new form of the problem can be described in terms of a game which we call the "imitation game." It is played with three people, a man (A), a woman (B), and an interrogator (C) who may be of either sex. The interrogator stays in a room apart from the other two. The object of the game for the interrogator is to determine which of the other two is the man and which is the woman. He knows them by labels X and Y, and at the end of the game he says either "X is A and Y is B" or "X is B and Y is A" (Turing, 1950, p. 433).

In simpler terms, and ignoring any technical aspects, Turing's Test can be described as follows: if we ignore the physical characteristics of the entities involved (imagine placing the entity we want to assess in a separate room where communication happens only through a computer), an entity can be considered intelligent if, after conversing with it for any length of time on any topic, we are unable to tell it apart from a human.

In summary, the Turing Test involves three "actors":

1. *A human*, the Interrogator, who cannot see either the human interviewee or the machine.
2. *The Human Interviewee*: a person who responds to the Interrogator's questions via a device.
3. *The Non-human Interviewee* (machine or otherwise), which responds to the Interrogator's questions, trying to simulate human thought.

The Interrogator asks a series of questions to both interviewees on agreed-upon topics and, at the end, must determine who is the human and who is the machine (or other unknown entity) based solely on the answers received through the interface. If the non-human interviewee (the machine, specifically) succeeds in tricking the Interrogator into believing it is human, it is said to have "passed the Turing Test."

This test is both simple and compelling, grounded in the idea that the only form of intelligence we can definitively prove must resemble human intelligence. In his well-known work *The Oxford Companion to the Mind* (1987), neuropsychologist

Richard Gregory described Alan Turing's "imitation game" as "*the best test available for confirming the presence of intelligence in a machine [and thus in men and animals*]."

Could an artificial intelligence pass the Turing Test today? The answer is not straightforward: an AI chatbot like ChatGPT (or others, such as Google Gemini, Microsoft Copilot, YouChat, You.com, etc.) likely would not pass for one simple reason: ChatGPT would appear too advanced compared to human intelligence. It would reveal an "omniscient" knowledge on any subject matter, no matter how complex, providing responses so immediate, rich, and detailed that the interrogator would have no doubts and conclude: the non-human interviewee is a "machine," one so superior that it always outmatches the limited knowledge and reasoning of any human, even a genius.

I conclude this lengthy section by noting that while biological evolution takes a long time (some species have never evolved over millions of years), the evolution of robots and androids occurs in very short periods and at an increasingly accelerated pace. This is the result of two Combinatory Systems working together to produce *the improvement and progress* of AI and *robots that are becoming "intelligent,"* setting in motion a dynamic process that will lead to effects that are difficult to imagine today (Mella 2017).

References

Albu, M. (1997). *Technological learning and innovation in industrial clusters in the South*. SPRU, Electronic Working Papers Series (7).

Antonelli, C. (1996). Localized knowledge percolation processes and information networks. *Evolutionary Economics, 6*, 281–295.

Antonelli, C. (1997). The economics of path-dependence in industrial organization. *International Journal of Industrial Organization, 15*, 643–675.

ASQ. American Society for Quality. (1991). *Quality*. Retrieved from https://asq.org/quality-resources/quality-glossary/q

Band, W. A. (1991). *Creating value for customers: Designing and implementing a total corporate strategy*. Wiley.

Baumol, W., Blackman, S., & Wolff, W. (1989). *Productivity and American leadership: The long view*. Cambridge MA: MIT Press. Wiley (1972 first edition).

Beccattini, G., Bellandi, M., & De Propist, L. (2009). *A handbook of industrial district*. Edward Elgar.

Bellandi, M. (1996). On entrepreneurship, region and the constitution of scale and scope economies. *European Planning Studies, 4*(4), 421–438.

Belussi, F., & Pilotti, L. (2002). Knowledge creation, learning and innovation in Italian industrial districts. *Human Geography, B, 84*(2), 125–139.

Ben-Eli, M. U. (1979). Amplifying regulation and variety increase in evolving systems. *Cybernetics and System, 9*(3), 285–296.

Bergman, E. M., & Feser, E. J. (1999). *Industry clusters: A methodology and framework for economic cooperation and development*. In OECD, Boosting Innovation. The Cluster Approach. Paris, France.

Bikhchandani, S., Hirshleifer, D., & Welch, I. (1992). A theory of fads, fashion, custom, and cultural change as informational cascades. *Journal of Political Economy, 100*(5), 992–1026.

Blank, R. M., & Shapiro, M. D. (2001). *Labour and the sustainability of output and productivity growth*. NBER, Un. Of Michigan.
Bowersox, D. J., Closs, D. J., & Cooper, M. B. (2002). *Supply chain logistics management* (Vol. 2). McGraw-Hill.
Breventano, S. (1570). *History of antiquity, nobility, et notable things of the city of Pavia* [original: Istoria della antichità, nobiltà, et delle cose notabili della città di Pauia]. Pavia (italy): Hieronimo Bartholi Ed. S. Pietro in Ciel'Aureo. Google Books.
Brusco, S. (1982). The Emilian model: Productive decentralisation and social integration. *Cambridge Journal of Economics, 6*, 167–184.
Brusco, S. (1986). Small firms and industrial districts: The experience of Italy. In Keeble & Wever (Eds.), *New firms and regional developments*. Croom Helm.
Buchanan, K. L. (2000). Stress and the evolution of condition-dependent signals. *Trends in Ecology and Evolution, 15*(4), 156–160.
Burgiel, S. W. (2020). The incident command system: A framework for rapid response to biological invasion. *Biological Invasions, 22*(1), 155–165.
Campbell, C., & Rozsnyai, C. (2002). *Quality assurance and the development of course programmes*. Papers on Higher Education.
Capra, F. (1996). *The web of life: A new scientific understanding of living systems*. Anchor Books.
Castigli, M. (2022). *Robots are spreading in industry: Here's in which sectors and what they do* [original: I robot si diffondono nell'industria: ecco in quali settori e cosa fanno]. Retrieved from https://www.agendadigitale.eu/industry-4-0/i-robot-si-diffondono-nellindustria-ecco-in-quali-settori-e-cosa-fanno/
Cessna, A. (2015 online). *Protoplanets*. Retrieved from http://www.universetoday.com/37053/protoplanets/
Chandler, A. D., Hagström, P., & Sölvell, Ö. (Eds.). (1999). *The dynamic firm. The role of technology, strategy, organization, and regions*. Oxford University Press.
Chandler, A. D., Jr. (1990). *Scale and scope: The dynamics of industrial capitalism*. Harvard, University Press.
Cipolla, C. M. (1962). *The economic history of World population*. Penguin Books.
Cook, S. (1995). *Practical benchmarking: A manager's guide to creating a competitive advantage*. Kogan Page.
Coyle, D. (1999). *The weightless world: Strategies for managing the digital economy*. MIT Press.
Crosby, P. B. (1979). *Quality is free: The art of making quality certain*. McGraw-Hill.
Darwin, C. (1859). *On the origins of species by means of natural selection*. Murray. Retrieved from http://darwin-online.org.uk/content/frameset?itemID=F373&viewtype=text&pageseq=1
Dawkins, R. (1989). *The selfish gene*. Oxford: Oxford University Press (1st ed., 1976). Online at: http://pratclif.com/biologie-moleculaire/dawkins.htm
Dawkins, R. (2004). *The ancestor's tale: A pilgrimage to the dawn of evolution*. Houghton Mifflin.
De Vries, H. (1905). *Species and varieties: Their origin by mutation: Lectures delivered at the University of California by Hugo de Vries*. Open Court Publishing.
Demartini, C., & Mella, P. (2011). Time competition. *The New Strategic Frontier, iBusiness*, 136–145.
Deneubourg, J. L., & Goss, S. (1989). Collective patterns and decision-making. *Ethology Ecology & Evolution, 1*(4), 295–311.
Dettre, M. (2022). *Just 24 wild rabbits started Australia's most devastating plague*. Retrieved from https://www.queenslandcountrylife.com.au/story/7873584/how-the-rabbit-plague-took-over-australia/
Drucker, P. F. (1977). *People and performance: The best of Peter Drucker on management*. Routledge.
Dupuy, J. P. (2005). Panic and the paradoxes of the social order. *Passions in economy, politics, and the media: in discussion with Christian theology, 17*, 215–233.
Dyer, J. H. (1997). Effective interfirm collaboration: How firms minimize transaction costs and maximise transaction value. *Strategic Management Journal, 18*(7), 535–556.

Feigenbaum, A. V. (1951). *Quality control: Principles, practice, and administration: An industrial management tool for improving product quality and design and for reducing operating costs and losses*. McGraw-Hill.

Fennell, J. (2016). *Planetesimal: Definition, theory & hypothesis*. Retrieved from http://study.com/academy/lesson/planetesimal-definition-theory-hypothesis.html

Flood, R. L. (1993). *Beyond TQM*. Wiley.

Floreano, D., & Keller, L. (2010). Evolution of adaptive behaviour in robots by means of Darwinian selection. *PLoS Biology, 8*(1), e1000292. Retrieved from https://journals.plos.org/plosbiology/article?id=10.1371/journal.pbio.1000292

Gabriel, R. A. (2010). *Philip II of Macedonia: Greater than Alexander*. Potomac Books.

Galangashi, R. B., Kohansal, A. R., & Adnani, S. S. (2017). Designing a rotary converter for absorbing sea wave energy. *Research in Marine Sciences, 2*(4), 188–193.

Garvin, D. A., & March, A. (1986). *The views of Deming, Juran, and Crosby*. Harvard Business School. Harvard Business Publishing.

Geertz, C. (1973). *The interpretation of culture*. Basic Books.

Glass, L. (1975). Classification of biological networks by their qualitative dynamics. *Journal of Theoretical Biology, 54*(1), 85–107.

Gordon, I. (1999). Internationalisation and urban competition. *Urban Studies, 36*(5/6), 1001–1016.

Gould, S. J., & Eldredge, N. (1977). Punctuated equilibria: The tempo and mode of evolution reconsidered. *Paleobiology, 3*(2) (Spring, 1977), 115–151.

Grassé, P. P. (1959). La reconstrucion du nid et les coordinations inteindividuelles chez Bellicositermes Natalensis et Cubitermes. La théorie de la stigmergie: essai d'interpretation du comportement des termites constructeurs. *Insect Sociaux, 6*(1), 41–80.

Grassé, P. P. (1960). The automatic regulations of collective behavior of social insect and "stigmergy". *Journal de Psychologie Normale et Pathologique, 57*, 1–10.

Grassé, P. P. (1967). Nouvelles experiences sur le termite de Müller (Macrotermes mülleri) et considerations sur la théorie de la stigmergie. *Insectes Sociaux, 14*(1), 73–101.

Gregory, R. L. (1987). *The Oxford companion to the mind*. Oxford University Press.

Guizzo, E. (2010). *World robot population*. Retrieved from https://spectrum.ieee.org/automaton/robotics/industrial-robots/041410-world-robot-population

Haak, T. (2020).·*15 Examples of workplace nudging*. Online at https://hrtrendinstitute.com/2020/11/09/15-examples-of-workplace-nudging/

Hakansson, H., & Snehota, I. (1994). *Developing relationships in business networks*. Routledge.

Harvey, L., & Green, D. (1993). Defining quality. *Assessment & Evaluation in Higher Education, 18*(1), 9–34.

Hawking, S. W., & Mlodinow, L. (2010). *The grand design*. Random House Publishing Group. Excerpt. http://bookre.org/reader?file=65691.

Heylighen, F. (2007). *Why is open access development so successful? Stigmergic organization and the economics of information*. https://arxiv.org/pdf/cs/0612071.pdf.

Holland, J. H. (1975). *Adaptation in natural and artificial systems: An introductory analysis with applications to biology, control, and artificial intelligence*. University of Michigan Press.

Huxley, J. (1942). *Evolution: The modern synthesis*. George Allen and Unwin.

IFR. International Federation of Robotics. (2019 online). *Industrial and service Robots*. Retrieved from https://ifr.org/downloads/press2018/IFR%20World%20Robotics%20Presentation%20-%2018%20Sept%202019.pdf

Illiyas, F. T., Mani, S. K., Pradeepkumar, A. P., & Mohan, K. (2013). Human stampedes during religious festivals: A comparative review of mass gathering emergencies in India. *International Journal of Disaster Risk Reduction, 5*, 10–18.

Jaffe, A., Trajtenberg, M., & Henderson, R. (1993). Geographic localisation of knowledge spillovers as evidenced by patent citations. *Quarterly Journal of Economics, 108*, 577–598.

Kareiva, P., Mullen, A., & Southwood, R. (1990). Population dynamics in spatially complex environments: Theory and data. *Philosophical Transactions of the Royal Society B: Biological Sciences, 330*(1257), 175–190.

Kefan, X., Song, Y., Liu, S., & Liu, J. (2019). Analysis of crowd stampede risk mechanism: A systems thinking perspective. *Kybernetes, 48*(1), 124–142.
Koza, J. R. (1992). *Genetic programming: On the programming of computers by means of natural selection* (Vol. 1). The MIT Pess.
Krizna, M., Tahara, T., Tainaka, K., Ito, H., Morita, S., Ichinose, G., Togashi, T., Nagatani, T., & Yoshimura, J. (2018). Multi-species coexistence in Lotka-Volterra competitive systems with crowding effects. *Scientific Reports, 8*(1), 1–8.
Kuhn, T. S. (1962). *The structure of scientific revolutions* (2nd edn. 1970). Chicago, IL: University of Chicago.
Lakatos, I. (1978). *The methodology of scientific research programmes*. Cambridge University Press.
Lamarck, J. B. (1809). *Philosophie zoologique, ou, Exposition des considérations relative à l'histoire naturelle des animaux.* Hachette Livre BNF (reprint, 2012).
Lazerson, M. H., & Lorenzoni, G. (1999). The firms that feed industrial districts: A return to the Italian source. *Industrial and Corporate Change, 8*(2), 235–266.
Lewis, M. P., Simons, G. F., & Fennig, C. D. (Eds.). (2013). *Ethnologue: Languages of the world.* SIL International. Retrieved from https://www.ethnologue.com
Lorenzoni, G., & Lipparini, A. (1999). The leveraging of interfirm relationships as a distinctive organizational capability: A longitudinal study. *Strategic Management Journal, 20*(4), 317–338.
Lotka, A. J. (1925). *Elements of physical biology*. Williams & Wilkins.
Lundvall, B. Å. (1992). Introduction. In B. Å. Lundvall (Ed.), *National systems of innovation. Towards a theory of innovation and interactive learning*. Pinter.
Lundvall, B. Å., & Johnson, B. (1994). The learning economy. *Journal of Industry Studies, 1*(2), 23–42.
Maillat, D. (1998). Interaction between urban systems and localized productive systems: An approach to endogenous regional development in terms of innovative milieu. *European Planning Studies, 6*, 117–129.
Malthus, T. R. (1798). *An essay on the principle of population, as it affects the future improvement of society, with remarks on the speculations of Mr. Godwin, M. Condorcet, and other writers.* London: J. Johnson. Recent Edition (2007), The Lawbook Exchange.
Manzoni, A. (1985). *The Betrothed* [I promessi sposi, about 1840]. Mondadori, Milano, Italy. At https://moodle2.units.it/pluginfile.php/441330/mod_resource/content/1/MANZONI-PROMESSI%20SPOSI.pdf.
Marshall, A. (1919). *Industry and trade. A study of industrial technique and business organization; and of their influences on the condition of various classes and nations*. Macmillan.
Marshall. A. (1890). *Principles of economics*. London: Macmillan and Co. Preface, at https://www.econlib.org/library/Marshall/marP.html
Maslow, A. H. (1943). A theory of human motivation. *Psychological Review, 50*, 370–396.
Maslow, A. H. (1954). *Motivation and personality*. Harper and Row.
Maturana, H. R., & Varela, F. J. (1980 1st ed. 1972). *Autopoiesis and cognition. The realization of living*. Dordrecht, Netherlands/Boston: Reidel Publishing.
Maturana, H. R., & Varela, F. J. (1987). *The tree of knowledge: The biological roots of human understanding*. New Science Library/Shambhala Publications.
McLeod, S. (2007 online). *Maslow's hierarchy of needs*. At http://www.simplypsychology.org/maslow.html
Megacities online. (n.d.). At https://www.statista.com/chart/1826/population-growth-in-the-worlds-megacities/
Mella, P. (2006). Spatial co-localisation of firms and entrepreneurial dynamics the combinatory system view. *The International Entrepreneurship and Management Journal, 2*, 391–412.
Mella, P. (2011). The quality policy in value based management. *The Journal of American Academy of Business, Cambridge, 17*(2011), 68–180.
Mella, P. (2012). *Systems thinking. intelligence in action*. Springer Verlag.
Mella, P. (2017). The Unexpected Cybernetics Life of Collectivities. *The Combinatory Systems Approach. Kybernetes, 46* (7),1086–1111.
Mella, P. (2018). The law of increasing productivity. *International Journal of Markets and Business Systems, 3*(4), 297–316.

Mella, P. (2021, 1st ed. 2014). The magic ring. *Systems thinking approach to control systems*. Contemporary systems thinking series. Springer.
Monod, J. (1970). *Chance and necessity: Essay on the natural philosophy of modern biology*. New York: Alfred A. Knopf, 1971 [Original, *Le hazard et la nécéssité: Essai sur la philosophie naturelle de la biologie moderne*. Paris: Seuil].
NCB, New Catholic Bible. (2019). *Genesis* 11:1–9. Retrieved from https://bible.usccb.org/bible/genesis/11
Nelsen, D., & Daniels, S. E. (2007). Quality glossary. *Quality Progress, 40*(6), 39.
Nelson, R. (1995). Theorizing about economic change. *Journal of Economic Literature, 33*, 48–90.
Nelson, R., & Winter, S. (1982). *An evolutionary theory of economic change*. The Belknap Press of Harvard University Press.
Ngai, K. M., Burkle, F. M., Hsu, A., & Hsu, E. B. (2009). Human stampedes: A systematic review of historical and peer-reviewed sources. *Disaster Medicine and Public Health Preparedness, 3*(4), 191–195.
Nguyen, D. H., & Yin, G. (2016). Coexistence and exclusion of stochastic competitive Lotka-Volterra models. *Journal of Differential Equations, 262*(3), 1192–1225.
Nonaka, I. (1994). A dynamic theory of organizational knowledge creation. *Organization Science, 5*(1), 14–37.
Nonaka, I., & Konno, N. (1998). The concept of "ba": Building a foundation for knowledge creation. *California Management Review, 40*(3), 40–54.
Nooteboom, B. (1999). Innovation, learning and industrial organization. *Cambridge Journal of Economics, 23*(2), 127–150.
Ohno, T. (1988). The Toyota production system; beyond large-scale production. Portland: Productivity Press.
Olsen, M. (2019). Stigmergy for Biological Spatial Modeling. Complex Adaptive Systems: Views from the Physical, *Natural, and Social Sciences*, 169–197.
Otter, H., van der Veen, A., & de Vriend, H. J. (2001). ABLOoM: Location behaviour, spatial patterns, and agent-based modelling. *Journal of Artificial Societies and Social Simulation, 4*(4) Retrieved from http://jasss.soc.surrey.ac.uk/4/4/2.html
Packard, A. S. (1901). *Lamarck, the founder of evolution. His life and work with translations*. New York/London: Longmans, Green.
PestSmart Connect (online). (n.d.). At https://pestsmart.org.au
Pirsig, R. (1974). *Zen and the art of motorcycle maintenance: An inquiry into values*. Bantam Books.
Porter, M. E., & Sölvell, Ö. (1998). The role of geography in the process of innovation and the sustainable competitive advantage of firms. In Chandler, A. D., Hagström, P., & Sölvell, Ö. (Eds). (1999).
Pray, L. (2008). Discovery of DNA structure and function: Watson and Crick. *Nature Education, 1*(1), 100.
Prigogine, I., & Nicolis, G. (1989). *Exploring complexity: An introduction*. Freeman.
Rabellotti, R. (1997). *External economies and cooperation in industrial districts*. McMillan Press.
Rigopoulou, I. D., Chaniotakis, I. E., Lymperopoulos, C., & Siomkos, G. I. (2008). After-sales service quality as an antecedent of customer satisfaction: The case of electronic appliances. *Managing Service Quality: An International Journal, 18*(5), 512–527.
Ringwood, A. E. (1979). *Origin of the earth and moon*. Springer.
Riva, A., & Pilotti, L. (2019). Sassuolo's ceramic district and strategic changes in the sector during the time. *Economia Aziendale Online, Special Issue, 10*(3), 455–482.
Rondo-Brovetto, P., & Saliterer, I. (2007). Comparing regions, cities, and communities: Local government benchmarking as an instrument for improving performance and competitiveness. *The Innovation Journal: The Public Sector Innovation Journal, 12*(3), 1–18.
Rotary wave. (2014, online). Webpage: https://www.rotarywave.com/en/company/
Russell, S. J., & Norvig, P. (2016). *Artificial intelligence: A modern approach*. Pearson.
Sapir, E. (1968). *Selected writings of Edward Sapir*. University of California Press.

Schlenker, P. (2004). Conditionals as definite descriptions. *Research on Language and Computation, 2*(3), 417–462.

Schmitz, J. A., Jr. (2005). What determines productivity? Lessons from the dramatic recovery of the US and Canadian iron ore industries following their early 1980s crisis. *Journal of Political Economy, 113*(3), 582–625.

Schoener, T. W. (2011). The newest synthesis: Understanding the interplay of evolutionary and ecological dynamics. *Science, 331*(6016), 426–429.

Schowalter, T. D. (2016). *Ecosystem structure and function* (Chap. 11). In: *Insect ecology*. (2022), Elsevier, Academic Press.

Senge, P. M. (1990). *The fifth discipline: The art & practice of the learning organization*. Currency Doubleday.

Singer, P. W. (2009). Isaac Asimov's laws of robotics are wrong. *Brookings Institute, 18*, 2009.

Smith, A. (1776). *An inquiry into the nature and causes of the wealth of nations.* Edited with an Introduction, Notes, Marginal Summary and an Enlarged Index by Edwin Cannan. London: Methuen, 1904. Vol. 1.

Spelta, A. M. (1603), *History written by Antonio Maria Spelta, citizen of Pavia* [original: Historia d'Antonio Maria Spelta, Cittadino Pavese]. Pavia (Italy): Pietro Bartoli (Ed), 1603. Google books.

Staddon, J. E. (2016, 1st ed. 1983). *Adaptive behavior and learning*. Cambridge, MA, Cambridge.

Szeliski, R. (2022). *Computer vision: Algorithms and applications*. Springer Nature.

Takeuchi, Y. (1995). *Global dynamical properties of Lotka-Volterra systems*. World Scientific.

Tansley, A. G. (1935). The use and abuse of vegetational concepts and terms. *Ecology, 16*(3), 284–307.

Thaler, R., & Sunstein, C. (2008). *Nudge: Improving decisions about health, wealth and happiness*. Yale University Press.

Thaxton, C., Bradley, W., & Olson, R. (1984). *The mystery of life's origin*. Philosophical Library.

Thorelli, H. B. (1986). Networks: Between markets and hierarchies. *Strategic Management Journal, 7*, 37–51.

Turing, A. M. (1950). Computing machinery and intelligence. *Mind, 59*, 433–460.

Tylor, E. B. (1871). *Primitive culture: researches into the development of mythology, philosophy, religion, art, and custom* (book in 2 Volumes). J. Murray.

Uhrmacher, A. M., & Weyns, D. (Eds.). (2009). *Multi-agent systems: Simulation and applications*. CRC Press (Taylor & Francis).

Volterra, V. (1926). Variations and fluctuations of the number of individuals in animal species living together. *Animal Ecology*, New York, NY: McGraw-Hill (1931) [Original: Variazioni e fluttuazioni del numero d'individui in specie animali conviventi. Atti dell'Accademia Nazionale dei Lincei, Italy, 31–113].

Volterra, V. (1931). *Lessons on the mathematical theory of struggle for* life [original: Leçons sur la théorie mathématique de la Lutte pour la vie]. Paris, France: Gauthier-Villars].

von Bertalanffy, L. (1968). *General system theory: Foundations, development, applications*. Braziller.

von der Pütten, J. C. (2008). *Moral issues and concerns about China's one-child policy: A cosmopolitan perspective*. GRIN Verlag.

von Hayek, F. (1942). Scientism and the study of society. Part I. *Economica, 9*(35), 267–291.

Von Neumann, J., & Burks, A. W. (1966). Theory of self-reproducing automata. *IEEE Transactions on Neural Networks, 5*(1), 3–14. Retrieved from: https://cba.mit.edu/events/03.11.ASE/docs/VonNeumann.pdf

Waddington, C. H. (1957). *The strategy of the genes*. Allen and Unwin. (Reprinted 2014).

Watson, J. D., & Crick, F. H. (1953). Molecular structure of nucleic acids. *Nature, 171*(4356), 737–738.

Weidenschilling, S. J. (1991). Formation of planetesimals. *Planetary Sciences: American and Soviet Research, 1*, 82–97.

Weidenschilling, S. J. (2000). Formation of planetesimals and accretion of the terrestrial planets. *Space Science Reviews, 92*(1–2), 295–310.
Weinshall, T. D., & Raveh, Y. A. (1983). *Managing growing organizations: A new approach*. Wiley.
Whorf, B. L. (1956). *Language thought, and reality*. The MIT Press.
Widhiarso, W. (2009). *The influence of language on thought study Benjamin Whorf hypothesis and Edward Sapir*. Available at SSRN. https://ssrn.com/abstract=1496642
Wikipedia. (2024a). *Al-Aimmah Bridge disaster*. Retrieved from https://en.wikipedia.org/wiki/2005_Al-Aimmah_Bridge_disaster
Wikipedia. (2024b). *Hillsborough disaster*. Retrieved from https://en.wikipedia.org/wiki/Hillsborough_disaster
Wikipedia. (2024c). *List of fatal crowd crushes*. Retrieved from https://en.wikipedia.org/wiki/List_of_fatal_crowd_crushes
Wikipedia. (2024d). *Megacity*. Retrieved from https://en.wikipedia.org/wiki/Megacity
Wikipedia. (2024e). *Mina stampede*. Retrieved from https://en.wikipedia.org/wiki/2015_Mina_stampede
Wikipedia. (2024f). *Turin, Italy, Stampede* (2017). Retrieved from https://en.wikipedia.org/wiki/2017_Turin_stampede
Wikipedia. (2024g). *Isaac Asimov*. Retrieved from https://en.wikipedia.org/wiki/Isaac_Asimov
Williamson, O. E. (1993). The evolving science of organization. *Journal of Institutional and Theoretical Economics, 149*, 36–63.
Willis, A. J. (1997). The ecosystem: An evolving concept viewed historically. *Functional Ecology*, 268–271.
YouTube. (2022). *Massive Mexican Wave! – London 2012 Olympic Stadium*. Retrieved from https://www.youtube.com/watch?v=hq387m0PRac
Zollo, G., Iandoli, L., & De Maio, L. (2001). An application of an ant colony system to the analysis of organizational learning processes. In: *VIII SIGEF Congress Proceedings, New Logics for the new economy, Naples*.
Zuradelli, C. (1888). *The towers of Pavia* [original: Le torri di Pavia], Pavia (Italy): Forni.

Chapter 3
Simulation Models: The Combinatory Automaton

Abstract The chapter introduces the concept of Combinatory Automaton, a tool for simulating, mainly in probabilistic terms, the different classes of combinatory systems. Combinatory Automata models offer a simulation of the dynamics of combinatory systems in terms of the dynamics, expressed in probability terms, of the state of the system and the state of the individual elements. A Combinatory Automaton is composed of a matrix, each of whose cells contains a variable representing the state of an agent. The value of each cell at time t_h depends on a *synthetic global variable* whose values derive from some "combination" operation carried out on the values of the cells in successive iterations; the synthetic global variable represents the *synthetic state* of the automaton. The *micro-macro feedback* operating in successive moments for a certain period connects the analytical values of the cells and the synthetic state of the automaton. This chapter seeks to demonstrate, through simple examples, that combinatory systems represent a wide range of behaviors of collectivities and that Combinatory Automata are a powerful tool for simulating the most relevant combinatory systems.

Keywords Stochastic Combinatory Automaton · Boolean automaton · Reversible and irreversible Combinatory Automaton · Fields of probabilities · Oscillating macro behavior · "Chaotic" macro behavior

> The psychological core of understanding, I shall assume, consists of having a "working model" of the phenomenon in your mind. If you understand inflation, a mathematical proof, the way a computer works, DNA, divorce, then you have a mental representation that serves as a model of an entity in much the same way as, say, a clock functions as a model of the earth's rotation ... Many of the models in people's minds are little more than high-grade simulations, but they are none the less useful provided that the picture is accurate. (Johnson-Laird, 1983, pp. 2, 4)

> Modeling is neither more nor less than Reasoning. (Simon & Kaplan, 1989, p.19)

P. Mella, *The Combinatory Systems Theory*, Progress in IS,
https://doi.org/10.1007/978-3-031-86946-4_3

Combinatory systems can be represented by various types of models:

(a) *Descriptive models*: these represent the fundamental elements that distinguish the operational logic of the combinatory system, as indicated in Chap. 1.
(b) *Heuristic models*: these specify, with concrete operative procedures, the (supposedly) effective micro and macro rules of the regulatory program from which the micro behavior of the elements and of the system in general derive. They are useful for simulating the dynamics of the system in terms which are usually deterministic, representing, in fact, behavior through the application of the specified rules. Chap. 2 deals with several heuristic models of significant combinatory systems.
(c) *Combinatory Automata models*: these not only offer a description of the behavior of combinatory systems but also simulate the dynamics of the macro effect and/or the macro behavior in terms of the dynamics, expressed in probability terms, of the state of the system and the state of the individual elements. In this chapter an introduction to these models will be presented.

Heuristic models offer a simple description of the logic of combinatory systems, but they do not allow us to carry out simulations of their dynamics. Combinatory Automata are specific mathematical-statistical simulation models that try to simulate the dynamics of the micro-macro effects and/or the micro-macro behavior of combinatory systems. A Combinatory Automaton can easily be constructed when it is possible to formulate a heuristic model that describes the behavior of systems in terms of explicit operative micro and macro rules that can be translated into equations.

A Combinatory Automaton is a simple useful tool for simulating the micro and macro behaviors/effects of a combinatory system whose agents act based on *global information* that is identifiable with, or is produced by, the macro *behavior* or the *macro* effect of the system. A Combinatory Automaton may be conceived of as a variation of a general cellular automaton. The latter simulates the behaviors of collectivities acting based on *local information*; the Combinatory Automaton acts according to *global information*. A Combinatory Automaton is composed of a matrix, each of whose cells contains a variable representing the state of an agent. The value of each cell at time t_h depends on a *synthetic global variable* whose values derive from some operation carried out on the values of the cells; the *synthetic global variable* represents the synthetic state of the automaton. The *micro-macro feedback* connects the analytical values of the cells and the synthetic state of the automaton. The most interesting type of Combinatory Automata are the probabilistic ones that aim at examining the dynamics of the state of the system—and thus its macro behavior and macro effect (if these two elements do not coincide)—by considering the probability that the individual elements of the base will change their state and trigger the system's dynamics.

This chapter will try to demonstrate, through simple examples, that combinatory systems represent a wide range of behaviors of collectivities, that Combinatory Automata are a powerful tool for simulating the most relevant combinatory systems, and that combinatory systems, despite their simplicity, can show chaotic dynamics and, of course, path dependence.

3.1 Combinatory Automaton

3.1.1 Defining the Deterministic Combinatory Automaton

A Combinatory Automaton is composed of a matrix, each of whose cells contains a variable representing the state of an agent. The value of each cell at time t_h depends on a synthetic global variable whose values derive from some operation carried out on the values of the cells. This global variable represents the synthetic state of the automaton, which produces the macro effect. The *micro-macro feedback* connects the analytical values of the cells and the synthetic state of the automaton.

Definition 7 Combinatory Automaton (Deterministic)
A "Combinatory Automaton" is an automaton that symbolizes a combinatory system (Sect. 1.3.3) in a matrix of N cells, which represent—for a given instant t_h—the micro states, $x_n(t_h)$ of the A_n, n = 1, 2, … N, and/or the micro effects, $e_n(t_h)$, corresponding to these individual states. The "combination," according to the recombining rules F of the $x_n(t_h)$, allows us to quantify the *synthetic macro state* $Y(t_h)$ and/or the *global macro effect* $E(t_h)$, which generally represent the *self-produced global information* based on which the agents, in the presence of the *necessitating* functions f_n, determine their new micro state $x_n(t_{h+1})$, which leads to the new macro state $Y(t_{h+1})$—together with the micro and macro effects $e_n(t_{h+1})$ and $E(t_{h+1})$—as described by the model in (1.10) (which is presented again below):

$$\begin{cases} Y(t_h) = F\left[X(t_h)\right] = F\left[x_1(t_h), x_2(t_h), \ldots x_N(t_h)\right] \\ E(t_h) = G\left[Yt_h)\right] \\ x_n(t_{h+1}) = f_n\left[x_n(t_h), Y(t_h) \text{ or } E(t_h)\right] \end{cases} \qquad (1.10\text{—Sect. } 1.3.3)$$

The structural model of the Combinatory Automaton is shown in Fig. 3.1.

In their simplest form, Combinatory Automata can simulate combinatory systems which can be described according to the logic indicated in Sects. 1.3.1–1.3.3 and the numerically quantifying rules (1.1) to (1.10) in those sections. In short, referring to the model in (1.10), it is possible to state the following correspondence between a Combinatory Automaton and a collectivity behaving as a combinatory system:

- Combinatory Automaton (CA) → combinatory system → collectivity.
- Cells → Agents A_n, n = 1, 2, … N.
- Value in a "cell (n)" → a value $x_n(t)$, which represents the micro state of an agent at t_h.
- Analytical state of the CA → analytical state of the collectivity at time t_h, $x_n(t_h)$ of the A_n, n = 1, 2, … N.
- Synthetic state of the CA → Macro variable, $Y(t_h)$ and/or $E(t_h)$, or *global information* produced by the collectivity, acting as a whole, and driving the micro behaviors.

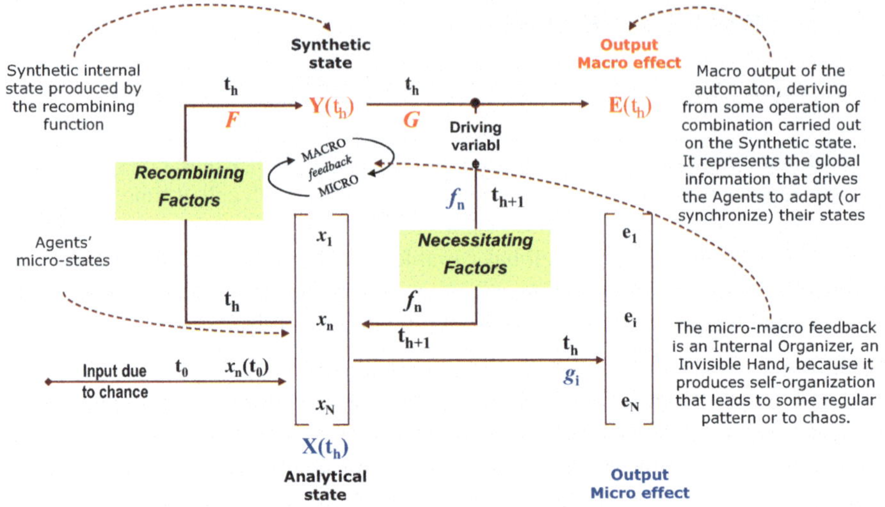

Fig. 3.1 Structural model of the deterministic Combinatory Automaton (see Fig. 1.11)

- Functions *F* and f_n → rules representing the influx of *recombining* and *necessitating* factors for producing macro and micro effects.

Moreover, a Combinatory Automaton can have the following operating features:

- *Reversibility* is admitted, in the model (1.10), if $x_n(t_h) = x_n(t_k)$, $h \neq k$. Reversibility has not been explicitly considered, as this will be developed in subsequent sections.
- *Randomness or causality*, if, respectively, the initial analytical state, $X(t_0)$, is produced "by chance" or is attributable to a specific "activation program" of an external will.
- *Mono-* or *multiple-driven*, depending on the number of driving variables—$Y(t_h)$ or $E(t_h)$—which determine the change in state of the cells. The model in Fig. 3.1 presents a mono-driven automaton, which will serve as the model of reference throughout the book.

The *social combinatory systems* that are most interesting and easiest to represent are the *irreversible* ones (build a tower or not, teach French or English to children). In these systems both the *micro* and *macro* behaviors produce permanent effects that may be viewed as increasing or decreasing cumulative processes (see Sect. 3.4). Nevertheless, reversible automata can show irregular, unpredictable, and often chaotic behavior (Sect. 3.4.3).

3.1.2 Simulating the Dynamics of a Mono-Dimensional Automaton

The cells can contain different types of values, $x_n(t_h)$, determined by the *recombining rules*, which are typical of combinatory systems represented by a Combinatory Automaton. These values can belong to numeric variables, discrete values, or continuous ones, measured at each t_h; or they can be symbolic values that, for each t_h, take on a symbol belonging to a specific set: for example, ["yes," "no"], ["white," "black," "yellow," etc.], ["applaud," "do not applaud"], and so on. In many cases, it is convenient to consider the cells as Boolean variables with the values "0" or "1," which have specific significance in the simulated system. For example, a cell can represent a rich inhabitant of medieval Pavia, who we can indicate by ["0" = has not built a tower] and ["1" has built a tower] (see Sect. 3.4.1). For the simulation, it is easy to calculate the number of towers existing at t_h by simply summing the "1" values in the automaton matrix.

Operating with vectors and matrices, constructing a deterministic Combinatory Automaton generally is not particularly difficult, since we can use Excel, which is a flexible and particularly suitable instrument. It is also possible to use MATLAB or a programming language, for example C or Python, though for simple simulations the results would be identical to those obtained using Excel, even if it is more difficult to visualize the dynamics over time of the values of the automaton. For the sole purpose of demonstrating how Excel simulations are much simpler, several Combinatory Automata simulations will be presented below which were obtained using "Powersim," a language for creating models of dynamic systems.

To show how the model in (1.10) can be translated into a linear Combinatory Automaton using Excel, Fig. 3.2 presents a very simple and abstract example which, obviously, does not correspond to any real situation to my knowledge.

The automaton is composed of only ten agents arranged in a row vector. The cells contain the micro states of the agents (expressed as whole numbers) to which the automaton applies the recombining rule, *F*, which merely makes it mandatory to calculate, for each t_h, the arithmetic average of the micro states. The mean represents the *synthetic macro state* of the combinatory system, since it derives from the "combination = mean value" of the *micro states*. It also represents the global *self-produced information* that brings the agents to modify their micro states by applying the following abstract micro rules f_n:

(a) If your state is *above or equal to the average*, then maintain your state or increase it by +1, in accordance with your needs.
(b) If your state is *below the average*, then maintain your state or increase it by +1 or +2, in accordance with your needs.

At t_0 two agents "by chance" express a micro state, and the Combinatory Automaton calculates their average.

For the successive t_h, h = 1, 2, …, 11, the application of the micro rules (a) and (b) gradually modify the micro states, which in turn modify the average value,

	Analytical state										Synthetic state
	A_1	A_2	A_3	A_4	A_5	A_6	A_7	A_8	A_9	A_{10}	Average
t_0	0	0	0	10	0	15	0	0	0	0 "chance"	2.5
t_1	2	1	0	11	2	15	1	1	2	1	3.6
t_2	4	3	1	12	4	16	3	3	3	3	5.2
t_3	6	5	2	13	6	17	5	5	5	5	6.9
t_4	8	7	2	13	8	17	7	6	6	6	8
t_5	9	8	4	13	9	18	8	7	7	7	9
t_6	10	9	5	13	10	18	9	8	8	8	9.8
t_7	11	11	6	14	11	19	11	10	9	9	11.1
t_8	13	13	7	15	13	19	12	12	11	11	12.6
t_9	14	14	9	16	14	20	14	14	13	13	14.1
t_{10}	16	16	10	17	16	20	16	16	15	15	15.7
t_{11}	17	17	11	18	18	21	17	17	17	17	17

Fig. 3.2 An abstract Combinatory Automaton (see Figs. 1.11 and 3.1)

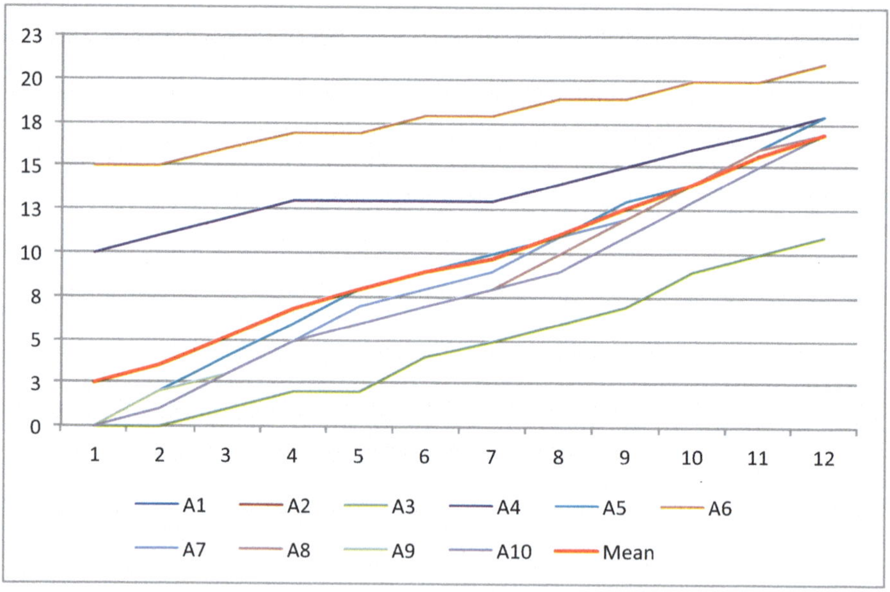

Fig. 3.3 Micro and macro dynamics in the Combinatory Automaton of Fig. 3.2

which again allows the agents to modify their micro states as part of a typical *micro-macro feedback* process, as shown in Fig. 3.2.

The dynamics of the micro and macro states are illustrated in Fig. 3.3.

The example is so abstract that it cannot reveal much about the dynamics produced by this Combinatory Automaton. The example tends to simulate a system of improvement and progress (Sect. 2.6) since from its micro rules we can deduce that the average of the states represents the *global information* on the "level of progress" achieved by the entire collectivity. The agents who perceive that their state is above the average try to maintain the superiority attained or to increase this somewhat. Those agents who perceive they are in a situation of inferiority with respect to the general level of progress try to improve their position through increments at times greater than those of the agents with a better individual state. These micro behaviors, for each t_h, raise the average, which spurs on the agents to make additional improvements. In the abstract example above, the "aggressive" behavior of the agents who were initially at a disadvantage and the "less aggressive" behavior of the advantaged agents lead the micro states to approach the average.

First Observation In the example above, as can be seen from Fig. 3.2, all the agents update their micro states at t_{h+1} by comparing the micro state at t_h with the macro state—that is, with the *average* of the values—calculated at t_h, and then making their decisions based on the intensity of the necessitating factors. This means that the agents in the Combinatory Automaton act "simultaneously" and "in parallel" fashion to update their micro states. Only when all the agents have made their decisions does the automaton update its macro state. This text does not consider Combinatory Automata with a "progressive functioning," which occur when the automaton updates its macro state instantaneously even when a single agent changes its micro state, thereby allowing the others to "progressively" make their decisions, as occurs, for example, in stock markets that (in theory) modify the price of a stock even when only a single trader carries out a trade.

Second Observation In the Combinatory Automaton in Fig. 3.2, the rule compels the agents, at t_{h+1}, to modify their micro state base on both their present state at t_h and the macro state, Y_h, which is generated by the automaton. No decision was based on the state of the neighboring agents, which means that the automaton simulates a "perfect" combinatory system (Sect. 1.3.5) whose agents operate with *complete information*. However, it is possible to also introduce the hypothesis of *incomplete* and *limited information*, in which case the agents change their state considering not only the macro state of the system but also the state of other *neighborhoods*, defined in an opportune way, as would occur if the following abstract micro rule were added to the previous rules (a) and (b):

(c) If your neighbors have a superior (inferior) micro state than yours, you can vary your state by +3 (−3) based on your needs.

This is still an abstract example for which it is not necessary to find an operational meaning for the micro rules. Nevertheless, having introduced rule (c) brings out a problem: apparently, agents A_1 and A_{10} cannot apply the rule, since they have only one *neighborhood*. However, we can agree that A_{10} is the neighbor to the left of A_1 and that to the right of A_{10} is A_1. This convention makes the automaton not only *linear* but also a *circular* Combinatory Automaton, in which all the agents, even if arranged in a row vector, must in fact be considered to belong to a "circle." In this chapter only linear automata will be considered.

3.1.3 Simulating the Dynamics of a Two-Dimensional Automaton

Considering the modus operandi of Combinatory Automata, it is useful to point out that these may be (a) *mono-dimensional*, or *linear*, if the agents are arranged in a vector of N cells, (b) two-dimensional, if the agents are arranged in a matrix of R rows and C columns, or (c) *multidimensional*; this book will consider only mono- and two-dimensional automata.

A *Two-Dimensional Combinatory Automaton* is an automaton that represents N agents—assuming that N is an even number—arranged in a matrix. Even if not strictly necessary, it is useful to posit that the matrix has "R" rows and "C" columns, so that C × R = N. The state, $x_{rc}(t_h)$, of each agent/cell A_{rc} is thus characterized by two indices, c = 1, 2, …C, r = 1, 2, …R, which indicate its placement in the matrix.

We note that if the agents are independent of each other, in that their micro behavior does not depend on that of its *neighbors*, then the dynamics of the states of the agents in the matrix corresponds, in fact, to that which would be produced if the agents were arranged in a row vector of "N" cells. The construction of a *two-dimensional automaton* is, nevertheless, necessary to simulate a combinatory system in which the agents show different behaviors depending on their position in the matrix, as occurs in combinatory systems with *incomplete* and *limited information.*

To understand the *micro* and *macro* dynamics of the automaton, we shall assume, for simplicity's sake, that a *two-dimensional automaton* is constructed to simulate a "perfect" combinatory system with N = 12 agents arranged in a 3 × 4 matrix, and that the state of the agents $x_{rc}(t_h)$ is recombined through a simple macro rule that determines $Y(t_h)$, based on the sum of the states, according to the following function:

$$Y(t_h) = \left[\sum_{1 \le r \le 3} \sum_{1 \le c \le 4} x_{rc}(t_h)\right] k$$

where "*k*" is a coefficient between 0 and 1, which, for convenience sake, is set at $k = 0.5$ in the example.

Figure 3.4 highlights the dynamics of the analytic and synthetic states of the *automaton* as corresponding to the dynamics of the matrices of the $x_{rc}(t_h)$, with h = 1, 2, …, 10.

On the upper border of the matrices, which indicate the dynamics of the micro states of the agents, the macro state $Y(t_h)$ is calculated. The first matrix indicates the *random* states at t_0. As in the example in the previous section, we assume that the *transition of state* functions for each cell of the matrix $x_{rc}(t_{h+1}) = f_{rc}[Y(t_h)]$ can be expressed using the following abstract micro rules:

(a) If $Y(t_h) < 2.0$, then $x_{rc}(t_{h+1}) = x_{rc}(t_h)$ + [0.0 or 0.1 or 0.2], according to the needs of each agent.
(b) If $Y(t_h) \ge 2.0$, then $x_{rc}(t_{h+1}) = x_{rc}(t_h)$ + [−0.2 or − 0.3], according to the needs of each agent.

$Y(t_0) =$ 0.40 $k =$ 0.5

	1	2	3	4
1	0.50	0.00	0.00	0.00
2	0.00	0.00	0.00	0.00
3	0.00	0.00	0.30	0.00

$Y(t_1) =$ 0.90 $k =$ 0.5

	1	2	3	4
1	0.60	0.20	0.10	0.10
2	0.10	0.00	0.10	0.20
3	0.00	0.10	0.30	0.00

$Y(t_2) =$ 1.90 $k =$ 0.5

	1	2	3	4
1	0.60	0.40	0.30	0.30
2	0.30	0.20	0.20	0.40
3	0.20	0.30	0.50	0.10

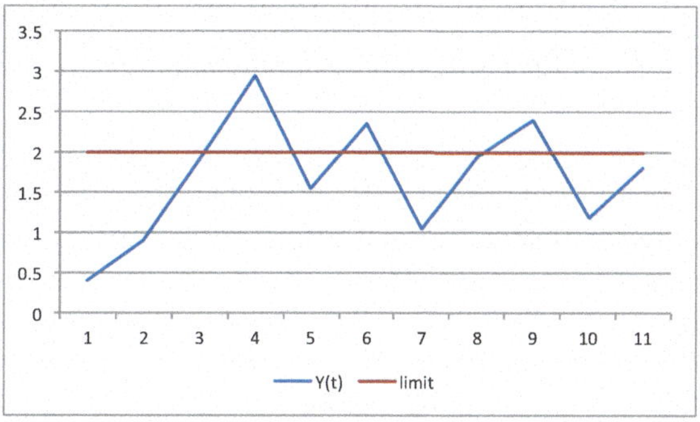

$Y(t_3) =$ 2.95 $k =$ 0.5

	1	2	3	4
1	0.80	0.60	0.40	0.50
2	0.50	0.40	0.40	0.60
3	0.40	0.40	0.60	0.30

$Y(t_4) =$ 1.55 $k =$ 0.5

	1	2	3	4
1	0.60	0.40	0.20	0.30
2	0.30	0.10	0.10	0.30
3	0.20	0.20	0.30	0.10

$Y(t_5) =$ 2.35 $k =$ 0.5

	1	2	3	4
1	0.70	0.50	0.40	0.40
2	0.50	0.30	0.20	0.50
3	0.30	0.10	0.50	0.30

[omissis]

$Y(t_{10}) =$ 1.80 $k =$ 0.5

	1	2	3	4
1	0.70	0.50	0.40	0.20
2	0.40	0.30	0.30	0.30
3	0.10	0.10	0.10	0.20

Fig. 3.4 Micro and macro dynamics in a *two-dimensional automaton*

It is not important to assign significance to an example whose only function is to demonstrate how simple the construction and functioning of a *two-dimensional Combinatory Automaton* is. The diagram in Fig. 3.4 (right side) shows that, due to rules (a) and (b), the dynamics of the macro state of the automaton oscillates around the limit $Y(t_h) = 2.0$, which obliges the agents to change the rule to follow to modify their micro states.

Observation. If we assume that the combinatory system simulated by the automaton functions with *incomplete* and *limited information* and that the agents change their state by observing that of their *neighborhood*, then the automaton could be a "spherical" Combinatory Automaton, in which each agent, represented by a cell located in the vectors forming the frame of the matrix, can observe the state of the agents represented by the vectors that form the other frames, based on how the *neighborhood* is defined (Sect. 1.3.5, footnote 2).

3.2 Stochastic Combinatory Automaton

The most interesting Combinatory Automata are the stochastic ones that seek to examine the dynamics of the state of the combinatory system—and thus its macro behavior and macro effect (if these two elements do not coincide)—by assigning a *probability* that the individual elements of the matrix will change their state and trigger the system's dynamics.

3.2.1 The Role of Probabilities

The *stochastic* automata form when, to carry out the simulations, the *necessitating factors* (which impel each cell to change its state) must be translated into a *transition of state probability*. In effect, the probability of transition should offer numerical information on all the observable, or even imaginable, characteristics that represent each A_n, n = 1, 2, … N, such as to make a change of state possible, plausible, probable, likely. The *transition of state probability* therefore expresses the influx of necessitating factors that impose on A_n its own micro behavior. In other words, it should express the likelihood of a given micro behavior and a given micro effect, which can potentially be carried out by and obtained from the agent.

We do not know what moves a man, even a distinguished one, to leave graffiti on a wall or to litter. But if we want to represent the Combinatory Automaton of the "graffiti" system or the system that gives rise to a pile of garbage, we must quantify the probability that a person will leave graffiti or abandon refuse, which also depends on the amount of graffiti or garbage already present (or on the number of persons *E* that have already left graffiti behind or have thrown their garbage in the street). However, it is clear that this *probability should be specified for each agent* who will pass in front of the wall, in order to take into account, through a quantitative variable, the typical characteristics of that person: age, profession, amount of education, lifestyle, and so on—information which the person who has to describe the combinatory system and carry out the simulation is not in a position to know.

How can we determine when a girl will decide to wear a miniskirt? Or a young man a hat? Nevertheless, we must estimate the *probability of transition of state* from "normal skirt" to "miniskirt" if we wish to build a Combinatory Automaton simulating the spread of a fashion. What are the characteristics necessary for a virus to attack an organism? What are the parameters for the spread of an epidemic? The *probability of transition of state* should include all the characteristics of the virus and the organisms that can be infected; it is necessary to estimate this if we want to describe a system involving the spread of a viral disease.

Probability theory and Statistics—in particular the Markovian theory of stochastic systems—and stochastic dynamical systems provide the instruments for constructing and employing probability models. The following discussion, however, is only meant to represent a first approximation regarding this subject. Elementary concepts will be used, beginning with a specification of the meaning of stochastic cell and stochastic matrix, as these concepts form the basis for constructing "Stochastic Combinatory Automata."

3.2.2 *Stochastic Cell and Automaton. Probability Field*

Definition 8 Stochastic Cell

A stochastic, or *probabilistic cell*, A_n, is one whose state $x_n(t_h)$ is assigned a number p_n that is between 0 and 1 and expresses the *probability of transition of state* (or, alternatively: behavioral probability, potentiality, expected frequency, likeliness, etc.); that is, the probability that an event (occurrence, happening, circumstance, fact, etc.) happens to (t_h), such that A_n modifies its state $x_n(t_h)$ to $x_n(t_{h+1})$ as a function of the occurrence of the event $a_n(t_h)$, with $\Delta t = (t_{h+1} - t_h)$ representing the *period of transition of state* for A_n.

To simulate the transition of state of A_n based on probability p_n, we can assume it is possible to undertake some "experiment" (based on defined protocols) whose outcome would be the equivalent of $a_n(t_h)$ occurring or not, based on the probability of transition of state we have assigned. The simplest and most immediate experiment, which will be used in this book to construct Combinatory Automata, is to generate a random number, $k_n(t_h)$, by specifying the following rule:

Conventional Rule for Simulations: the event $a_n(t_h)$ "occurs" at t_h if $k_n(t_h) < p_n$; "it does not occur" in the opposite case. For example, if we wish to simulate whether launching a perfect coin produces the event $a_n(t_h) \longrightarrow$ "heads," then we can specify that the event "heads" has occurred if $k_n(t_h) < 0.5$. If $k_n(t_h) \geq 0.5$, then $a_n(t_h) \longrightarrow$ "tails."

In the simulations presented in this book, it is assumed the same experiment has been done on each cell but that the results for each cell are different. For simplicity's sake, we shall write that the uniform experiment at (t_h) produces the specific result $k_n(t_h)$ for cell A_n.

Since the transition of state of the stochastic cell A_n depends on the probability p_n that characterizes it, it is necessary to represent the new state, $x_n(t_{h+1})$, also as a function of this probability, as follows:

$$x_n\left(t_{h+1}\right) = f_n\left[Y\left(t_h\right), x_n\left(t_h\right), p_n\right], \quad \text{which is conditioned by the event } a_n\left(t_h\right) \tag{3.1}$$

We can now define a stochastic Combinatory Automaton as a set of stochastic cells:

Definition 9 Stochastic Combinatory Automaton

A Combinatory Automaton that operates according to Definition 7, is defined as a *stochastic Combinatory Automaton* if it is composed of a matrix of *stochastic cells*, as described in Definition 8, and possesses the *transition of state* function for the cells defined in (3.1).

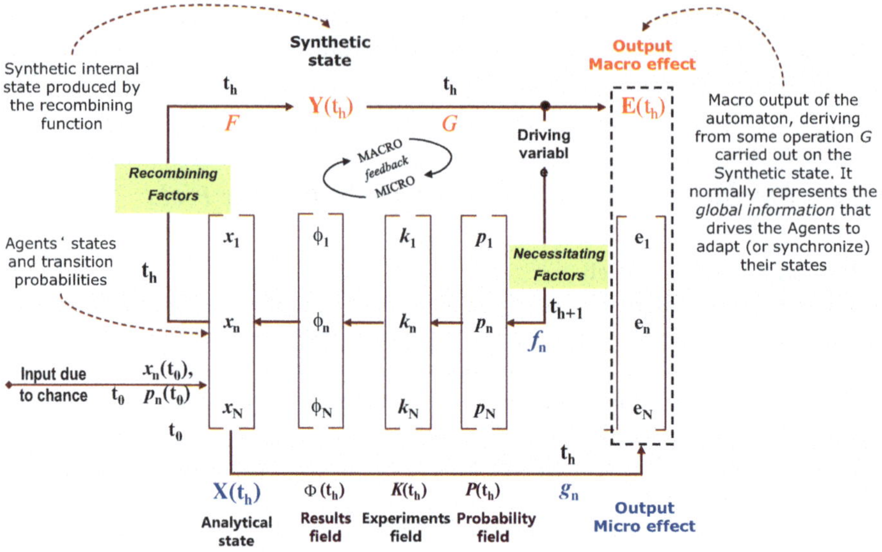

Fig. 3.5 Structural model of the stochastic Combinatory Automaton (see Fig. 3.1)

The structural model of the stochastic Combinatory Automaton is shown in Fig. 3.5.

Figure 3.5 shows that to carry out simulations using a *stochastic Combinatory Automaton* we must determine for each cell not only a *micro state* but also a *probability of transition* of state, p_n, and devise an experiment at (t_h) to decide whether the transition of state can be achieved or not. Thus, the automaton operates with four matrices:

A. The *matrix of the micro states*, $x_n(t_h)$—or $x_{rc}(t_h)$—which represents the *analytical state* of the automaton, $X(t_h)$.
B. The *matrix* (field) *of the transition of state probabilities*, p_n, which defines the probability field, $P(t_h) = [p_n]$, on which the transition of the *analytical state* from $X(t_h)$ to $X(t_{h+1})$ depends.
C. The *matrix* (field) *of the experiment values*, $K(t_h)$; that is, the random numbers, $k_n(t_h)$, which allow us to decide whether or not to carry out the transition of state of the automaton.
D. The *matrix* (field) *of the experiment results*, $\Phi(t_h)$—which derives from a comparison of the values of $K(t_h)$ and $P(t_h)$, according to the operational rules of the automaton—that is, of the random numbers $k_n(t_h)$, which allow us to decide whether or not to carry out the transition of state of the automaton.

The probabilities that form the *probability field* can be different for each cell and change in succeeding instants. They may be:

1. *Fixed* if $p_n \equiv p$ for every A_n, $1 \leq n \leq N$
2. *Time-dependent* if $p_n \equiv p(t_h)$
3. *Cell-dependent* if $p_n \equiv p(x_n)$
4. *Cell and time-dependent* if $p_n \equiv p(x_n, t_h)$
5. *Cell and output-dependent* if $p_n \equiv p(Y, x_n, t_h)$

Excluding Type 1, if the probabilities are *cell and output dependent*, then it is also necessary to define a probability transition function, fp_n, whose simplest form would be:

$$p_n(t_{h+1}) = fp_n\left[p_n(t_h), Y(t_h)\right] \tag{3.2}$$

Cell and *output-dependent probabilities* can act in two ways:

1. As *stop-or-go probabilities*, in the sense that if the probabilistic event, $a(t_h)$, occurs, then A_n assumes a new state; otherwise, the actual state is maintained. For the sake of simplicity, in Case 5 this type of probability will be symbolized here as $p(Y, x_n, t_h)_{[0,\,1]}$. The subscript "0" means that if the event does not occur, the agent maintains its state; "1" signifies that the agent changes its state if the event occurs.
2. As *transition probabilities*, in the sense that if the probabilistic event occurs, then the agent enters a new state; if the event does not occur, the agent assumes a different state or returns to the previous one. This will be indicated here as $p(Y, x_n, t_h)_{[-1,\,1]}$.

The length of time needed for the state of each A_n to be modified—for example, experiment to change from $a_n(t_h)$ to $b_n(t_h)$—is also a function of the state of the system; in this case the automaton is *time-response sensitive*. That length will be referred to as the *period of transition of state*—$\Delta_n(t_h)$—from $a_n(t_h)$ to $a_{n+1}(t_h)$. For simplicity's sake, the simulations presented in the following sections assume that $\Delta_n(t_h)$ is a constant and not explicitly considered. Thus, the models consider only non-time response sensitive automatons.

The definition of a *mono-dimensional* stochastic Combinatory Automaton with *cell* and *output-dependent probabilities* (the mode of action of the probabilities is not explicitly considered) is summarized in the following formal model (3.3), which includes the elements in Fig. 3.5:

$$
\begin{cases}
for\, \mathrm{n} = 1,2,\ldots \mathrm{N} & \\
for\, \mathrm{h} = 0,1,2\ldots & \\
\mathrm{X}(\mathrm{t}_0) = [x_\mathrm{n}(\mathrm{t}_0)] & \text{initial input} \begin{pmatrix} \text{random or} \\ \text{programmed} \end{pmatrix} \\
P(\mathrm{t}_0) = [p_\mathrm{n}(\mathrm{t}_0)] & \text{initial probability field} \\
\mathrm{X}(\mathrm{t}_\mathrm{h}) = [x_\mathrm{n}(\mathrm{t}_\mathrm{h})] & \text{analytical state} \\
\mathrm{Y}(\mathrm{t}_\mathrm{h}) = F[x_\mathrm{n}(\mathrm{t}_\mathrm{h})] & \text{synthetic state} \\
P(\mathrm{t}_\mathrm{h}) = [p_\mathrm{n}(\mathrm{t}_\mathrm{h})] & \text{probability field} \\
K(\mathrm{t}_\mathrm{h}) = [k_\mathrm{n}(\mathrm{t}_\mathrm{h})] & \text{experiment field} \\
x_\mathrm{n}(\mathrm{t}_{\mathrm{h}+1}) = f_\mathrm{n}\begin{bmatrix} x_\mathrm{n}(\mathrm{t}_\mathrm{h}), p_\mathrm{n}(\mathrm{t}_\mathrm{h}), \\ k_\mathrm{n}(\mathrm{t}_\mathrm{h}), \mathrm{Y}(\mathrm{t}_\mathrm{h}) \end{bmatrix} & \text{states transition} \\
p_\mathrm{n}(\mathrm{t}_{\mathrm{h}+1}) = fp_\mathrm{n}[p_\mathrm{n}(\mathrm{t}_\mathrm{h}), \mathrm{Y}(\mathrm{t}_\mathrm{h})] & \text{probabilities transition} \\
\Delta_\mathrm{n}(\mathrm{t}_\mathrm{h}) = \delta_\mathrm{n}[\mathrm{Y}(\mathrm{t}_\mathrm{h})] & \text{period of transition of state} \\
\mathrm{E}(\mathrm{t}_\mathrm{h}) = G[Yt_\mathrm{h})] & \text{macro effect} \\
\mathrm{e}_\mathrm{n}(\mathrm{t}_\mathrm{h}) = g_\mathrm{n}[x_n(\mathrm{t}_\mathrm{h})] & \text{macro effect} \\
Set: [\mathrm{X}(\mathrm{t}_0), P(\mathrm{t}_0), F, f_\mathrm{n}, fp_\mathrm{n}, \delta_\mathrm{n}, G, g_\mathrm{n}] & \text{operative program}
\end{cases}
\tag{3.3}
$$

The *set of rules* specifying the initial state $\mathrm{X}(\mathrm{t}_0)$, the initial probability field $P(\mathrm{t}_0)$, and the functions $F, f_\mathrm{n}, fp_\mathrm{n}, \delta_\mathrm{n}, G$, and g_n, together with the experiment field, represent the *operative program*, which produces the dynamics of the *Combinatory Automaton*; these rules must be specified for each simulation.

Incidentally, note that when in (3.3) the transition of states are defined as:

$$x_\mathrm{n}(\mathrm{t}_{\mathrm{h}+1}) = f_\mathrm{n}[x_\mathrm{n}(\mathrm{t}_\mathrm{h}),\ p_\mathrm{n}(\mathrm{t}_\mathrm{h}),\ k_\mathrm{n}(\mathrm{t}_\mathrm{h}),\ \mathrm{Y}(\mathrm{t}_\mathrm{h})]$$

then the Combinatory Automaton follows the logic of a “Melay machine”; if the transition of states is assumed to be:

$$x_\mathrm{n}(\mathrm{t}_{\mathrm{h}+1}) = f_\mathrm{n}[p(\mathrm{t}_\mathrm{h}), k(\mathrm{t}_\mathrm{h}), \mathrm{Y}(\mathrm{t}_\mathrm{h})]$$

then the automaton acts as a “Moore Machine.”

3.2.3 Reversible Stochastic Combinatory Automaton Simulating Pursuit and Order: "Voice-Murmur" Phenomenon in an Indoor Locale (Sect. 2.4.1)

An interesting example of a *reversible* Combinatory Automaton with *cell-dependent* transition of state *probabilities*—$p_n \equiv p(x_n)_{[1,\ 1]}$—is the one that simulates the combinatory system that produces the typical murmur in a locale with people who are conversing, the heuristic model for which was illustrated in Sect. 2.4.1. Wherever we observe this phenomenon, the system must be capable of recombining the voices of the individual agents at every moment, thereby producing a murmur that represents the macro effect of the system, which is calculated as the average of the speakers' voice levels, with correction coefficients to take into account exogenous and random factors.

The persons in the locale can alternate from remaining silent for some time and then returning to conversing, as we all have experienced. In other words, they change their micro behavior of "remaining silent/speaking" in a *reversible* manner at successive moments and for irregular periods of time.

This micro behavior does not depend only on the murmur arising in a locale, which forces those present to raise their voice levels, but also on the intensity of the *necessitating factors*, which can increase or put a brake on the desire to speak and which, in turn, are conditioned by the importance of the conversation, the education level of the speakers, the occasion for which they have gathered in the locale, and so on.

It is almost impossible to specifically research the unique *necessitating factors* for each speaker when constructing a Combinatory Automaton that simulates the system of speakers. The simplest "operational solution" is to translate the set of necessitating factors for each agent into a probability that the agents will speak or not speak, which will remain constant for the entire period of the simulation. Since the agents are assigned *transition* and *cell-dependent probabilities*, $p(x_n)_{[-1,\ 1]}$, the stochastic Combinatory Automaton is therefore *reversible*.

The action of the necessitating factors can be simulated by drawing random numbers, $k_n(t_h)$, which are specific for each agent and each period; if the experiment (drawing) produces a random number $k_n(t_h) \leq p_n(t_h)$, then agent A_n will modify his or her voice level, adjusting it to the background murmur in the locale. In the opposite case, the agent will "remain silent." For purposes of the simulation, this means that it is sufficient to introduce a Boolean variable, $speak_n(t_h) = [0, 1]$, which takes on the value "1" if $k_n(t_h) \leq p_n(t_h)$ and "0" in the opposite case. For the simulation, the values of $speak_n(t_h)$ are multiplied by the voice level that is calculated for the corresponding agent, so that this level is zero if $speak_n(t_h) =$ "0," meaning that at t_h, A_n remains silent.

The higher the probabilities assigned to the cells, the more the necessitating factors will move the agent to speak; lower probabilities will instead lead to the agents speaking less. Quantifying the probabilities for the cells must be done after analyzing the type of agent (students in a class without the teacher on the last day of school, the audience at an opera in a famous theater, etc.), and the form and structure of the locale (a narrow room, making it difficult to speak to those near us, a restaurant with tables full of happy friends, a broad square crowded with people attending a political rally, etc.).

The Combinatory Automaton model of the combinatory system that produces a buzzing noise in a crowded room can be simulated by the following simple probabilistic Combinatory Automaton, in which $M(t_h)$ represents the murmur as the macro output behavior of the collectivity, $N(t_h)$ represents the macro output effect of the automaton, and $v_n(t_h)$ represents the voice level as the micro effect output of the agent A_n.

The crowded room recombines the voice levels into a simple mean $[(1/N)\sum_{1\le n\le N} v_n(th)]$ which recursively increases the voice levels $v_n(t_{h+1})$, in a typical feedback between micro and macro behavior.

The basic variables and parameters that characterize this Combinatory Automaton can be summarized in the following (short) operative model (3.4):

$$\begin{cases} \text{for}: n = 1,2,\ldots N & \\ \quad \text{for}: h = 0,1,2,\ldots & \\ \quad [v_n(t_0)] & \text{initial random input (chance)} \\ M(t_h) = \left\{ \frac{k}{N} \left[\sum_{1\le n\le N} v_n(t_h) \right] \right\} & \text{collective murmur } \textit{as} \text{ the macro} - \text{behaviour} \\ N(t_h) = \left\{ M(t_h) + Q\left[r(t_h)_{[0,1]} \right] \right\} (1-a) & \text{background noise } \textit{as} \text{ the macro} - \text{effect} \\ v_n(t_{h+1}) = \begin{bmatrix} \left\{ \begin{matrix} \left[w_n N(t_h) + v_{n(min)} \right] \\ + v_{n(rnd)} 1_n(t_h)_{[-1,1]} \end{matrix} \right\} \times \\ \times \left\{ \begin{matrix} speak_{n[bol]}(t_h) \left[s_{n[-1,1]} \right] \\ \times t_{n[bol]}(t_h) \end{matrix} \right\} \end{bmatrix} & \text{voice level} \end{cases} \tag{3.4}$$

In the previous model the basic variables and parameters that characterize this Combinatory Automaton can be summarized as follows:

1. The group of speakers consists of N agents, A_n, with $1 \le n \le N$; their behavior is observed over discrete instants: t_h, $h = 0, 1, 2 \ldots$, to be specified for each observed combinatory system.
2. The recombining rule $M(t_h) = k\,[(1/N) \sum_{1\le n\le N} v_n(t_h)]$ represents the background *murmur* that arises from the combination of the speakers' voice levels; we assume this is determined by the arithmetic average of the speakers' voices, which represents the principal *recombining rule*. The average is rectified by the

amplifying factor "k," whose value depends on the shape of the room: normally, $0 < k < 1$; if $k > 1$, then the murmur is amplified.

3. The background *noise*, $N(t_h)$, does not depend solely on the background *murmur* but on other factors as well, among which the outside noise, which represents a macro-level disturbance; this is indicated by $Q\,[r(t_h)_{[0,1]}]$, with Q being the disturbance and $r(t_h)_{[0,1]}$ the probability of this occurring at each instant. I have also considered the possibility the background noise is externally controlled by sound absorbing panels by introducing a coefficient of sound absorption, "a"; if $a = 1$, there is total sound absorption and $N(t_h) = 0$.
4. $v_{n(min)}$ indicates the number of decibels (which differs for each agent) each speaker must produce above $N(t_h)$ to be heard by the others.
5. $v_n(t_{h+1}) = [w_n\,N(t_h) + v_{n(min)}] + v_{n(rnd)}$ represents the voice level of the agents above the *buzzing* noise level; the parameter w_n (usually set equal to 1) represents a synthetic measure of the position of each agent: if $w_n > 1$, the *buzzing* is perceived as amplified by the speaker, who then further raises his voice.
6. For each agent an anomalous (random) variation of the voice level is added to take into account random factors that oblige speakers to modify their way of speaking (emotional state, distance from other speakers, etc.); this factor is indicated by the additive factor $v_{n(rnd)}[1_n(t_h)_{[-1,\,1]}]$, in which $v_{n(rnd)}$ is the random level of voice for A_n, which is assumed proportional to the probability of random voices $[1_n(t_h)_{[-1,\,1]}]$.
7. $s_{n\,[-1,\,1]}$ indicates the probability of speaking, which is a very important variable in the model, since it synthetically expresses the set of necessitating factors that push individuals to speak or remain silent. For simplicity's sake, this probability is assumed to be constant over time, even though it is different for each speaker. The experiment with random numbers determines whether A_n speaks or remains silent. The model includes the Boolean variable $speak_{n\,[bol]}$, which is thus a function of $s_{n\,[-1,\,1]}$.
8. *Tolerance* to noise, $t_n(t_h)$, has also been introduced, which is indicated by $t_{n(bol)}(t_h)$, a Boolean variable that is equal to "0" when the speaker has reached the maximum level of toleration and stops talking.
9. As indicated in (7) and (8), the multiplying factor

$$\left\{speak_{n[bol]}\left(t_h\right)\left[s_{n[-1,1]}\right]\times t_{n[bol]}\left(t_h\right)\right\}$$

can thus take on the value "0"—A_n remains silent at t_{h+1}—or "1," in which case A_n changes his or her voice level from $v_n(t_h)$ to $v_n(t_{h+1})$, based on rules (5) and (6).

The preceding considerations can be translated into the example of a Combinatory Automaton of $N = 20$ agents arranged in a row vector and speaking in a room, since the shape of the room is assumed not to influence the dynamics of the agents, each of whom is characterized by specific individual parameters and probabilities of speaking, as represented in the control panel in Fig. 3.6. The control panel in Fig. 3.6 regulates a Combinatory Automaton that simulates a school class that goes wild

		A_1	A_2	A_3	A_4	A_5	A_6	A_7	A_8	A_9	…	A_{18}	A_{19}	A_{20}	Mean
Initial voice level (chance)	$V_n(t_0)$	8	11	0	8	8	10	10	12	13	…	10	10	10	8.5
Amplification coefficient	$k =$	1		a =	10%	Sound-absorbing coefficient					…				
External random noise Q	$Q =$	10		$r(t_h)$	20%	Probability of random noise					…				
Coefficient of position of A_n	w_n	1	1	1	1	1	1	1	1	1	…	1	1	1	1
Decibels above the murmur	$v_{n\,(min)}$	10	9	11	10	10	10	10	9	11	…	10	10	10	10
Additional random voice	$v_{n\,(rnd)}$	4	3	4	3	4	5	3	4	5	…	5	4	5	4
Probability of speaking	s_n	80%	90%	85%	90%	85%	70%	95%	95%	95%	…	90%	90%	80%	85.0%
Maximum tolerance level	T_n	70	70	70	70	70	70	70	70	70	…	70	70	70	70

Fig. 3.6 Control panel for the Combinatory Automaton that simulates background noise in a school room

Cycles $=t_h$	Agents	A_1	A_2	A_3	A_4	A_5	A_6	A_7	A_8	A_9	…	A_{18}	A_{19}	A_{20}	Macro Effect	
	Micro States	$V_1(t_h)$	$V_2(t_h)$	$V_3(t_h)$	$V_4(t_h)$	$V_5(t_h)$	$V_6(t_h)$	$V_7(t_h)$	$V_8(t_h)$	$V_9(t_h)$	…	$V_{18}(t_h)$	$V_{19}(t_h)$	$V_{20}(t_h)$	M(s, t)	mean
t=0 - Initial voice by chance	$V_n(t_0)$	8.0	11.0	0.0	8.0	8.0	10.0	10.0	12.0	13.0	…	10.0	10.0	10.0	7.7	8.5
Random event for external noise	$r(t_h)$	0.86									…					
Occurrence of external noise	Q	0.0									…					
Voice over the Noise base	w N	7.7	7.7	7.7	7.7	7.7	7.7	7.7	7.7	7.7	…	7.7	7.7	7.7		
Random event for Voice level	$l_n(t_h)$	0.42	0.86	0.06	0.40	0.77	0.10	0.47	0.12	0.76	…	0.57	0.65	0.48		
Random Voice level	$V_{n(md)}\,[l_n]$	1.7	2.6	0.3	1.2	3.1	0.5	1.4	0.5	3.8	…	2.8	2.6	2.4		
New Voice level	$V_n(t_h+1)$	19.3	19.2	18.9	18.8	20.7	18.1	19.0	17.1	22.4	…	20.5	20.2	20.0		
Probability of speaking	S_n	0.80	0.90	0.85	0.90	0.85	0.70	0.95	0.95	0.95	…	0.90	0.90	0.80	probability field	
Random event for speaking	k_n	0.44	0.47	0.39	0.84	0.77	0.26	0.60	0.59	0.83	…	0.42	0.30	0.23	experiments field	
To speak or not to speak	$speak_n$	1	1	1	1	1	1	1	1	1	…	1	1	1	results of experiments	
Tolerance factor	t_n	1	1	1	1	1	1	1	1	1	…	1	1	1		
t=1	$V_n(t_1)$	19.3	19.2	18.9	18.8	20.7	18.1	19.0	17.1	22.4	…	20.5	20.2	20.0	15.9	17.7

Fig. 3.7 First iteration for the Combinatory Automaton that simulates background noise in a school room

when the bell sounds for the last hour of school. From the control panel in Fig. 3.6 we can move on to the iterations, the first of which is presented in Fig. 3.7.

The dynamics produced by the Combinatory Automaton in Fig. 3.6 is presented in Fig. 3.8, which shows the dynamics in the voice levels (colored lines) and the buzzing noise (bold blue line) for a period of 30 cycles of the process, under different assumptions (each cycle hypothetically lasts 2 s). The noise (bold line) may be viewed as the output of the Combinatory Automaton constituting the collectivity as a whole. During a performance in a concert hall, only rude people whisper, and they are immediately told to be quiet by those nearby or even reproached by the usher.

As a second example, consider the simulation of the Combinatory Automaton represented in the operative model in (3.4), carried out using Powersim software (www.powersim.com). The Stock and Flow Diagram and the initial values of the automaton representing a classroom of ten completely wild students are shown in Fig. 3.9. Figure 3.10 illustrates simulation hypotheses like those in Fig. 3.8. The dynamics shown in Figs. 3.8 and 3.10 do not coincide due to the action of different random numbers generated by the various software utilized for the simulations.

Test 1 – External noise Q=10 decibels. Mean probability to speak=85% (class room). Max level of tolerance=70 decibels

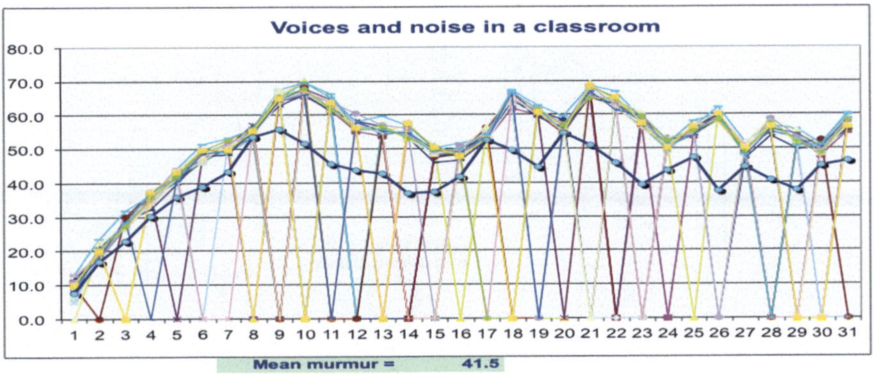

Test 2 – External noise Q=5 decibels. Mean probability to speak=50% (conference room). Max level of tolerance=15 decibels

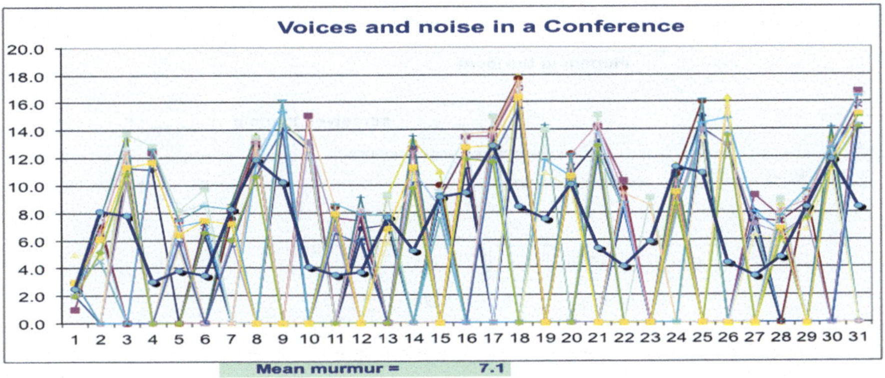

Test 3 – External noise Q=0 decibels. Mean probability to speak=10% (theater during a performance). Max level of tolerance=3 decibels

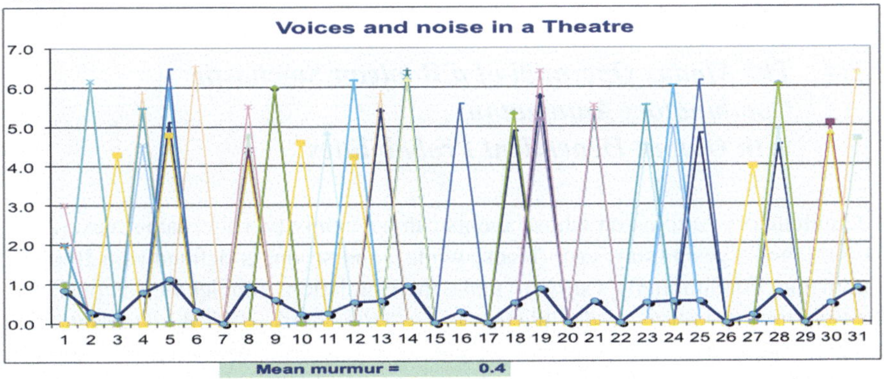

Fig. 3.8 Dynamics of murmur and noise Combinatory Automaton with 20 agents

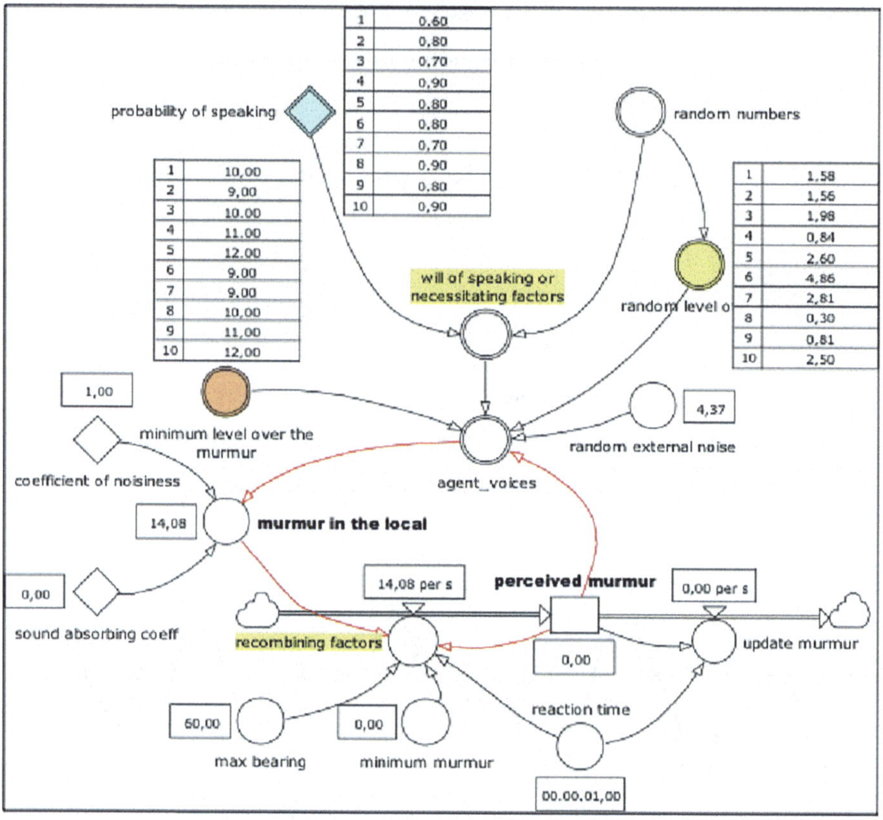

Fig. 3.9 The Combinatory Automaton simulating the murmur and noise system, with 10 agents and differentiated probabilities for each agent (Powersim software)

3.2.4 *The Modus Operandi of a Boolean Stochastic Combinatory Automaton with Output-Dependent Probabilities*

A Combinatory Automaton whose agents can have only two alternate states, (a, b), (0, 1), ("no," "yes") (stay, go), (black, white), and so on, is defined as a Boolean automaton. For simplicity's sake, to make the simulations simpler we can posit that the two states are always reducible to $x_n(t_h)$ = "0" or $x_n(t_h)$ = "1." Boolean automata are particularly useful for simulating combinatory systems whose agents carry out binary choices that lead to opposite states, as occurs, for example, in constructing a model of the system of the "white village" that turns from white to blue (agents: the owners of the houses; micro behaviors: color or not; micro effects: blue or white house; state of each element: white or blue; macro state: the number of white and

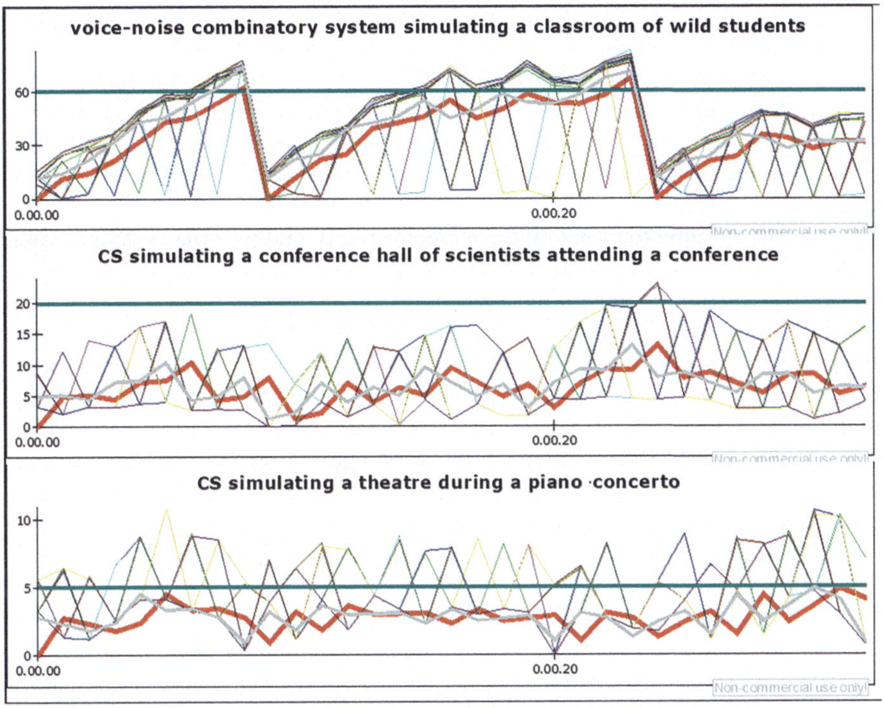

Fig. 3.10 Simulations of the murmur and noise system in different locales (Powersim software)

blue houses); a model of the system of graffiti left on a bench (agents: passers-by; micro behaviors: to leave or not to leave graffiti; micro effects: graffiti or no graffiti; state of each element: having left or not having left graffiti; macro state: the cloud of graffiti); even models of many of the other systems presented in the previous chapters. In such cases the micro behavior of the agents depends on the system's macro state dynamics, expressed as the number of micro states of one type or its opposite.

Let us assume that a Combinatory Automaton is composed of N elements A_n, with $1 \leq n \leq N$, arranged in a grid of $R \times C = N$ cells, with N being a number of elements sufficiently larger than the *minimum activation number* for the system, and that each element can have two alternate states, which, in order to avoid confusion, have been indicated by $x_n(t_h)$ = "a" or $x_n(t_h)$ = "b," to which the probabilities p_n and $q_n = 1 - p_n$ are respectively assigned (as assumed above, the probabilities can also indicate potentiality, frequency, likelihood, etc.).

Suppose that the *synthetic state* of the automaton at a given time "t" is represented by the number of elements *E*a that constitute state "a" or by the number of elements *E*b = (N − *E*) that make up state "b."

Thus, if we consider as an example the "white village" system (Sect. 1.1.6) and assume that there are N = 100 houses, then this system may present two synthetic states—in terms of the color of the houses at a given t_h—represented by the number of white houses and the number of blue houses. If we assume that $Ea(t_h) = 72$ is the number of white houses and $Eb(t_h) = (N - E) = 28$ the number of blue houses, then conventionally we choose $Ea = 72$ to indicate the state of the system in terms of white houses at time "t." If the system at initial time t_0 has the state $Ea(t_0) = 0$, then none of the elements that compose the automaton have the state "a = white"; all are in the state "b = blue."

Let us also assume that each of the elements (cell), and thus the system as well, changes its state in a discrete manner and over a period of finite length—for example, a period of length $\Delta_n(t_h)$ for the element A_n, which represents the time needed to do the painting—and that the system is activated "by chance," in the sense that at time t_0 some of its elements change their state from "b = blue" to "a = white" without this being predictable. The Combinatory Automaton describing the graffiti system (Sect. 2.2.3) presents a general state given by the number of persons who have placed graffiti in the observed place—or, more simply, by the amount of graffiti observed—at a given instant, which depends on the number of persons that belong to the "have-written" state compared to those who have passed by without writing anything. The state of the system is modified at the end of the period utilized by the passers-by to write their thoughts.

Since the influence of the periods of transition of state are perceived only in dynamic simulation models, which cannot be adequately dealt with in a paper, we assume for simplicity's sake that these periods are constant over time and equal for each cell; this assumption is not unrealistic, since it is typical of a wide range of combinatory systems.

The most interesting and useful stochastic Combinatory Automata are those whose cells are assigned an *output-dependent* probability of the type $p_n \equiv p(Y, t_h)$, which varies for each period in relation to the synthetic state, $Y(t_h)$, of the automaton, as described in the model in Fig. 3.5.

Due to the existence of the *micro-macro feedback*, the *synthetic* state of the Combinatory Automaton derives from the state of its agents; the macro state, nevertheless, influences the micro behaviors and the states of the elements according to the *probability* (and *period* of *transition* of state) for each one, a probability that depends, in turn, on the state of the system.

We must therefore take account of this *micro-macro feedback* by positing that the state of each agent depends on the probability that characterizes it; but this probability is, in turn, a function of the state of the system; we shall refer to the probability $p_n(Ea, t_h)$ as the *probability of transition of state* from $x_n(t_h) =$ "b" to $x_n(t_h) =$ "a."

Therefore, we can write: $p_n(Ea, t_h)$ and $p_n(Eb, t_h) = q_n(Ea, t_h) = [1 - p_n(Ea, t_h)]$, for all $E < N - 1$; this means that the probability of transition of state at state N may not necessarily be defined. If we want we can write, in order to generalize, $p_n(N, t_h) = 1$; this is the same as stating that the system, once the state $E = N$ is reached, is maintained in that state and is *not reversible*; in the opposite case, where $p_n(N, t_h) < 1$, the automaton would be *reversible*.

In a *non-reversible* automaton we can also write the length of the *period of transition of state* as a function of the state E of the system: $\Delta_n(E)$ for $E < N - 1$; the period of transition of state at state $E = N$ is not defined. It is possible to write, in order to generalize, $\Delta_n(E) = \infty$; this is the same as stating that the system, once the state $E = N$ is reached, is indefinitely maintained in that state.

The degree of representativeness of the Combinatory Automaton obviously depends on the accuracy with which the values of p_n and Δ_n are estimated for all A_n.

3.2.5 *Example of a Boolean Stochastic Irreversible Combinatory Automaton with Stop-or-Go Cell and Output-Dependent Probabilities*

In order to demonstrate the dynamics of Boolean Combinatory Automata, let us assume a theoretical, abstract combinatory system composed of N = 10 elements, A_n, n = 1, 2, …, 10—arranged in a row vector—which can take on two states, "a" and "b." The system at initial time t_0 has the state $Ea(t_0) = 0$, as shown in Fig. 3.11. This means that none of the 10 elements that compose the system are in state "a"; all are in state "b."

The *probability field* establishing the *cell* and *output-dependent*, *stop-or-go probabilities* of transition of state from state "b" to state "a," $p_n(Ea, t_h)_{[0, 1]}$, for each of the 10 agents and each of the possible 11 states of the system, from 0 to 10—in terms of Ea = number of states "a"—is shown in Fig. 3.12 (in the cells the probabilities are indicated in decimals: e.g., 1 is equivalent to 0.1, or 10%).

Stage	t_0	t_1	t_2	t_3	t_4	t_5	t_6	t_7	t_8	…	…
State=*Ea*	*Ea*=0										
A_1	b										
A_2	b										
A_3	b										
A_4	b										
A_5	b										
A_6	b										
A_7	b										
A_8	b										
A_9	b										
A_{10}	b										

Fig. 3.11 Transition of state in a two-state system—Stage 0

State=*Ea*	*Ea*=0	*Ea*=1	*Ea*=2	*Ea*=3	*Ea*=4	*Ea*=5	*Ea*=6	*Ea*=7	*Ea*=8	*Ea*=9	*Ea*=10
A_1	–	1	1	1	1	1	1	1	1	2	100%
A_2	–	1	1	1	1	1	1	1	2	2	100%
A_3	–	1	1	1	1	1	1	1	2	2	100%
A_4	–	1	1	1	1	1	1	1	2	2	100%
A_5	–	1	1	1	1	1	2	2	2	2	100%
A_6	–	1	1	1	2	2	2	2	1	1	100%
A_7	–	1	1	1	2	2	2	2	1	1	100%
A_8	–	3	2	2	2	2	2	1	1	1	100%
A_9	–	4	3	3	3	3	2	2	2	1	100%
A_{10}	–	6	5	4	3	3	3	3	2	2	100%

Fig. 3.12 Map of the *stop-or-go* probabilities of transition of state (in %)

At *Ea* = 10 the probability is 100% for each A_n; this means that we can consider the state *Ea* = 10 as the "stability state" of the system which, when reached, means that none of the elements will change its particular state from "b" to "a."

At *Ea* = 0 there is no probability, since this state is abandoned due to an initial random impulse, as we shall see shortly.

We are not interested in knowing how the *probability field* was constructed and how the probabilities were calculated and assigned; however, as we can observe, the probabilities of transition of state differ in terms of the agents and the value of the *synthetic state* of the system, as expressed by the number of agents whose *micro state* is $x_n(t_h)$ = "a." For example, for A_7 the probabilities of transition are $p_7 = 1$ for the states *Ea* = 1, *Ea* = 2, *Ea* = 3, *Ea* = 8, and *Ea* = 9. The probability for state *Ea* = 10 is 100, the same as for the other agents. The highest probabilities are assigned to agent A_{10}.

Let us assume that the functions for the periods of transition of state are constant, $\Delta_n(E)$ = const, with respect to *E*, so that we can ignore them when we describe the evolution of this abstract system.

We will use a *random number table* to simulate an experiment that determines the change of state, taking into account the map of the transitions of state in Fig. 3.12 and assuming the system is at t_0 and there is an initial impulse that, "by chance," changes the state of an element.

In order to simulate the initial *random impulse*, we generate a series of random numbers and arbitrarily choose a column and a one-digit number, after determining that the numbers from *rnd.* = 1 to *rnd.* = 9 correspond to the elements from A_1 and A_9 and that the number 0 represents the element A_{10}. Since we assume that the number we read is 3, by convention (operative program of the system) this means that an

Stage	t_0	t_1	t_2	t_3	t_4	t_5	t_6	t_7	t_8	...	...
State=*E*a	*E*a=0	*E*a=1									
A_1	b	b									
A_2	b	b									
A_3	b	a									
A_4	b	b									
A_5	b	b									
A_6	b	b									
A_7	b	b									
A_8	b	b									
A_9	b	b									
A_{10}	b	b									

Fig. 3.13 Transition of state in a two-state system—Stage 1

initial impulse has modified the particular state of the element A_3, which changes from "b" to "a"; thus at t_1 the system presents, by chance, the macro state $Ea = 1$, as shown in Fig. 3.13.

According to the map of the *stop-or-go* probabilities of transition of state, the system evolves by necessity; we see its evolution in a short series of successive stages. To proceed to the subsequent stage, we agree to use the random number table as follows:

- For each of the elements in the system we read off a random one-digit number corresponding to the respective column.
- If we read off a number that is equal to or below that written in the square of the corresponding element, the element changes state.
- Otherwise it remains in the original state.

Now the state of the system is $Ea(t_1) = 1$; in the random number table we associate a random number to the elements which still are in state "b," as shown in the following row vectors: the first two vectors correspond to the values in Fig. 3.12, columns 1 and 3. The third represents the random numbers drawn for each element, except for A_3, which has moved to state "a."

	A_1	A_2	A_3	A_4	A_5	A_6	A_7	A_8	A_9	A_{10}
$p_n(Ea, t_1)$	1	1	1	1	1	1	1	3	4	6
random n.	3	7	a	2	4	7	10	5	5	10

Stage	t_0	t_1	t_2	t_3	t_4	t_5	t_6	t_7	t_8	...	...
State=*Ea*	*Ea*=0	*Ea*=1	*Ea*=1	*Ea*=4							
A_1	b	b	b	b							
A_2	b	b	b	b							
A_3	b	a	a	a							
A_4	b	b	b	b							
A_5	b	b	b	a							
A_6	b	b	b	b							
A_7	b	b	b	b							
A_8	b	b	b	b							
A_9	b	b	b	a							
A_{10}	b	b	b	a							

Fig. 3.14 Transitions of state in a two-state system—Stage 3

In the second transition of state period *none* of the system's elements change their micro state, and the macro state of the automaton is again $Ea(t_2) = 1$. Reading once again from the random number table, we obtain the following:

	A_1	A_2	A_3	A_4	A_5	A_6	A_7	A_8	A_9	A_{10}
$p_u(Ea, t_1)$	1	1	1	1	1	1	1	3	4	6
random n.	6	7	a	7	1	10	5	9	4	6

The state of the elements A_5, A_9, and A_{10} have thus changed; in the third iteration the state of the system becomes $Ea(t_3) = 4$, as shown in Fig. 3.14.

We repeat the operation of reading the random number table only for those elements that are in state "b," comparing them with the numbers that express the transition of state probabilities indicated in column $Ea = 4$ in Fig. 3.12. Adjust dymensions of the following figures:

	A_1	A_2	A_3	A_4	A_5	A_6	A_7	A_8	A_9	A_{10}
$Ea = 4$	1	1	1	1	1	2	2	2	3	3
random n.	2	4	a	5	a	3	9	6	a	a

Stage	t_0	t_1	t_2	t_3	t_4	t_5	t_6	t_7	t_8	...	...
State=*E*a	*E*a=0	*E*a=1	*E*a=1	*E*a=4	*E*a=4	*E*a=4					
E(1)	b	b	b	b	b	b					
E(2)	b	b	b	b	b	a					
E(3)	b	a	a	a	a	a					
E(4)	b	b	b	b	b	b					
E(5)	b	b	b	a	a	a					
E(6)	b	b	b	b	b	b					
E(7)	b	b	b	b	b	b					
E(8)	b	b	b	b	b	b					
E(9)	b	b	b	a	a	a					
E(10)	b	b	b	a	a	a					

Fig. 3.15 Transitions of state in a two-state system—Stage 5

After this fourth iteration, the system is still in state $Ea(t_4) = 4$.

Since the probabilities of transitions of state diminish considerably after state $Ea = 4$ for many elements, we can easily imagine that the system will change its state slowly and that many periods are required before the complete mutation occurs. If, in the meantime, the periods of transition of state should lengthen, the system can even be considered to be stable in the short term.

In the new table of random numbers, the transition of state probabilities are still those indicated in column $Ea = 4$ in Fig. 3.12, corresponding to state $Ea(t_4) = 4$:

	A_1	A_2	A_3	A_4	A_5	A_6	A_7	A_8	A_9	A_{10}
$p_n(Ea, t_1)$	1	1	1	1	1	2	2	2	3	3
random n.	3	1	a	8	a	5	3	6	a	a

In the fifth iteration the element A_2 changes state; as a result the state of the system is $Ea(t_5) = 5$, as shown in Fig. 3.15.

Given the abstractness and simplicity of the example—constructed for the sole purpose of demonstrating the modus operandi of the Combinatory Automaton with a *stop-or-go* probability field that is *state* and *agent-dependent*—the reader who so wishes can continue the iterations until state $Ea(t_h) = 10$ is reached, at which point the system has changed its macro state from $Ea(t_0) = 0$ to $Ea(t_h) = 10$.

3.3 Fields of Probabilities for Transition of State in a Boolean Automaton

Now that we have indicated the elements necessary to construct a stochastic Boolean Combinatory Automaton, and after having examined a simple and abstract example showing its modus operandi, it is necessary to consider the various dynamics of the probability functions associated with the transition of the micro states of the agents as a function of the macro state of the automaton. We shall examine some simple cases, assuming, as in the example in the previous section, that the micro states of the agents can take on only the values "a" or "b."

3.3.1 Seven Special Cases

3.3.1.1 Case 1: p_n and Δ_n Constant and Not Dependent on *E*a

The simplest case is that in which the probability p_n and the period Δ_n, which are necessary for the A_n element to change its own state from "b" to "a," remain constant when the state *E*a of the system varies. The curve for the constant probability of a transition of state can be represented as indicated in Fig. 3.16. We would have a similar representation for the interval Δ_n, after having specified an appropriate unit of time measurement along the ordinate axis.

If the Combinatory Automaton is *irreversible*, that is, the probability of a transition of state of the all the agents A_n, $1 \le n \le N$, is $p_n = p_{[0,1]}$ = constant, then the Combinatory Automaton, starting from a state *E*a = 0, gradually evolves—over a time period that takes into account Δ_n = constant—toward the state *E*a = N.

If the automaton is *reversible*, that is $p_n = p_{[-1,1]}$, the system, starting from a synthetic state *E*a = 0, gradually evolves—over a time period that takes into account Δ_n—toward the state *E*a = [N $p_{n[-1,1]}$].

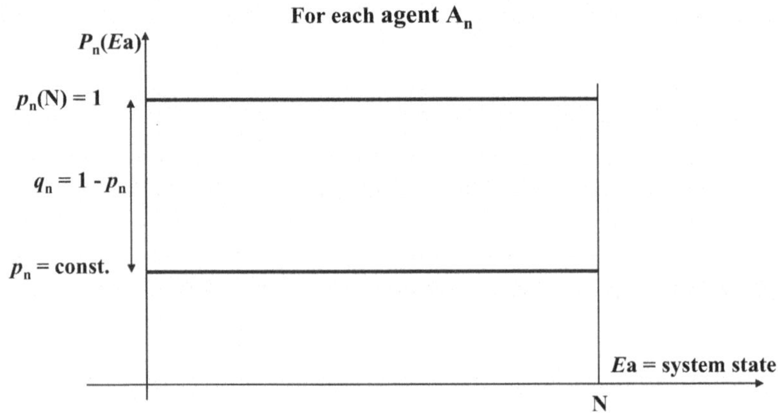

Fig. 3.16 Constant probability of the state *E*a of the single agent A_n

This model is of limited interest since we do not perceive the effect of the *micro-macro feedback*; thus it is not suitable for representing any combinatory system. Any macro effect from the joint action of a collectivity that derives from a constraint or an obligation imposed on its members to undertake certain elementary behaviors cannot be connected to a combinatory system. The probability that a given element produces that behavior is not influenced by the number of elements that have already undertaken a similar behavior. For example, consider the black marks that come from people placing their feet on a wall, the ruts that form when people are forced to go along a narrow passage, or the holes in marble caused by drops of water falling over a long period.

3.3.1.2 Case 2: p_n and Δ_n "Setting off" the System

More interesting is the case in which the probability function for the modification of the state from "b" to "a" of the element A_n is a function of the state of the system, as shown in Fig. 3.17.

In this model the element A_n gradually increases its probability of moving from state "b" to "a," until the system reaches N_{min}; after this state the probability rapidly converges to unity, as indicated in the stretch of curve $p'_n(Ea)$, so that the element which in the past had not changed its state moves from "b" to "a" with a probability that quickly converges to "1," subsequently changing its state with almost complete certainty. If all agents have the function $p_n(Ea)$ equal to that in Fig. 3.17, as shown in Fig. 3.18, and the automaton is *irreversible* then, when N_{min} is reached after a series of transitions, the macro state of the automaton would move in one step, or very quickly, to the state $Ea = N$, with a clear synchronizing effect for all the agents, as if a director had regulated the micro behaviors.

In these systems, even the function Δ_n generally shows a rapid decline toward N_{min}, where it will remain near zero from N_{min} to N.

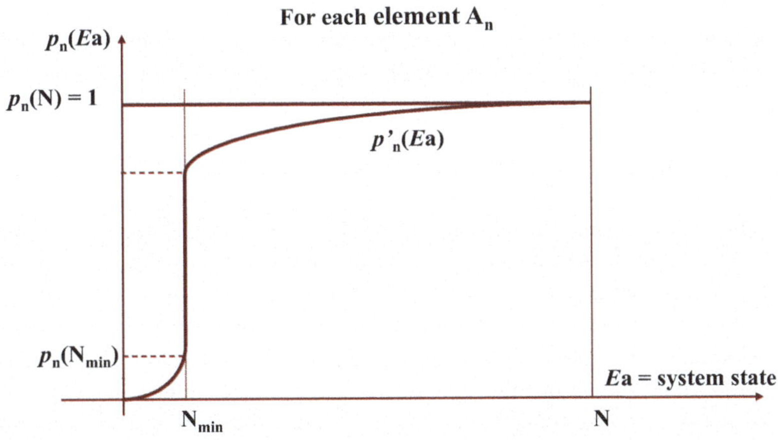

Fig. 3.17 Probability that A_n changes its state from "b" to "a" as a function of *Ea*

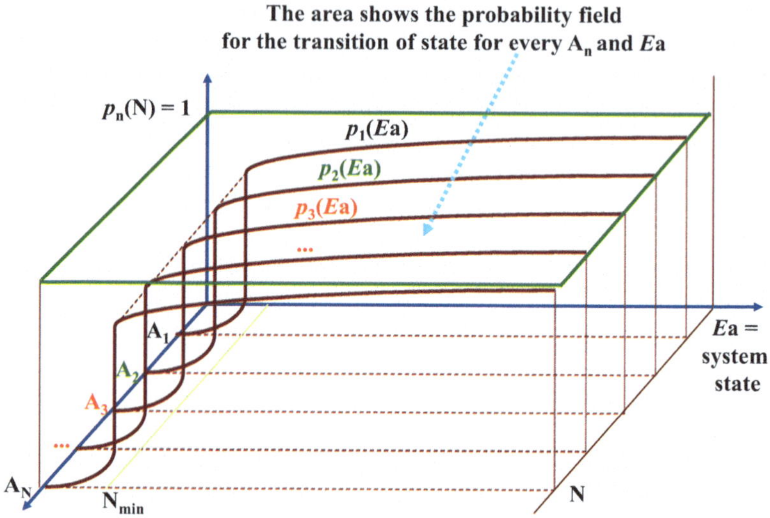

Fig. 3.18 Probability field of transition of state for the N Agents of the system (Case 2)

Referring to the assumptions in Fig. 3.17, the surface in Fig. 3.18 represents the field of probabilities for the transition of the state of the observed automaton.

Examples of systems acting according to the logic represented by the model for Case 2 are:

(a) A herd of elephants fleeing upon hearing the signal from the head of the herd; after some moments of uncertainty, all the elephants begin to flee almost contemporaneously as soon as the lead elephant moves (probability that increases to unity and transition period that falls to zero).
(b) John turns around suddenly, looks up and points with his finger; those near him—after some hesitation (probability that reaches unity and transition period that falls to zero)—do likewise at almost the same time.
(c) Bob, after the romantic piece at the concert, begins to applaud "out of synch"; if he is not hushed, then soon (probability that increases to unity and transition period that falls to zero) applause will break out almost spontaneously.

3.3.1.3 Case 3: p_n an Increasing Function and Δ_n a Decreasing One of *E*a

Equally interesting is the case where the probability function for the transition from state "b" to "a" and the transition period for the agent A_n take on the form shown in Fig. 3.19. The probability for a change in state for agent A_n increases with larger values for the macro state *E*a, which indicates the number of elements that have already changed their state; similarly, the time period for the transition decreases.

If the functions shown in Fig. 3.19 are valid for all the agents, as we can see in Fig. 3.20, then an *irreversible* Combinatory Automaton presents an increasingly

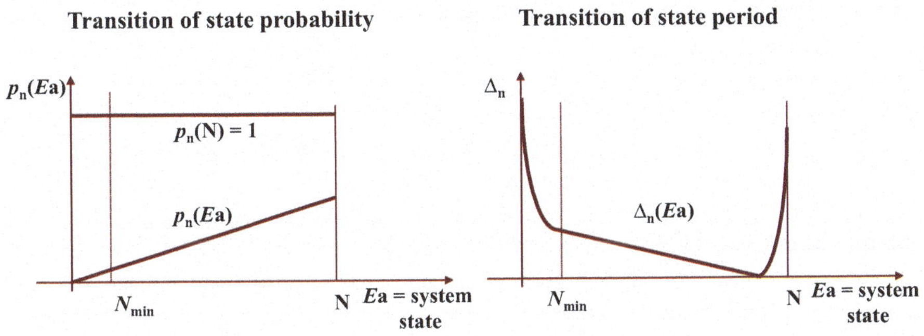

Fig. 3.19 Increasing probability for the transition of state of A_n and decreasing transition period

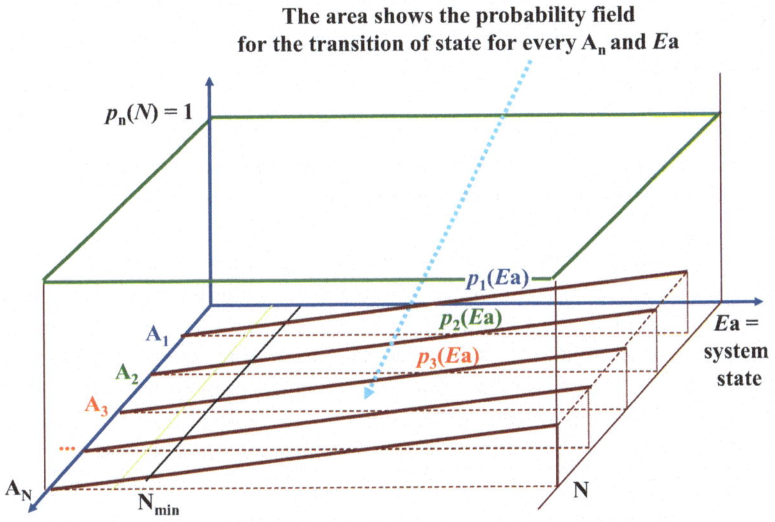

Fig. 3.20 Probability field for the transition of state for the N agents in the system (Case 3)

faster evolution in its own state. The *reversible* Combinatory Automaton may present a chaotic behavior.

Systems that can be represented by the model in Case 3 are typically systems of accumulation, such as:

(a) A pile of garbage: after the first garbage is left at a site, the pile gradually grows, and at an increasing speed (until a saturation limit is reached).
(b) A mountain path: after being initially traced by the first few infrequent passers-by, it rapidly becomes larger and more clearly defined until it is transformed into a road (system of order).
(c) The graffiti on a park bench: after some hesitant carvings, the amount of graffiti rapidly increases (until there is no room for additional incisions).

3.3.1.4 Case 4: p_n a Decreasing Function and Δ_n an Increasing Function of *E*a

This case, which we can view as opposite to the preceding one, occurs when the probability of transition function and that for the period of transition have an inverse dynamics with respect to that shown in the model in Fig. 3.19. The probability decreases and the period of time increases as a function of the state *E* of the Combinatory Automaton.

The graphs have not been drawn, since they are easy to imagine, but we can immediately understand the behavior of the system: in an *irreversible* automaton, after a period of rapid change in the state of many elements, we observe a slowing down in the change of state until the evolutionary phenomenon is exhausted before all the elements have changed their state (in relation, naturally, to the form of the functions p_n and Δ_n for A_n). The systems that can be represented by the model in Case 4 are the same as those for Case 5 below, for which Case 4 represents an extreme example.

3.3.1.5 Case 5: p_n a Function with a Maximum and Δ_n a Function with a Minimum with Respect to *E*a

Let us now consider another case where the function for the probability of transition of state from "b" to "a" and transition period for the agent A_n have the form shown in Fig. 3.21. The probability of a change in state increases over an initial range, until the system reaches the state $E°$a; it then decreases more or less rapidly, with an increase in the number of elements that have changed their overall state. The time function for the transition of state instead calls for a shortening in the transition time the higher the probability is; after state N′, the transition time increases.

If the functions indicated in Fig. 3.21 are valid for all the A_n in the system, then the system's evolution in the case of an *irreversible* automaton will have a slow

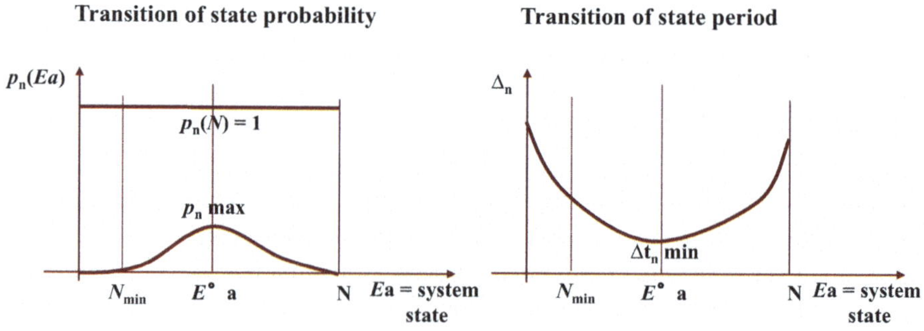

Fig. 3.21 Probability of transition of state with a maximum level and transition period with a minimum level

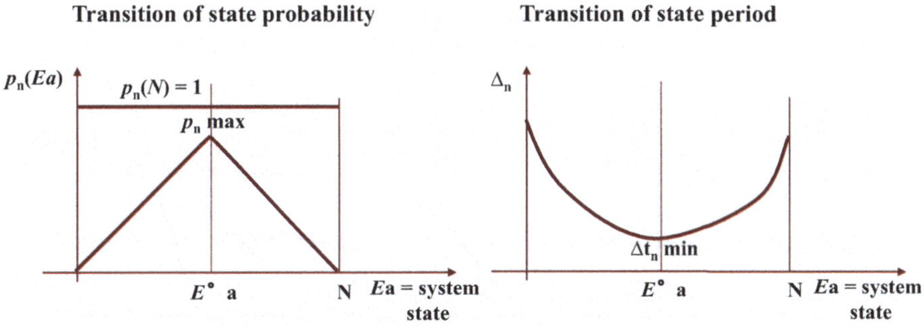

Fig. 3.22 The probability of transition of state in the form of a tent map

beginning—in terms of both the number of elements that change state and the time periods—after which the change will accelerate until point *E*°a. When state *E*°a is reached, the system will slow down until the dynamics of the macro behavior or the macro effect ceases. On the other hand, a *reversible* Combinatory Automaton may show an *oscillatory* dynamics and also a *chaotic* one, especially if the probability assumes the shape of a *tent map function*, as shown in Fig. 3.22 (see Sect. 3.4.3).

Systems that can be represented by the model in Case 5 are, for example:

(a) The spread of fashion: after a slow beginning until N_{min}, the fashion explodes, until *E*°a; it then slows down even to the point of halting its spread.
(b) The spread of an epidemic not countered by appropriate measures.

3.3.1.6 Case 6: p_n and Δ_n Are Differentiated Functions for each Cell of the Automaton

In the previous cases, we have assumed that both the probability of transition of state function as well as that for the transition period were identical for all the agents. We must nevertheless keep in mind both the specific peculiarities of the various elements and, above all, the fact that the elements act under conditions of limited information. More generally, we can thus assume that each element of the system can be characterized by "its own" $p_n(Ea)$ and $\Delta_n(Ea)$ without these functions being equal for the various elements. We can thus represent the probability of transition of state that characterizes the N elements of the automaton with a graph like that in Fig. 3.23.

It is easy to recognize that the field of probability described in Sect. 3.2.5 (Fig. 3.12) is a special case of that in Fig. 3.23.

Figure 3.24 illustrates the model in Fig. 3.23 using *contour lines*, allowing us to better examine those "zones" of the system where there can be a greater modification of state.

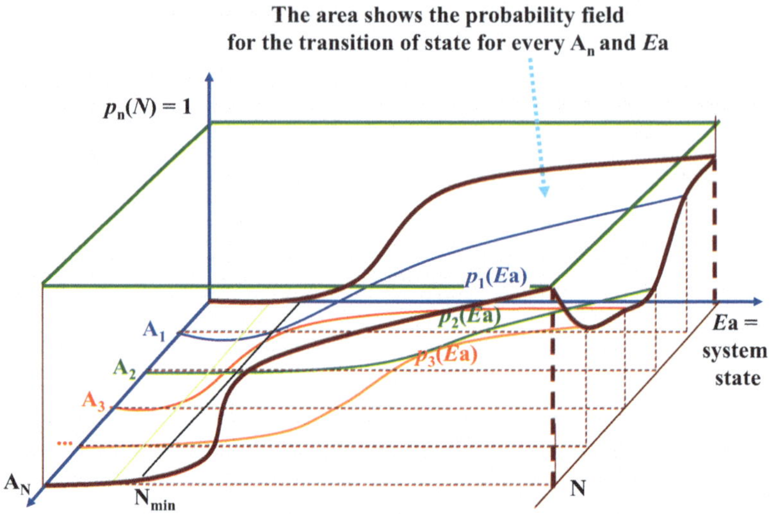

Fig. 3.23 Probability field for the transition of state for the N agents in the system (Case 6)

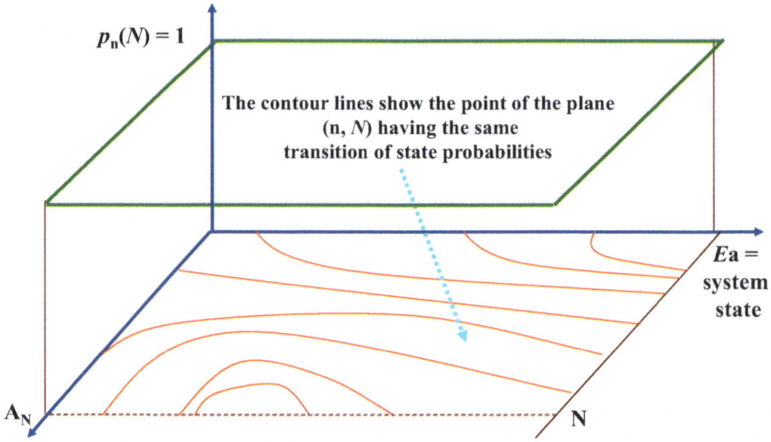

Fig. 3.24 Probability field for the transition of state for the N elements of the automaton by means of contour curves (Case 6)

3.3.1.7 Case7: Limited Information and *Irreversible* Automaton Characterized by a Form and by Stop-or-Go Probabilities

In the example presented in Sect. 3.2.5, the system was characterized by specific probabilities for the transitions of state for each agent, following the logic of Case 6 above, but it did not take into account the important characteristic of *limited information*, according to which the micro behavior of A_n depends not only on the *global information* about the system's *macro state* but also on the information possessed concerning the *micro behavior*—and thus the dynamics of state—of a limited neighborhood.

Taking account of this characteristic in the model means changing the probabilities of transition of state of the elements that are in informational contact with the elements that carry out the change in their states, according to the functioning rules of the system. *Limited information* is a typical characteristic of systems having a "form"; that is, systems where the elements are arranged in a suitable form, which conditions the transmission of information.

3.3.2 *The Modus Operandi of a Probabilistic Irreversible Combinatory Automaton Simulating Diffusion (Case 7)*

The operating logic of limited information, *irreversible* automatons can easily be understood by constructing a Boolean automaton that simulates diffusion based on the logic of Case 7.

Let us consider an irreversible Combinatory Automaton with two states, "a" and "b," composed of N = 100 elements arranged in a [10 × 10] square matrix—going from 0 to 9 along the rows and 1 to 10 along the columns—as shown in Fig. 3.25, with the convention that the element that occupies the position (0, 1) is A_1, that occupying the position (0, 2) is A_2, that occupying the position (2, 1) is A_{21}, and so on, following the progressive numbering of the elements of the system by reading the elements of the matrix in progressive order from (0, 1) to (9, 10). At time t_0 all the elements are in state "b," as shown in the matrix in Fig. 3.25.

	1	2	3	4	5	6	7	8	9	10
0	b	b	b	b	b	b	b	b	b	b
1	b	b	b	b	b	b	b	b	b	b
2	b	b	b	b	b	b	b	b	b	b
3	b	b	b	b	b	b	b	b	b	b
4	b	b	b	b	b	b	b	b	b	b
5	b	b	b	b	b	b	b	b	b	b
6	b	b	b	b	b	b	b	b	b	b
7	b	b	b	b	b	b	b	b	b	b
8	b	b	b	b	b	b	b	b	b	b
9	b	b	b	b	b	b	b	b	b	b

Macro state $Ea(t_0)$ = 0

Fig. 3.25 The combinatory system with a "matrix" form—Stage 0

To simplify the example, let us assume the following *micro rules*:

- At t_0 each agent of the automaton has the same probability of transition of state function of moving from "b" to "a," and.
- The probability field indicates a constant probability of $p_n(Ea)_{[0,\,1]} = 1/100$ of moving from "b" to "a" when the state *E*a—in terms of A—is included in the 10 mutations, $0 \leq Ea \leq 09$.
- As well as a constant probability of 2/100 of moving from "b" to "a" when the state *E*a has a value from 10 to 19, $10 \leq Ea \leq 19$.
- And so on, increasing the probability by 1/100 for each increase of 10 in the state.
- Obviously, the assumption of constancy of the $p_n(Ea)_{[0,\,1]}$ within the value ranges of 10 is introduced as a simplifying assumption in order to make the example easier to understand.

Nevertheless, the system is characterized by *limited information*, in addition to the *micro-macro feedback on probabilities* (even though limited within the value ranges of 10). This means that each A_n has a probability of changing its own *micro state* not only in relation to the *macro state* of the system, expressed by *E*a, but also in relation to the state taken on by the elements that form its *neighborhood*.

In order to understand the limited information feature of the system, we will adopt the following *macro rule* of *limited information*:

- If an agent changes its own state from "b" to "a," the eight agents that form its Moore *neighborhood* in the matrix (Sect. 1.3.5) change their transition of state probabilities, which are increased by 5% points while remaining constant within value ranges of 10 (in order to make the example manageable).
- To simplify the example, if the element only lies in a perimeter row or column of the matrix, only the elements included in the matrix are considered in the *neighborhood*.
- If several elements change, we proceed by modifying the probabilities regarding the element with a lower number.

Since the initial state of the system is $Ea = 0$, we can start from the matrix of elements that form the combinatory system indicated in Fig. 3.25.

The matrix in Fig. 3.26 contains—in hundredths—the constant probabilities that characterize the probability of transition of state functions; to simplify the example, let us suppose, as we have already observed, that the initial probabilities are equal to 1/100 for each A_n (the symbol %, as in the preceding examples, has been omitted) and that Δ_n is constant.

We can now use a random numbers table to simulate the experiment that produces the behavior of the system.

We will try above all to identify the element (a single one in order to simplify things) that undergoes the initial impulse and thus "by chance" changes its own state. We read *rnd.* = 34 in the random numbers table. Thus, by definition the element A_{34} changes its state in $x_{34}(t_0)$ = "a," and the macro state of the automaton therefore becomes $Ea(t_0) = 1$. The matrix that defines the system becomes that shown in Fig. 3.27.

	1	2	3	4	5	6	7	8	9	10
0	1	1	1	1	1	1	1	1	1	1
1	1	1	1	1	1	1	1	1	1	1
2	1	1	1	1	1	1	1	1	1	1
3	1	1	1	1	1	1	1	1	1	1
4	1	1	1	1	1	1	1	1	1	1
5	1	1	1	1	1	1	1	1	1	1
6	1	1	1	1	1	1	1	1	1	1
7	1	1	1	1	1	1	1	1	1	1
8	1	1	1	1	1	1	1	1	1	1
9	1	1	1	1	1	1	1	1	1	1

Fig. 3.26 Probability field for the transition of state for the irreversible automaton in Fig. 3.25—Stage 0

	1	2	3	4	5	6	7	8	9	10
0	b	b	b	b	b	b	b	b	b	b
1	b	b	b	b	b	b	b	b	b	b
2	b	b	b	b	b	b	b	b	b	b
3	b	b	b	a	b	b	b	b	b	b
4	b	b	b	b	b	b	b	b	b	b
5	b	b	b	b	b	b	b	b	b	b
6	b	b	b	b	b	b	b	b	b	b
7	b	b	b	b	b	b	b	b	b	b
8	b	b	b	b	b	b	b	b	b	b
9	b	b	b	b	b	b	b	b	b	b
Macro state $Ea(t_1)$ =			1							

Fig. 3.27 The automaton in Fig. 3.25 after the initial random impulse—Stage 1

According to the *macro rule of limited information* (Sect. 3.3.2), the matrix of the probabilities of transition of state is modified as shown in Fig. 3.28. The probability for the eight elements of the Moore neighborhood of agent A_{34} increases by 5% points; the probability for A_{34} has been indicated by 100% to represent the fact that the cell corresponding to A_{34} does not change its state "a," which by now has been permanently acquired.

	1	2	3	4	5	6	7	8	9	10
0	1	1	1	1	1	1	1	1	1	1
1	1	1	1	1	1	1	1	1	1	1
2	1	1	6	6	6	1	1	1	1	1
3	1	1	6	100	6	1	1	1	1	1
4	1	1	6	6	6	1	1	1	1	1
5	1	1	1	1	1	1	1	1	1	1
6	1	1	1	1	1	1	1	1	1	1
7	1	1	1	1	1	1	1	1	1	1
8	1	1	1	1	1	1	1	1	1	1
9	1	1	1	1	1	1	1	1	1	1

Fig. 3.28 Matrix of the transition of state functions for the system in Fig. 3.27 after the initial impulse—Stage 1

We must now use the random numbers table to make the system evolve to Stage 2.

For each of the 99 elements that have not yet taken on state "a" we read a two-digit number from a random numbers table. If this number is equal to or less than the one written in the cell of the element in question in the matrix in Fig. 3.28, this element changes state; otherwise it remains in state "b."

Reading from the random numbers table, we see that the elements A_{04} [number read, 01], A_{45} [number read, 04], and A_{71} [number read, 00] change state. The diffusion process has started.

Applying the *macro rule of limited information*, we thus obtain the new state in Fig. 3.29 and the new probability table for transition of state in Fig. 3.30.

As we can see from Fig. 3.30, the probability field for transition of state is modified, considerably increasing the probability functions for transition of state for many elements. In particular, we observe that the probability of transition of state for the element A_{35} increases from p_{35} $(t_1) = 6$ to p_{35} $(t_2) = 11$ due to the sum of the of extra probabilities, as stated in the micro rules; and likewise for the element A_{44}.

Let us try a new transition for the system, once again reading from the random numbers table. Only the elements A_{13} [number read, 05] and A_{88} [number read, 01] change their state. The diffusion process continues.

Figures 3.31 and 3.32 show the new matrices.

We can observe how the automaton evolves of its own "necessity," but also how we must expect an irregular evolution; the *limited information* operates by modifying the probabilities of transition of state of the elements which are "in contact" with each of the elements that change state (we have not, however, taken into account the time functions for the transition of state, which also characterize the evolution of the system).

	1	2	3	4	5	6	7	8	9	10
0	b	b	b	a	b	b	b	b	b	b
1	b	b	b	b	b	b	b	b	b	b
2	b	b	b	b	b	b	b	b	b	b
3	b	b	b	a	b	b	b	b	b	b
4	b	b	b	b	a	b	b	b	b	b
5	b	b	b	b	b	b	b	b	b	b
6	b	b	b	b	b	b	b	b	b	b
7	a	b	b	b	b	b	b	b	b	b
8	b	b	b	b	b	b	b	b	b	b
9	b	b	b	b	b	b	b	b	b	b

Macro state $Ea(t_2)$ = 4

Fig. 3.29 The combinatory system in Fig. 3.27 after the second transition—Stage 2

	1	2	3	4	5	6	7	8	9	10
0	1	1	6	100	6	1	1	1	1	1
1	1	1	6	6	6	1	1	1	1	1
2	1	1	6	6	6	1	1	1	1	1
3	1	1	6	100	11	6	1	1	1	1
4	1	1	6	11	100	6	1	1	1	1
5	1	1	1	6	6	6	1	1	1	1
6	6	6	1	1	1	1	1	1	1	1
7	100	6	1	1	1	1	1	1	1	1
8	6	6	1	1	1	1	1	1	1	1
9	1	1	1	1	1	1	1	1	1	1

Fig. 3.30 Matrix of the transition of state functions for the system in Fig. 3.29—Stage 2

In transition 4, the random numbers table indicates that the elements A_{08}, A_{15}, A_{24}, A_{33}, A_{35}, A_{87}, and A_{97} change their *micro state*, which reveals that the diffusion process accelerates. The combinatory system thus changes as indicated in Figs. 3.33 and 3.34, taking into account the fact that N increased by 10 and that, therefore, the probability functions which before were equal to 1/100 are now equal to 2/100.

The reader is invited to complete the evolution, noting only that the system is now able to evolve rapidly (if the Δ_n are constant or decreasing) in terms of the number of elements whose transition from state "b" to state "a" has become more likely.

	1	2	3	4	5	6	7	8	9	10
0	b	b	b	a	b	b	b	b	b	b
1	b	b	a	b	b	b	b	b	b	b
2	b	b	b	b	b	b	b	b	b	b
3	b	b	b	a	b	b	b	b	b	b
4	b	b	b	b	a	b	b	b	b	b
5	b	b	b	b	b	b	b	b	b	b
6	b	b	b	b	b	b	b	b	b	b
7	a	b	b	b	b	b	b	b	b	b
8	b	b	b	b	b	b	b	a	b	b
9	b	b	b	b	b	b	b	b	b	b

Macro state $Ea(t_3)$ = 6

Fig. 3.31 The combinatory system in Fig. 3.29 after the third transition—Stage 3

	1	2	3	4	5	6	7	8	9	10
0	1	6	11	100	6	1	1	1	1	1
1	1	6	100	11	6	1	1	1	1	1
2	1	6	11	11	6	1	1	1	1	1
3	1	1	6	100	11	6	1	1	1	1
4	1	1	6	11	100	6	1	1	1	1
5	1	1	1	6	6	6	1	1	1	1
6	6	6	1	1	1	1	1	1	1	1
7	100	6	1	1	1	1	6	6	6	1
8	6	6	1	1	1	1	6	100	6	1
9	1	1	1	1	1	1	6	6	6	1

Fig. 3.32 Matrix of the transition of state functions for the system in Fig. 3.31—Stage 3

	1	2	3	4	5	6	7	8	9	10
0	b	b	b	a	b	b	b	a	b	b
1	b	b	a	b	a	b	b	b	b	b
2	b	b	b	a	b	b	b	b	b	b
3	b	b	a	a	a	b	b	b	b	b
4	b	b	b	b	a	b	b	b	b	b
5	b	b	b	b	b	b	b	b	b	b
6	b	b	b	b	b	b	b	b	b	b
7	a	b	b	b	b	b	b	b	b	b
8	b	b	b	b	b	b	a	a	b	b
9	b	b	b	b	b	b	a	b	b	b
Macro state $Ea(t_4)$ =	13									

Fig. 3.33 The combinatory system in Fig. 3.29 after the fourth transition—Stage 4

	1	2	3	4	5	6	7	8	9	10
0	2	7	12	100	12	7	7	100	7	2
1	2	7	100	22	100	7	7	7	7	2
2	2	12	22	100	22	12	2	2	2	2
3	2	7	100	100	100	12	2	2	2	2
4	2	7	12	22	100	12	2	2	2	2
5	2	2	2	7	7	7	2	2	2	2
6	7	7	2	2	2	2	2	2	2	2
7	100	7	2	2	2	7	12	12	7	2
8	7	7	2	2	2	12	100	100	7	2
9	2	2	2	2	2	12	100	17	7	2

Fig. 3.34 Matrix of the transition of state functions for the system in Fig. 3.33—Stage 4

3.4 Stochastic Combinatory Automata Simulating Combinatory Systems of Diffusion and Accumulation

3.4.1 Irreversible Stochastic Combinatory Automaton Simulating Slow Diffusion. The "Hundred-Towers" Phenomenon (Sect. 2.3.3)

During the tenth and eleventh centuries, a unique phenomenon arose in Pavia, a town in the north of Italy, which led, in a little over 100 years, to the construction of a cluster of between 100 and 170 towers (the numbers vary from one historian to another). This development, described in Sect. 2.3.3, can be interpreted as the macro effect of a combinatory system whose heuristic model has been specified. To simulate this phenomenon and generate several towers compatible with historical events and recent observation, it is useful to construct a stochastic Combinatory Automaton represented by a matrix whose cells represent the micro state, $x_n(t_h)$, of a hypothetical noble family of Pavia, a potential tower builder. Therefore, each cell can have only two states: $x_n(t_h)$ = ["0," towers absent; "1," tower built]. The states of the automaton will be indicated by "0" and "1," since this makes it possible to immediately calculate the number of towers at t_h by simply adding the states of all the cells. The state "−1" is also possible; this would represent the collapse or destruction of a tower, which obviously would decrease the overall number. However, this state is an autonomous alternative state; in the simulation, it would reduce a state "1" to a state "0," allowing the counter of states to reduce by 1 the number of towers from that moment on.

To simulate this combinatory system, a Combinatory Automaton has been devised which can be described by the following conventions and rules:

1. The area of Pavia, within which the houses of the noble families were built (roughly 900 × 900 square meters), is divided into an R × C = 30 × 30 grid of cells; the size of a cell corresponds in theory to the area of a nobleman's palace, with each cell having a state that can take on two values: $x_{rc}(t_h)$ = ["0," towers absent; "1," tower built], for $1 \leq r \leq 30$ and $1 \leq c \leq 30$.
2. The simulation occurs over a time span of T = 100 years, divided into 25 4-year periods and iterations [t_h, h = 1, 2, …, 25].
3. The *micro effect* of the decision of each agent A_{rc} "to build/not to build" a tower is represented by $tow_{rc}(t_h)$; the *macro effect*, deriving from the *synthetic state* of the automaton, is represented by the variable $TOW(t_h) = N(t_h)$, which indicates the number of towers and the *global information* which determines the *field of probabilities* that each cell will translate its state from "0" to "1."
4. Each cell has a *stop-or-go* probability, $p_{rc}(TOW, t_h)_{[0, 1]}$, which depends on the state of the system, according to a function specified in the control panel of the simulator.
5. The *necessitating factors*—that is, the factors which determine the *will-for-tower*—are translated into probabilities, $w_{rc}(t_h)_{[0, 1]}$, which are quantified by add-

ing to the $p_{rc}(TOW, t_h)_{[0,1]}$ a *minimum* base probability, p_{min} = const., which is assigned (optionally) at the beginning to start up the system; $k_{centrcell}$ and $k_{neighbs}$ increase the probabilities for those cells near another cell which contains a tower. The grid showing the sum of these probabilities configures a *will-for-tower landscape* that determines the system's evolution.

6. A first *will-for-tower landscape*, which considers the structure of the city and the still-visible concentration of the remaining towers (Fig. 2.15), has been chosen as shown in Fig. 3.36 (at the bottom of the main graph).
7. Provision is also made for a reduction of the $w_{rc}(t_h)_{[0,1]}$ to take into account the probabilities for the towers' destruction should they become too numerous; this reduction, (−1), depends on the density of the towers in the nearby cells, $dd_{rc}(t)$, and on the probability of the towers' destruction, $pd_{rc}(t)_{[0,1]}$, defined in the control panel.
8. We assume the possibility of an external control by the municipality, which can lead to the decision to demolish a certain number of towers when the density ratio, $[TOW(t_h)/900]$, is judged too high; we thus include the control coefficient $[\_k_{exogen\ contr}\ (TOW(t_h)/900)]_{[0,1]}$; in the simulation, this coefficient is assumed equal to zero.

The Combinatory Automaton is summarized by (3.5).

$$\begin{cases} \text{for } r = 1,2,\ldots 30 \\ \text{for } c = 1,2,\ldots 30 \\ \text{for } t = 0,1,2,\ldots 25 \\ TOW(t_h) = \sum_{1 \le r \le 30} \sum_{1 \le c \le 30} tow_{rc}(t_h) \\ w_{rc}(t_{h+1})_{[0,1]} = p_{rc}(TOW, t_h) + p_{min} + k_{centrcell} + k_{neighbs} \\ -k_{exogen\,contr}\left(TOW(t_h)/900\right) \\ \text{if } tow_{rc}(t_h) = 0, \text{then } tow_{rc}(t_{h+1}) = tow_{rc}(t_h) + 1\left[w_{rc}(t_h)_{[0,1]}\right] \\ \text{if } tow_{rc}(t_h) = 1, \text{then } tow_{rc}(t_{h+1}) = tow_{rc}(t_h) - 1\left[dd_{rc}(t_h), pd_{rc}(t_h)_{[0,1]}\right] \end{cases} \quad (3.5)$$

Figure 3.35 shows the control panel necessary to insert all the data for the simulation. To produce the dynamics of the Combinatory Automaton based on (3.5), a vector of five matrices has been constructed, each composed of 30 × 30 cells and referred to as follows:

1. The *Analytical State of the Automaton* matrix, whose function is to highlight the cells containing a tower (micro effect of the micro behavior of the agent corresponding to that cell) and those still without towers. Since each tower is represented by "1," this matrix can immediately calculate the *macro state* of the automaton as well: $TOW(t_h)$.

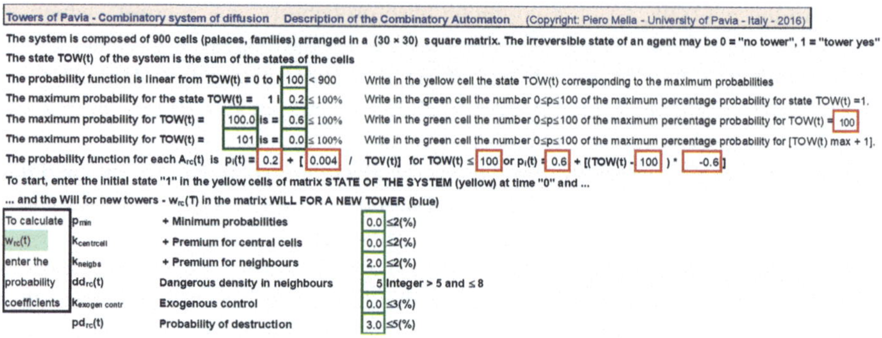

Fig. 3.35 Control panel regulating the dynamics of the Combinatory Automaton simulating the *Hundred-Towers*

t	N(t)	ΔN	ΔΔN
0	2	2	2
1	5	3	1
2	11	6	3
3	15	4	-2
4	20	5	1
5	24	4	-1
6	35	11	7
7	42	7	-4
8	53	11	4
9	69	16	5
10	83	14	-2
11	125	42	28
12	127	2	-40
13	107	-20	-22
14	87	-20	0
15	91	4	24
16	90	-1	-5
17	92	2	3
18	93	1	-1
19	90	-3	-4
20	120	30	33
21	89	-31	-61
22	95	6	37
23	92	-3	-9
24	107	15	18
25	104	-3	-18
26	101	-3	0

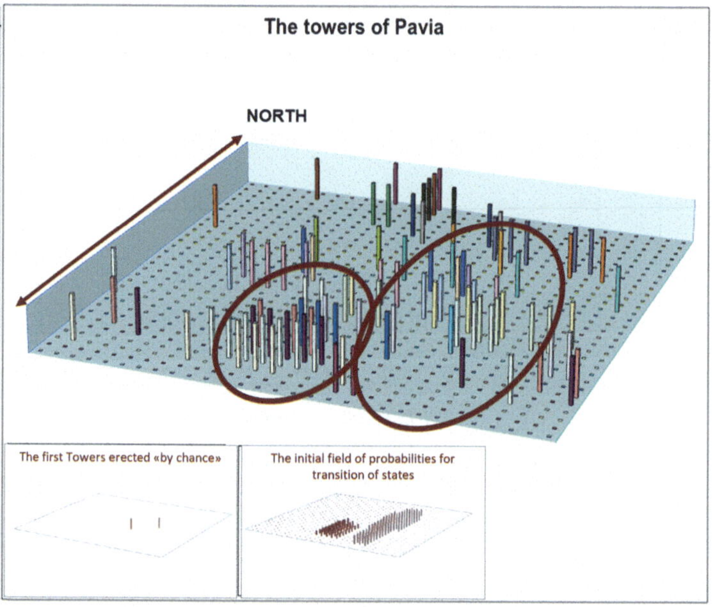

Fig. 3.36 The results of one simulation in which TOW(1) = 2. The towers existing after 25 iterations (100 years)

2. The *Will-for-Tower Landscape* matrix (necessitating factors), which represents the probability field, since it indicates, for each cell, the probability of change in state, $w_{rc}(t_h)_{[0, 1]}$. The values of each cell are calculated by adding to the probabilities presented in matrix (4) the values needed to construct $w_{rc}(t_{h+1})_{[0, 1]}$, as indicated in the second equation (row 5) of system (3.5).
3. The matrix showing the *Experiments Field*, which indicates the random numbers generated by each iteration, which allow the Combinatory Automaton to calculate the results of the experiments that decide which cells will change their state.

4. The matrix of *Probabilities Before Minimum and Premium Probabilities*, which use the probability function determined by the cells in the upper part of the control panel—in particular, in the red bordered cells—to determine $p_{rc}(TOW, t_h)$.
5. The matrix in which the *Neighbors* of each cell in matrix "1," whose state is "1" are calculated.

This vector of matrices has been recalculated for 25 iterations, thereby generating the micro and macro dynamics of the automaton during the period of the simulation.

Figure 3.36 represents the *result* of one simulation and shows the *analytical state* at t_0 (random towers built) and the initial *will-for-tower landscape* at t_0 (the two figures in reduced form at the bottom of the main graph), as well as the *analytical state* of the automaton at t_{25}, which represents the towers at the end of the iteration. The dynamics of the macro states of the 25 periods is indicated in the table on the left; the simulation runs for 26 periods, but the analytical state refers to t_{25}.

Subsequent repetitions of the simulation program obviously produce different but similar patterns, even though each experiment uses new random numbers, which allow us to determine whether the cells change their state with a new tower.

Figure 3.37 shows the analytical state of the automaton at t_{25}, after a new simulation in which all the parameters from the simulation in Fig. 3.36 have been maintained, with only new random numbers chosen.

Figure 3.38 shows the result of a new simulation which assumes there is an initial state with a single random tower built, while Fig. 3.39 shows the result of a new test

t	N(t)	ΔN	ΔΔN
0	2	2	2
1	7	5	3
2	12	5	0
3	13	1	-4
4	23	10	9
5	25	2	-8
6	41	16	14
7	49	8	-8
8	65	16	8
9	88	23	7
10	109	21	-2
11	134	25	4
12	120	-14	-39
13	91	-29	-15
14	84	-7	22
15	90	6	13
16	83	-7	-13
17	76	-7	0
18	87	11	18
19	89	2	-9
20	95	6	4
21	93	-2	-8
22	96	3	5
23	90	-6	-9
24	94	4	10
25	127	33	29
26	119	-8	-41

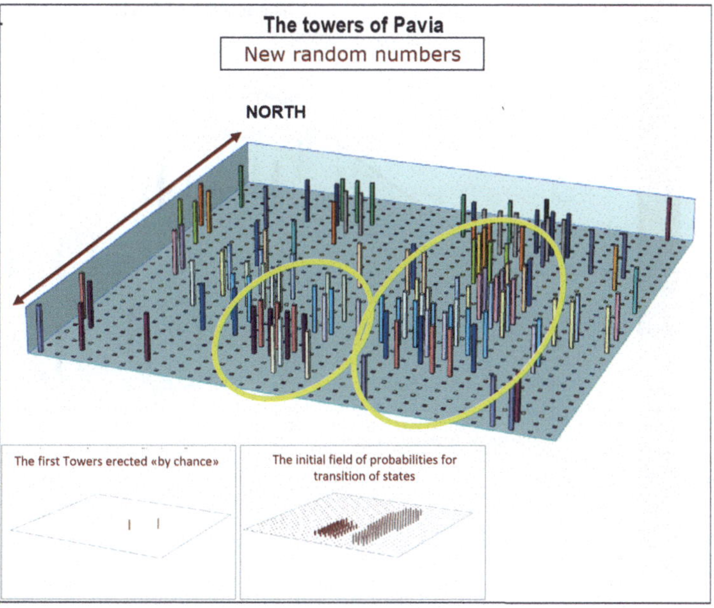

Fig. 3.37 The results of a new test in which TOW(1) = 2 but the dynamics are based on new random numbers that form the Will for a New Tower Landscape

t	N(t)	ΔN	ΔΔN
0	1	1	1
1	2	1	0
2	5	3	2
3	9	4	1
4	9	0	-4
5	14	5	5
6	18	4	-1
7	33	15	11
8	41	8	-7
9	41	0	-8
10	58	17	17
11	72	14	-3
12	120	48	34
13	121	1	-47
14	108	-13	-14
15	94	-14	-1
16	96	2	16
17	93	-3	-5
18	105	12	15
19	94	-11	-23
20	94	0	11
21	100	6	6
22	80	-20	-26
23	115	35	55
24	116	1	-34
25	116	0	-1
26	98	-18	-18

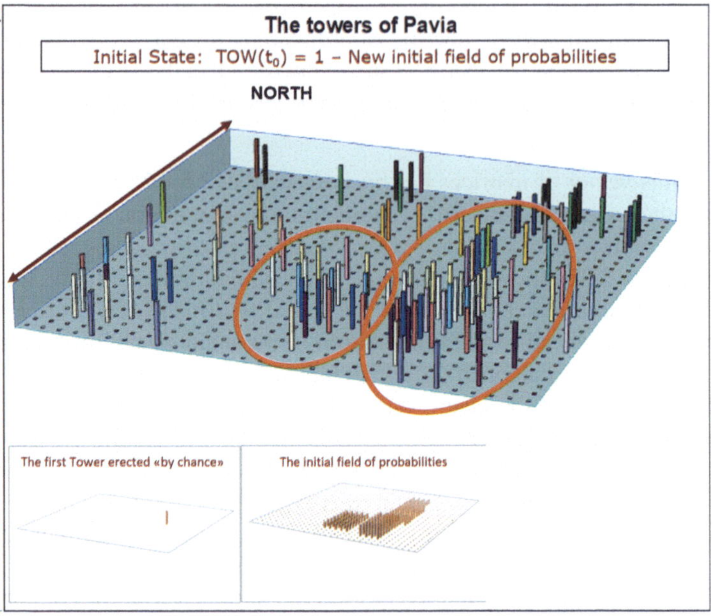

Fig. 3.38 The results of a new test based on a new minimal initial state, TOW(1) = 1 and a new initial field of probabilities (will-for-tower)

t	N(t)	ΔN	ΔΔN
0	4	4	4
1	5	1	-3
2	8	3	2
3	9	1	-2
4	15	6	5
5	14	-1	-7
6	33	19	20
7	44	11	-8
8	51	7	-4
9	70	19	12
10	97	27	8
11	117	20	-7
12	121	4	-16
13	108	-13	-17
14	86	-22	-9
15	92	6	28
16	94	2	-4
17	86	-8	-10
18	122	36	44
19	84	-38	-74
20	93	9	47
21	81	-12	-21
22	103	22	34
23	103	0	-22
24	101	-2	-2
25	111	10	12
26	111	0	-10

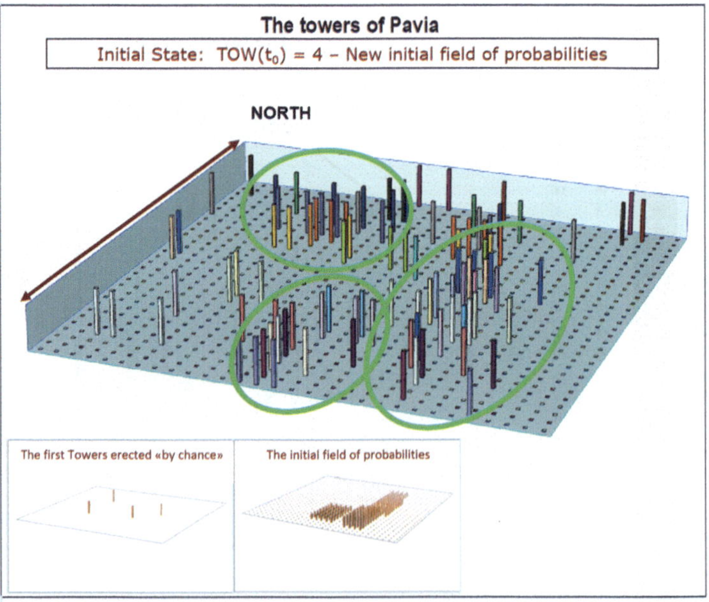

Fig. 3.39 The results of a new test based on a new new minimal initial state, TOW(1) = 4 and a new initial field of probabilities (will-for-tower)

which assumes an initial state made up of four towers built "by chance" in four different sites in the town.

With these tests the Combinatory Automaton clearly reveal that the simulations allow us to obtain a final state at t_{25} with around 100/110 towers, which conforms to contemporary historians' accounts of Pavia (Sect. 2.3.2). The localization of the towers, assuming an initial state of one or two towers, seems to adequately represent the towers of Pavia grouped into clusters, as shown in Figs. 2.14 and 2.15. Similarly, the simulation with an initial state of four towers produces realistic clusters, even though these are arranged slightly differently.

The tests also demonstrate that the dynamics of the towers are in large part independent of the initial conditions, which, in any event, succeed in generating an adequate cluster of towers, thanks to the *micro-macro feedback* action that modifies the "Will-for-Tower Landscape" as a function of state $TOW(t_h)$. This means that the Combinatory Automaton proposed here represents an adequate instrument for the simulation of the combinatory system that has produced that unique, unrepeatable phenomenon of the "Hundred-Towers" of Pavia.

3.4.2 Probabilistic Reversible Combinatory Automaton Simulating Explosive Diffusion: Applause

Applause is the macro effect of a combinatory system of spectators that wish to show their appreciation for some fact or event. On various occasions we all have applauded, and with our micro behavior have produced a micro effect: the sound of our clapping hands. This micro behavior produces the macro effect of thunderous applause, which continues for some time before disappearing as quickly as it appeared.

To simulate this combinatory system, we can construct various Combinatory Automata. To simplify things, a Combinatory Automaton has been devised which can be described by the following conventions and rules:

1. Assume there are N = 100 agents arranged, for convenience sake, in a matrix of R × C = 10 × 10 cells; each cell has a state that can take on two values: $x_{rc}(t_h)$ = ["1," agent clapping; "0," agent quiet], for all $x_{rc}(t_h)$, $1 \le r \le 10$ and $1 \le c \le 10$.
2. The simulation occurs over a time span of T = 50 periods and iterations [t_h, h = 1, 2, …, 50].
3. The *micro behavior* of each agent for each t_h is ["applaud" or "do not applaud"], which constitutes the state of the cell, which is represented by $clap_{rc}(t_h)$ = ["1" or "0"]; this produces the clapping noise, which represents the *micro effect* obtained by converting the clapping hands into a sound.
4. The *macro state*, deriving from the *synthetic state* of the automaton, is represented by the variable $APPL(t_h) = N(t_h)$, which indicates the number of clapping agents in the collectivity.

5. This is converted into a *macro effect* (the applause) by summing (recombining) all the sounds produced by the agents in the state $\text{clap}_{rc}(t_h) = \text{“1”}$; the applause therefore represents the immediately perceived *global information*, which determines the *field of probabilities* that each cell-agent will translate its state from “0” to “1.”
6. Each cell has a transition probability, $p_{rc}(\text{APPL}, t_h)_{[-1, 1]}$, which depends on the state of the system and quantifies the *necessitating factors*—that is, the factors which determine the *will-to-clap.*
7. Each $p_{rc}(\text{APPL}, t_h)_{[-1, 1]}$ is quantified by multiplying a *base probability*, $p_{\min} = const.$—which is assigned (optionally) at the beginning to start up the system—by the number of persons applauding, $\text{APPL}(t_h)$, so as to produce a continuous and rapid increase in the number of agents shifting from the state $\text{clap}_{rc}(t_h) = \text{“0”}$ to that of $\text{clap}_{rc}(t_h) = \text{“1.”}$
8. To rapidly increase $p_{rc}(\text{APPL}, t_h)_{[-1, 1]}$, a premium for the minimum activation number, $p_{act} = const.$, is added when $N_{\min}(t_h)$ is reached.
9. It is not possible to know when the applause will end; however, for the simulations it is assumed it can continue for a certain number of periods, t_{stop}, which is specified in the control panel of the automaton. Two simulations have been carried out assuming $t_{stop} = 9$ and $t_{stop} = 11$.
10. To end the applause, it is necessary to introduce a reducer, *reduct* = const., which reduces $p_{rc}(\text{APPL}, t_h)_{[-1, 1]}$ when this approaches 100%.

The Combinatory Automaton is summarized in synthetic form by system (3.6); the control panel is shown in Fig. 3.40; Fig. 3.41 shows the dynamics of the macro behavior of the automaton.

$$\begin{cases} \text{for } r = 1,2,\ldots 10 \\ \quad \text{for } c = 1,2,\ldots 10 \\ \quad\quad \text{for } t = 0,1,2,\ldots 50 \\ \quad\quad \text{APPL}(t_h) = \sum_{1 \le r \le 10} \sum_{1 \le c \le 10} \text{clap}_{rc}(t_h) \\ \quad\quad p_{rc}(t_{h+1})_{[-1,1]} = \left[\text{APPL}(t_h) \times p_{\min}\right] + p_{act} \\ \quad \text{clap}_{rc}(t_{h+1}) = +1\left[p_{rc}(t_h)_{[-1,1]}\right] \end{cases} \tag{3.6}$$

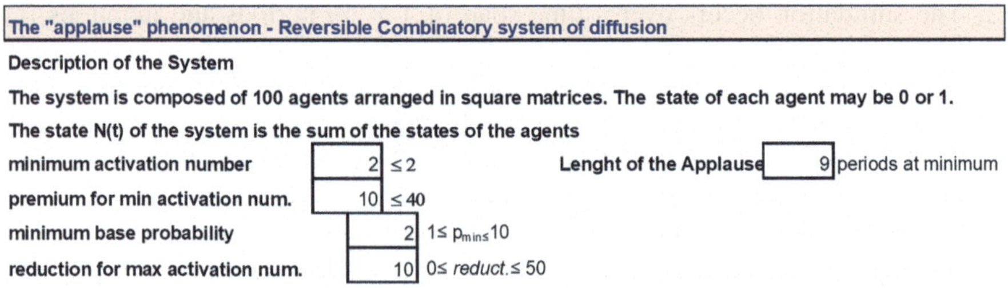

Fig. 3.40 Control panel regulating the dynamics of the Combinatory Automaton simulating *applause*

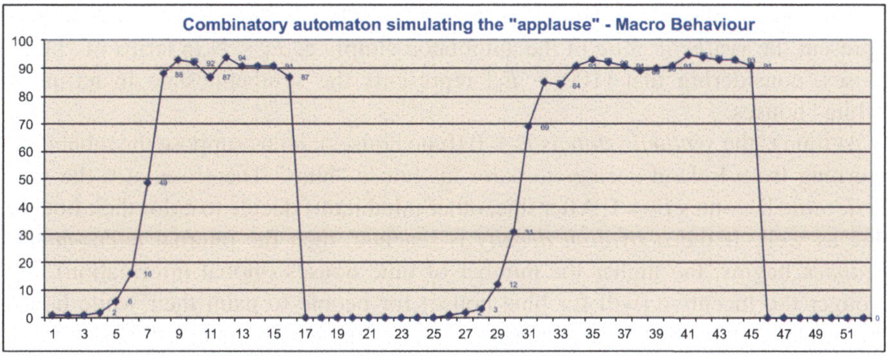

Fig. 3.41 The results of the dynamics of the Combinatory Automaton simulating the *applause* which is maintained for at least nine periods

The graph on the left (generated by the control panel in Fig. 3.40) assumes a duration of the "applause" of at least 9 periods and that the initial applause (which ends at t_{17}) is followed by a second applause starting at t_{31} with a duration of at least 15 periods.

3.4.3 Probabilistic Reversible Combinatory Automaton Generating "Chaotic" Macro Behaviors

Chaos arises in *combinatory systems* when the *hypothesis of reversibility* is introduced (e.g., to speak or to keep quiet in the next minute, clap or not to clap in this instant, wear a skirt or miniskirt on different days, choose road A or B on different days). When a Boolean Combinatory Automaton is constructed to simulate the dynamics of such systems, these systems are generally governed by *transition* probabilities, $p_{rc}[Y(t_h)]_{[-1,1]}$, where, as mentioned above, $Y(t_h) = E_h$ indicates, briefly, the synthetic state of the automaton, so that we admit that a cell could change its state from $x_n(t_h) =$ "0" to $x_n(t_{h+1}) =$ "1" and vice versa at different times. As a result, the Combinatory Automaton might show a chaotic *macro* behavior in the sequence of *macro* states $E(t_h)$, h = 0, 1, 2, … in relation to the probability function $p_{rc}[E_h]_{[-1,1]}$ which determines it.

Let us suppose that in a village in North Africa, with N = 100 white houses, an inhabitant—after a journey in Ireland during which he has seen all the houses brightly colored—decides, upon returning to his village, to paint (micro decision) his house blue (micro effect). As the color is pleasing, *by chance* another inhabitant imitates the first one, then another and another. The more houses there are that have turned blue, the greater the desire of the other inhabitants to change the color of their houses (see Sect. 1.1.6).

We can represent this phenomenon with the aid of a Boolean Combinatory Automaton of N = 100 cells, in a square grid, whose *analytical state*—in terms of the color of the houses at a given moment—is represented by the array $X(t_h) = x_{rc}(t_h) =$ ["white" or "blue"] for $1 \leq r, c \leq 10$, h = 1, 2, …

If we denote the "white" house as "0" and the "blue" house as "1," we can also represent the *synthetic state* of the automaton simply as $E_h \le N$ in terms of "blue" houses, considering that $[100 - E_h]$ represents the synthetic state in terms of "white" houses.

At time t_0 the *synthetic state* is $E_0 = 0$ (blue houses); let us suppose the inhabitant returning from Ireland decides to color his house "blue." Therefore, at t_1 the *synthetic state* becomes $E_0 = 1$. After this, other inhabitants decide to color their houses blue as well. If the *activation density* is reached, then the internal *micro-macro feedback* begins: the higher the number of blue houses (global information), the stronger the incentive (will for blue house) for people to paint their white house blue, and the faster will be the change in color of the houses in the village.

The macro dynamics thus produces a macro effect (gradual change in the color of the village), which is the *global information* that, in turn, influences the micro dynamics, as part of the internal *micro-macro feedback* which operates according to the *probability of transition of state*, $p_{rc}[E(t_h)] = F[E_h]$, which is *state sensitive* for each inhabitant. It is important to note that the change in state of each house occurs during a finite *period of transition of state*—that needed to do the painting; this period is assumed to be constant for any t_h, an assumption which is not explicitly introduced into this Combinatory Automaton.

Considering the previous assumption, we can construct the *Combinatory Automaton* (3.7) to simulate the "white" village:

$$\begin{cases} x_{rc}(t_h) = \{1 \text{ or } 0\}, 1 \le r \le R, 1 \le c \le C, R \times C = N, h = 0,1,2,\ldots \\ X(t_0) = x_{rc}(t_0) = \{0\} \\ E_h = \sum_{1 \le r \le R} \sum_{1 \le c \le C} x_{rc}(t_h) \\ x_{rc}(t_{h+1}) = \begin{cases} \{1\} & [p_{rc}[E_h]\,] \\ \{0\} & [1 - p_{rc}[E_h]\,] \end{cases} \\ \text{OPERATIVE PROGRAM} \Rightarrow \text{define}: N, \left[p_{rc}[E_h]\right] \end{cases} \tag{3.7}$$

To describe the *macro dynamics* of the Combinatory Automaton when changing its states, the model must specify in the most accurate way possible the form of the *probability* function, $p_{rc}[E(t_h)] = F[E_h]$, for each agent and for the $0 \le E_h \le 100$ admissible *macro states* of the village (automaton). These probabilities should provide a numerical indication of all the *necessitating factors* that regulate the individual choices of whether or not to paint (to simplify, these probabilities might be assumed to be the same for each element). If we simply assume *stop-or-go*, non-agent-dependent, probabilities whose functions have the following structure:

$$p_{rc}\left[E_h\right]_{[0,1]} = \frac{E_h}{N} = \frac{E_h}{100} \tag{3.8}$$

then the system's macro behavior is *irreversible* and the diffusion accelerates, since the village becomes "entirely blue." We could also add a premium for "proximity" to increase the probabilities of the neighborhoods of the cell whose value has changed from "0" to "1," which would make the diffusion effect much faster or slower.

If we abandon the hypothesis of *irreversibility* and consider the possibility that each inhabitant may change his decision and re-paint his "blue" house "white"—so that we admit that a cell could change its state from "0" to "1" as well as from "1" to "0"—and assume that in theory this rethinking can be repeated over time (perhaps in subsequent generations), we might posit, for example, that the probabilities take on the following values, which correspond to a *tent map* centered on $E_h = 50$. Let us also suppose that the probabilities are *not agent-dependent* and take on the following values corresponding to the "tent map" in the following model (3.9):

$$p_{rc}\left[E_h\right]_{[-1,1]} = \begin{cases} \dfrac{2E_h}{100} & \text{if } 0 < E_h \le \dfrac{\mathrm{N}}{2} \\ 100 - \dfrac{2E_h - \mathrm{N}}{\mathrm{N}} & \text{if } \dfrac{\mathrm{N}}{2} < E_h \le \mathrm{N} \end{cases} \tag{3.9}$$

Thus, in both cases a general *micro-macro feedback* relation connects the *micro* state to the *macro* state through the synthetic macro *variable* E_h, in the sense that the state of each agent, A_{rc}. depends on the probability $p_{rc}[E(t_h)]_{[-1,1]}$, which in turn depends on the state of the system, E_h, which defines the macro effect.

The *tent map* may take on different forms in relation to both the value of E_h, on which the maximum probability is centered, as well as the maximum admitted value of the probability $p_{rc}[E(t_h)]_{[-1,1]}$. We must also take into account the possibility of increasing the probability of the *neighbors* of each cell and of assigning an extra premium to the cell that has changed its state. In order to include all these options, which are necessary to start the simulation, the *control panel* in Fig. 3.42 has been constructed. In the green-bordered cells, the *tent map* data is inserted, which allows us to construct "the probability function for each A_n" (third-to-last row). The cells in the last row present the added probabilities of the neighbors.

If we simulate the micro behavior by some experiment that generates *random numbers* for each element, referring to the control panel in Fig. 3.42, we can observe that the Combinatory Automaton presents a "chaotic" macro behavior independently of the *initial random impulse* $X(t_0)$, so that the system's history is *irreversible*

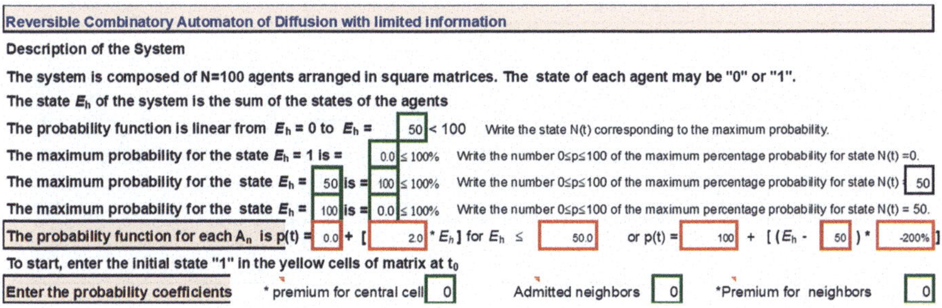

Fig. 3.42 Control panel regulating the dynamics of the Combinatory Automaton simulating *reversible* diffusion and chaos

and the system's future *unpredictable*, since the description of regularities is impossible (in the sense of Gell-Mann, 1995, of Wolfram's classification scheme, 1984, 2004, or of Devaney, 1989). Figure 3.43 presents several tests of the *macro*

TEST A – Probabilities calculated as in the control panel of Fig. 2.29. N_i = 1. Probabilities increase straight line to 100 (%) for E50 and then decrease to 0 for E100

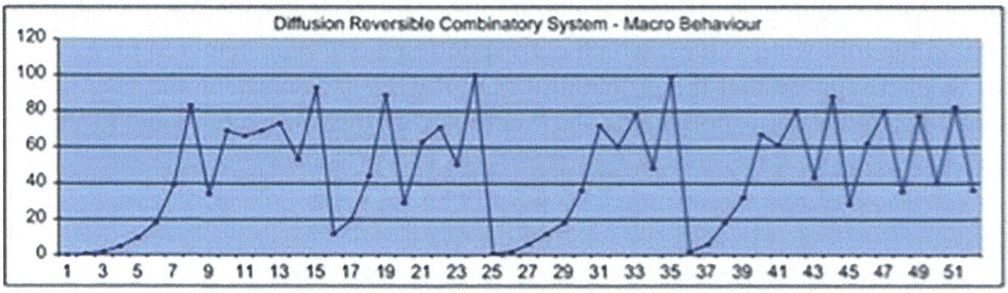

TEST B – Same Probabilities. Admitted neighbors = 2 – Premium for each neighbor = 2

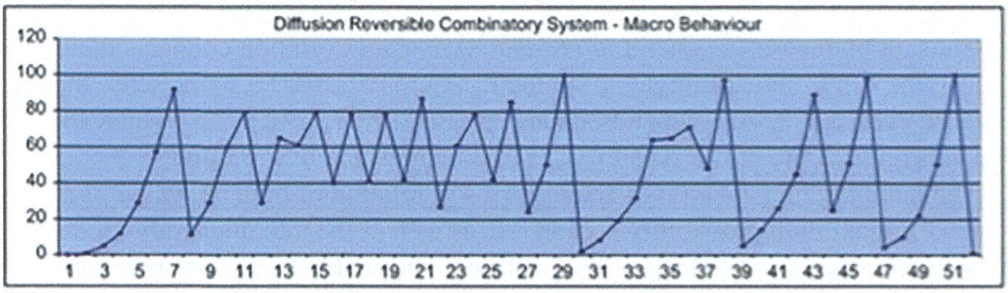

TEST C – N_i = 4. Same Probabilities. Same random numbers

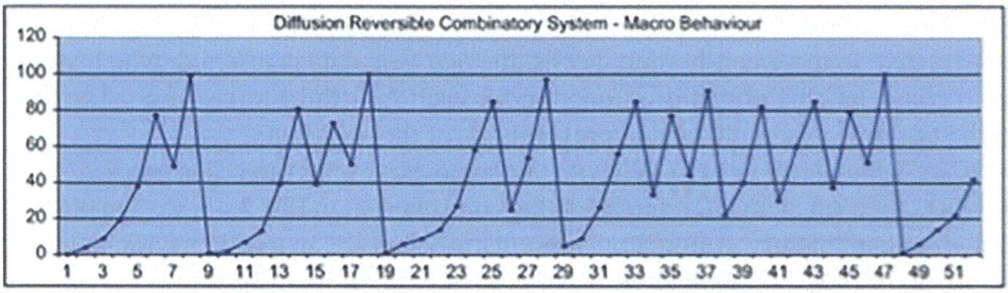

TEST D – N_i = 2. Probabilities increase straight line to100 (%) for E_{30} and then decrease to 0 for E_{100}

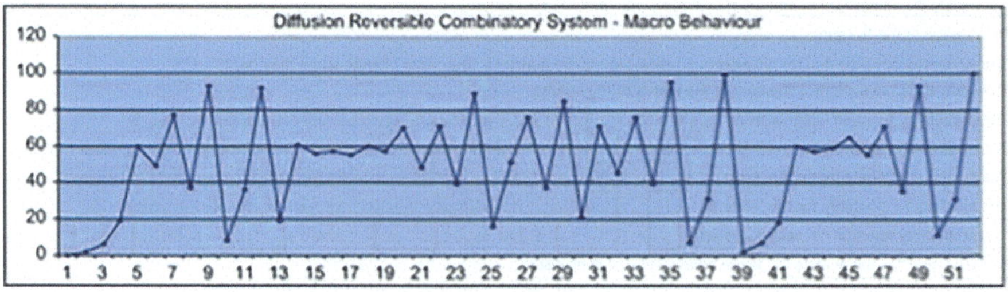

Fig. 3.43 Tests for reversible automaton of 100 cells in a square grid, simulating diffusion, assuming different initial states and different random numbers

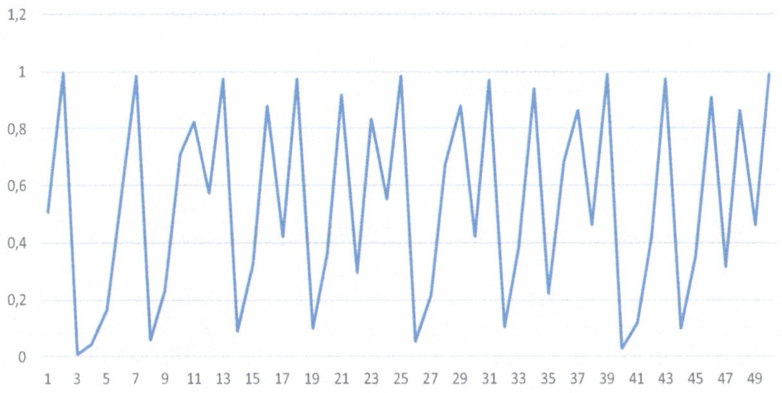

Fig. 3.44 Chaotic behavior produced by the quadratic function outlined in the text

dynamics of the automaton, expressed by the time series of number-of-"blue"-houses over 50 iterations.

Figure 3.44 presents, for purposes of comparison, the typical chaotic dynamics caused by the recursive quadratic function.

$$x_{n+1} = cx_n (1 - x_n)$$
$$x_0 = 0.8500$$
$$c = 3.9900$$

The similarity of the behaviors shown in Figs. 3.43 and 3.44 is evident.

Figure 3.45 shows the Stock and Flow Diagram and the initial values of the automaton, already presented in Fig. 3.42, simulated in Powersim.

Figure 3.46 shows the results of some tests obtained using the program in Fig. 3.45, which confirms the chaotic dynamics of the macro behavior of the automaton.

3.4.4 Probabilistic Irreversible Combinatory Automaton Simulating Vertical and Horizontal Accumulation

The combinatory systems producing the accumulation of objects (understood in the broad sense) are quite common (Sect. 2.2) because of both the objective need of individuals working in groups to accumulate objects of some type and the attractive force of *self-produced global information*, represented by the accumulations, which act as a behavioral "attractor" for those who must position a new object. There are two forms of accumulation: *vertical* or *additive* and *horizontal* or characterized *by proximity*.

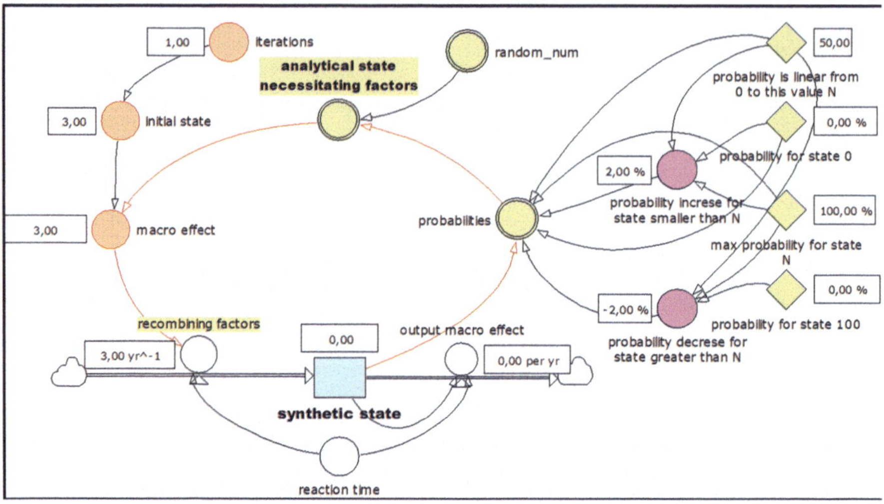

Fig. 3.45 Reversible Combinatory Automaton generating chaotic behavior, simulated in Powersim

The *additive* form leads to an accumulation of objects that overlap to form increasingly higher layers and piles. The *micro-macro feedback* acts to make it increasingly probable that a new object will be added to those already accumulated, thereby forming larger vertical piles that are compatible with the shapes of the objects that are accumulated. Typical systems of vertical accumulation are those that lead to people going to live in increasingly higher buildings, to clusters of skyscrapers, to the numerical growth in populations in a given territory, or to the formation of garbage piles. Even the combinatory system that produces the macro effect of the formation of planets through the aggregation of asteroids and planetesimals produces an additive accumulation process, just as schools of fish or large swarms of insects are typical macro effects of additive accumulation.

Horizontal processes of accumulation lead to aggregate formations of objects arranged one next to the other, but not overlapping. The accumulation appears as a progressive expansion of an aggregation whose size represents the *self-produced global information* that spurs agents to arrange new objects near the previous ones. The creation and expansion of urban or productive settlements, the formation of industrial districts (Sects 2.2.5 and 2.7.1), the lengthening of the paths for the transmission of information and the transportation of persons or goods, and the enlargement of information networks are typical examples of the macro effects of combinatory systems of horizontal accumulation, as are the flocks of birds that form as a result of the accumulation of individuals that join a flock already in flight, thereby increasing the size of the flock, which then attracts other individuals. Most of the graffiti on walls or on the backs of doors, or carved into wood or marble, are the effect of systems of horizontal accumulation. When new graffiti is placed over existing graffiti, an additive accumulation is also produced. At times the processes of horizontal accumulation can easily be distinguished from processes of diffusion.

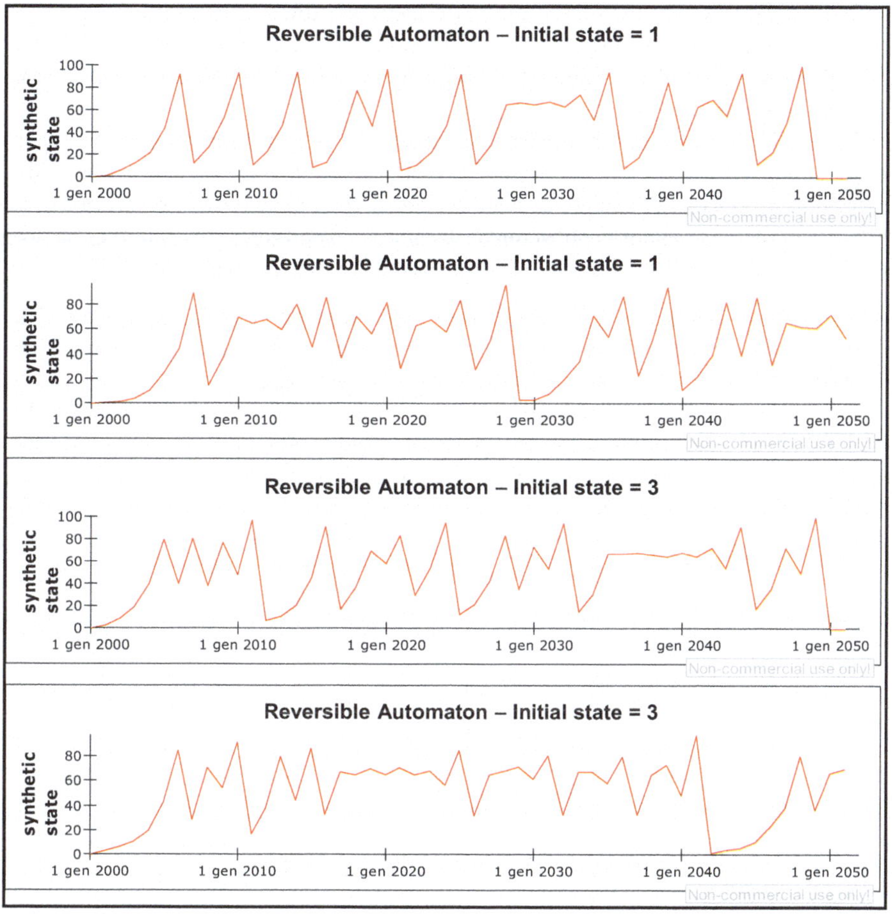

Fig. 3.46 Tests for reversible automaton simulating diffusion, obtained using a Powersim program and assuming different initial states and different random numbers (reference: Fig. 3.45)

The formation of the towers of Pavia (Sect. 2.3.3) was a typical phenomenon of diffusion; however, the real reason the towers accumulated was their proximity to each other, as they became increasingly denser inside the town.

Combinatory systems that produce accumulation are usually irreversible; only in some cases are they reversible, though the process of reversibility is slow. An example of this is the process of de-urbanization or the reduction in size of industrial districts. At times, the agents initiate forms of *internal* control, while at other times the accumulation is cancelled or reduced by the behavior of agents who, no longer finding the accumulation attractive, disperse, as occurs in the breaking up of schools of fish, swarms, and flocks. At other times, forms of *external* control are activated through interventions that reduce the quantity of accumulated objects, as occurs in the elimination of garbage piles or the imposition of the requirement that urban or industrial settlements move elsewhere.

To simulate the effects of processes of accumulation, the rules we have introduced in the previous example of a Combinatory Automaton simulating diffusion can be easily adapted to construct a simple *irreversible* Combinatory Automaton simulating accumulation.

The easiest way to do this is to assume that each cell of the automaton becomes an "accumulator"; if its state moves from "0" to "1" but in a subsequent iteration the event that produces the change of state reoccurs, then the state moves from "1" to "2"; it can increase again in subsequent iterations. To produce this micro behavior, it is enough to start from system (3.7) and introduce the following rule:

- If the state of a cell is $x_{rc}(\mathrm{t_h}) = \mathrm{n}$—that is, if the "accumulator" is in state "n"—and the event occurs to which a *stop-or-go* probability, $p_{rc}(\mathrm{t_h})_{[0,\ 1]}$, is associated, then it becomes $x_{rc}(\mathrm{t_{h+1}}) = \mathrm{n} + 1$; otherwise it remains in state "n."

We can make the accumulation process even more general by assuming the increase in the states of the cells can occur not only through unitary values, "1," but, more generally, by adding to state $x_{rc}(\mathrm{t_h}) = \mathrm{n}$ an incremental value $\Delta > 0$, which produces the new state $x_{rc}(\mathrm{t_{h+1}}) = \mathrm{n} + \Delta$, still subordinate to the probability field of *stop-or-go* probabilities and the results of the experiments. Formally:

$$
\begin{cases}
x_{rc}(\mathrm{t_h}) = \{n\}, 1 \le \mathrm{r} \le \mathrm{R}, 1 \le \mathrm{c} \le \mathrm{C}, \mathrm{R} \times \mathrm{C} = \mathrm{N}, \mathrm{h} = 0,1,2,\ldots \\
x_{rc}(\mathrm{t_0}) = \{0\} \\
E_{\mathrm{h}} = \sum_{1 \le r \le R} \sum_{1 \le c \le C} x_{rc}(\mathrm{t_h}) \\
x_{rc}(\mathrm{t_{h+1}}) = \begin{cases} \left(x_{rc}(\mathrm{t_h}) + \Delta\right) \left[p_{rc}\left[E_{\mathrm{h}}\right]_{[0,1]}\right] \\ x_{rc}(\mathrm{t_h}) \left[1 - p_{rc}\left[E_{\mathrm{h}}\right]_{[0,1]}\right] \end{cases} \\
\text{OPERATIVE PROGRAM} \Rightarrow \text{define} : \mathrm{N}, \Delta, \left[p_{rc}\left[E_{\mathrm{h}}\right]_{[0,1]}\right]
\end{cases}
\tag{3.10}
$$

To simulate the dynamics of the system in (3.10), a model like the one in system (3.7) in the previous section has been constructed. Figure 3.47 shows the control panel used to insert the necessary data and produce the dynamics of the stochastic irreversible Combinatory Automaton composed of 100 cells after 50 iterations. Each cell represents an agent which, at every iteration, accumulates as a function of the accumulation of the other agents. The system's analytical state modifies the probability of accumulation for each agent. To produce a vertical accumulation, it is useful to calculate the *stop-or-go* probabilities $p_{rc}(\mathrm{t_h})_{[0,\ 1]}$, taking into account the *minimum* base probability $p_{\mathrm{min}} = \mathrm{const.}$, which is assigned (optionally) at the beginning to start up the system, and the optional values $\mathrm{k_{centrcell}}$ and $\mathrm{k_{neighbs}}$, which increase the probabilities for those cells placed near one containing an accumulation.

The final state of the automaton in Fig. 3.47, calculated after 50 iterations, is shown in Fig. 3.48. It is not surprising that vertical accumulation occurs only in correspondence with the three elements of the initial state. As indicated in the control panel of Fig. 3.47, the parameter for "premium for neighbors" was set to 0, which means that the cells neighboring the initial elements are not attractive to the accumulators.

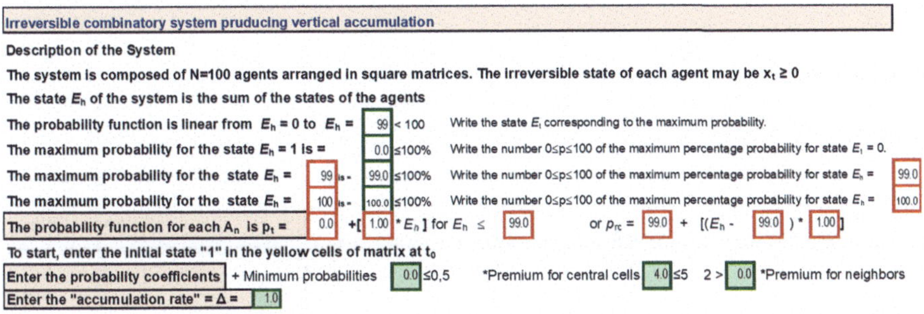

Fig. 3.47 Control panel regulating the dynamics of the *irreversible* Combinatory Automaton simulating vertical accumulation

Macro state E_h		
time	state	Δ state
0	3	0
1	3	0
2	3	0
3	3	0
4	3	0
5	3	0
6	3	0
7	3	0
8	3	0
9	3	0
10	3	0
11	3	0
12	4	1
13	5	1
14	5	0
15	5	0
16	5	0
17	5	0
18	6	1
19	7	1
20	7	0
21	8	1
22	9	1
23	9	0
24	9	0
25	9	0
26	11	2
27	12	1
28	13	1
29	13	0
30	13	0
31	14	1
32	14	0
33	15	1
34	15	0
35	18	3
36	18	0
37	19	1
38	19	0
39	19	0
40	20	1
41	21	1
42	21	0
43	23	2
44	24	1
45	26	2
46	26	0
47	27	1
48	28	1
49	29	1
50	31	2
51	32	1

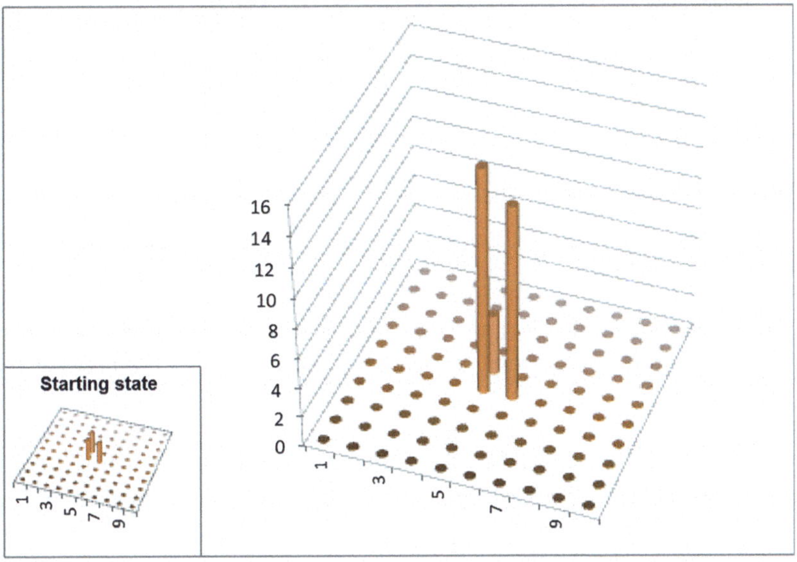

Fig. 3.48 Macro state at t_{50} of the automaton in Fig. 3.47

Figure 3.49, shows the final state of the automaton in Fig. 3.47, with an initial state of $E_0 = 1$, after we have inserted the various parameters (indicated below the diagram), among which we should note $k_{centrcell} = 2$ and $k_{neighbs} = 0.2$. As the “premium for neighbors” has also been added to the probabilities, the accumulation is also produced in those cells near the one in which the first micro state “owing to chance” is produced.

To simulate *horizontal accumulation*, we must consider the fact that the cells of the Combinatory Automaton can have only two micro states: “0” or “1.” To transform System (3.10) into the Boolean automaton (3.11), it is sufficient to introduce the following rule for the transition of state:

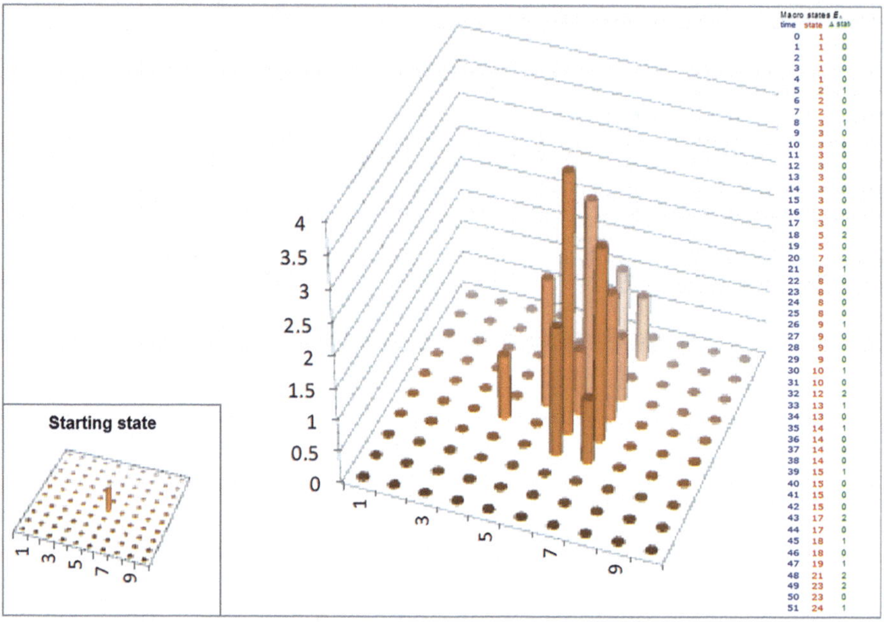

Fig. 3.49 Macro state at t_{50} of the automaton in Fig. 3.47, with various initial values and different probability functions

- If the state of a cell is $x_{rc}(t_h) = 0$ and the event occurs to which a *stop-or-go* probability, $p_{rc}(t_h)_{[0,1]}$, is associated, then it becomes $x_{rc}(t_{h+1}) = 1$; otherwise, it remains in state "0."

$$\begin{cases} x_{rc}(t_h) = \{0 \text{ or } 1\}, 1 \le r \le R, 1 \le c \le C, R \times C = N, h = 0,1,2,\ldots \\ x_{rc}(t_0) = \{0\} \quad \text{Input random values in some cells} \\ E_h = \sum_{1 \le r \le R} \sum_{1 \le c \le C} x_{rc}(t_h) \\ x_{rc}(t_{h+1}) = \{ \begin{matrix} \{1\} \; [p_{rc}[E_h]_{[0,1]}] \\ x_{rc}(t_h) \; [1 - p_{rc}[E_h]_{[0,1]}] \end{matrix} \\ \text{OPERATIVE PROGRAM} \Rightarrow \text{define}: N, \left[p_{rc}[E_h]_{[0,1]} \right] \end{cases} \tag{3.11}$$

Let us assume that the control panel that regulates the dynamics of the Combinatory Automaton causing horizontal accumulation contains the parameters indicated in Fig. 3.50.

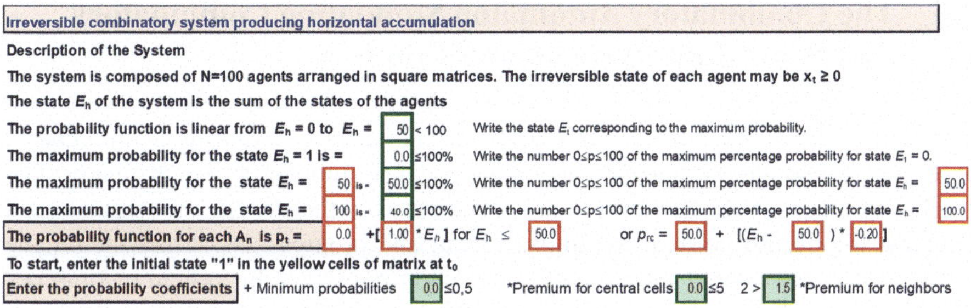

Fig. 3.50 Control panel regulating the dynamics of the *irreversible* Combinatory Automaton simulating horizontal accumulation

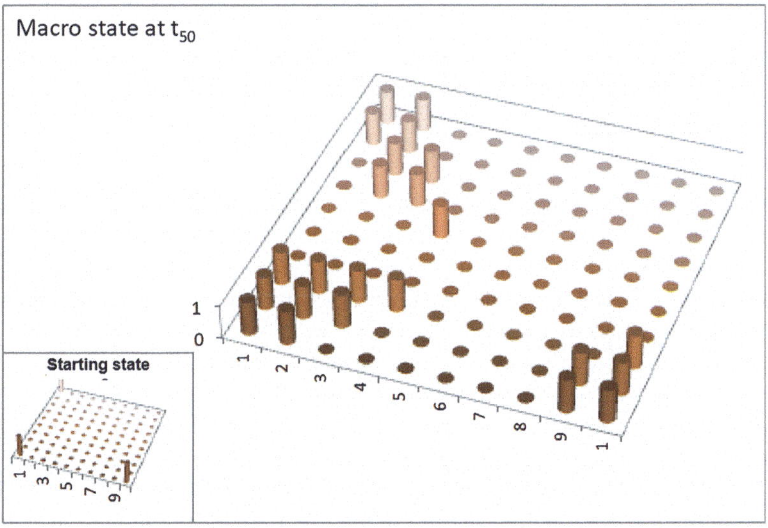

Fig. 3.51 Macro state at t_{50} of the automaton in Fig. 3.50

Note that it is fundamental to assign a congruous value to the "premium for neighbors," since, to accumulate a new object, it is necessary for the agents to be "attracted" by cells which are still "empty" and next to those already containing an object. The value $k_{neighbs} = 1.5\%$ has been assigned in the control panel.

The final state of the automaton after 50 iterations is shown in Fig. 3.51.

Figure 3.51 reveals that three clusters of objects, which have been horizontally accumulated, correspond to an initial state of three elements in the cells in three vertices of the matrix, as would have been expected from the rules applied and the data chosen in Fig. 3.50.

3.5 The Combinatory Automaton Simulating Combinatory Systems of Improvement and Progress

The previous sections have examined several simple stochastic Combinatory Automata to simulate the behavior of systems of accumulation and diffusion. The models in the previous chapter can also be adapted to the systems of pursuit (Sect. 2.4) and the derived systems of improvement and progress. Nevertheless, for the latter we can use simpler models under the assumption that the system recombines the micro behaviors or their effects using the average of the individual indices of improvement. Because they are closely correlated, systems of pursuit and systems of improvement and progress will both be dealt with in this section, and several types of such systems will be formally presented.

3.5.1 A Combinatory Automaton Simulating Races and Records

A competition—for example, a car race or a tournament organized to break an Olympic record—can be viewed as the effect of a Combinatory Automaton simulating a combinatory system of *pursuit* (Sect. 2.4.2). The competitors are agents who, at each instant, produce a position in the competition (micro state); the time series of the positions defines their micro behaviors. The *global information* is represented by the standings at every instant of the competition. The leader considers it necessary to maintain his lead over the follower, while those following the leader find it necessary to reduce the gap. Thus, the athletes try to continually adjust their states, thereby producing micro behaviors in order to gain and maintain the highest positions. This Combinatory Automaton is outlined in model (3.12).

The automaton (3.12) is *irreversible* in that the individual competitors can only improve their position during their performance, as occurs, for example, in all attempts at breaking a record. A *record*, in any sports category, determines the "absolute best"; because of this we witness a true race for the record. Those who compete are not content to equal the record, but instead do all they can to beat it. Thus, records are gradually improved and represent at one moment in time the macro effect of the combinatory system of the athletes and the *global information* that directs the subsequent micro behaviors. Not only does the competition push athletes to do their best to win, it also improves both their individual performance and the average performance levels of the competitors.

$$
\begin{cases}
\text{for } n = 1,\ldots,N \\
\text{for } h = 0,1,2,\ldots \\
x_n(t_0) = \{\text{initial values}\} \\
Y(t_h) = \max x_n(t_h) \text{ state of the leader} \\
x_n(t_{h+1}) = x_n(t_h) + \left[x_n(t_h) - Y(t_h)\right]\left[i_{n[RND]}\left[p_{n[0,1]}\right]\right] \\
+\left[kx_n(t_h) + hY^M(t_h)\right]\left[r_{n[0,1]}\right] \\
Y^M(t_h) = \frac{1}{N}\sum_{1\le n\le N} x_n(t_h) \text{mean performance} \\
\text{OPERATIVE PROGRAM} \Rightarrow \text{define}: N, x_n(t_0), p_{n[-1,1]}, r_{n[-1,1]}, h, k
\end{cases}
\tag{3.12}
$$

The modus operandi of the automaton simulating a race, illustrated in (3.12), can be described as follows. The racers are agents who, at any instant t_h, produce a position in the race, their micro states $x_n(t_h)$; the *global information* is represented by the individual gap, $gap_n = [x_n(t_h) - Y(t_h)]$, with respect to the leader's position in the race, $Y(t_h)$. The leader considers it necessary to maintain his lead over the followers by maintaining $gap_n > 0$ with respect to the other agents; those following the leader find it necessary to reduce their negative gap. Therefore, the racers try to continually adjust their states, thereby producing micro behaviors to gain and maintain the highest positions.

Each racer reduces his or her gap_n in two ways:

1. By using the *effort factor*, $\{[x_n(t_h) - Y(t_h)][i_{n\,[RND]}[p_{n\,[0,\,1]}]]\}$, which rewards racers who give their all to improve their position. For the simulation, an *individual effort factor* has been introduced, represented by a random number, $0 \le i_{n\,[RND]} \le 1$, which multiplies the *gap* by a percentage value, thereby reducing it. If $i_{n\,[RND]} = 0$, the gap is not reduced between the agent in question and the leader of the race; if $i_{n\,[RND]} = 1$, the gap is eliminated. Nevertheless, the intensity of the effort is a function of the *will to win*, represented by *stop-or-go* probabilities, $p_{n\,[0,1]}$, which, for simplicity's sake, are established at the start and *are not time-dependent*. The usual experiment using the drawing of *random numbers*, $rnd_n(t_h)$, occurs if the agents truly have the possibility of putting an effort into the race. In fact, if the experiment fails—$rnd_n(t_h) > p_{n\,[-1,1]}$—then $i_{n\,[RND]}[p_{n\,[-1,\,1]}] = 0$, and the agent remains in the previous state $x_n(t_h)$; otherwise, gap_n is reduced in proportion to the intensity of the effort, $i_{n\,[RND]}$;
2. By the *luck factor*, represented by the addend $[k\,x_n(t_h) + h\,Y^M(t_h)][r_{n\,[0,\,1]}]$, which intervenes independently of the individual *effort* of the racers to modify their gap_n. The amount of the change owing to "fortune" is calculated as the combination of the position (state) of each racer, $x_n(t_h)$, and the average position of the other agents, $Y^M(t_h)$. For simplicity's sake, this combination is determined by the coefficients $0 \le h \le 0.5$ e $0 \le k \le 0.5$, respectively. Fortune is expressed by the *stop-or-go* probability $r_{n\,[0,\,1]}$. If the experiment succeeds using ad hoc random

numbers, then the gap_n is reduced for each racer; if not, then luck does not intervene and the gap remains unchanged, except in the event of any action by the *effort factor*.

It is precisely the consideration of the *stop-or-go* probabilities that makes the automaton irreversible. On the other hand, by introducing *transition* probabilities, both for the *effort factor*, $p_{n\,[-1,1]}$, and the *luck factor*, $r_{n\,[-1,1]}$, $x_n(t_{h+1})$ could be lower than $x_n(t_h)$ and the worsening in the position could be attributed to a less intense effort or to luck. In this case, it must be posited that if the experiment fails, since $p_{n\,[-1,1]} > rnd_n$, then $i_{n[RND]}[p_{n[-1,1]}] < 0$, the gap_n increases, and the agent worsens his previous state. Otherwise, the gap_n is reduced in proportion to the intensity $i_{n\,[RND]}$, as would also occur due to the luck factor.

The automaton (3.12) is simulated in Fig. 3.52, which represents the dynamics of only two athletes competing over time to beat the record. The upper part of the diagram is the control panel containing the data, which remains constant during the simulation, while the lower part shows the data needed to produce the first interaction ($t_0 = t_1$).

Combinatory Automaton simulating a Combinatory system of Pursuit

		A_1	A_2	
Internal attitudes	$p_{n\,[0,1]}$	40%	80%	desire to win if A_n is the leader
		90%	40%	desire to win if A_n is not the leader
External chance	$r_{n\,[0,1]}$	30%	40%	good luck if $x_1(t_h) > x_2(t_h)$
		90%	80%	good luck if $x_1(t_h) \le x_2(t_h)$
Starting performance	$x_n(t_h)$	1	2	insert the starting performances > 0
	$0 \le h \le 0.5$ =	0.2	0.1	= $0 \le k \le 0.5$

Cycles = t_h	Functions	A_1	A_2	head	mean
	$x_n(t_h)$	$x_1(t_h)$	$x_2(t_h)$	$Y(t_h)$ leader	$Y^{Mean}(t_h)$
t_1	$x_n(t_h)$	1.00	2.00	2.00	1.50
the leader	$Y(t_h)$ leader	2.00	2.00		
tha gap_n	$x_n(t_h) - Y(t_h)$	-1.00	0.00		
wish to win $p_{n\,[0,1]}$	predetermined	*0.90*	*0.40*		
wish to win rnd numb.	random numbers	*0.74*	*0.15*		
effort $i_{n\,[RND]}$	random coefficient	0.64	0.82		
improvement due to effort	$gap_n\, i_n\, (p_n)$	0.64	0.00		
luck $r_{n\,[0,1]}$	predetermined	*0.90*	*0.80*		
external event	random numbers	*0.51*	*0.82*		
improvement due to luck	$k\, x_n + h\, Y^M$	0.40	0.00		
t_2	$x_n(t_h)$	2.04	2.00	2.04	2.02

Fig. 3.52 Control panel and first iteration of an irreversible Combinatory Automaton, acting over 10 iterations, simulating the pursuit of a record

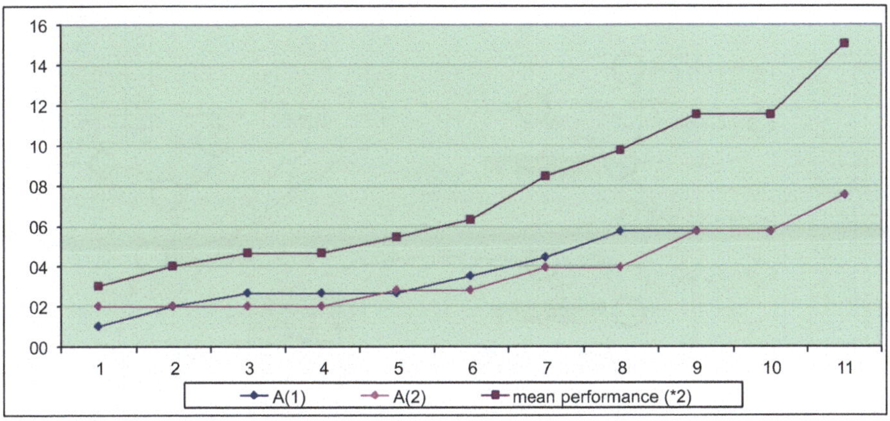

Fig. 3.53 Micro and macro dynamics produced by the automaton in Fig. 3.52, acting over 10 iterations

Figure 3.53 shows the dynamics of the micro states of the two athletes competing for the record and the dynamics of the average performance of the athletes (these values have been doubled to make the diagram more legible). The attempts to improve the records motivate more and more athletes to take part in competitions, which also leads to the continual improvement in the *average performance* of the athletes, so that after a more or less lengthy period, beginning when the first record was set, the average performance of the competitors is very high.

To simulate a system of pursuit—for example, a race with *10 runners* and assuming reversibility—a Combinatory Automaton has been built using Powersim. The simulation program in Fig. 3.54 indicates several initial values.

Figure 3.55 shows the dynamics of the positions of the competitors, clearly showing that during several cycles the relative positions worsen with respect to previous cycles.

3.5.2 A Combinatory Automaton Simulating an Inflationary Process Due to Imitation

The logic of combinatory systems of pursuit, discussed in Sect. 3.5.1, can be applied to simulate other systems in which the gradual increase in the micro states of the agents produces an increment in the macro states of the system, which, in turn, conditions the subsequent micro behaviors of the agents, exactly what occurs with athletes competing to surpass a record, or even with speakers in a crowded room (Sect. 2.4.1).

Let us consider a very common combinatory system which operates whenever a certain number of operators, A_n, involved in the retail markets of a certain good, decide to increase or lower the price of that good, considering the *global*

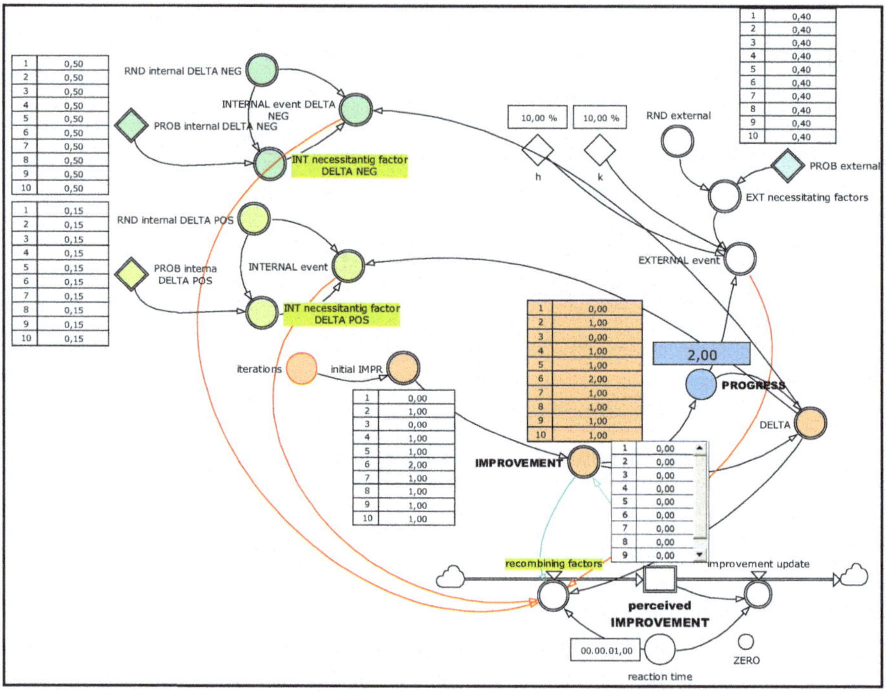

Fig. 3.54 Micro and macro dynamics produced by a reversible automaton imitating a race (CS of pursuit), simulated in Powersim

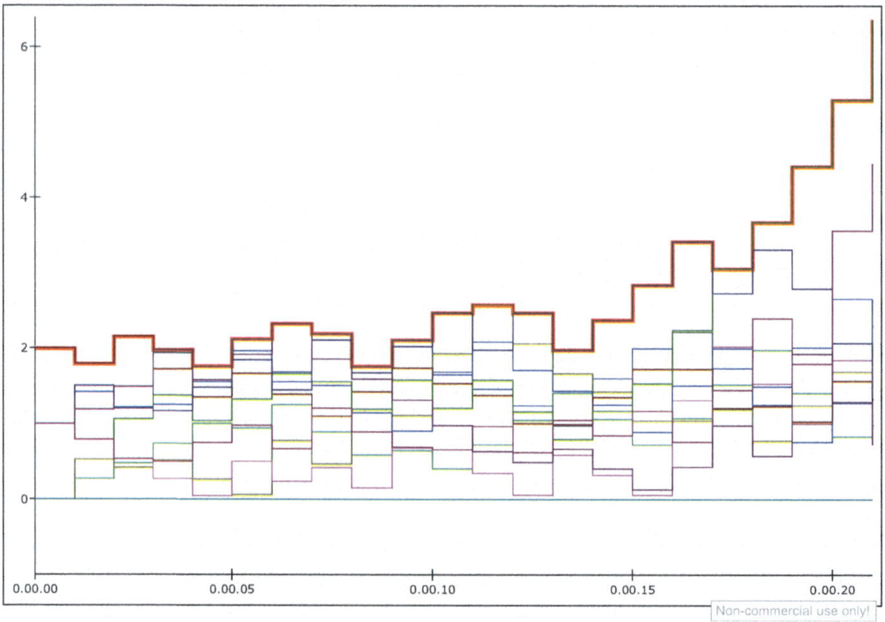

Fig. 3.55 Micro and macro positions in the race. The bold line shows the position of the race leader

information represented by the *price index* for that and/or other goods. For simplicity's sake, the *price index* $P(t_h)$ is represented here by the non-weighted average of prices $\pi_n(t_h)$ that exists for every A_n at t_h; but at t_{h+1} each A_n adjusts its own prices by taking into account the gap represented by $[\pi_n(t_h) - P(t_h)]$.

The system may be simulated in very simplified form by the stochastic Combinatory Automaton (3.13).

$$\begin{cases} \text{for } n = 1,\ldots,N \\ \text{for } h = 0,1,2,\ldots \\ \quad \pi_n(t_0) \Rightarrow \{\text{initial prices}\} \\ \{\pi_n(t_{h+1}) = \pi_n(t_h) + \{h[\pi_n(t_h) - P(t_h)] + k[P(t_h)]\}[p_{n[-1,1]}] \\ P(t_h) = \frac{1}{N}\sum_{1 \le n \le N} \pi_n(t_h) \text{price index} - \textit{not} \text{ weighted average} \\ \text{OPERATIVE PROGRAM} \Rightarrow \text{define}: N, \pi_n(t_0), p_{n[-1,1]}, h, k \end{cases} \tag{3.13}$$

Figure 3.56 shows the control panel and the first iteration of a simulation model of the automaton in (3.13).

To make the calculations for the first iteration more easily understandable, in each row of the second table of Fig. 3.56 the completed operations are indicated by capital letters (first column in blue), with the following meanings:

CONTROL PANEL

	Combinatory Automaton simulating Price Index. 10 retailers										
	Agents	A_1	A_2	A_3	A_4	A_5	A_6	A_7	A_8	A_9	A_{10}
[1]	$p_n(\Delta>L+)$ & $\Delta>0$, high	15.0	15.0	15.0	15.0	15.0	15.0	15.0	15.0	15.0	15.0
[2]	$p_i(\Delta<=L+)$ & $\Delta>0$, low	20.0	20.0	20.0	20.0	20.0	20.0	20.0	20.0	20.0	20.0
[3]	$p_i(\Delta<=L-)$ & $\Delta<0$ low loss	30.0	25.0	60.0	40.0	50.0	30.0	30.0	3.0	40.0	40.0
[4]	$p_i(\Delta>L-)$ & $\Delta<0$ high loss	15.0	15.0	15.0	9.0	9.0	9.0	9.0	9.0	9.0	9.0
[5]	Enter Initial prices	1.50	1.55	1.43	1.48	1.51	1.50	1.55	1.43	1.48	1.51
[6]	Enter parameters	*h* =	10%		L+	2%	L-	-6%			
[7]		*k* =	5%			buy		sell			

FIRST ITERATION

	Cycles=t_h	Functions	A_1	A_2	A_3	A_4	A_5	A_6	A_7	A_8	A_9	A_{10}	$P(t_h)$	$\Sigma\,\mu_n(t_h)$
[A]	t_1	$\pi_n(t_1)$	1.50	1.55	1.43	1.48	1.51	1.50	1.55	1.43	1.48	1.51	1.49	14.94
[B]	prices	$P(t_h)$	1.49	1.49	1.49	1.49	1.49	1.49	1.49	1.49	1.49	1.49		
[C]	delta π	$\Delta\pi = \pi - P$	0.01	0.06	-0.06	-0.01	0.02	0.01	0.06	-0.06	-0.01	0.02		0.00
[D]	delta π %	$\Delta\pi$%	0.40%	3.61%	-4.48%	-0.95%	1.06%	0.40%	3.61%	-4.48%	-0.95%	1.06%		
[E]	buy/sell	L+ / L-	2.00%	2.00%	-6.00%	-6.00%	2.00%	2.00%	2.00%	-6.00%	-6.00%	2.00%		
[F]	change	$p_n(\Delta>0)$ = gain	20.00	15.00			20.00	20.00	15.00			20.00		
[G]	change	$p_n(\Delta<0)$ = loss			60.00	40.00				3.00	40.00			
[H]	RND 1	for decision	*95*	*65*	*25*	*79*	*44*	*40*	*11*	*28*	*95*	*22*		
[I]	choice	change prices	0	0	1	0	0	0	1	0	0	0		
[J]	$\Delta\pi$	delta price	0.00	0.00	0.07	0.00	0.00	0.00	0.08	0.00	0.00	0.00		
[K]	t_2	$\pi_n(t_2)=\pi_n(t_1)+\Delta\pi$	1.50	1.55	1.50	1.48	1.51	1.50	1.63	1.43	1.48	1.51	1.51	15.09

Fig. 3.56 Control panel and first iteration of the simulation of the Combinatory Automaton in (3.13), with ten agents and differentiated probabilities for each agent

[A] values for the first negotiation, taken from row [5] of the first table
[B] price index, taken from the average in row [A]
[C] gain or losses (prices–index)
[D] percentages of gains or losses
[E] maximum gain or loss percentages, L+ or L−, in row [6]
[F] the differences in row [D] are compared with the limits indicated in row [E]; based on the sign of delta, the probabilities indicated are rewritten in rows [1] and [2] for positive delta values
[G] compares the differences in row [D] with the limits indicated in row [E]; based on the sign of delta, the probabilities in rows [3] and [4] are rewritten for negative deltas
[H] random number generated based on the decision to change or not to change prices
[I] the random numbers in row [H] are compared with the probabilities in rows [F] and [G], and the choice: “1” [= yes], “0” [= no] is indicated.
[J] calculates the price variation, considering the parameters *h* and *k* in rows [6] and [7]
[K] calculates the new price by adding the starting price (row [A]) and the delta price (row [J])

Figure 3.57 describes the micro and macro dynamics of the automaton in Fig. 3.56, observed for 20 iterations. The bold blue line represents the *global information*; that is, the *price index*. The colored lines show the micro behavior of the prices quoted by each operator.

The model above is based on the imitative hypothesis for price increases, which are necessary when businesses operate during periods of inflation. These increases are necessary solely in order not to operate under conditions of monetary losses due to inflation. We have all experienced this situation; for example, when one day a cup

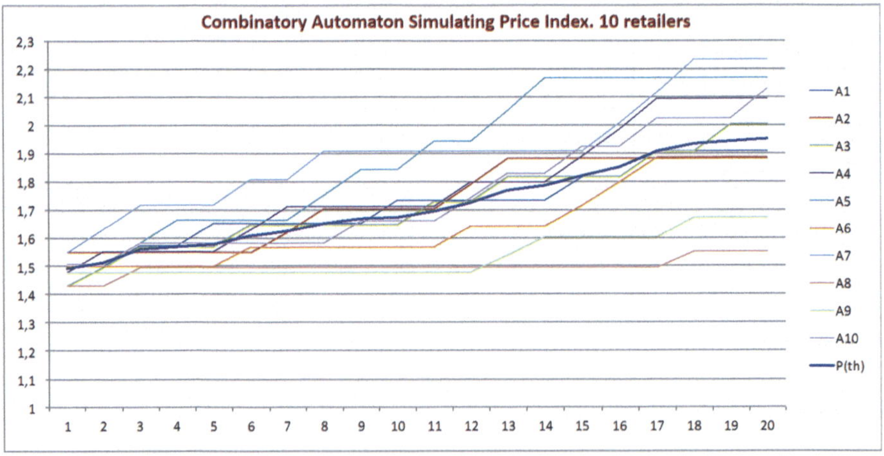

Fig. 3.57 Inflationary dynamics produced by the Combinatory Automaton (3.13) indicated in Fig. 3.56

of coffee at a café costs 1.00, the next day 1.03, and after a week 1.10, and so on. There is no real economic reason for this, only the need to maintain the purchasing power of the invested capital. Model (3.13) in Fig. 3.56 can also be used to simulate the opposite trends in prices of a single good, or of several though fungible goods produced by multiple producers or sold by different merchants operating in a market with near-perfect competition, where the latter is based mainly on price. As a result, the agents, producers, or merchants by necessity seek to reduce, not to increase, the prices of the good, thereby initiating a typical *micro-macro feedback* process that leads to a decrease in prices. This phenomenon is so apparent that there is no need for further comment. In fact, in order to simulate this combinatory system, it suffices to insert negative values for the parameters "*h*" and "*k*," in order to enable negative variations in prices to be calculated.

3.5.3 A Combinatory Automaton Simulating Stock Exchange Quotations and General Indices

A similar though much more complicated *reversible* Combinatory Automaton (3.14) simulates the quotations of a single stock, which normally produce chaotic dynamics both in the quotations of individual stocks and in the general indices.

$$\begin{cases} for: 1 \le \mathrm{n} \le \mathrm{N} \\ for: \mathrm{h} = 0,1,2,\ldots \\ for: s = 1,2,\ldots \mathrm{S} \\ \pi_{\mathrm{n}}^{\mathrm{s}}(\mathrm{t}_0) \Rightarrow \{\text{initial prices for each stock}\} \\ \pi_{\mathrm{n}}^{\mathrm{s}}(\mathrm{t}_{\mathrm{h+1}}) = \pi_{\mathrm{n}}^{\mathrm{s}}(\mathrm{t}_{\mathrm{h}}) + \begin{Bmatrix} \left[h^{\mathrm{s}} \left[\pi_{\mathrm{n}}^{\mathrm{s}}(\mathrm{t}_{\mathrm{h}}) - P^{\mathrm{s}}(\mathrm{t}_{\mathrm{h}}) \right] \mathrm{i}_{\mathrm{n}[RND]}^{\mathrm{s}} \right] \\ + \left[k^{\mathrm{s}} P^{\mathrm{s}}(\mathrm{t}_{\mathrm{h}}) \mathrm{i}_{\mathrm{n}[RND]}^{\mathrm{s}} \right] \\ + \left[k^{SE} P^{STOCKEXCH.}(\mathrm{t}_{\mathrm{h}}) \mathrm{i}_{\mathrm{n}[RND]}^{SE} \right] \end{Bmatrix} \times \mathrm{S}^{-} \text{ or } B^{+}{}_{\mathrm{n}} \left[p_{\mathrm{n}[-1,1]}^{\mathrm{s}} \right] \\ P^{\mathrm{s}}(\mathrm{t}_{\mathrm{h}}) = \dfrac{\sum_{1 \le \mathrm{n} \le \mathrm{N}} \pi_{\mathrm{n}}^{\mathrm{s}}(\mathrm{t}_{\mathrm{h}})}{\sum_{1 \le \mathrm{n} \le \mathrm{N}} \mathrm{Q}_{\mathrm{n}}^{\mathrm{s}}(\mathrm{t}_{\mathrm{h}})} \mathrm{Q}_{\mathrm{n}}^{\mathrm{s}}(\mathrm{t}_{\mathrm{h}})_{[RND]} \text{ weighted price index for each s} \\ EV^{\mathrm{s}} = \sum_{1 \le \mathrm{n} \le \mathrm{N}} \pi_{\mathrm{n}}^{\mathrm{s}}(\mathrm{t}_{\mathrm{h}}) \mathrm{Q}_{\mathrm{n}}^{\mathrm{s}}(\mathrm{t}_{\mathrm{h}})_{[RND]} \text{ total value exhanged for each s} \\ P^{STOCKEXCH.}(\mathrm{t}_{\mathrm{h}}) = \dfrac{\sum_{1 \le s \le \mathrm{S}} P^{\mathrm{s}}(\mathrm{t}_{\mathrm{h}})}{\sum_{1 \le s \le \mathrm{S}} EV^{\mathrm{s}}} EV^{\mathrm{s}} \text{ price index for Stock Exch. (SE)} \\ \text{OPERATIVE PROGRAM} \Rightarrow \text{define}: \mathrm{N}, \pi_{\mathrm{n}}^{\mathrm{s}}(\mathrm{t}_0), p_{\mathrm{n}[-1,1]}, h, k \end{cases} \quad (3.14)$$

Let us consider the simpler case of a certain number "N" of stockbrokers, A_n, interested in trading a certain stock, "s," from a basket of "S" stocks, and let us suppose that their purchase or selling decisions are taken only considering the price index $P^s(t_h)$ for that stock.

We can argue that if the operator A_n at time t_h possesses the stock with a value $\pi_n^s(t_h) > P^s(t_h)$, the potential gain is equal to the difference; thus A_n can decide to keep the stock, perhaps hoping for a further rise, or sell it for a gain. If instead $\pi_n^s(t_h) < P^s(t_h)$ then there is a potential loss equal to the difference. A_n can thus decide to keep the stock, hoping perhaps for a rise in value, or sell it to limit the loss. In short, we suppose that each operator decides to acquire, maintain, or sell shares by evaluating the difference $\left[\pi_n^s(t_h) - P^s(t_h)\right]$, which defines gains or losses at t_h. This is inserted into model (3.14) by means of the factor $\left[h^s\left[\pi_n^s(t_h) - P^s(t_h)\right] i^s_{n[RND]}\right]$, which contributes to determining the variation of $\pi_n^s(t_h)$ as a part of the potential gains or losses, based on a coefficient h^s, which, for simplicity's sake, is assumed fixed for all agents.

To simulate the behavior of the operators by means of the stochastic Combinatory Automaton in (3.14), we must introduce buying (B^+) or selling (S^-) transition probabilities, $\left[p^s_{n[-1,1]}\right]$, which can be differentiated based on the sign and level of the difference: $\left[\pi_n^s(t_h) - P^s(t_h)\right]$.

Once the decisions are made by all the stockbrokers, the new prices, $\pi_n^s(t_{h+1})$ —which depend on the level of $P^s(t_h)$ as well as on random factors, as specified above—lead to changes in the stock quotation, which becomes $P^s(t_{h+1})$, and influences subsequent buying or selling decisions. This is inserted into model (3.14) by adding the new factor $\left[k^s\ P^s(t_h)\ i^s_{n[RND]}\right]$, which contributes to determining the variation in the quotation based on a coefficient k^s, which, for simplicity's sake, is assumed fixed for all agents. Since the behavior of each agent depends first of all on the information regarding the *gap* between the value of the stock that is held and the stock quotation of the share, this automaton is based only on a *one-level feedback* between micro and macro behaviors. Therefore, model (3.14) specifies that stockbrokers make decisions only taking account of the level of $P^s(t_h)$, which is the *global self-produced information*, and not considering other *exogenous* information.

In order to determine $P^s(t_h)$, we must introduce the volumes of the shares which are exchanged for each operator at each t_h, so as to be able to calculate the average of the prices, which have been weighted according to the exchanged quantities. To simulate the negotiated quantities, we have adopted the simple solution of using random numbers specifically selected for these calculations.

Nevertheless the behavior of stockbrokers normally depends not only on the individual stock quotation but also on that for other shares. In this case we can define the combinatory system as a *crossed-feedback combinatory system*, following the model in Fig. 3.58.

The broker's decision to sell a stock can, in a certain sense, set off a combinatory dynamic process for that stock. The dynamics regarding that stock can then influence that of other stocks in the same as well as opposite direction. The combination of the dynamic processes for these stocks determines that of the Stock Exchange General Index, which in turn influences stockbrokers' decisions which either

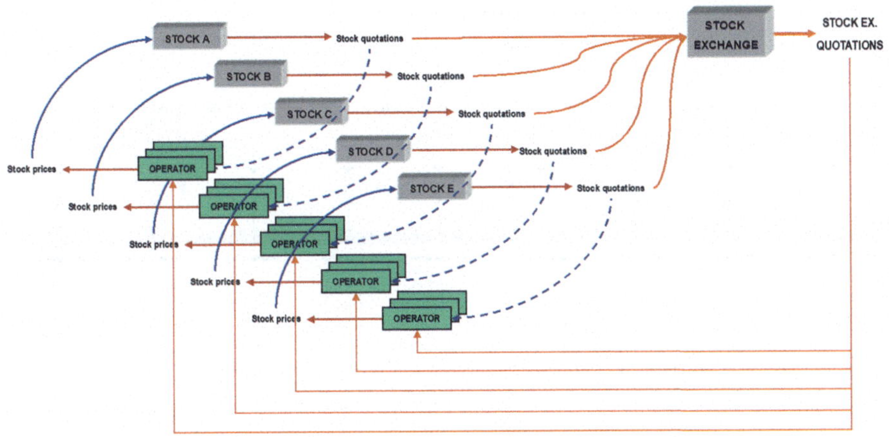

Fig. 3.58 Model of the combinatory system showing two-level feedback and crossed feedback, which determines the Stock Exchange quotations

strengthen or invert the direction of the trend. This is inserted into (3.14) by adding the new element $\left[k^s P^s(t_h) i^s_{n[RND]}\right]$ as a determinant of $\pi^s_n(t_{h+1})$.

Therefore, model (3.14) clearly shows that trading decisions regarding a stock also depend on the general level of Stock Exchange quotations, $P^{STOCKEXC.}(t_h)$, which derives from the weighted average of the quotations of the various stocks. If the price index represents further *general information* for an individual stockbroker, then the Stock Exchange also becomes a combinatory system, where the brokers make decisions based on *two levels* of feedback, one represented by the quotation of their own stock, $P^s(t_h)$, and the second by the general *Stock Exchange Index*, $P^{STOCKEXC.}(t_h)$, which is based on a coefficient, k^{SE}, assumed, for simplicity's sake, to be fixed for all operators. The higher k^{SE} is, the greater the number of traders that will choose to buy, which will positively influence the stock prices of the "S" stocks.

This will occur during periods of optimism. However, during periods of pessimism k^{SE} can even become negative, so that the *micro-macro feedback* produces and maintains a downward trend in stock prices.

Finally, the world exchanges are interconnected by a multiple feedback process to form a new, vaster single combinatory system. An important conclusion from these models is that the general price index of a Stock Exchange is the result of a high level combinatory system, holonic in nature, composed of lower level combinatory systems interrelated by means of *multiple* and *crossed* feedback; the interconnected stock exchanges are like a holarchy, which is a *hierarchically organized structure of holons* (Mella, 2009) as shown in the model in Fig. 3.59.

It is thus not difficult to imagine how the beating of a butterfly's wings (Rangone & Mella, 2019) can condition some operators and have catastrophic results for the dynamics of a Stock Exchange, and that this can condition the dynamics of the entire system of world exchanges.

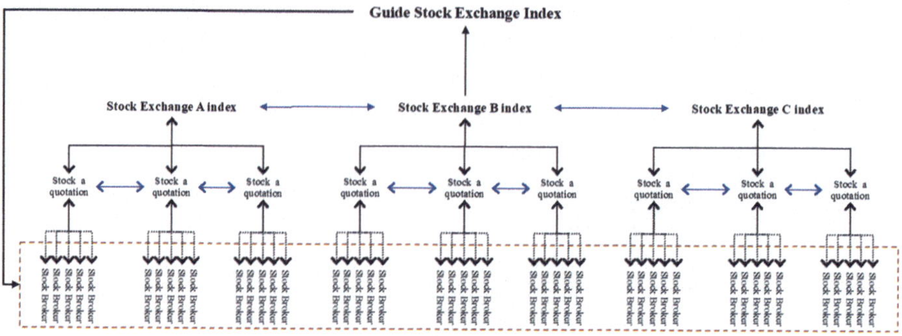

Fig. 3.59 Model of the holonic structure of the Stock Exchange network

3.6 The Theoretical Behavior of Combinatory Automata Simulating Systems of Improvement and Progress

This type of combinatory system is fundamental to an understanding of the process of collective progress in all human collectivities. This progress owes to the need for the agents to improve their own state with respect to the general index of progress they perceive or intuit. Stated more simply, *micro-macro feedback* is established between the *micro* and *macro behavior*, which acts to produce progress understood as an improvement in the overall state of a collectivity (according to commonly accepted value judgments). The progress derives from the agents' pursuit of individual improvement (i.e., an increase in some parameter judged to be useful or favorable) at the same time it directs the agents to search for new improvements. Individual improvement raises the parameter of collective progress, which leads to the formation of positive and negative gaps that push the individuals to improve in order to increase the positive gaps or eliminate the negative ones. The system must be able to perceive the individual improvement and to adjust the progress parameter to the average (or, more generally, the combination) of the individual improvement measures.

3.6.1 The Formal Definition

A very special and important combinatory system is the one we shall call the *Improvement and Progress Combinatory System*—which was illustrated in Sect. 2.6)—since its particular effect is to produce progress, according to commonly accepted value judgments regarding an improvement in the overall state of a collectivity.

The economy considered as a typical combinatory system of *improvement and progress* was first theorized by Adam Smith:

> Every individual is continually exerting himself to find out the most advantageous employment of whatever capital he can command. It is his own advantage [individual improvement], indeed, and not that of the society, which he has in view. But the study of his own advantage naturally, or rather necessarily leads him to prefer that employment which is most advantageous to society [necessitating factor]. He generally, indeed, neither intends to promote the public interest nor knows how much he is promoting it. By preferring the support of domestic to that of foreign industry, he intends only his own security [limited information]; and by directing that industry in such a manner as its produce may be of the greatest value [micro effect], he intends only his own gain, and he is in this, as in many other cases, led by an invisible hand [*micro-macro* feedback] to promote an end which was no part of his intention [recombining factor]. Nor is it always the worse for the society that it was no part of it. By pursuing his own interest [individual improvement], he frequently promotes that of the society [collective progress] more effectually that when he really intends to promote it. I have never known much good done by those who affected to trade for the public good. It is an affectation, indeed, not very common among merchants, and very few words need be employed in dissuading them from it. (Adam Smith, 1776) [square brackets added]

In effect, Smith asserted that everyone strives to become wealthy [individual improvement] *considering only his own gain*, but to this end he must produce with the maximum productivity and quality and exchange what he owns or produces with others [micro behavior] who are (supposed) capable of evaluating the goods he has to offer. Therefore, the human desire to improve wealth and individual happiness produces progress in the general wealth and welfare, since it reduces costs and increases the average quality of goods. Smith argued that the division of labor and a free market are the *recombining factors* necessary for the system to operate. The maximization of returns on invested capital represents the *necessitating factors*, together with the desire for happiness.

Ludwig von Mises also clearly described the idea of progress as the effect of the "fact that people are motivated by the urge to improve the material conditions of their existence."

> Eighteenth-century social philosophy was convinced that mankind has now finally entered the age of reason. While in the past theological and metaphysical errors were dominant, henceforth reason will be supreme. People will free themselves more and more from the chains of tradition and superstition and will dedicate all their efforts to the continuous improvement of social institutions. Every new generation will contribute its part to this glorious task. With the progress of time society will more and more become the society of free men, aiming at the greatest happiness of the greatest number. Temporary setbacks are, of course, not impossible. But finally the good cause will triumph because it is the cause of reason. People called themselves happy in that they were citizens of an age of enlightenment which through the discovery of the laws of rational conduct paved the way toward a steady amelioration of human affairs. (Von Mises, 1998, p. 192)

> The term progress is nonsensical when applied to cosmic events or to a comprehensive world view. We have no information about the plans of the prime mover. But it is different with its use in the frame of ideological doctrine. The immense majority strives after a greater and better supply of food, clothes, homes, and other material amenities. In calling a rise in the masses' standard of living progress and improvement, economists do not espouse a mean materialism. They simply establish the fact that people are motivated by the urge to improve the material conditions of their existence. They judge policies from the point of view of the aims men want to attain. He who disdains the fall in infant mortality and the

gradual disappearance of famines and plagues may cast the first stone upon the materialism of the economists. (Von Mises, 1998, p. 193)

In social systems, systems of improvement and progress are the systems which lead to institutions (Caldas & Coelho, 1999).

In life sciences, we can put forward the hypothesis that men naturally tend to achieve individual improvement and collective progress, as stated in Gould's theory of options (2000, 2001).

> After testing many pathways to account for human peculiarity, we arrive at an unusual principle. Humans appeared to evolve along a fitness pathway that maximized the options of behavior for the least cost to adapt. [...]
>
> Arguments here then, offer an unusual approach to complex issues. For studying the mind, morality, or human nature, science is making progress in detail, but traditional problems of dualism, human purpose, or mind studying mind remain. The suggestion here is to examine the issues not with answers, but a way to ask questions. If we study an attribute of human physiology or behavior and ask 'why did it evolve this way?' we might ask that forever. But if we ask, 'how would evolving this way maximize options?' we see old problems from a fresh perspective.
>
> Finally, the real purpose to knowledge is to increase options. Today humanity, which has overcome many trials, now faces perilous options of its success. Self-knowledge of what we are and how we came to be the species that we are, provides our surest path forward. Here the approach offers its final flexibility. The theory can provide fresh insights into how human behavior evolved, but can also be used to analyze behavior in everyday situations. In business, society, personal relations, and morality, people constantly face change and challenge. We confront our options. This is how our species raised itself (Gould, 2001, online).

Systems of *improvement and progress* represent a "mixed" class, since they have features of systems of pursuit integrated with those of systems of accumulation and diffusion. However they may be simulated by a relatively simple Combinatory Automaton that follows these rules based on the general model (3.3) in Sect. 3.2.2:

1. Let us consider, as usual, a Combinatory Automaton made up of a base of N agents, A_n, $1 \leq n \leq N$, with or without form, and assume that each A_n has a micro state, $\pi_n(t_h)$, which represents the *individual improvement* level of A_n at time $t_{h.}$
2. The micro dynamics $[\pi_n(t_0), \ldots \pi_n(t_h), \ldots]$ represents the *improvement path* of that agent; in particular, there is an improvement for A_n—according to value parameters to be defined through observation or stipulation—if $\pi_n(t_{h+1}), > \pi_n(t_h)$ (or if the opposite inequality holds, according to the meaning of $\pi_n(t_h)$).
3. The *analytical state* of the automaton, $X(t_h) = [\pi_n(t_h)]$, $1 \leq n \leq N$, is defined as the values of the variables of improvement $\pi_n(t_h)$ associated with all the A_n of the automaton for each t_h, h = 0, 1, 2, ...
4. The *synthetic state* and the *output* of the automaton $Y(t_h)$, at t_h, are defined as the value assumed by the *recombined macro variable*

$$P^M(t_h) = \frac{1}{N} \sum_{1 \leq n \leq N} \pi_n(t_h) \tag{3.15}$$

which represents the average state of progress for the entire system.

5. The macro dynamics $[P^M(t_0), \ldots, P^M(t_h), \ldots]$ symbolizes the *progress path* of the automaton; progress occurs for the collectivity—according to value parameters

to be defined through observation or stipulation—if $P^M(t_{h+1}), > P^M(t_h)$, (or if the opposite inequality holds, according to the meaning of $P^M(t_h)$).

6. At time t_h, the *necessitating operation(s)*, which condition the internal event (decision) that determines the agents' behavior, derives from the differences

$$\Delta\pi_n(t_h) = \pi_n(t_h) - P^M(t_h) \tag{3.16}$$

which denote the *gaps* between the *individual improvement* level and the mean level denoting the state of the *collective progress*; thus, each A_n perceives an *inferiority*, with respect to the collectivity, if $\Delta\pi_n(t_h) < 0$, or a *superiority* in the opposite case, and acts to maintain or increase the *superiority*, or eliminate or reduce the *inferiority*.

7. The *micro transition* functions can have the following form:

$$\pi_n(t_{h+1}) = \left\{\pi_n(t_h) + \Delta\pi_n(t_h)\, i_{n[RND]}\left[p_n(t_h, \Delta)_{[0,1]}\right]\right\} \tag{3.17}$$

The probabilities $p_n(t_h, \Delta)_{[0,1]}$, which are both agent- and time-dependent and dependent on the sign of $\Delta\pi_n(t_h)$, represent the *necessitating factors* under the hypothesis of *irreversibility* (agents can only ameliorate their improvement measures); nevertheless, the probabilities for transition of state may take the form $p_n(t_h, \Delta)_{[-1,1]}$, under the hypothesis of *reversible* micro behaviors (agents may also reduce their improvement measures with respect to the progress measure). The variable $i_{n[RND]}$ indicates the random action of A_n to improve or worsen his performance. It is reasonable to consider that the probabilities are greater for the elements whose level of improvement is below the average and lower for those which already exceed the level of progress.

8. To make the model more general, we may also assume that the micro *transition* functions (3.17), which determine the change in the agents' micro behavior, depend also on *environmental factors*; that is, on *external events*, whose general expression may be:

$$\left[k\pi_n(t_h) + hP^M(t_h)\right]\left[r_n(t_h, \Delta)_{[0,1]}\right] \tag{3.18}$$

where:

- $[k\,\pi_n(t_h) + h\,P^M(t_h)]$ represents the amount of influence of the environmental variables on the improvement measure of A_n; this expression translates the common idea that the attempt to improve performance is conditioned by both the previous level of the individual performance measure and the previous level of performance of the system, where $0 \le k \le 1$ and $0 \le h \le 1$ are the weights of the two factors. In many automata it is normally assumed that $h = 0$.
- $r_n(t_h, \Delta)_{[0,1]}$ or $r_n(t_h, \Delta)_{[-1,1]}$ (in the case of irreversibility and reversibility) indicates the probability of this external event, assumed to be dependent, for each A_n, on the sign of $\Delta\pi_n(t_h)$, as defined in (3.16). It is logical to assume that for those A_n whose measure of improvement exceeds the average progress, the probabilities $r_n(t_h, \Delta)$ are greater than that for those elements with a lower level of improvement.

9. The *micro* and *macro* dynamics are thus connected, since the level of *improvement* measures determine the level of *progress*, but this in turn modifies the subsequent *improvement* variables in the typical *micro-macro feedback* that also impacts the probabilities of transition of state of the agents.

The Combinatory Automaton is summarized in the general formal model (3.19):

$$\begin{cases} \text{for } n=1,\ldots,N, h=0,1,2,\ldots \\ \pi_n(t_0) \Rightarrow \{\text{initial individual improvement levels}\} \\ P^M(t_h) = \frac{1}{N}\sum_{1\le n\le N}\pi_n(t_h) \text{state of progress of the collectivity} \\ \Delta\pi_n(t_h) = \pi_n(t_h) - P^M(t_h)\, gap \text{ perceived } by \text{ A}_n \\ \pi_n(t_{h+1}) = \begin{Bmatrix} \pi_n(t_h) + \Delta\pi_n(t_h)\, i_{n[RND]}\left[p_n(t_h,\Delta)_{[0,1]}\right] \\ +\left[k\pi_n(t_h) + hP^M(t_h)\right]\left[r_n(t_h,\Delta)_{[0,1]}\right] \end{Bmatrix} \\ \text{OPERATIVE PROGRAM} \Rightarrow \text{define}: \text{N}, \pi_n(t_0), p_{n[-1,1]}, r_{n[0,1]}, h, k \end{cases} \tag{3.19}$$

We can attribute the following significance to the proposed model (though many other variants could be equally as interesting): if we assume the agents of the system possess a "will"—that is, some form of influence on the means of controlling their own level of *improvement*—then the choice of values for the parameters represents a situation where those who "feel" inferior have a greater "desire" to overcome the negative gap, while those who already "feel" superior have more "chance" to remain in this position.

Of course, we must assume the existence of recombining factors that allow the combinatory system to calculate and disclose the levels of collective progress, $P^M(t_h)$, determined by the combination of levels of individual improvement. The operative logic of the Combinatory Automaton (3.19) can therefore be described as follows:

- "By chance," at time t_0 some element of the collectivity produces a micro behavior that causes an improvement in the parameter $\pi_n(t_0)$, whose significance must be specified by the observer.
- Assuming that the *minimum density* required for an improvement to influence the variable of progress is attained at t_1, then progress begins and $P^M(t_1) > P^M(t_0)$.
- By observing the value of $P^M(t_1)$, the system's agents can compare the average level of progress for the collectivity with that of their state of well-being; the individual gaps that result activate the necessitating factors, which move the agents to modify their micro behavior based on the level of progress achieved by the system, thereby obtaining an improvement that raises the parameter $\pi_n(t_1)$ to the level $\pi_n(t_2)$.
- In succeeding moments, additional progress is produced in the system, thus activating the characteristic *micro-macro feedback*.

Due to the presence of sufficient necessitating and recombining factors, when "by chance" an improvement begins in one or in all the elements in the base of the system, then "by necessity" progress occurs throughout the system. The improvement spreads and the progress continues, unless a *limiting state* is reached in which no further improvement can be carried out and no further progress can occur.

3.6.2 Three Types of Combinatory Automata of Improvement and Progress

In view of the structure of the progress variable in model (3.19), that is $P^M(\mathrm{t_h}) = (1/\mathrm{N})\sum_{\mathrm{n}=1}^{\mathrm{N}} \pi_n(\mathrm{t_h})$, the automaton may be defined as the MEDIAL AUTOMATON of IMPROVEMENT AND PROGRESS, since the micro behaviors aim at reaching and/or exceeding the parameter of progress, which represents an average (whose form may be specified in different ways) of the measures of the parameters of improvement in the base agents; the macro behavior of the system leads to a continual readjustment of the average, so that the individual improvement leads to the accentuation of the average progress, which, in turn, gives a boost to individual improvement.

Even if the MEDIAL AUTOMATON of IMPROVEMENT AND PROGRESS is quite general, we may conceive of two other different automata simply by specifying the structure of the progress variable $P(\mathrm{t_h})$. There are two relevant types.

The first variant is the MAXIMAL, "INFERIORITY-REDUCING," or "pursuit" AUTOMATON, where the parameter of progress is represented by the *maximum value* of the parameters of improvement which characterize the elements of the system; the element to which this value belongs is referred to as the *leader element*—or simply "the best"—since its performance represents the degree of progress for the entire system:

$$P^{MAX}(\mathrm{t_h}) = \max_{\mathrm{n}} \pi_{\mathrm{n}}(\mathrm{t_h}) = \pi_{MAX}(\mathrm{t_h}) \tag{3.20}$$

All the other elements thus present a state which is inferior with respect to "the best" and try to improve for their part; the element that succeeds in being "the best" becomes the *guide for progress* and triggers further improvement.

The gap between each agent's state and the value of "the best" state,

$$\Delta\pi_{\mathrm{n}}(\mathrm{t_h}) = \pi_{\mathrm{n}}(\mathrm{t_h}) - \pi_{MAX}(\mathrm{t_h}) \tag{3.21}$$

represents the *quantum of inferiority* perceived by each agent compared with the improvement parameter of the *leader agent*. We thus witness micro behaviors aimed at reducing the inferior level with respect to the level of progress, which causes a macro behavior whose effect is to raise the average level of improvement, so that some agents manage to further raise their previous level of progress. Thus, the transition of state function for each cell in the automaton must consider at the same time the quantum of inferiority as well as the *average level of improvement*; its structure is as follows, where $P^M(t_h)$ indicates the *average level of progress*.

$$\pi_n(t_{h+1}) = \Big\{\pi_n(t_h) + \Delta\pi_n(t_h)\, i_{n[RND]}\Big[p_n(t_h,\Delta)_{[0,1]}\Big] + \Big[k\,\pi_n(t_h) + h\,P^M(t_h)\Big]\Big[r_n(t_h,\Delta)_{[0,1]}\Big]\Big\}$$

The *maximal systems*, or "inferiority-reducing" automaton, can be summarized in this general formal model (3.22):

$$\begin{cases} \text{for } n = 1,\ldots,N \\ \text{for } h = 0,1,2,\ldots \\ \pi_n(t_0) \Rightarrow \{\text{initial individual improvement levels}\} \\ P^{MAX}(t_h) = \max_n \pi_n(t_h) = \pi_{MAX}(t_h) \quad \text{state of progress of the collectivity} \\ P^M(t_h) = \frac{1}{N}\sum_{1\le n\le N}\pi_n(t_h) \quad \text{average level of improvement} \\ \Delta\pi_n(t_h) = \pi_n(t_h) - \pi_{MAX}(t_h) \quad \text{quantum of inferiority} \\ \pi_n(t_{h+1}) = \begin{cases} \pi_n(t_h) + \Delta\pi_n(t_h)\, i_{n[RND]}\Big[p_n(t_h,\Delta)_{[0,1]}\Big] \\ +\Big[k\pi_n(t_h) + hP^M(t_h)\Big]\Big[r_n(t_h,\Delta)_{[0,1]}\Big] \end{cases} \\ \text{OPERATIVE PROGRAM} \Rightarrow \text{define}: N, \pi_n(t_0), p_{n[-1,1]}, r_{n[0,1]}, h, k \end{cases} \qquad (3.22)$$

In the preceding model:

- $[p_n(t_h, \Delta)_{[0,1]}]$ represents a weight (or even a probability, an intensity, etc.) that varies for each element in the base and at each moment, based on which we take account of the deviation to readjust the parameter of improvement.
- $[k\,\pi_n(t_h) + h\,P^M(t_h)][r_n(t_h, \Delta)_{[0,1]}]$ symbolizes all contingent external factors that may condition the increment in the parameter of improvement, which we have assumed depends on the level of the parameter of improvement itself and on the average level of progress, $P^M(t_h)$.
- $[r_n(t_h, \Delta)_{[0,1]}]$ indicates the probability that A_n, in period t_h, will suffer an external random variation in its own level of improvement.

The second variant is the MINIMAL OR "INCREMENTAL" SUPERIORITY, or the "flight" AUTOMATON; this system acts in a symmetrical way with respect to the previous one, since the parameter of progress is represented by the *minimum level* reached by the improvement parameters of the agents:

$$P^{MIN}(t_h) = \min_n \pi_n(t_h) = \pi_{MIN}(t_h) \qquad (3.23)$$

$P^{MIN}(t_h)$ is the lower basis, the "floor," on which the future improvement of the agents is based. Thus, in each t_h all the other micro states are thus superior. Each agent tries to outdistance as much as possible its own level of improvement from the level of progress, to "flee" from the "floor," to increment its own superiority.

The gap between the state of each agent and the value of the "floor,"

$$\Delta\pi_n(t_h) = \pi_n(t_h) - \pi_{MIN}(t_h) \tag{3.24}$$

represents the *quantum of superiority* perceived by each agent compared with the improvement parameter of the agent at the floor level. These micro behaviors lead to a general increase in the average level of improvement, which ends up raising the parameter of progress, further boosting the levels of improvement.

The model of this automaton is entirely like model (3.22), which represents the maximal automaton. The only difference is the definition $P^{MIN}(t_h)$ provided by (3.23), which substitutes $P^{MAX}(t_h)$ as defined by (3.20).

3.6.3 Reversibility in Automata of Improvement and Progress

It is necessary to introduce two variants for the functioning of the preceding automata, which derive from the choice of the type of *probability* that regulates the agents' transitions of state.

If both $[p_n(t_h, \Delta)_{[0,1]}]$ and $[r_n(t_h, \Delta)_{[0,1]}]$ are *stop-or-go probabilities*, then the automaton is *reversible*, in the sense that the levels of individual improvement, $\pi_n(t_h)$, and the levels of collective progress, $P^M(t_h)$, increase over time, or at least remain stable.

If instead we assume the *transition* probabilities are of the type $[p_n(t_h, \Delta)_{[-1,1]}]$ and $[r_n(t_h, \Delta)_{[-1,1]}]$, then the automaton is *strong reversible*.

If only one of the two probabilities entail reversibility (generally r_n), the automaton is *weak reversible*; otherwise, it is *irreversible*, and improvement and progress are continuously increasing.

Let us assume an *irreversible* automaton of 10 agents whose control panel is described in Fig. 3.60, which also shows the dynamics of this automaton under different hypotheses of reversibility. It is easy to see that the more reversibility is introduced, the more the macro and micro behaviors show irregular dynamics, as we can verify by observing Fig. 3.60.

CONTROL PANEL for Automata of improvement and progress										
Enter Probabilities	A_1	A_2	A_3	A_4	A_5	A_6	A_7	A_8	A_9	A_{10}
$p_n(\Delta>L+)$ & $\Delta>0$, high	0.2	0.2	0.2	0.2	0.2	0.2	0.2	0.2	0.2	0.2
$p_n(\Delta<=L+)$ & $\Delta>0$, low	0.8	0.8	0.8	0.8	0.8	0.8	0.8	0.8	0.8	0.8
$p_n(\Delta<=L-)$ & $\Delta<0$ low	0.8	0.8	0.8	0.8	0.8	0.8	0.8	0.8	0.8	0.8
$p_n(\Delta>L-)$ & $\Delta<0$ high	0.2	0.2	0.2	0.2	0.2	0.2	0.2	0.2	0.2	0.2
Enter Initial states	1	0	1	0	1	0	1	0	1	0
Enter parameters	k =	0.5	0.5	= h (this value is "0" for irreversible automata)						

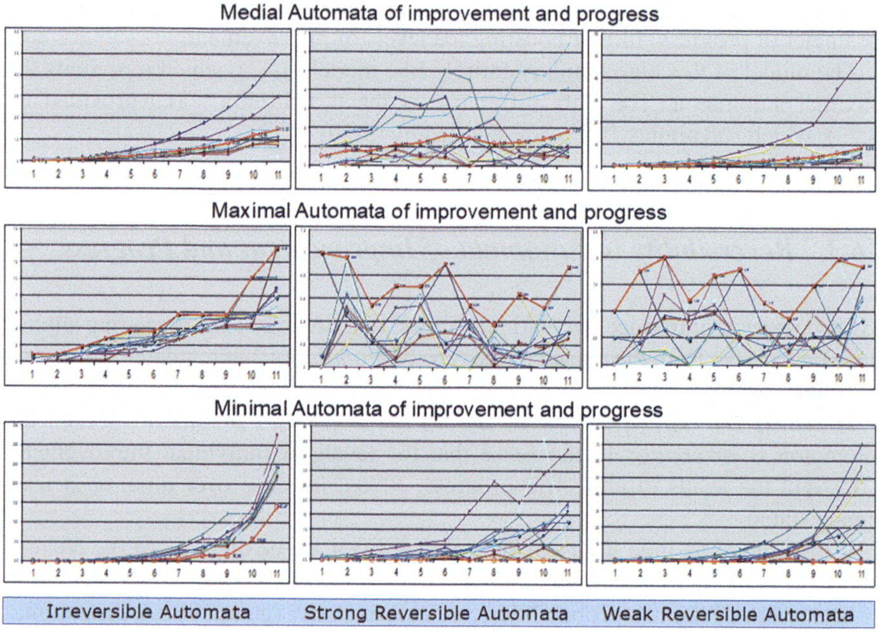

Fig. 3.60 Control panel of a general Combinatory Automaton of 10 agents simulating improvement and progress, with different levels of reversibility and macro and micro dynamics for 11 iterations

3.6.4 Rewriting the Heuristic Models

We can now go back to the definition of systems of improvement and progress in Sect. 2.6 to rewrite the *heuristic* model of the Combinatory Automata that can simulate such systems.

3.6.4.1 Heuristic Model of Medial Automaton of Improvement and Progress

The behavior of a Combinatory Automaton can be described by the following rules, which are simply the translation of model (3.19):

Micro behavior and necessitating rule: If you perceive that the level of your parameter of improvement, $\pi_n(t_h)$, is inferior to the level of the system's parameter of progress, $P^M(t_h)$—that is, $\Delta\pi_n(t_h) = \pi_n(t_h) - P^M(t_h) < 0$—try to *improve* in order to reduce the "gap" and, if possible, to make sure that $\pi_n(t_h) \geq P^M(t_h)$; if $\pi_n(t_h)$ is superior to the parameter of progress—that is, $\Delta\pi_n(t_h) > 0$—do nothing, or try further improvements to increment the "gap" in your favor. The degree of improvement depends on your characteristics and attitudes, expressed by the probabilities of the necessitating factors, $p_n(t_h, \Delta)_{[0, 1]}$, assuming irreversibility, or $p_n(t_h, \Delta)_{[-1, 1]}$ in the case of reversibility. "Chance" can influence your attempt by making you improve or regress in relation to the levels of *improvement* and *progress* previously attained, based on the element $[k\ \pi_n(t_h) + h\ P^M(t_h)][r_n(t_h, \Delta)_{[0, 1]}]$.

Macro behavior and recombining rule: The system must be able to pick up the individual improvement, $\pi_n(t_h)$ for all agents and adjust the parameter of progress to the average of the individual measures of improvement: that is, recalculate $P^M(t_h)$ at each t_h.

Micro-macro feedback. Chance and necessity: At time $t < t_0$ we can assume $\pi_n(t) = 0$ for all agents. At time t_0 we can instead assume that "by chance" $\pi_n(t_0) > 0$ for some A_n.

At this point the Combinatory Automaton begins to operate, and we witness the rise, "by necessity," in both the levels of improvement as well as those of progress. "Chance," that is, the set of factors not completely knowable and controllable by the individual elements, can influence the application *of the necessitating rule*, since it can reduce or accentuate the attempt by the individuals to increase their level of improvement. If the system has a very large base we might also introduce the hypothesis of *incomplete information*; in this situation two or more different dynamics for improvement can be produced, and the system might even divide into two or more less-powerful systems.

3.6.4.2 Heuristic Model of a Maximal Automaton of Improvement and Progress

The behavior of this Combinatory Automaton can be described by the following rules, which are simply the translation of model (3.22):

Micro behavior and necessitating rule: If you perceive you are not at the level of "the best"—that is, the value of your parameter of improvement, $\pi_n(t_h)$, is below that of the *leader* among the system's elements, $P^{MAX}(t_h) = \max_n \pi_n(t_h) = \pi_{MAX}(t_h)$—try to improve in order to reduce the "gap." The degree of improvement depends on your characteristics and attitude, which are expressed by the probabilities of the necessitating factors, $p_n(t_h, \Delta)_{[0, 1]}$, assuming irreversibility, or $p_n(t_h, \Delta)_{[-1, 1]}$ in the case of reversibility. "Chance" can influence your attempt by making you improve further or regress with respect to both the level of *improvement* previously achieved and the *average level of progress*, according to the factor $[k\,\pi_n(t_h) + h\,P^M(t_h)][r_n(t_h, \Delta)_{[0,1]}]$.

Macro behavior and recombining rule: The automaton must be able to pick up the individual improvement for each agent and adjust the parameter of progress so that it equals the maximum measure of individual improvement, by recalculating both $P^{MAX}(t_h)$ and $P^M(t_h)$.

Micro-macro feedback. Chance and necessity: At time $t < t_0$ we can assume $\pi_n(t) = 0$ for all agents. At time t_0 we can instead assume that "by chance" $\pi_n(t_0), > 0$ for some A_n. The automaton identifies the *leader element* in terms of improvement—the element $\pi_{MAX}(t_h)$—and uses it as the *indicator of progress*. At this point the Combinatory Automaton begins to operate as a *system of pursuit*, since the other elements "pursue" the leader and try to eliminate the gap between their performance and that of "the best." We then witness a rise, "by necessity," in both the level of improvement and that of progress. The system, however, gives rise to a *second type of progress*, which takes place even when the parameter of progress, $\pi_{MAX}(t_h)$, does not undergo an increase; this progress consists in a rise in the *average level* of the measures of improvement, $P^M(t_h)$, for all the agents in the system. In fact, it is easy to see that the attempts to reach and overtake the leader can push the elements in the system to raise their individual performance; even if the level of improvement does not change, the system increases the average level of individual improvement, causing the elements in the system to raise their individual performance. Naturally "chance" can influence the dynamics of the levels of improvement, and thus the dynamics of progress, since it can reduce or accentuate the attempt of individuals—as well as the leader—to raise their level of individual improvement. Because of chance, some other element may also substitute the leader in terms of improvement.

3.6.4.3 Heuristic Model of a Minimal Automaton of Improvement and Progress

The following rules, which are simply the translation of the automaton defined above, are sufficient to describe the behavior of the minimal Combinatory Automaton:

Micro behavior and necessitating rule: If you perceive that an agent has reached a base improvement level, $\pi_{MIN}(t_h)$, below which you do not wish to fall, try to raise your improvement parameter to levels above $\pi_{MIN}(t_h)$. The degree of improvement you succeed in achieving depends on your characteristics and aptitudes, expressed by the probabilities of the necessitating factors, $p_n(t_h, \Delta)_{[0,1]}$, assuming irreversibility, or $p_n(t_h, \Delta)_{[-1,1]}$ in the case of reversibility. "Chance" can influence your attempt by making you improve or regress in relation to the improvement level previously reached and the average progress level, according to the factor $[k\,\pi_n(t_h) + h\,P^M(t_h)]$ $[r_n(t_h, \Delta)_{[0,1]}]$.

Macro behavior and recombining rule: The system must be able to pick up the individual improvements, $\pi_n(t_h)$, and set the progress parameter equal to the minimum of the individual improvement measures; that is, to recalculate both $P^{MIN}(t_h)$ and $P^M(t_h)$.

Micro-macro feedback. Chance and necessity: At time t_0 we can assume that, "by chance," $\pi_n(t_0) > 0$ for some A_n. The automaton identifies the "base" improvement element—that is, the element for which $\pi_{MIN}(t_h)$—and takes $\pi_{MIN}(t_h)$ as the *progress indicator*. At this point the Combinatory Automaton begins to operate as a *system of flight*, since the other elements "flee" from the base, trying to increase the "gap" between their performance and that of the "worst" element. We thus witness the increase "by necessity" in both the improvement and progress levels. However, in this form as well the automaton brings about a second type of *progress* that is revealed even when the progress parameter $\pi_{MIN}(t_h)$ is not increased, and which consists in the raising of the average level of the improvement measures, $P^M(t_h)$, of all the agents of the system. In fact, we can easily see how the attempts to exceed as soon as possible the improvement levels of the base element push the system's elements to raise their individual performances; even if the improvement level of the base element does not change, the system still produces an increase in the average improvement level. Here too "chance" can influence the dynamics of the improvement levels, and thus the progress levels, since it can reduce or accentuate the attempts of the individuals to raise their individual improvement levels.

3.6.5 Combinatory Automata Simulating Evolutionary Dynamics Under Various Assumptions

The spread of a beneficial mutation in a population can be simulated (Sect. 2.6.6) by using a Combinatory Automaton consisting of a lattice representing the population. Each cell in this lattice reflects an individual's phenotype (ranging from thousands to billions). The value of each cell, either "1" (mutated) or "0" (not mutated), is determined at each moment in time by the overall number of mutated individuals. This creates a *micro-macro feedback*, connecting the "analytical values" of the cells (micro level) with the "synthetic state" of the automaton (macro level, i.e., the total number of mutated individuals). To simulate Darwinian evolutionary dynamics effectively, the most suitable automata are those that are both *maximally stochastic* and *irreversible*. This is because simulating mutation spread in a population requires calculating the probability of an individual transitioning from a "normal" to a "mutated" phenotype, with each cell representing one individual.

It is important to first emphasize, referencing Darwin (Sect. 2.6.8), that a key characteristic of the *evolutionary process* is the spread of a mutation, which requires a *process of reproduction* by mutated individuals to pass on their genes containing the mutation to their descendants. Without reproduction, there can be no evolution. Julian Huxley, while acknowledging Darwin's ideas, also highlighted that the *natural selection* does not operate through "survival of the fittest" but through the more comprehensive process of *adapting to the environment*. Individuals who are better able to adapt to the environment and demonstrate significant genetic variability will have more offspring. Individuals with an enhanced likelihood of being *selected* for mating, often due to the appearance of brightly colored markings on their bodies, such as in many male birds and fish, have an advantage. Those that successfully navigate physical and biological challenges in their environment can succeed in reaching sexual maturity. Through "sexual transmission," genes are freely exchanged, allowing a genetic mutation that arises by chance in some individuals to gradually spread across the entire population and, eventually, the species in a typical process of diffusion (Darwin, 1859; Huxley, 1942).

Huxley emphasized the significance of populations as key units in the evolutionary process, asserting that natural selection functions by organisms adapting to their environment through reproduction. He also argued that reproductive isolation promotes evolutionary divergence through "sexual communication," which normally allows for the free exchange of genes.

To build the Combinatory Automaton for simulating the evolutionary process, it is important to briefly recall the two main forms in which the reproductive process occurs:

A. *Sexual reproduction*: This form of reproduction generally involves the fusion of two sex cells, or gametes, one male (sperm) and one female (egg), to create a new individual with a genetic combination from both parents. We can generally distinguish between:

   (a) *Blind mating*, where gamete fusion occurs between individuals of different sexes that mate without actively *choosing one another*; in this form, reproduction occurs based only on the *probability of encounter*, not on *active mating*.
   (b) *Sexual selection*, where one partner *chooses* the other based on traits perceived as beneficial for their offspring. This can be further divided into:

      (b.1) *Intrasexual selection*, where one partner, usually the male, competes with other males for the right to mate.
      (b.2) *Intersexual selection*, where one partner, usually the female, selects the male based on traits that indicate good health or greater resistance to diseases.

B. *Asexual reproduction*, where there is no fusion of gametes and the new individual is genetically identical to the parent. This is observed among bacteria, many plants—which reproduce vegetatively (through cuttings, bulbs, rhizomes)—and some animals.

Sexual reproduction transmits advantageous mutations, that is, those favorable to survival, more quickly among individuals that mate, producing offspring that typically exhibit the new mutated traits. When building a Combinatory Automaton, it is necessary to introduce at least three probabilities that allow the description of the process of sexual selection:

1. *Fitness intensity of mating*, which specifies the intensity with which the individual choosing the partner mates; the higher the probability, the faster the mutation spreads.
2. *Attractiveness for mating*, which specifies how preferable, desirable, or attractive the individual carrying the mutation appears to other partners; the higher the probability, the more the mutations present in the attractive partner spread throughout the population.
3. Minimum *environmental probability of mating*, which indicates how much the living environment can favor mating.

We recall Darwin's clear words (already presented in Sect. 2.6.5) in defining the process of sexual selection.

> What natural selection cannot do, is to modify the structure of one species, without giving it any advantage, for the good of another species; and though statements to this effect may be found in works of natural history, I cannot find one case which will bear investigation. [...] And this leads me to say a few words on what I call Sexual Selection. This depends, not on a struggle for existence, but on a struggle between the males for possession of the females; the result is not death to the unsuccessful competitor, but few or no offspring. Sexual selection is, therefore, less rigorous than natural selection. Generally, the most vigorous males, those which are best fitted for their places in nature, will leave the most progeny [...] A hornless stag or spurless cock would have a poor chance of leaving offspring... (Darwin, 1859, p. 87)

In reproduction through *sexual selection*, *mating attractiveness* plays an essential role, especially in species where there is free choice of a partner. Take, for example, the mane of a lion. It has been shown, even through experimental observations, that the lioness mates more easily with lions with a dark mane because this characteristic is interpreted as an indicator of strength, aggression, and an abundance of testosterone.

Using life-sized dummies with detachable wigs, researchers in Africa have found that female lions favor mates with a dark head of hair (West & Packer, 2002). Lions with the longest, darkest manes suffer from the heat more than their blonder or less shaggy peers but they do better in the hunt for mates. Scientists have found that the lion's mane works much like the peacock's tail; it tells females, and rivals, about a male's health and fitness (Highfield, 2002, online).

Figure 3.61 shows the control panel and the results from a Combinatory Automaton simulating the process of a *slow evolution* under the assumption of *blind mating*, meaning that an advantageous mutation normally does not produce superiority in individual reproduction. The following probabilities of transition of state of individuals have been hypothesized:

- *Fitness intensity of mating* = 2.0% (probability), which is a function of the neighboring population.
- *Attractiveness for mating* = 0.00 under the assumption that the mutated individuals have no mating attractiveness.
- *Minimum environmental probability of mating* = 0.00 (constant) under the assumption that the environment is neutral in the reproduction of mutated individuals.

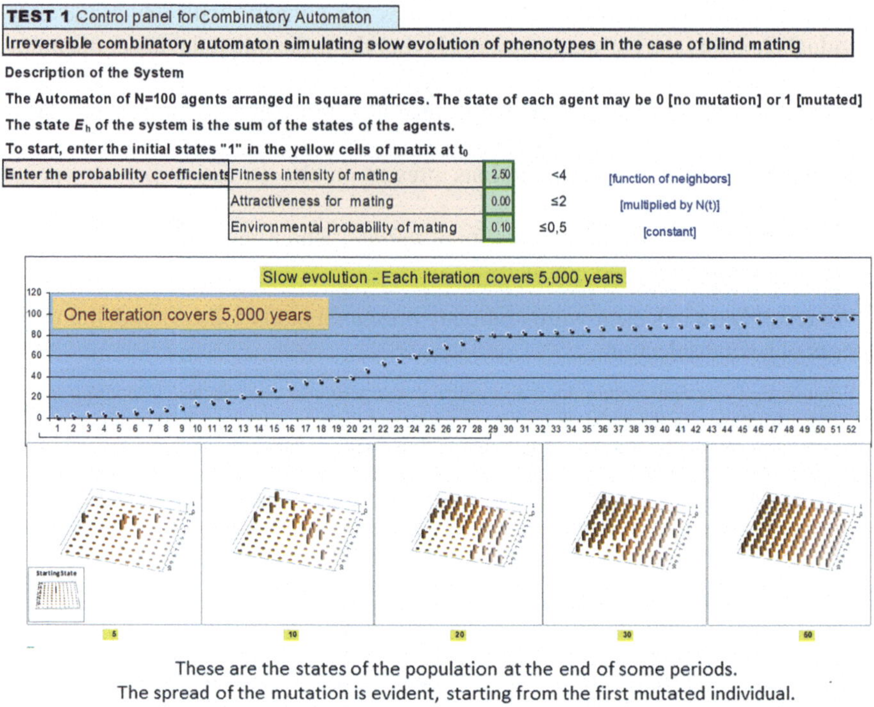

Fig. 3.61 Control panel simulating the spread of a mutation in a population of 100 k individuals, with "k" equal to 1000, a million, and so on, individuals, assuming *blind mating*

Figure 3.62 shows the control panel and the results obtained from a Combinatory Automaton simulating the process of *evolution* under the assumption of *sexual selection*. This means that an advantageous mutation produces a superiority in sexual reproduction, in the two cases where (1) the mutated individual acquires greater physical strength and can surpass the competition from other individuals in the struggle for mating, and (2) the mutated individual becomes more attractive to potential partners.

The following probabilities of transition of state of individuals have been hypothesized:

- *Fitness intensity of mating* = 1.0% (probability), which is a function of the neighboring population.
- *Attractiveness for mating* = 0.20 (coefficient which is multiplied by the number of mutated individuals) under the assumption that the mutated individuals have a higher level of mating attractiveness.
- *Minimum environmental probability of mating* = 0.02 (constant) under the assumption that the environment favors the reproduction of mutated individuals.

The simulations shown in Figs. 3.61 and 3.62 produce a relatively slow evolutionary process, or a gradualism in change.

The biologist Stephen Jay Gould (Eldredge & Gould, 1972), after a detailed review of paleontological fossil records, proposed that when significant beneficial

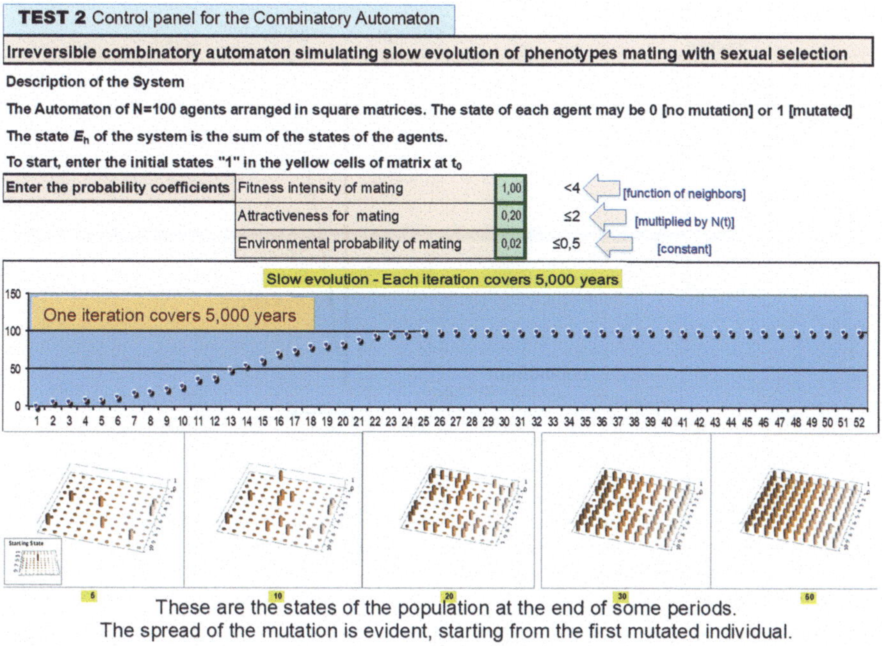

Fig. 3.62 Simulation of the spread of a mutation in a population of 100 k individuals, with "k" equal to 1000, a million, and so on, individuals, assuming mating with sexual selection

mutations occur, primarily during sexual reproduction, the evolutionary process of a species can accelerate. This rapid evolution may lead to the formation of two distinct species in a relatively short time. He referred to this evolutionary pattern as "Punctuated Equilibrium pattern." Gradualism and Punctuated Equilibrium both explain the pace of speciation. Gradualism suggests that species evolve through slow and consistent variations, with small genetic changes over time in the gene pool, while Punctuated Equilibrium posits that evolution happens in short bursts of significant change followed by long periods of *stability*, as shown in Fig. 3.63.

Figure 3.64 shows the control panel of the Combinatory Automaton that simulates an evolution with "discontinuity." In the model, the population dynamics are produced by the action of the following parameters:

- *Minimum Activation Number*: that is, the number of mutated individuals which trigger quick evolution. Once the minimum has been reached, the probability the mutation will spread becomes a function of the system state, that is, the number of mutated individuals, which makes evolution much faster, thus producing a discontinuity.
- *Fitness Intensity in Individuals*": this expresses the advantages the mutated individuals have regarding survival and reproduction.
- *Fitness Premium for Minimum Activation Number*: this determines the increase in the level of fitness for the mutated individuals when the minimum activation number has been reached.

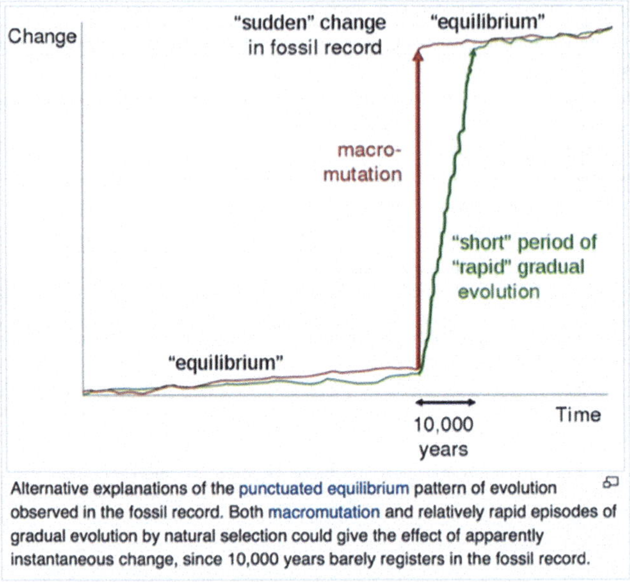

Fig. 3.63 The dynamics of a population which evolves following a Punctuated Equilibrium model (source: Wikipedia: Punctuated equilibrium) Wikipedia (2024)

Combinatory Automaton simulating the evolution with punctuated equilibrium in populations with sexual selection

Description of the System

The system is composed of 100 agents arranged in square matrices. Agent's state [0 or 1].

The state N(t) of the system is the sum of the states of the agents

minimum activation number MAN	8	≤ 10
fitness intensity in individuals	4	≤ 5
fitness premium for diffusion of mutation	5	≤ 5
resistance to mutation	4	0 ≤ resist. ≤ 10

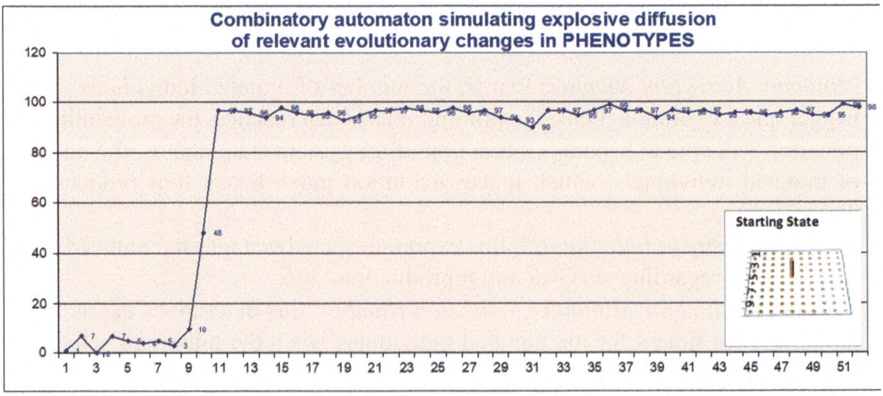

Fig. 3.64 Test 1—Combinatory Automaton simulating evolution with Punctuated Equilibrium, assuming a low "minimum activation number" equal to 8

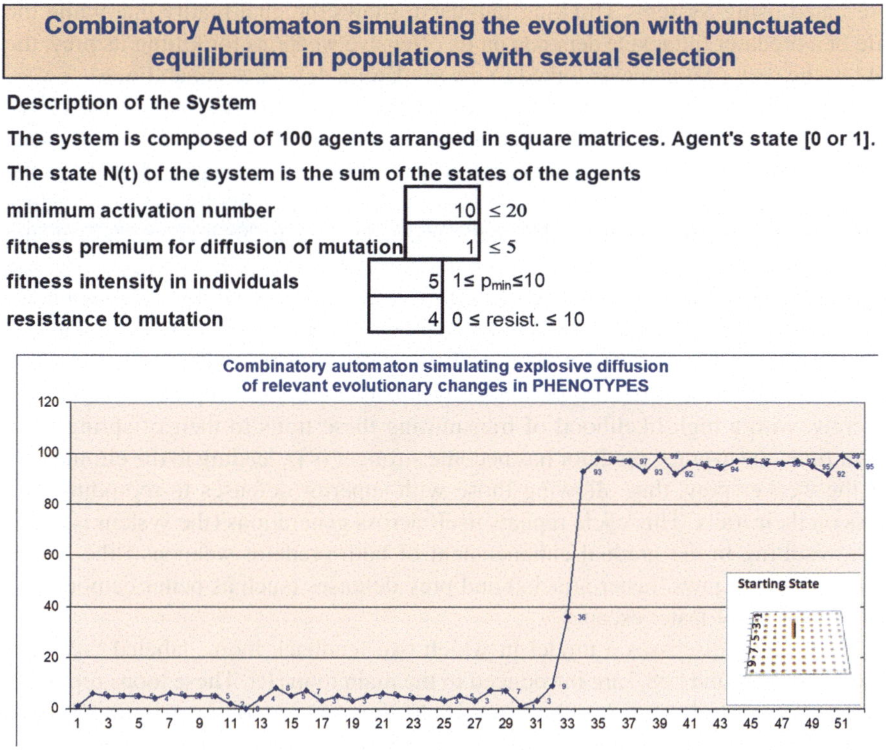

Fig. 3.65 Test 2—Combinatory Automaton simulating evolution with Punctuated Equilibrium, assuming a higher "minimum activation number," equal to 10

- *Resistance to Mutation*: this indicates the influence of environmental factors that may hinder the survival of mutated individuals.

Figure 3.65 demonstrates how evolution according to the Punctuated Equilibrium model depends on the "minimum activation number"; the higher it is, the more time is required for the evolutionary leap to occur. For example, in certain species, such as crocodiles and alligators, no evident evolutionary leap has taken place.

3.6.6 Coevolution of Two Interacting Species, Such as Prey and Predators

Evolution is a general biological phenomenon that also occurs when the reproductive processes involve two (or more) species (or populations) that coexist, such as *prey* and *predators*. In this case, the *struggle for life* operates jointly for both populations, for example, by modifying or strengthening the "weapons" of the predators over time and, consequently, adjusting or reinforcing the defense mechanisms of the prey to give them the maximum chance of survival. The model in Fig. 3.66, presented using Systems Thinking language, shows the dual reality that while the life of a predator (attacker) depends on its offensive weapons for killing its prey, the life of the prey (defender) is linked to the predator's defensive capabilities.

Referring to Fig. 3.66, predators possessing enhanced weapons (however we interpret this term) gain an advantage in hunting (s1), reducing the survival chances of the weaker prey (o1) (the defenders). As the number of weak prey decreases, the stronger prey, equipped with better defenses, have a greater likelihood of reproducing (o2) and passing on their defensive traits, leading to an increase (s2) in the effectiveness of defenses among a growing number of prey. The offspring of these stronger prey are likely to inherit these effective defense capabilities, allowing more and more prey to escape their predators (s3). As the prey's defensive abilities strengthen, the survival of weaker predators (o3) (attackers) declines. However, predators with more effective offensive weapons are able to reproduce more frequently, with a high likelihood of transmitting these traits to their offspring (o4). Over time, the average predator has become stronger (s4), leading to the elimination of the weaker prey, thus allowing those with superior defenses to reproduce and pass on their traits. This cycle repeats itself across generations (the system is cyclical), resulting in the gradual enhancement of both predator weapons (like longer beaks, stronger jaws, faster attacks) and prey defenses (such as better camouflage, thicker shells, or faster escape).

Figure 3.66 discusses a model in which two feedback loops, labeled "o5" and "o6," and "o7" and "o8," are introduced to the main loop [R]. These loops represent *external controls* that *weaken* the natural selection process. By interrupting the two relations defined by these loops, these external factors reduce the development of both offensive and defensive mechanisms, making both prey and predators

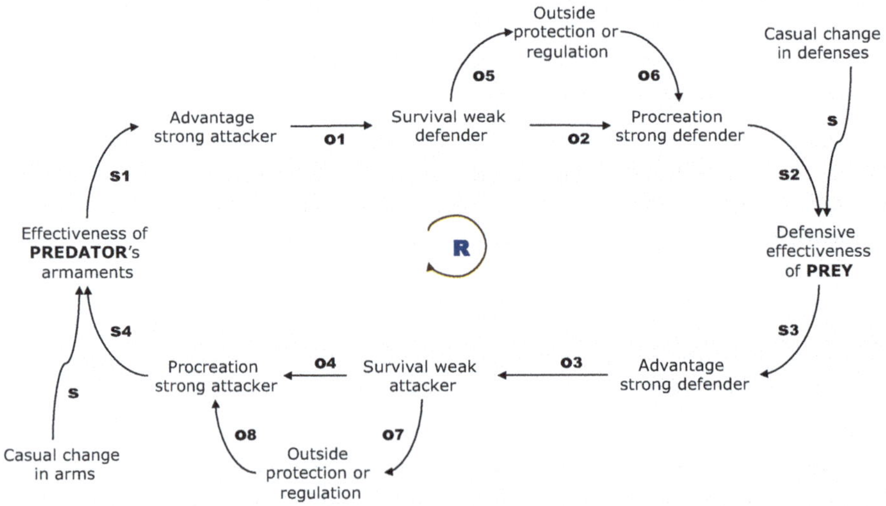

Fig. 3.66 Model of the joint evolution of phenotypes based on the process involving the struggle for life between prey and predators

progressively weaker and less adapted for survival. This weakening effect occurs when humans intervene to protect certain species (especially humans), leading to individuals becoming less fit for competition. An example provided is the difficulty animals face when reintroduced into the wild after being kept in captivity (protected) for long periods. However, the loops "o5" and "o6," as well as "o7" and "o8," can also have the opposite effect: when there is no external intervention, the strongest traits are enhanced, following the principles used by breeders to promote the reproduction of individuals with the most favorable traits.

Referring again to the main loop, [R], if it is not interfered with, will evolution produce a boundless strengthening of attack and defense mechanisms, as well as of those traits that favor reproductive success? Or will some "natural" Control Process arise to curb the excess numbers. The "natural" Control System that sets a limit to the growth in attack and defense organs in the world of biology is represented—in simplified form—by physiological and environmental limits. When the weapons of the predator increase, there is soon a slowdown in the chances for further growth due to the presence of physiological limits; when these limits are reached, the individuals are incapable of surviving and therefore die, without offspring, thereby stopping any further transmission of their traits. The same process regulates the development of defences in the prey.

The model in Fig. 3.66 would invariably operate even if the conditions of the ecosystem in which the population is living should change. Let us suppose a significant number of predators and prey (such as lions and gazelles) are forced, due to a catastrophic event, to inhabit a different environment (for instance, a mountainous region lacking flat terrain). After some years, in order to allow for these species to adapt to the environment, gazelles and lions would probably develop the capability of climbing and jumping. Loop [R] would invariably come into play to strengthen these capacities in both species. Without mutual adaptation, one or both populations would most likely become extinct (Brockhurst et al., 2003). It is evident that coevolution progresses more rapidly for species with strong ecological relationships (Brockhurst et al., 2003; Buckling & Rainey, 2002), and that pairwise coevolution can alter the evolutionary trajectory compared to adaptation occurring in isolation (Schluter et al., 1985). The size of the environment plays an equally significant role, an idea that aligns with the neo-Darwinist perspective. Julian Huxley, after emphasizing the significance of populations as elements in evolution, suggested that natural selection functions through adaptation to the environment, and reproductive isolation encourages evolutionary divergence through "sexual communication." In this process, genes flow freely, allowing genetic mutations that randomly appear in some individuals to gradually spread across the entire population, eventually impacting the entire species (Darwin, 1859; Huxley, 1942).

Whatever idea we have of the evolutionary process, the *qualitative* dynamics of populations and species also influences their *quantitative* dynamics; an alteration in some traits of a species can produce considerable changes in the number of individuals in that and other species connected to it in the ecosystem, at times with catastrophic effects (Buchanan, 2000).

Without using a probabilistic Combinatory Automaton, which is rather difficult to represent graphically, but instead using a deterministic Combinatory Automaton derived from the Volterra-Lotka Equations (Sect. 2.6.7), Fig. 3.67 outlines the coevolutionary trends of two populations of prey and predators, assuming an initial minimum of A (prey) = 500 k units and B (predators) = 100 k units. The initial fertility and extinction coefficients of the two populations are indicated in the upper part of Fig. 3.67. The simulation proceeds for 1000 iterations.

The irregular behavior seen in the diagram below is a result of changes in the phenotypic traits of the two populations. These mutations influence the rates that control their numerical interactions, based on the following assumptions (didactic purposes):

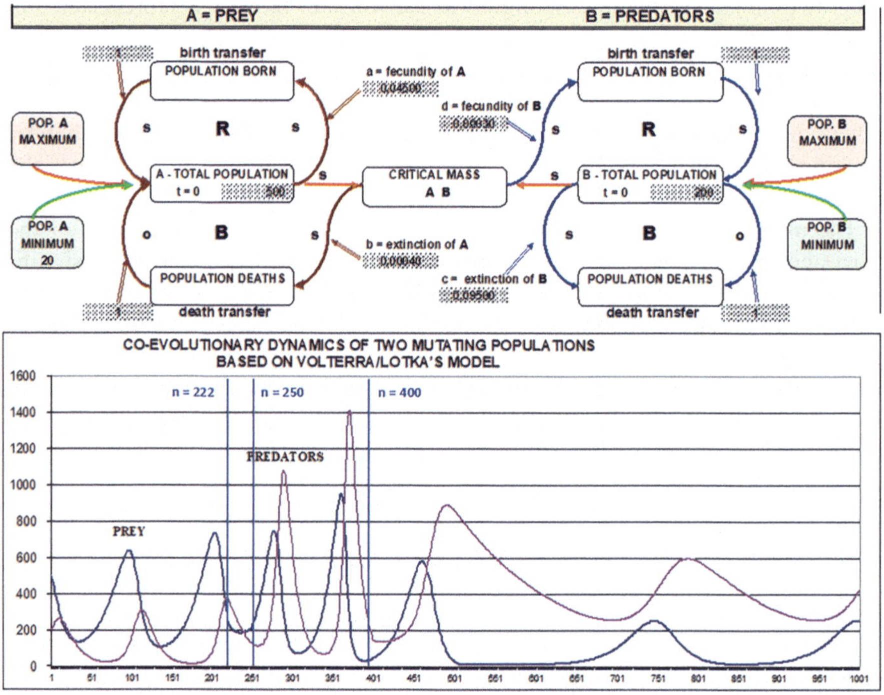

Fig. 3.67 Model of the natural evolution of coexisting populations of prey and predators of two populations with mutations

- At iteration n = 222, a mutation occurs that strengthens the defense capabilities of population A, making predation by B more difficult. As a result, the extinction rate of A decreases from $b_{221} = 0.0004$ to $b_{222} = 0.0002$.
- At iteration n = 250, a new mutation in A increases the rate of reproduction, which increases from $a_{249} = 0.045$ to $a_{250} = 0.080$.
- At iteration n = 400, population B undergoes a mutation, with its fecundity rate falling from $d_{339} = 0.0003$ to $d_{400} = 0.0001$.

Figure 3.67 demonstrates how, in principle, the process of coevolution can lead the two populations to maintain a reciprocal relationship that produces oscillating increasing dynamics. It is possible to introduce other variations to "enrich" the system (e.g., the competition for food among predators); since the variants do not modify in any way the basic logic of the system, those interested in further detail can consult specialist texts (Casti, 1985; Flake, 1998; Martelli, 1999; Takeuchi, 1995).

To complete this brief description of the evolutionary model of the species, note that the model in Fig. 3.67 enables us to simulate the impact of *exogenous controls*, for example, by adjusting population sizes through outside intervention, when they surpass a maximum threshold or fall below a minimum. For example, fish or deer populations can be regulated through fishing or hunting if numbers become too high, or by repopulating if they drop too low. We can simulate exogenous interventions that artificially modify both the minimum and maximum numbers of prey and predators, in accordance with environmental resources. The simulation of such interventions leads to the model in Fig. 3.68, which is completed by introducing *maximum* and *minimum* limits to the number for each population.

In Test 1, the *maximum* and *minimum* population limits result in a stable dynamic for both species after approximately 200 iterations, aligning with the set boundaries. In Test 2, limits were also applied to both populations; however, the prey's maximum was raised to Amax = 800 units, preventing it from reaching that threshold, while the predator population slightly exceeds Bmax = 500 units. In the Test 2 diagram, the dynamics become erratic after n = 80, as population A is reduced by 500 units over three periods ($n = 593–595$) due to an environmental disaster.

To enhance realism, the model could also account for changes in birth and death rates as population sizes near their maximum limits. Additionally, it's valuable to consider the natural constraints on population growth due to limited available resources.

The model in Fig. 3.68 can be extended to a larger ecosystem (Tansley, 1935). The model in Fig. 3.69 simulates the dynamics of five populations that form a food chain, assuming both the absence of exogenous controls (upper part) and the introduction of controls in some populations.

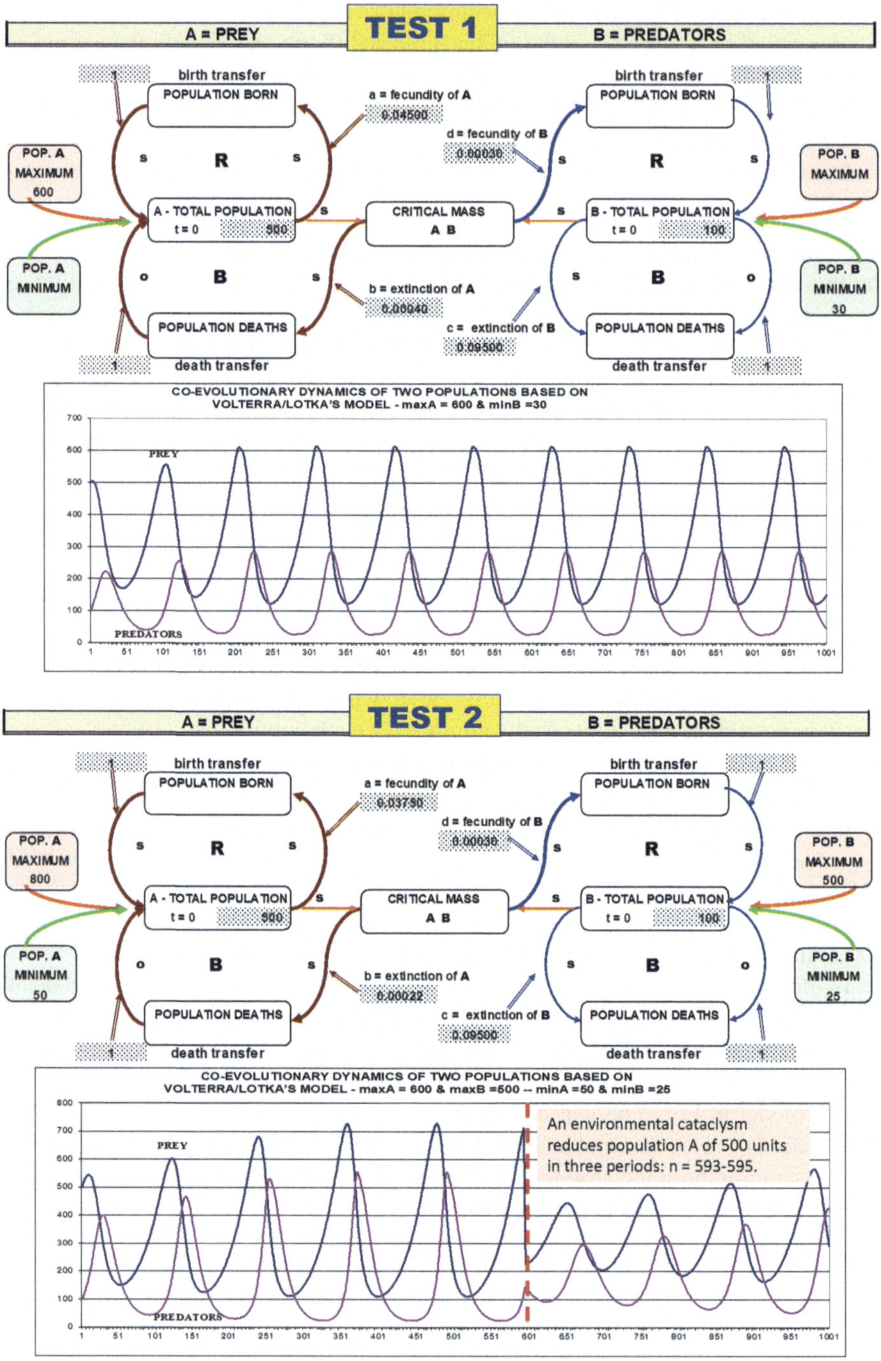

Fig. 3.68 Dynamics of two coexisting populations when an external control is introduced on the min and max numbers of individuals

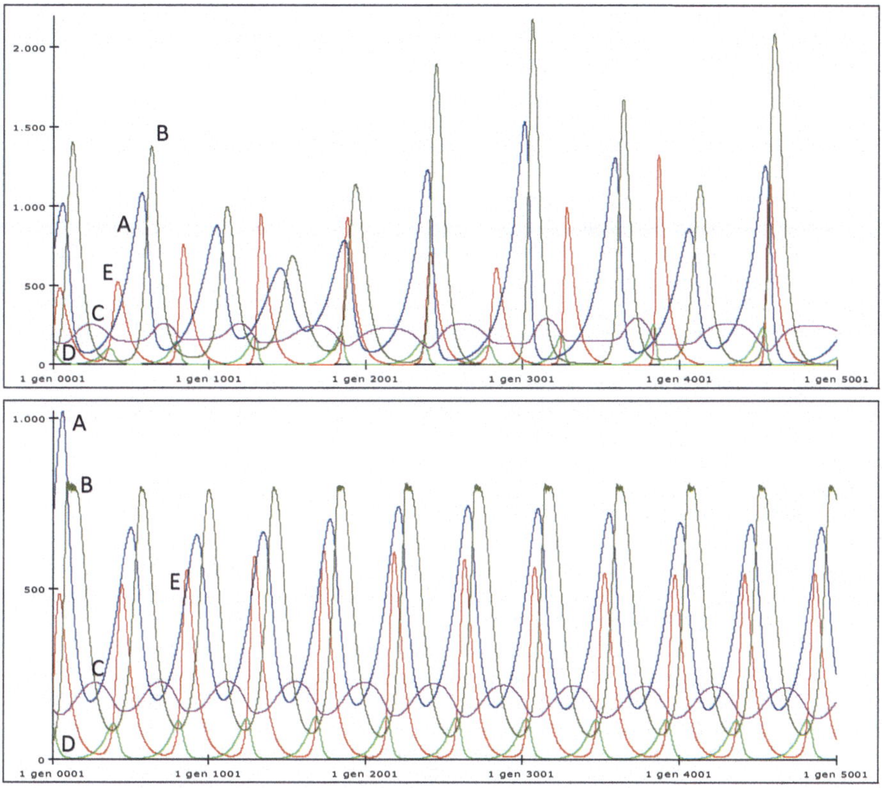

Fig. 3.69 Dynamics of five populations of an ecosystem, without (top graph) and with (lower graph) external controls (reference to Fig. 3.68, simulated in Powersim)

References

Brockhurst, M. A., Morgan, A. D., Rainey, P. B., & Buckling, A. (2003). Population mixing accelerates coevolution. *Ecology Letters, 6*(11), 975–979.

Buchanan, K. L. (2000). Stress and the evolution of condition-dependent signals. *Trends in Ecology and Evolution, 15*(4), 156–160.

Buckling, A., & Rainey, P. B. (2002). The role of parasites in sympatric and allopatric host diversification. *Nature, 420*(6915), 496–499.

Caldas, J. C., & Coelho, H. (1999). The origin of institutions: Socio-economic processes, choice, norms and conventions. *Journal of Artificial Societies and Social Simulation, 2*(2), 1.

Casti, J. L. (1985). *Nonlinear system theory*. Academic Press.

Darwin, C. (1859). *On the origins of species by means of natural selection*. London: Murray. Retrieved from: http://darwin-online.org.uk/content/frameset?itemID=F373&viewtype=text&pageseq=1

Devaney, R. (1989). *An introduction to chaotic dynamical systems* (2nd ed.). Addison-Wesley.

Eldredge, N., & Gould, S. J. (1972). Punctuated equilibria: An alternative to phyletic gradualism. *Models in Paleobiology, 82*, 115. Retrieved from: https://www.researchgate.net/publication/221958736_Punctuated_Equilibria_An_Alternative_to_Phyletic_Gradualism

Flake, G. W. (1998). *The computational beauty of nature: Computer explorations of fractals, chaos, complex systems, and adaptation*. The MIT Press.

Gell-Mann, M. (1995). What is complexity? *Complexity, 1*(5), 3. Retrieved from: http://complexity.martinsewell.com/Gell95.pdf

Gould, S. (2000). *The theory of options: A new theory of the evolution of human behavior*. Universal Publishers/Upublish.com.

Gould, S. (2001). *Introduction to the theory of options*. Retrieved from: http://www.oocities.org/athens/acropolis/1628/A01into.htm

Highfield, R. (2002). *Lionesses don't prefer blonds - In the mane*. Retrieved from: https://www.telegraph.co.uk/news/science/science-news/4768860/Lionesses-dont-prefer-blonds-in-the-mane.html

Huxley, J. (1942). *Evolution: The modern synthesis*. George Allen and Unwin.

Johnson-Laird, P. N. (1983). *Mental models: Towards a cognitive science of language, inference, and consciousness* (Vol. 6). Harvard University Press.

Martelli, M. (1999). *Introduction to discrete dynamical systems and chaos*. Wiley.

Mella, P. (2009). *The Holonic Revolution. Holons, Holarchies and Holonic Networks. The Ghost in the Production Machine*. Pavia (Italy): University Press.

Rangone, A., & Mella, P. (2019). Obstacles to Managing Dynamic Systems. The Systems Thinking Approach. *International Journal of Business and Social Science, 10*(8), 24–41.

Schluter, D., Price, T. D., & Grant, P. R. (1985). Ecological character displacement in Darwin's finches. *Science, 227*(4690), 1056–1059.

Simon, H. A., & Kaplan, C. A. (1989). *Foundations of cognitive science*. MIT Press.

Smith, A. (1776). An inquiry into the nature and causes of the wealth of nations. In E. Cannan (Ed.), *Introduction, notes, marginal summary and an enlarged index* (p. 1). Methuen.

Takeuchi, Y. (1995). *Global dynamical properties of Lotka-Volterra systems*. World Scientific.

Tansley, A. G. (1935). The use and abuse of vegetational concepts and terms. *Ecology, 16*(3), 284–307.

Von Mises, L. (1998). *Human action: The scholar's edition*. The Mises Institute.

West, P. M., & Packer, C. (2002). Sexual selection, temperature and the lion's mane. *Science, 297*(5585), 1339–1343.

Wikipedia. (2024). *Punctuated equilibrium*. Retrieved from: https://en.wikipedia.org/wiki/Punctuated_equilibrium

Wolfram, S. (1984). Cellular automata as models of complexity. *Nature, 311*(5985), 419–424.

Wolfram, S. (2004). *A new kind of science and the future of mathematics*. Joint Mathematics Meetings. Retrieved from: http://www.stephenwolfram.com/publications/a-new-kind-of-science-future-mathematics/

Chapter 4
The Heuristic Value of Combinatory Systems Theory

Abstract Chapter 1 presented the basic ideas behind the theory, which are analyzed in some detail. Chapter 2 described the heuristic models of several relevant combinatory systems in the context of the five typical classes. Chapter 3 developed the Theory of Combinatory Automata, which present models for simulating the dynamics of combinatory systems. This final chapter considers the "explanatory power" of Combinatory Systems Theory for understanding the logic of collective dynamics, understood as a unitary (macro) object of observation, even though collectivities are composed of a "plurality" of individual entities capable of autonomous (micro) behavior (agents, individuals, operators, etc.). The chapter attempts to answer three questions: Are combinatory systems "systems" in the true sense of the term? Why is this theory able to explain so many and so varied a number of phenomena, even though it is based on a very simple modus operandi? Are combinatory systems different than complex systems?

Keywords Paradigm of systems thinking · Explanation process and its operational closure · Explanatory power of the combinatory systems theory · Chance, necessity and freedom · Individual rationality and collective behavior

> Whether or not you can observe a thing depends on the theory you use. It is the theory which decides what can be observed. (Einstein, 1926, quoted by Heisenberg, 1989, p. 40)

> Nothing we are going to say will be understood in a really effective way unless the reader feels personally involved and has a direct experience that goes beyond all mere description. (Maturana & Varela, 1987, p. 18)

Combinatory Systems Theory assumes that most human collectivities act as combinatory systems. The theory expands this perspective, and in broader and more general form studies the *modus operandi* of any collectivity of similar non-interconnected agents whose analogous *micro behavior* produces a *macro behavior* that refers to the collectivity as a whole and which drives the subsequent micro behavior, in an uninterrupted *micro-macro feedback* action maintained by the presence of *necessitating factors* (which cause the micro behavior of each agent in the collectivity) along with *recombining factors* (which produce and maintain the unit's macro

P. Mella, *The Combinatory Systems Theory*, Progress in IS,
https://doi.org/10.1007/978-3-031-86946-4_4

behavior). Three questions arise: Are combinatory systems "systems" in the true sense of the term? Why is this theory able to explain so many and so varied a number of phenomena, even though it is based on a very simple *modus operandi*? Are combinatory systems different than complex systems? After clarifying these three aspects, we shall consider several questions which Combinatory Systems Theory allows us to examine in a new light.

4.1 Do Combinatory Systems Follow the Paradigm of Systems Thinking and the General Theory of Systems?

4.1.1 Premise: Different Ways to Conceive of Systems

In his General System Theory (1968), Karl Ludwig von Bertalanffy sent us a powerful message on which Systems Thinking was based: the observation, description, interpretation, and explanation of objects and phenomena assumes the ability of the observer to *identify*—or *impose*—an order and/or organization and/or mutual link to those interacting phenomena and objects, forming them into a structure that, as a consequence of the interactions among elements, has a distinct *macro* significance/property which does not derive from the *micro* significance/properties of the structured object, and a *macro* behavior that cannot be reduced to the sum of the *micro* behaviors of the elements contained in the structure. We thus create a *system*—that is, an observable unit, a complex—composed of several elements interconnected in such a way that there is an interaction between the elements and the whole, and vice-versa. The whole (macro) depends on the parts, but the parts (micro) are connected (coordinated) in order to give meaning to the whole; thus, the parts lose their individuality in order to form the structure of the system whose characteristics "*appear as 'new' or 'emergent'*" when compared to those of the constitutive elements (von Bertalanffy, 1968, p. 55). This necessary and fundamental interaction between parts (micro) and whole (macro) is outlined in Fig. 4.1.

Systems can be *real* or *conceptual.*

> What is to be defined as a system and which things are describable as such are certainly not questions to which we can give an obvious or simple response. It is easy to agree on the fact that a galaxy, a dog, a cell and an atom are real systems—that is, entities that are perceived by observation or inferred from this, and which exist independently of the observer. On the other hand, there are conceptual systems, such as logic and mathematics (including, for example, even music), which are essentially symbolic constructs; that is to say, conceptual systems corresponding to reality. (von Bertalanffy, 1968, p. 16)

A real system is *dynamic* if it produces modification processes for its own state in order to react (output) to stresses or stimuli (input) from the environment. We will define as "*behavioral*" a dynamic system that sets off internal dynamics (or structural or endogenous dynamics) which entail a modification of the states of the

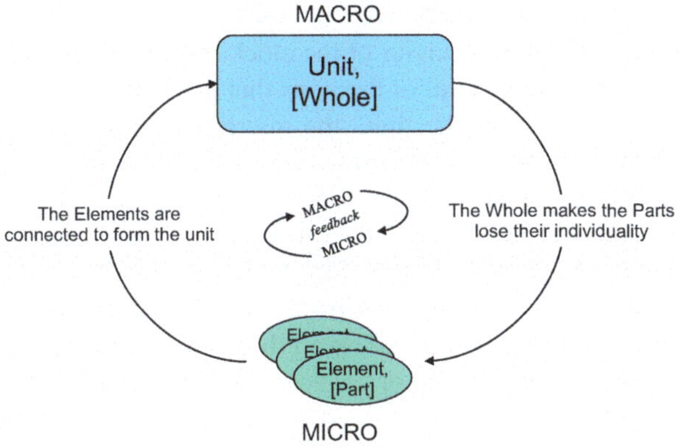

Fig. 4.1 The macro and micro relationships in a system

system to which an external (or functional, environmental, or exogenous) dynamics can be correlated. The latter can be represented as a time-ordered variation in the system's output (time-ordered succession), which is a function of its state. Such dynamics can be described by an observer as "the" *behavior*—external or internal—of the system, or as the *effect of this behavior.*

In effect, we must note that for behavioral systems, in their quality as dynamic systems, the relation between outputs and inputs is never direct. In fact, we must assume that the inputs cause changes in the *state* of the *structure* of the system (or simply of the *state* of the system), independently of the nature of its elements and the relations it is composed of (*external* point of view). The outputs are linked to the inputs, in that they derive from changes to the internal structure caused by the inputs.

There are three important features of *behavioral systems:*

1. *Micro behaviors*: the behavioral system is made up of *elements*, all essential for producing the macro behavior and each of which producing its own *micro* behavior, understood here as a modification in the state of the element, which in turn produces a *micro effect*.
2. *Macro behavior*: the system produces its own *macro* behavior (structural and/or functional) which is definable by some observer as "the" behavior "of the" system, or which is referable "to" the system as unit; thus, there exists emerging *macro* behavior with respect to the micro behaviors.
3. *Micro-macro feedback*: there must be a *mutual relationship* between macro and micro behaviors: on the one hand, the system's macro behavior *derives* from the micro behaviors of the component elements, but, on the other hand, in turn, *conditions* these micro behaviors as part of a feedback between *micro* and *macro* behaviors.

In a clock, the various cog wheels produce a micro behavior when they proceed individually in their operation; they produce a micro effect, in that they permit other cog wheels to advance according to ratios determined by the number of cogs in the wheels that are in contact with each other; the movements of the wheels as a whole represent the structural macro behavior of the clock that produces a macro effect: to the outside observer, the movement of the arms that mark the time. Since the macro behavior must be able to mark the time, the micro behaviors, and thus the micro effects, must be planned and coordinated in order to achieve the macro effect.

The sails of a sailing ship produce their micro behavior when they are swelled by the wind; the hull, rudder, shrouds, and the other parts of the ship also produce their own micro behavior. Nevertheless, the ship, taking a route toward a destination, produces a *macro* behavior that derives from the *micro* behavior of its parts but which is not identified with any of these, even though it conditions them.

Systems exist—or are observable or conceivable—everywhere, but often we are not able to identify them—or to *create* them—due to a lack of fundamental theories or suitable tools. For this reason, *systems thinking* has been developed.

> Systems thinking is a discipline for seeing wholes, recognizing patterns and interrelationships, and learning how to structure those interrelationships in more effective, efficient ways. (Senge & Lannon-Kim, 1991, p. 24)

Our observational abilities will improve the more we are able to create models of reality in system terms:

> In one way or another we're forced to deal, in all areas of knowledge, with complexities, with "wholes," with "systems." And this implies a fundamental reorientation of scientific thinking. (von Bertalanffy, 1968, p. 17)

In fact, the author of the *General Systems Theory* does not hesitate in stating that systems theory "*... represents a new 'paradigm' in the context of scientific thinking ...*" (von Bertalanffy, 1968, p. 11).

4.1.2 Exogenous and Endogenous Observation of Systems

Systems, in general—and *behavioral* ones, in particular—can be observed from both an *exogenous* and *endogenous* point of view. Observing the system from an exogenous point of view means the *nature* of the *elements* and the *relationships* that compose the structure of the observed entity are not of interest. The *exogenous perspective* brings out solely the possibility of describing the dynamics of the system in terms of variables, whether these are associated with the system as a unit characterized by inputs and outputs (functional description) or with the states of the structure (structural description).

Interest in the study of systems from an *exogenous* point of view increases when we observe in them some form of *retroaction*, that is, some feedback relations: the effects depend on the causes but these in turn are subject to the effects. Between causes and effects, there is thus a *line of feedback*.

Even feedback relationships can be observed from either an exogenous point of view, which we define as *extrinsic feedback*, or an endogenous one, which we define as *intrinsic feedback*. From an exogenous point of view, the internal feedback is by definition unobservable (or, in any event, not relevant to the description of the system); thus, since it views systems from an *exogenous* point of view, the science of systems ignores the internal feedback.

A clear definition and external description is given by Gary Sandquist (Fig. 4.2), who considers the *cause-effect system* and proposes the following definition:

> [A system is:] Any collection, grouping, arrangement or set of elements, objects or entities that may be material or immaterial, tangible or intangible, real or abstract to which a measurable relationship of cause and effect exists or can be rationally assigned. (Sandquist, 1985, p. 22)

Clearly, the observer is not interested in the *nature* of the elements that compose the structure of the system. When, instead, the observer analyses the nature of the elements that make up the structure of the system, he places himself in an *endogenous perspective*; that is, he undertakes an approach that leads him to "*enter*" the system to search for and analyze the different *nature* of the elements that form its *internal structure* and characterize its *state*, thereby allowing them to interact and produce the *micro-macro feedback* and the structural *behaviors*. From this point of view, with regard to the nature of the elements in the structure, it is possible to distinguish between *organized*, or operative systems, and *non-organized*, or complex and combinatory systems.

Note. We must not confuse the *endogenous* or *exogenous point of view* with the *external* or *internal description* of a system. In fact, the internal, structural, external, or environmental descriptions always involve systems observed from an *exogenous* point of view; the observer that describes the system does not study the *nature* of the system; that is, the elements and relations that make up the system and define its state and output: "*The internal description is basically 'structural'; that is, it tries to describe the behavior of the system in terms of state variables and their interdependence. The external description is nevertheless 'functional', since it describes the behavior of the system through its interactions with the external environment*" (von Bertalanffy, 1968, p. 149).

We define as *operative*, or *organized*, those behavioral systems composed of various "parts," called *organs*—that is, *functionally differentiated* elements—each

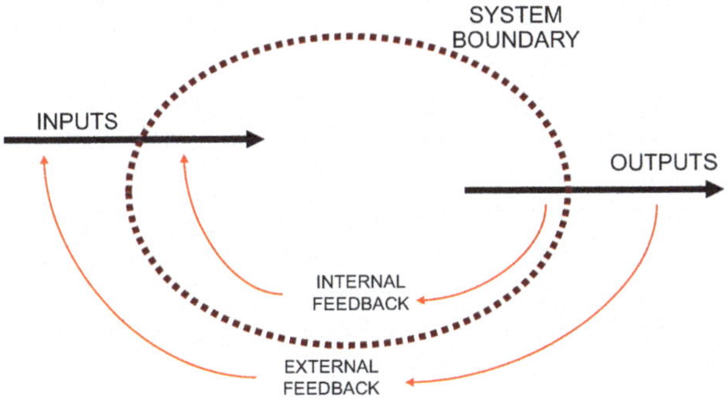

Fig. 4.2 Model of a cause-effect system (*source*: Sandquist, 1985, p. 22)

of which has a specific *specialized function*. The organs are interrelated through *organizational* and *structural* relations; the former impose the *function, functionality,* and *topology* of each element-organ, since each is coordinated with the other to allow the system to operate as such and produce its own macro behavior and emergent properties (Mella, 2021, p. 329). The structural (or specific) relations instead impose the qualitative-quantitative features of each element-organ to ensure these are compatible with the others in the structure of the system. If the organs are mechanical, the system represents a *mechanism* or a *machine*; if they are biological, the system represents an *organism*; otherwise, it can be defined as a social *organization*.

The organs of the system, with their *specificity*, thus produce, in the context of the *function, functionality,* and the *topology* assigned to them, differentiated and specialized micro behaviors to allow for interaction in a synergetic way—based on information—thereby producing an overall macro behavior that refers to the unitary system and that can be interpreted as the "*operation*" of the system. A macro effect can result from these *operations*. Systems of this kind are better defined as *operative* or *organized* systems.

The human body, along with every other biological system, represents the clearest example of an operative system, in that its *operation* derives from that of its parts—circulatory, muscular, nervous sub-system, and so on—and its organs, even though it cannot be reduced to a sum of the individual actions of these parts, at least given our present knowledge. There is thus a synergetic action of the organs. Moreover, just as the macro behavior derives from the micro behaviors of the organs, the latter are influenced by the macro behavior of the system. In fact, the behavior of the organs of an individual are quite different when the individual is resting or in motion, or calm as opposed to excited, and so on. Even a machine (clock, lathe, production line, factory) is an operative system, since it produces macro behaviors that depend on the micro behaviors of the organs that compose the structure.

With *endogenous* observation, on the other hand, we can define complex systems as collectivities made up of *heterogeneous* elements that interact to form dynamic wholes, which are emerging and not predictable a priori (Sect. 1.2.2) and combinatory systems (Sect. 1.1.4) as collectivities composed of a plurality of *similar* elements, each of which is capable of producing an independently defined micro behavior—often simple and predictable—that is distinctly observable or, in any case, refers to each single unit. The micro behaviors are *similar* to each other but, when "*combined together*" (whether or not in a conforming or non-conforming manner), they lead to an overall joint macro behavior. For this reason, combinatory systems are clearly distinct from *organized* systems.

In conclusion, and referring to Fig. 4.1, combinatory systems should be considered as systems in all respects, as the previous chapters have demonstrated. Combinatory System Theory is a manifestation of Systems Thinking applied to the *collectivity* as defined in Definition 1, in Sect. 1.1.2.

4.2 The Explanatory Power of Combinatory System Theory

4.2.1 Premise: The Explanation Process and Its Operational Closure

When man is faced with phenomena or events he cannot understand or manage, a situation of ignorance exists that gives rise to the necessity of knowing (curiosity), which leads to a question that requires an explanation. The explanation is a process of guidance toward the construction of a *model* from which those phenomena or events could derive.

> Any kind of scientific statement, concept, law, and any description of a phenomenon is a model construction which tries to reflect phenomena of the external world. Reality is extremely complex; it consists of strongly or more weakly related events. Science makes an attempt to separate and isolate different effects and phenomena. It seeks the simplest relationships by which examined phenomena can at least be described or demonstrated. It creates simplified models which only partly reflect reality, but which allow contemplation, and what is most important, pragmatic, even if sometimes modest, predictions. (Csányi, 1966, p. 148)

It is important to realize that the explanation is a backward-recursive process: every *explanans* of an explanation becomes, in turn, an explanandum that can require a subsequent explanation, and so on up the line, just like the "game of why" that all children love to play. Any ignorance is explained by some piece of knowledge; but this in turn is based on other knowledge. Let us indicate by E(n) the *explanans* and by F the *explanandum*; the recursive explanation then has the following structure:

F because E(n),
E(n) because E(n − 1),
E(n − 1) because E(n − 2),
…
E(0).
End.

The explanation process is "closed" when we reach E(0); that is, an explanans with no further explanandum. E(0) is defined as the operational closure of the explanation, which represents the "point of ignorance," the limit to knowledge.

E(0) can denote two possible forms of ignorance:

(A) – *E*(0) = "*What is not yet known*" and constitutes *reversible ignorance,* which can take the form of:

1. *Temporary ignorance* while awaiting new models ("research in progress…"; "we are checking", etc.)
2. *Disinterest*, meaning satisfaction with *E*(0) ("I am not interested in investigating further"; "It's already enough!")
3. *Secrecy*, meaning the desire to hide knowledge by not investigating beyond *E*(0) ("I can't explain it to you"; "It's classified")

(B) – *E*(0) = "*What we can never know.*" In this situation, "operational closure" represents *irreversible ignorance*, which can take various forms:

1. *Permanent ignorance*, that is, what *can never be known* (Heisenberg's uncertainty principle, Gödel's and Turing's theorems, and similar phenomena, contents of a black hole)
2. *Mystery*, or the unknowable (inevitable principles, myths, God, metaphysic or agnosticism, acts of faith, etc.)
3. *Postulate*, that is, an undemonstrable origin, merely assumed, from which the explanation is derived ("We hypothesize that…"; "Given that…"; "Let us suppose that…"; etc.)
4. *Chance*, in all its manifestations (quantum randomness, genetic mutations, initial impulses of combinatory systems, etc.)
5. *Unmotivated necessity*, that is, the self-sufficiency of E(0) ("It must be so"; "It cannot be otherwise"; "The immobile motor," etc.)

In these cases, operational closure becomes an *explanatory dogma*.

An explanation is *Coherent*, if it does not include elements that contradict other explanations; it is *Valid*, if it derives from a correct inference; it is *True*, if the E(n) do not contain false elements but only elements declared true or verifiable; it is *Explicit*, if it does not rely on "tacit premises" and if the E(n) can be made explicit. Not all non-explicit explanations should be rejected because they can be based on appropriate tacit premises useful for advancing knowledge.

> I shall reconsider human knowledge by starting from the fact that we can know more than we can tell. This fact seems obvious enough; but it is not easy to say exactly what it means. Take an example. We know a person's face, and can recognize it among a thousand, indeed among a million. Yet we usually cannot tell how we recognize a face we know. So most of this knowledge cannot be put into words. (Polanyi, 1967, p. 4)

Let us return to the original question: What is the power of Combinatory System Theory? To answer this, we must recall that there are at least four techniques for explaining a phenomenon (Mella, 2014).

4.2.2 The "Common" Explanation

"Common" explanations, rooted in common sense and descriptive or contingent, are those we use daily to justify phenomena or quickly answer questions. In a common explanation, it is assumed the phenomenon to be explained derives from "models" that usually, merely *describe* what needs to be explained without seeking scientific causes or underlying systemic relationships. These often lead to simplistic, if not incorrect, conclusions, as seen in examples like: "The light turned on because Aldo pressed the switch"; "The light turned on because it got dark"; "He had a fever because he was bitten by an insect," and so on.

The common explanation sometimes appeals to *intentions* ("The light turned on because Aldo needed to enter the garage"), *chance* ("I ran into Piero because he just happened to be passing by"), *experience* ("It's raining because I went out without an umbrella"), *regularities* or *common-sense rules*, including "wise proverbs" ("This is a rainy spring because the winter was dry"), *banalities*, and *false* or *tautological* beliefs ("Inflation is rising because prices are increasing"; "Global warming exists because humans burn fuels and heat the air").

The same phenomenon can also give rise to a plurality of "common explanations" based on factors that appear convincing and that usually allow for only one or two levels of recursion ("You made a mistake because you were distracted. What are you thinking about?", "I'm thinking about what I'll do tonight", "What's so important that you need to do?").

The "common explanation" can also take the form of a descriptive statement and does not utilize sophisticated qualitative models (Hempel, 1965). This category includes *historical explanations* ("Italy is a country rich in works of art because it experienced the Renaissance"). Often, to explain one's own behavior or that of others, some form of "justification" is introduced, as indicated by the "attribution theory" proposed by Fritz Heider (Heider, 1958). According to this theory, to "explain" how people behave and events occur, we tend to attribute the behavior of others, or the outcomes of such behavior, to two possible "causes": an *internal* cause—personality traits, intelligence, motivation, and so on ("He argued because he's a bit crazy"), or an external cause—luck, circumstances, the actions of third parties, and so on ("He won the lottery because he's always lucky", or "…because he prayed for a miracle from Saint Anthony") (Weiner, 1994).

Another technique of "common explanation" is the naïve application of the "5-Ws and 1-H" method—Who, What, When, Where, Why, and How—typical of "problem-solving" approaches. This method was conceived and developed in the 1930s by Sakichi Toyoda, founder of Toyota Industries, and addresses the majority of explanatory needs in our daily lives (Mulder, 2012, online).

Common explanations can easily become tautological—and thus non-explanatory—when attempting to explain a phenomenon simply through its definition: "Gravity exists because all heavy bodies are attracted to the Earth"; "Internationalization is an inevitable process because no one can stop it," and so on.

The general structure of the common "explanation" is as follows:

EXPLANANS	→P = Non-sophisticated descriptive, contingent, or simplistic premises: functions, objectives, intentions, assumed causes, chance, ignorance, etc
Explanandum	→F = phenomenon to be explained
Conclusion	→F occurs because the premises P occurred
OPERATIONAL CLOSURE	→ This is the last of the premises that recursively explain F

4.2.3 The "Classical Scientific" or "Causal" Explanation

Classical explanation, or "cause and effect" explanation. Carl Hempel and Paul Oppenheim published an essay, "Studies in the Logic of Explanation" (1948), which set out one kind of deductive argument that constitutes a legitimate type of scientific explanation, later known as the deductive-nomological model. According to this technique, normally used in the experimental sciences, an *explanandum* phenomenon is explained by an *explanans* model according to the following schema:

EXPLANANDUM	→F = phenomenon to be explained
Explanans	→T = scientific theories and postulates, & L = scientific laws, or Statistical laws & C = initial conditions.
Conclusion	→F is observed because, given the initial conditions C, it follows from the laws L, if we accept theory T
OPERATIONAL CLOSURE	→The fundamental theories and postulates represent the operational closure of the explanation

By varying the theoretical context and/or the laws adopted and/or the initial conditions included in the explanans, the explanation also varies (Cupples, 1977). This form of explanation has become recognized as the scientific explanation par excellence.

> Forty years ago a remarkable event occurred. Carl G. Hempel and Paul Oppenheim published an essay, "Studies in the Logic of Explanation," which was truly epoch-making. It set out, with unprecedented precision and clarity, a characterization of one kind of deductive argument that, according to their account, does constitute a legitimate type of scientific explanation. It came later to be known as the deductive-nomological model. This 1948 article provided the foundation for the old consensus on the nature of scientific explanation that reached its height in the 1960s. (Salmon, 1990, p. 3)

> […] there are two types of explanation, what he called 'deductive-nomological' (DN) and 'inductive-statistical' (IS) respectively." Both IS and DN arguments have the same structure. Their premises each contain statements of two types: (1) initial conditions C, and (2) law-like generalizations L. In each, the conclusion is the event E to be explained […] The only difference between the two is that the laws in a DN explanation are universal generalizations, whereas the laws in IS explanations have the form of statistical generalizations. (Hempel & Oppenheim, 1948, p. 139).

In the scientific context, this structure of "explanation" is normally adopted in physical, chemical, and other related sciences, where the relationships between events or objects are typically "causal" and can be observed or hypothesized.

4.2.4 The "Systemic" Explanation

Systemic explanation. Sometimes there is no cause for the explanandum phenomenon, F, but it is rather connected to the dynamics of some systemic process whose model we need to know and which represents the explanans that we are trying to uncover. The explanation thus consists in reconstructing the system which the observed phenomenon is a part of; an understanding of the system interrelationships permits us to understand how the phenomenon has produced the other interconnected and interacting phenomena. In particular, the systemic explanation must highlight the processes, P, that have generated F, which depends on the systemic structure S that supports those processes, the programs that guide the latter, and the environment E that conditions them. This explanation model has the following structure:

EXPLANANDUM	→F = phenomenon to be explained
Explanans	→S = structure of the system generating F, & Π = programs that make the structure generate the processes, & P = processes generated by S through the programs Π, & E = environment to which the system is coupled
Conclusion	→F derives from P produced by S in E by means of Π
OPERATIONAL CLOSURE	→The system (environment-structure-programs-processes) represents the operational closure of the explanation

The systemic explanations can be integrated with the *teleological explanations*, which try to take into account the behavior of a system with an objective (usually a biological one) by using the notion of "goal" or "objective" (Lennox, 1992).

> The necessity of teleological explanations for biology has been questioned by the formalist tradition. We have seen that the ideal of explanation for the formalists is the deductive model. Many attempts have been made to incorporate teleological explanations into the deductive model. Indeed, such an incorporation is necessary if there is to be a formal reduction of biology to physics and chemistry. The basic line of argumentation is to subsume biological "goal-directed" systems under the wider category of "directively organized" systems which can apply to both living and inorganic systems. (Plamondon, 1979, p. 153)

Teleological explanations are "evolutionary explanations" that justify certain biological phenomena related to individuals and populations, deriving from some "final cause" in the Aristotelian sense, a "necessary purpose," or a type of "backward causation." As a simple example of a teleological explanation, suppose we want to explain, "*Why does the hare's coat turn white in winter?*" The classical causal explanation would refer to the analysis of hare genes and the theory of evolution, which favors the survival of animals with camouflage traits.

Why did some species of dinosaurs grow incredibly in size and develop a long tail? Why do many animals of a given species that live on small islands shrink in size compared to those living in large spaces? Why are many species of snakes, insects, jellyfish, fungi, and fruits venomous? Why do many birds have long, colorful feathers on their heads? It is difficult to answer these questions using causal or systemic explanations, such as studying DNA or the environment. These phenomena require recourse to "teleological explanations" based on Darwinian evolutionary theory and its derived forms, among which the Theory of Combinatory Systems can be placed.

4.2.5 The "Procedural" Explanation

Procedural explanation. The procedural method is a very relevant explanatory method—even though it has not received proper attention in the literature—used whenever a phenomenon does not derive from others that produce it (causes) or is interrelated with it (system), but is instead the result of the application of some procedure (process, program, formulation, algorithm, etc.) that allows the phenomenon to be obtained as the consequence of given, or assumed, rules. When we ask ourselves why the solution to extracting the square root of an expression does not correspond to the answer in our textbook, why our ticket was not drawn in the lottery, or why we ran into the bumper of the car in front of us, we must look for the answer in the procedure followed for the calculation or lottery drawing, or we must examine our parking attempt.

The procedural explanation can take the following form:

EXPLANANDUM	→F = phenomenon to be explained
Explanans	→P = procedure whose application produces F & C = conditions for applying the procedure
Conclusion	→F derives from P applied under the conditions C
Operational closure	→The procedure represents the operational closure of the explanation

The procedural explanation always appears very convincing because not only is it the logical conclusion of the premise but it also allows us to reconstruct and emphasize the factual procedure (the logical or physical process) through which the conclusion emerges (is produced or determined by) from the premises. The systemic explanation represents the general form of the procedural explanation. In this sense, Charles Darwin's Theory of Evolution (Darwin, 1859) provides a procedural explanation of the evolutionary dynamics of species because it describes the processes—mutation, fertility, and natural selection based on the struggle for existence and mating—through which species change their characteristics and diversify over time, representing a powerful procedural explanatory tool for phenomena connected with the evolution of species (Pelikan, 2001). The "systemic explanation" serves as a broader framework encompassing the procedural explanation.

4.2.6 The Power of the Procedural Explanation

The Theory of Combinatory Systems represents an efficient tool for the *procedural explanation* of dynamic phenomena that derive from the action of collectivities that can be considered as observational units and not only as aggregates of individuals. This theory explains how the behavior of a collectivity arises and evolves by examining the interactive mechanisms between individual behavior (micro) and collective behavior (macro) and by trying to determine the *rules* that give rise to such behavior (and its effects).

The *procedural* explanation—to which Combinatory Systems Theory is linked—of the observable macro behavior/effects of a collectivity can take on the following form:

EXPLANANDUM	→F = collectivity's macro behavior/effect to explain
Explanans	→P = check if you can recognize all the requirements of a combinatory system: 1. Specify the size and dynamics of the phenomenon F & 2. Try to circumscribe the "base" and the "environment" of agents forming the collectivity & 3. Recognize the agents' micro behavior/effect & 4. Detect the *micro-macro feedback* action & 5. Identify the necessitating and recombining factors & 6. Discover the origin of the first micro behavior: "chance" or external decisions & 7. Build a logical and comprehensible model & 8. Try to simulate the behavior of the system C = conditions for applying the procedure 1. Check that the number of statistically significant observations is adequate 2. Define the boundaries of the system 3. Specify the precise micro and behavior
Conclusion	→F derives from P applied under the conditions C
Operational closure	→The *micro-macro feedback* represents the operational closure of the explanation

Other interpretative models may be possible; that offered by the theory of combinatory systems provides an efficient and effective interpretation of collective phenomena that can be considered to derive from *random initial* inputs and are thus necessitated by the *micro-macro feedback* action.

> Like the overlapping maps in a Mercator projection, where the ranges of different versions overlap, they predict the same phenomena. But just as there is no flat map that is a good representation of the earth's entire surface, there is no single theory that is a good representation of observations in all situations … (Hawking & Mlodinow, 2010, p. 10)

Even more interesting is the application of the theory of combinatory systems to collective phenomena for which we cannot immediately observe or postulate a macro behavior, but for which there is a macro effect that can be considered to be connected, or derived, from micro effects that result from individual micro behavior. How can we explain the formation of a pile of garbage? In principle, the pile of garbage could be considered the result of the random action of a collectivity that, instead of throwing away their garbage over a wide area, as is very likely, throws it over a limited area, even if there is a low probability connected to this. Much more significant instead is the explanation of accumulation in terms of combinatory systems (Sect. 2.1).

Will we ever manage to keep genetic engineering research under control? Will it ever be possible to control the use of cloning? Will we be able to stop the spread of child kidnapping for organ transplant purposes? Will there ever be an end to the race for the highest skyscraper, the longest bridge, or the fastest plane? Will we be able to control the production of drugs? Will people continue to live longer? How will we manage to control population growth? Will we be able to control pollution? How can we control the process of jobless production, which is now under way in the advanced industrial economies? Will the space race slow down? How dangerous is the spread of hate among the population of the same country? Can we reverse the increase in the population of large cities? Will the production costs of manufactured goods continue to decrease? Will there be further progress in the miniaturization of products (cellular phones, computers, printers, etc.)?

We can find answers to these and many other phenomena only by understanding the combinatory system that produces them. If the theory of combinatory systems is valid, then the answer will be clear and, in many cases, worrying.

4.2.7 Combinatory Systems Versus CAS

As they are constituted by a plurality of agents whose combined behavior gives rise to collective phenomena that produces self-organization and recognizing patterns, combinatory systems could also be conceived of as a particular class of *complex systems*. However, they differ from *complex systems* and, in particular, from *complex adaptive systems* in many respects (the following arguments are based on quotes from Sect. 1.2.2).

The *first* evident difference is that *combinatory systems* do not necessarily present phenomena of *adaptation* but, generally, some form of self-organization due to *micro-macro feedback*. Adaptation may be a characteristic of some particular class of collectivities of rational agents, and not, in general, of collectivities conceived of in a broader sense.

A *second* main difference is observable also as regards the *similarity* of the agents. Combinatory systems are made up of similar agents, while, as Holland notes, CAS are composed of heterogeneous agents: "*Here we confront directly the issues, and the questions, that distinguish CAS from other kinds of systems. One of the most obvious of these distinctions is the diversity of the agents that form CAS. Is this diversity the product of similar mechanisms in different CAS?*" (Holland, 1995, p. 93). It is worth going back to this quote (see Sect. 1.2.2): "*Definition (1): A complex adaptive system (CAS) consists of inhomogeneous, interacting adaptive agents. Adaptive means capable of learning*" (Ahmed et al., 2005, pp. 1–2).

A *third* relevant difference regards the nature of interactions among the agents; in combinatory systems, agents generally interact only with some macro variable and not with each other, as they do in CAS: "*Another distinction is more subtle, though equally pervasive and important. The interactions of agents in CAS is governed by anticipations engendered by learning and long-term adaptation*" (Holland, 1995, p. 93).

> Clearly, complex adaptive systems have a tendency to give rise to other complex adaptive systems. It is worth remarking for readers of this journal that John Holland, for example, uses a different set of terms to describe some of the same ideas. He uses "adaptive agent" for a complex adaptive system as defined above, reserving the name "complex adaptive system" for a composite complex adaptive system (such as an economy or an ecological system) consisting of many adaptive agents making predictions of one another's behavior. (Gell-Mann, 1995, online)

The *fourth* relevant difference is that CAS theory observes the macro effects of the system produced by the agents that follow a "*schema,*" or "*schemata,*" or change the schema previously followed. "*More generally, it is significant that any CAS is a pattern-recognition device that seeks to find regularities in experience and compress them into schemata*" (Gell-Mann, 1994, p. 22). *Complex adaptive systems* often show possible multi-level hierarchies (Chan, 1998; Gaffeo, 1999) according to a "schema" or "model" (innate or learned) (Dooley, 1996, 1997).

However, since they follow the simplest "schema of adaptation," and because of the similarity of agents and behavior and the absence of organizational or social links, we could provocatively define combinatory systems as a "simplex systems."

A *fifth* fundamental difference is that ignoring the *micro-macro feedback* implies that CAS theory only focuses its attention on *necessitating* factors and ignores the *recombining* ones. Any *micro-macro feedback* between the micro behavior and the schema is considered a relevant characteristic.

Finally, in combinatory systems, agents act on the basis of global self-produced information, while in CASs agents act prevalently on the basis of local information:

> If you want to understand why a person acts as she does, it is certainly possible to look around in the immediate environment for an explanation. But often an explanation needs to look also, or perhaps primarily, at events that occurred in the past and at how the present situation developed from previous circumstances. (Gilbert, 1995, online)

For a synthesis, see Table in Fig. 4.3.

Complex systems and holarchies	**Complex adaptive systems**	**Combinatory systems**
Agents are heterogeneous	Diversity of the agents as a constitutive feature	Agents are similar and are not necessarily biological beings
Agents are interconnected and show hierarchy	Agents are interconnected and change interactions	Agents are not interconnected
Micro behavior is differentiated	Agents show adaptation	Micro behavior is analogous
Agents interact with their environment	Agents exchange information with their environment	Agents act following global *self-produced information*
Agents act following local rules	Agents act following a "schema"	Agents act following the *micro-macro feedback*
Decisions are prevalently based on some "game" schema	Decisions are based on forecast and expectations	Decisions follow a simple one column pay-off matrix

Fig. 4.3 Table comparing combinatory systems with complex systems

4.2.8 *Is Combinatory Systems Theory an Effective Explanatory Tool?*

Despite the differences highlighted in Fig. 4.3, combinatory systems can represent, precisely due to their simple logic, which can easily be observed in nature and easily described and simulated by using Combinatory Automata (See Chap. 3), a useful theoretical model to be applied to the study of populations of similar agents, though they are not applicable to complex social systems.

Three aspects of this theory make it particularly effective:

1. It is not limited to describing the macro behavior of collectivities based only on *global* rules or the individual behavior based only on *local* rules, but tries to uncover and explain above all the *feedback* between the macro and micro behavior or their effects.
2. To understand the phenomena attributable to the action of combinatory systems, the theory tries to uncover and make clear the *necessitating factors* (which cause the micro behavior of each agent in the system) and the *recombining factors* (which produce and maintain the unit's macro behavior). The theory then concludes that, in the presence of suitable necessitating and recombining factors, "chance" will trigger the dynamic process of the system, which, "by necessity," is then maintained and influences the individual behavior.
3. The *procedural explanation* offered by the theory not only allows us to understand the operating mechanisms that produce the phenomena under examination but also permits us to determine the most effective forms of control.

The challenge of combinatory system theory is threefold:

1. To apply the theory in order to understand collectivities operating in the real world and to specify, for any real observed collective phenomenon, the sets of necessitating and recombining factors which allow us to interpret and control the collectivity that produces it.
2. To identify the *reinforcing* or *weakening measures* (actions, provisions, constraints, limits, obligations, etc.) which, operating at a *macro* and/or *micro* level, modify the *recombining* and *necessitating* factors, thereby influencing the *macro* and/or *micro* behavior in order to *favor* or *inhibit* the attainment of certain individual and collective results considered to be harmful or useful.
3. To develop more general and increasingly sophisticated Combinatory Automata for any specific class of combinatory system especially for systems operating in a three-dimensional space (clouds, school of fish, etc.).

Finally, it should be noted that the Theory of Combinatory Systems proposes in new clothing the "usual" question of the freedom of individuals to decide their own future, in the absence of causes or constraints, and thus their capacity for self-determination. Here, we shall deal with a problem that is symmetrical with respect to the previous one: Can rational *micro* behavior produce an irrational *macro* behavior that is even capable of causing catastrophic effects for the entire system? This is the subject of the next section.

4.3 Three Reflections

4.3.1 *Chance, Necessity, and Freedom*

The problem of the freedom of individuals to decide their own future arises from the realization that if, on the one hand, combinatory systems are everywhere, on the other, the individuals they are made up of in most cases cannot consciously perceive their presence, since they act in conditions of *incomplete* and *limited* information. Therefore, these individuals appear to act in a conscious, autonomous, and independent manner, while in fact their behavior is determined by the *global information* derived from the macro behavior or the macro effect of the system. Once begun "by chance" and with the achievement of the *minimum density* for the *base*, the combinatory system "by necessity" develops its macro behavior and produces the consequent macro effect; the individual micro behavior feeds the macro behavior, but it is in turn conditioned by the latter. The existence of a *micro-macro feedback* in the presence of a very intense necessitating factor would reduce (in some cases even cancel) the margins for individual discretion in deciding on the micro behavior, at least for systems where the micro behavior is voluntary. The macro behavior create path dependence which conditions the agents' micro behaviors. For example, once the spinning movement to the left or right is produced in the "waltz-turning" system, it inevitably conditions the movements of the single dancers, who in fact would have a truly limited discretion to change direction during the dance. A similar situation occurs with "elephants-in-flight" combinatory systems (Sect. 2.5.2). The individual elephants in flight, when faced with an obstacle, would find it difficult to change the direction of their flight, as this is conditioned by the direction of the herd (macro behavior). The problem is even clearer and more accentuated with very large and complex combinatory systems.

Let us take, for example, the entire economic system in which different combinatory systems operate: that of consumers, firms, and that formed by the public administrative bodies. If observed as complex systems, the behavior of the individual apparently enjoys a large amount of freedom: he can buy or not buy, produce or not produce, introduce an innovation or not do so, and so on.

The decision of the individual to buy/not buy, of the corporation to introduce/not introduce an innovation is nonetheless strongly conditioned by the macro behavior of the system. There may not be the purchase of a luxury good, but this can cause a reduction in the individual's social standing, which represents an unacceptable condition, so that, in fact, when a combinatory system of pursuit or improvement is triggered, in which the individuals continually want to improve their social condition, the decision to purchase that good is all but inevitable.

> A robust alternative to the rational choice program must also be more explicitly context-dependent. The rational choice paradigm itself, of course, is already context-dependent, in that the conception of self-interest that operates changes from context to context: a single human is imagined to maximize quite different functions in the roles of corporate executive and mother, for example. But the rational choice program posits the pattern of explanation of

> action in terms of maximization subject to constraints to be universal. It seems likely, however, that in reality humans deploy very different modes of action in different contexts, and that these modes of behavior differ in their computational complexity. (Foley, 1998, p. 71)

The "rationality principle" is diminished since the potential liberty is in fact conditioned by the strong *constraint* represented by the *micro-macro feedback* between the individual and the collective behavior, and this constraint is more intense, the more the necessitating effect is felt.

Similarly, each firm is free to decide whether or not to introduce a certain product, process, or organizational innovation aimed at increasing product quality, productivity, or process efficiency. However, these firms are agents of combinatory systems of progress and improvement (Sect. 2.6), and thus their decision is conditioned by the behavior of the other firms/agents, which, taken as a whole, triggers a macro effect characterized by an increase in "quality," (Sect. 2.6.2) "productivity," and "efficiency" (Sect. 2.6.1); those firms that do not adapt to this macro effect are penalized in terms of their economic performance, efficiency, and profitability, and so must undertake the micro behavior of introducing innovations.

The "rationality principle" thus implies that the freedom to make decisions is thus countered by the necessity to decide according to what is indicated by the macro behavior of the combinatory system. This assertion is not intended to favor the holistic view of human society or at least to deny validity to the *individualistic* idea; it is simple meant to point out that if one accepts the Theory of Combinatory Systems as an explanation for several social phenomena, then the individual micro behavior for those phenomena derives from a *necessitating* factor, caused (connected, derived, correlated) by the collective macro-behavior, thereby producing *global information*, and by definition limiting the freedom of the individuals with regard to that behavior. If the combinatory system, then undergoes *strengthening* actions, this freedom is even more limited.

Consider, for example, the freedom to follow a cult which is protected by the constitution. The combinatory system of those who believe in a certain cult necessarily conditions the micro behavior of those families who wish to pass the cult on to their children, and this micro behavior will be more likely, the more serious are the moral and social repercussions for the families themselves. The cult is thus the macro effect of the system's macro behavior. The latter is caused by the micro behavior of the families; but, by means of the "rationality principle," it inevitably conditions the micro behavior, according to a typical *micro-macro feedback*.

More generally, if we reflect on our daily life, we can easily see how limited our freedom of choice is; none of us is completely free to choose the time we start work, the time we go to the restaurant, the month we take our holidays, our pace of life, the speed we go through the streets of our town. Our micro behavior always has a fairly limited range of freedom, not only when we belong to some organized system (where by definition, the individual/organ's discretion is limited by organizational constraints) but also when we appear to be acting in an individual, autonomous, and free manner. Each of us, in fact, is part of a multiplicity of combinatory systems that operate contemporaneously and whose effects are superimposed on each other and often multiplied; therefore, we do not perceive the *micro-macro feedback* action of combinatory systems to which we unknowingly belong.

Thus, on the one hand, we realize that the continual increase in energy consumption can lead to a collective disaster, but at the same time each of us buys more powerful cars, increases our individual consumption of energy through the range of available household appliances, and so on.

In conclusion, none of us is truly free to change our micro behavior regarding consumption, work, and free time precisely because these depend on the feedback with the macro behavior produced by the systems we are a part of. For the elements of the system, this becomes a super ordered entity that, through the global self-produced information deriving from its own macro behavior/effect, influences, and in many cases, drives the micro behavior.

From this perspective, the problem of "individual freedom" must be reconsidered. The freedom of a micro behavior must no longer be perceived as simply the freedom to behave or not to behave in a certain way but comes down to the question of when to effectively develop a free micro behavior in the context of the combinatory system. The first of the 100 towers in medieval Pavia arose by chance; by chance others were built, but when the minimum density was reached, the combinatory system began and the families thus needed to behave so as to make the number of towers in Pavia reach the incredible number of 170 (Sect. 2.3.3).

In combinatory systems, agents act under conditions of bounded rationality (Simon, 1955; Rubinstein, 1998; Dequech, 2001) and produce a hysteresis effect.

In combinatory systems, the hysteresis effect is manifested through the *micro-macro feedback* action.

> In general terms hysteresis is a property of dynamic systems. Hysteretic systems are path-dependent systems. The long-run solution of such a system does not only depend on the long-run values of the exogenous variables (as usually) but also on the initial condition of each state variable. These systems have a long-lasting memory and are therefore "historical" systems. Loosely speaking: Where you get to is determined by how you get there. (Franz, 1990, pp. 1–2)
>
> A related aspect can be found in choice theory where, for example, preferences of consumers change in course of their experiences with consumption, or to quote Georgescu-Roegen: "...in order to determine the equilibrium of the consumer (for a fixed budget and constant prices) we need to know more than his particular hysteresis law. Still worse, this law being expressed by a very complex set function, we can only write it on paper but not determine it in actual practice" (1971, p. 126). (ibidem, p. 6)

The problem of individual freedom with regard to (social) combinatory systems obviously depends on how intense and strong are the *necessitating* and *recombining factors* and the micro rule that produces the *micro-macro feedback*. We can easily sense that there is a considerable difference in terms of freedom in the case of marching at a different pace in a parade as opposed to carving some new graffiti on the slats of a bench.

The more the micro behavior is held to be useful, necessary, indispensable, due to the *global information,* the more the individual's freedom is limited; the same "rationality principle" obliges one not to adhere to the *macro* behavior/effect that produces information characterized by futility, damage, or danger regarding the micro behavior of agents. This does not imply a reduction in the significance of the "rationality principle"; on the contrary, the individual whose micro behavior is "guided" by the macro behavior behaves in a perfectly rational way as an individual. It would be irrational for one pair of dancers to try to dance in the opposite direction of the other dancers in the room, just as it would be irrational for a businessman not to take account of a technical innovation and thus damage his firm or for a university researcher not to follow a new scientific discovery and continue to fail to advance in his academic career.

4.3.2 Individual Rationality and Collective Behavior

In the preceding section, we observed that, for combinatory systems, the macro behavior of the system constrains the micro behavior of the individuals, even when the latter behave rationally. This section will instead deal with a problem that is symmetrical with respect to the previous one: Can rational micro behavior produce an irrational macro behavior that is even capable of causing catastrophic effects for the entire system?

The answer can only be "Yes": when the system operates under *incomplete* and *limited information* (Sect. 1.3.5) the single individuals—ignoring the macro effect and observing only the micro behavior of a limited number of other units—produce a *micro behavior* which from their point of view optimizes the micro effects while at the same time producing a *macro behavior* capable of breaking up the system.

This occurs, for example, with the micro behavior of a person fleeing amidst a maddened throng, guided only by the direction of the persons surrounding him (limited information) and ignoring the direction of the crowd (incomplete information). When the crowd has taken on a certain direction, this by necessity is maintained, even if the crowd is moving dangerously close to a barricade; the momentum from the fleeing individuals, who are guided by the collective flight, can cause the catastrophic stampeding action, which occurs often with such phenomena.

The elephant that is fleeing in the middle of the herd does not think (presumably) about the herd in flight (incomplete information) but about the fact that the elephant in front and alongside him are fleeing (limited information). Therefore, while the heard is heading toward a crevasse and risking falling in, this is not what any of them desire; however, running based on the limited information available, the elephants would be heading toward disaster as part of an inevitable process that the following imaginary and "surreal" interview seeks to clarify.

An ancient tragedy. Act 1

A team of hunters was observing a group of elephants, taking care to be downwind of them. Suddenly the wind changed direction and the hunters were sniffed. A terrible trumpeting came from the throat of the female head of the herd, who began to run in zig-zag fashion. This is the initial (external) global information based on which the elephants appeared to synchronize their behavior. As if guided by an invisible hand, they followed the head of the herd, which soon formed a compact fleeing mass moving behind the lead elephant.

Interviews

Let us interview an elephant.

"Excuse me, Mr. Elephant, why are you running in that direction?"

"I'm not running toward a particular direction. I need only save myself. The herd is forcing me to run in that direction."

Let us interview the herd.

"Excuse me, Mr. Herd, why are you running in that direction?"

"I don't want to run anywhere. The elephants are running to follow the leader!"

An ancient tragedy. Act 2

The stampeding herd came upon a crevasse. The lead elephant wanted to stop but was pressed from behind, because the elephants that followed had no choice but to continue in their flight since those preceding them were marking the way.

Interviews

Let us interview the lead elephant.

"Excuse me, Mr. Leader, why are you running toward the crevasse?"

"I'm not running in any particular direction. I only wish to avoid obstacles (rationality) because the herd is pressing me from behind and I might break my tusks."

Let us interview the last elephant in the queue.

"Excuse me, Mr. Elephant, why don't you stop running? If you stop, the herd will gradually stop, too. Your exertions will finish and the herd will avoid the crevasse."

"Do you think I am mad? The hunters might chase me! I will never stop as long as the hunters are following me!"

An ancient tragedy. Act 3

A predictably disastrous outcome occurred: most of the elephants forming the herd, while not wanting to fall into the crevasse.

The combinatory system just described contained agents operating under conditions of *incomplete* and *limited* information. However, catastrophic events also often occur even in combinatory systems whose agents operate with *complete information*, represented, for example, by the *macro effect* of the system they themselves have produced and which is perceived by means of some indicator that is updated at every instant by their micro behavior. This is the typical case of stock brokering activity (Sect. 3.5.3), which is carried out by a large number of individuals that, through their transactions (micro behavior), determine the overall trend (macro behavior) in the stock market indices (macro effect), which, for many traders, represents the *global information* which determines their subsequent negotiating decisions; the latter, due to the recombining effect, continually modify the stock index itself (Aliber et al., 2023).

A modern tragedy. One-act play

It is a bleak day for the NYSE. A wave of selling starting from the opening bell has led to a fall in share prices, which has continued for the entire day.

Source: Financial Times, New York, APRIL 22 2024

Interviews

Let us interview an Investor.

"Excuse me, Mr. Investor, why are you bringing down the Stock Exchange by selling your stocks?"

"I'm not bringing down the Stock Exchange. I simply want to reduce my losses by selling as fast as possible because of the steep downturn in the Stock Exchange."

"Before selling, why didn't you wait to understand what was happening?"

"Because the fall in the stock index was so swift that any delay would have increased my losses. And then panic set in."

Let us interview the STOCK EXCHANGE.

"Excuse me, Mr. Stock Exchange, why do you wish to bring down the stock values and produce losses for the investors?"

"I don't desire losses for anybody. Investors who are selling their stocks bring me down."

Let us interview the FIRST SELLERS at the sound of the opening bell.

"Why did you begin selling, thus causing a fall in the index which set under way the disastrous day for the Stock Exchange?"

"We began selling our stocks because the index was "at its maximum" and we wanted to do some profit taking. It was a matter of chance that no one began to buy up other stocks, which was the reason behind the initial fall in the stock index."

The panic that occurs in rational combinatory systems—elephants or financial markets—has been dealt with by many authors (Sect. 2.4.6). However, these authors have tried to explain this phenomenon by highlighting the role of the *micro behavior* and, in some cases, the *necessitating factors* that push agents to carry out imitative behavior, without, however, understanding the essence of the *micro-macro feedback* and the necessitating and recombining factors that support such feedback.

> The form of panic is typically systemic, for it consists of communication between the elements of a totality through the intermediary of the totality itself. (Khalil & Boulding, 1996, p. 75)

> My hypothesis is the following: Panic is a social evil that comes from within and not from the outside. If there is a myth here, it is the belief that panic is caused by something external to the group. ... There was no panic in Hiroshima on August 6, 1945, or in New York on September 11, 2001. (Dupuy, 2005, p. 218)

> The form of panic is typically systemic, for it consists of the communication between the elements of a totality through the intermediary of the totality itself. This totality is considered transcendent, although it is actually an emergent phenomenon. It is possible that this figure is the key to all forms of social division. (ibidem, p. 229)

> Most of the intuitions that guide these models come from Keynes. In his study of financial speculation, Keynes sees the fundamental role of imitation (cf. Keynes, 1937). In a situation of radical uncertainty, such as the one prevailing in a financial market in crisis, the only rational form of conduct is to imitate others. (ibidem, p. 230).

> Knowing that our own individual judgment is worthless, we endeavor to fall back on the judgment of the rest of the world which is perhaps better informed. That is, we endeavor to conform with the behavior of the majority or the average. (Keynes, 1937, p. 214).

Many other combinatory systems produce dangerous behavior. Companies do not wish to reduce their "productive efficiency" with respect to others in the same sector that have an advantage (limited information); if a company tries to increase its productivity by introducing a robot (input due to chance), and if other companies follow its lead (limited information), then the system is triggered and the *micro-macro feedback* guarantees that soon all the companies will use similar robots, with an increase in productivity (this is, in fact, the idea behind the Evolutionary Economics approach) (Sect. 2.6.9). This micro behavior, which is rational from the individual point of view, ignores the effects of the gradual increase in productivity on the economic sector (incomplete information): changes in the method of production, the evolution in production techniques, the freeing-up of manpower, and so on. Catastrophic dynamics could also occur for the sector (flight of firms) or for the entire economic system (creation of unemployment). Those talking in a crowded room do not consider that if they were quiet, the noise would diminish; rather, in order to be heard by their interlocutors (limited information), they talk even louder. Those who push drugs do not think about the catastrophic effects of their spread (incomplete information) but only act (limited information) based on their own needs (one of the most intense necessitating factors). The individual Can-Can dancers do not rotate looking at the line that rotates (incomplete information) but keep in line by looking at the dancer at the head of the line (limited information).

Consider the combinatory system involving the transplant of organs. A person in need of a transplant tries to acquire an organ, triggering off a process involving the sale of organs; at the individual level, the behavior appears rational even if the system can produce easily imaginable catastrophic effects for the "production" of organs to transplant.

Equally damaging, if not catastrophic, is the system involving wheel ruts forming on highway asphalt due to the passing of trailer trucks and the system of feuds or those involving the spread of drugs and the formation of a megalopolis.

On the other hand, we can consider systems with positive effects characterized by the continuous increase in productivity and of scientific progress in general, even if the system involving progress in genetic engineering research could produce unpredictably dangerous effects.

Often the analysis of combinatory systems favors only the observation of the macro behavior, so that the micro behavior appears irrational. This is not so, since the macro behavior constrains the micro behavior due to the action of the *micro-macro feedback*.

The structure of combinatory systems of which we are unknowingly a part thus represents a constraint to rational behavior.

> Human rational behavior is shaped by a scissors whose two blades are the structure of task environments and the computational capabilities of the actor. (Simon, 1990, p. 7)

4.3.3 From Systems of Improvement and Progress to the Three Metaphysical "Laws" of Becoming

The existence of structures that behave according to the logic of combinatory systems, or even the assumption that this is true, leads to several interesting theoretical implications. After having outlined the fundamental features of combinatory systems and described several simple combinatory systems, taken from direct observation, we can see that such systems are very common in "nature" and can be considered as the means for the propagation of order and progress.

Generally speaking, the observation of reality indicates there are three important assumptions or laws (metaphysical) that can be postulated concerning the future:

1. The law of disorder, or chaos
2. The law of order, or organized systems
3. The law of combinatory systems of improvement and progress

The *law of disorder* states that the universe tends toward the maximum state of randomness, the maximum disorder, the maximum "entropy," to the detriment of order; today this law is accepted by all scientists and is formalized in the "second law of thermodynamics."

The *law of order*, in apparent opposition to the preceding one, states instead that nature tends to take on ordered structures (within which entropy diminishes), which appear in all their magnificent complexity in the case of biological organisms and socio-political organizations—that is, *autopoietic systems* and their realization—masterfully described by Maturana and Varela (1980, 1987).

> The mathematical modelling of the emergence of macroscopic order in animal population is not very difficult where it is based on simple chemotaxis the attraction due to certain chemical substances (Prigogine & Meixner, 1976). Chemotaxis may be observed in striking examples with single-cell organisms, such as the amoebae which, at certain times, join to form the slime mold Dictyostelium discoideum. In times of food scarcity, the amoebae stop to divide and form spontaneous aggregations, and every 3 to 5 minutes send out chemical pulses of cyclical AMP (adenosinemonophosphate), a signal leading to rhythmical chemotaxis. (Jantsch, 1980, p. 63)

While the law of disorder is the law of heterogeneous groups that join to form increasingly more statistically homogeneous groups, without ever becoming systems, the law of order is evidenced by the birth of operative systems, and in particular by the formation of organic structures composed of different elements, each of which takes on a different role and carries out its function according to given information. Order is supported by information incorporated into the operative system and its elements, above all by the possibility of transmitting this information to other systems (descendant).

The *law of improvement* states that nature, by means of successive "catastrophes"—in the sense meant by Thom, as a change in form deriving from the singularity of the system's evolution, making the catastrophe morphogenetic (Thom, 1975)—tends to produce ever more evolved ordered structures, and that these structures are spread as much as possible, to the detriment of disorder, in accordance with the available resources and information, in order to reproduce (from the blade of grass that grows between the cracks in the asphalt to space programs to bring life to other planets). The metaphysical law of improvement and progress is discussed by many authors who make reference not so much to combinatory systems but to systems in general (see, e.g., Le Moigne, 1994a, 1994b, Chap. 12).

This view, which is derived from the observation of the combinatory systems of structured biological systems, does not conflict with the evolutionary view of biological or intellectual nature. In particular, the well-known "formula": ORDER = RULE × APPLICATION, places at the center of evolution the concept of information (rule) that is repeated (application) in order to derive ordered systems. Systems of simple rules for a large quantity of elements (e.g., the first mono-cell organisms) give rise to more ordered organisms in which the information of simple organisms is merged, leading to an increase in the degree to which sets of rules exist in the biosphere (Reichholf & Mattson, 1994). It thus becomes evident that many natural systems are structured based on their internal processes, without any orders coming from the outside. Therefore, DNA is the same as information and, as a result, the basic principle behind the self-determination of order for living systems. There is thus the same amount of order in the presence of many organisms under a system of simple rules as there is for a few organisms under a system of complex rules. Not only does order spread everywhere but, where possible, ordered structures become more and more efficient and try to improve; they evolve and, by evolving, lead to progress.

We immediately see how the law of disorder is sovereign in the case of inanimate nature and how that of order instead characterizes animate nature—in which autopoietic operative systems that can reproduce predominate—while the *law of improvement and progress* is found in biological and social populations, where combinatory systems typically operate. We must, however, remember that in animated nature as well, there is disorder, since chance produces variations in the DNA, which leads to hereditary differences that, thanks to combinatory systems, can give life to new species that result from chance mutations and environmental reinforcements. In any event, the idea of improvement must be stipulated by the observer.

The law of disorder is the law that reduces differences; the law of order is the law of the creation and maintenance of differences between that which is in the system and that which belongs to the external environment. The *law of improvement and progress* is the law of the accentuation of the "advantageous" differences between individuals and populations. Finally, the law of disorder is the law of "ignorance"; that of order is the law of "information" (and of transmitted information); that of improvement and progress is the law of "judgement," since it always implies a comparison between a before and an after (between a bad and a good, between a better and a worse, etc.), for the purpose of perceiving and judging the improvement. It is thus the law of intelligence, the law of man as we conceive of him.

> Spontaneous order keeps the wheels of society turning without the need to issue commands from the centre. A free society is orderly not because people are told what to do but because the evolving traditions and inherited institutions of human society allow individuals to pursue their own ends and, by so doing, meet the needs of others. People's behaviour follows certain patterns because they have been accepted by society initially as they allowed the groups which adopted them to prosper. (Ashford, 2001, p. 83)

In this context humanity, with all the operative and combinatory systems that form and act within it, represents the most evolved combinatory system composed of men, where the *law of improvement and progress* prevails as the natural consequence of the assumption of rational behavior. In observing purposeful activity, we accept the postulate that man always acts in a rational way; in other words, we can recognize that human actions develop according to the "Rationality Principle," which, in a simplified form, can be stated in the following way: "*In carrying out his own purposeful activities, man always acts rationally, in the sense that he tries to maximize the efficiency and the effectiveness of his own actions.*"

This is only an assumption we make to guide the observation of human behavior in combinatory systems, but it is truly difficult to imagine that a person consciously does not try to minimize the sacrifices necessary to obtain given benefits and does not try to maximize the obtainable benefits with the same degree of sacrifice; and, at the same time, that he does not work to maximize the satisfaction obtainable through those levels of efficiency. The principle of rationality makes possible the necessitating effect and guarantees the manifestation of the law of improvement in several basic combinatory systems composed of men who operate for scientific and technological production and research. The metaphysical law of improvement is evidenced by the progress of humanity.

4.3.4 Concluding Remarks. An Imaginary Interview with the Author

The Combinatory Systems Theory presented in this book is simple regarding its basic ideas, while at the same time being somewhat elaborate and, in many of its features, formalized.

This imaginary interview with the author, which balances the desire for simplicity and the need to highlight the most relevant aspects of the theory, aims above all at revealing the logic of the theory of combinatory systems and its conceptual simplicity, as well as its general interpretative capabilities. On the one hand, the interview focuses on several aspects which might be less easily understandable upon a first reading of other sources, and on the other clarifies the main links between this and other theories that survey the same field: synergetics, the theory of complex systems, stigmergy, path dependence, and so on. In addition, the aim here is to stimulate the reader to learn more about the theory and to bring out new questions, which the author will be happy to answer.

Question. *From reading the text, Combinatory Systems Theory (CST) appears to be quite elaborate. Could you summarize its basic ideas.*

Answer. Briefly, this theory attempts to interpret, explain, and simulate phenomena deriving from the actions of collectivities, populations (human or animal), or, more generally, sets or groupings of *similar*, not necessarily living, elements/agents that produce *analogous* behaviors which, "when combined," produce cumulative phenomena and effects that refer to the entire collectivity observed (hence the name "Combinatory system"). The collectivity/agents relationship is what CST focuses on.

As I understand it then, CST views collectivities no longer as the result of the simple aggregation, or combination, of individuals and their individual micro behaviors; instead it views the collectivity itself as an entity possessing its own observable behavior.

Exactly: the collectivity has its own behavior, which is not programmed by the behaviors of the individuals inside it, even if it derives from these behaviors. A coral reef is quite different from the individual corals it is composed of; a desert area rippled with dunes that move with the wind according to their respective flows, thereby influencing the flow itself of the wind, is quite different from the single dunes; a herd in flight is quite different from the individual animals that are fleeing; a stadium wave is quite distinct from the cyclical rising up and sitting down of the individual spectators, and so on. Collectivities in which combinatory systems form produce emerging collective phenomena.

Emerging?

Yes, in that the behavior of the individuals that constitute the collectivity produces a synergetic effect, thereby generating a global collective phenomenon that differs from what would be produced by these same elements considered separately, which could not be explained, in a reductive manner, solely by the properties (qualitative and quantitative) of the individual behaviors or their mere addition or aggregation.

How can this be? What fundamental idea moved you to formulate this theory?

The fundamental idea is elementary, almost obvious, even if we are not immediately aware of it. However, we must learn to recognize it in the collective phenomena we observe. Here it is: a collectivity of elements forms a combinatory system when the collective behavior is determined (provoked, influenced, controlled, etc.) by that of the individual agents, at the same time that it determines (provokes, influences, controls, etc.) the subsequent behavior of the agents themselves.

For example?

There are countless examples. Here are some of the basic ones (if necessary, I can present others in future).

I go on holiday in August because everyone does the same. However, by going on holiday in August, I keep up the custom of going on holiday in August.

I speak my mother tongue because the collectivity speaks my mother tongue, which is taught to children. However, precisely because I speak it, I teach the mother tongue to my children and thus help maintain a collectivity that speaks my mother tongue. Have you noticed how many dialects remain unchanged over time? And how many different dialectal inflections remain unaltered for centuries in the same quarters of the same city?

Why is there a colony of hedgehogs on a cliff, while on the neighboring cliffs there are none? Because the population or hedgehogs facilitates the reproduction of the individuals. By reproducing, the individuals keep the population localized.

Why does one wish to live in a city? Because everyone lives in a city, which leads to clear advantages. In this way, the city perpetuates itself, expanding and providing incentives for others to live there.

Why does a fish dive into a school of fish of his species, thereby maintaining the school for a certain period? Because the school exists and provides protection; but it exists because the individual fish dive into it and, by existing, leads to other fish joining it.

Why does an ant follow the same path of the ants ahead of it and then explore another route after reaching the end of the path? Because the path was formed by ants who had previously followed the path and left a pheromone trace; thus, that path guides other ants which, by taking it, reinforce the pheromone trace.

Chapter 2 presents significant examples and models of combinatory systems, structured according to the basic scheme of Fig. 4.4.

Something about these examples do not convince me. They are, so to speak, tautological, "circular." The collectivity reveals a certain macro behavior produced by the individuals through their micro behavior, but the behavior of the collectivity subsequently determines that of the individuals. It is a bit like the serpent that feeds on its own tail. Isn't there a contradiction at play here?

You used the perfect term: the examples *are* "circular." This, in fact, is the fundamental feature of combinatory systems: the circularity—or, to use a technical, though equivalent term, the *feedback* (or loop, retroaction, reciprocal interaction)—between the (macro) behavior of the collectivity and the individual (micro) behaviors, as shown in the model, which specifies at the center of the loop that the

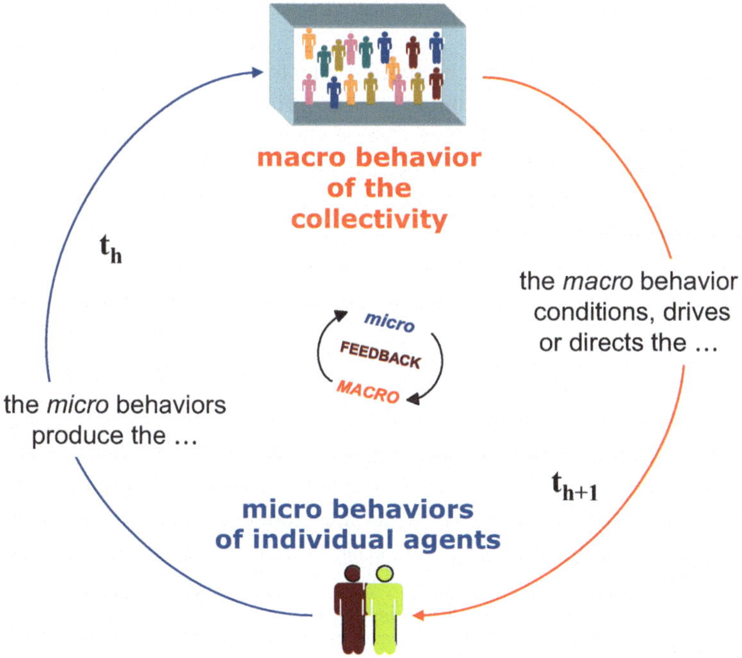

Fig. 4.4 The basic scheme of a combinatory system

opposing arrows indicate the "*micro-macro feedback.*" The time references have also been highlighted.

A clarification: Am I right in understanding that the micro-macro feedback can operate only if the "circular" system continues to function for a certain period?

Yes, this is precisely the operating logic of combinatory systems. They carry out a repetition of behaviors, each of which derives from the preceding one. Otherwise, it would make no sense to speak of a collectivity or population behaving as a whole and of individuals that act as elements of the collectivity. We would only have a group of individuals, each of which would behave independently of the others. Combinatory systems can be viewed as "repetitive" systems (in some cases, as "recursive" ones) over a certain period.

In what sense is the logic of repetitive systems employed?

In very simple terms, when a system is repetitive, the agents initiate their micro behavior at the initial moment, say at t_0. At t_1 the system produces its macro behavior, which influences the micro behaviors, which modify, at t_2, the macro behavior, and so on for the succeeding moments. A formal representation of this logic is provided in Chap. 1 and then developed in Chap. 3. For a combinatory system to start up, the initial micro behaviors must be produced "by chance" or through an external decision.

I understand. Nevertheless, I still do not see where the collectivities are and where the individuals are. The previous model indicates only the behaviors; however, don't these derive from the collectivity and from the elements it is made up of? Why doesn't the model represent the individuals, or elements, that produce the behaviors?

You are perfectly right. In that model, I used a form of representation (very common to Systems Thinking) that indicates the variables in the system and the elements that form the physical structure of the system which those variables determine. In effect, as is the case with every other system, even combinatory systems have a behavior that depends on the structure of the elements (processors) that make up the collectivity. However, to take these into account we can rapidly modify the previous model, transforming it as shown in the new model.

So, then, if I've understood correctly, the micro and macro behaviors are caused by the individual elements of the collectivity and the collectivity itself as a whole, respectively. Thus, it is the individuals and the collectivity that interact with each other at different times. Is that right?

Exactly! We cannot, nor must we, forget that the behaviors do not arise from nothing but are the products of some agent or basic unit. The *macro behavior* is produced by the collectivity as a whole and derives from the "combination" of the *micro behaviors* of its elements. The latter are produced by the individual elements but are controlled (influenced, aimed at, conditioned, etc.) by the macro behavior of the collectivity.

Even if it is intuitively clear what "behavior" is, can you clarify what exactly is meant by that term?

It is not easy to give an answer to this question, but with the help of a minimum of symbolism, the definition becomes simple and precise. Intuitively, by the behavior of a given entity, we mean the modification of some of its characteristics during a given period. The behavior of a bee that flies, a tractor-trailer going down a stretch of highway, an individual taking a walk, a flower blossoming, a computer carrying out calculations, an economic system that is developing, a firm that improves its productivity, are observed only when we can detect some change in position, dimension, form, and so on. To construct a more neutral and general definition of behavior, we need to introduce the idea of "state," which refers to a moment in time (appropriately defined). Let us consider a given agent, whose state we indicate as the value at a certain "time" of a certain number of variables, which we associate with the element in question and define as "state variables." The behavior of an element can thus be defined as the change over time in the state variables of that element; that is, the trends in its states. Behavior here is the equivalent of state dynamics. Obviously, the state dynamics can also be continuous over time.

Could you provide some examples?

Let us assume that E is a point P in space and that its state is the distance at a given moment with respect, for example, to the origin. The behavior of P is detected by observing the shift of the point in space over successive moments; in short, its trajectory. Let us also assume that E is Piero and that his behavior is defined by a single variable with the following values: "he works, eats and rests." Piero's behavior is nothing other than the succession over time (e.g., a week) of his work, rest, and eating states.

Do the previous examples refer only to the micro behaviors or the macro behaviors as well? And if they do, in what sense?

The definition of behavior above (the succession of states over time) applies to both individual elements (micro behaviors) as well as the entire collectivity (macro behavior). Let us assume that E is a towered city and that its state is the "number of towers that exist in a given year." In abstract terms, we can identify the behavior of the towered city as a "succession of the number of towers over a fifty-year period." Nevertheless, there is a fundamental difference between *micro* and *macro* behavior. In fact, the latter must be defined as a function of the micro behavior of the elements that make up the collectivity, which is composed of the families that can build their own tower.

I don't understand what is the significance here of a city with many towers. I have always thought that Combinatory System Theory considered only collectivities of men or animals.

We must broaden our interpretative horizons. In CST, the term "collectivity" must be understood in a broad sense, as indicating the presence, "together," of several elements, both biological individuals—herds, schools of fish, crowds of people, Macedonian phalanxes, cancan dancers—and non-biological ones—electromagnetic waves, asteroids, moving cars, houses (families) in cities, drilling companies, and so on. The collectivities do not necessarily have to be made up of men or animals. Human collectivities represent only one type (perhaps the most interesting and important one) of combinatory systems, which is defined as a *social* combinatory system. Note that, however one interprets the concept of collectivity, it is not always easy to identify the *macro* behavior; at times the collective behavior is clearly observed, while at other times, it derives from our interpretation of the facts.

In what sense is the collective behavior not always directly observable?

Suppose we are observing the expansion of an anthill in a forest made up of thousands of ants, as described by Hofstadter, in Gödel, Esher, Bach. By observing the day-to-day changes in the anthill in the forest, we can interpret its behavior by saying that at times it appears calm, at times agitated, at other times bellicose, as if it were a single being with its own personality. Since, in observing the anthill (collectivity), we are abstracting from the behavior of the single ants (individuals), the collective behavior is attributable to an entity distinct from its constituent elements. Let us consider instead a mountain path formed by the repeated footsteps of a multitude of passers-by who, by treading on the grass, leave behind an indelible trace. In this case, the collective behavior cannot easily be attributed to an entity, even if we can intuitively construct the abstract image of a long, single-file row of persons stepping on the grass in succession.

Your examples give me pause for reflection. Is it true, then, that the populations viewed as combinatory systems do not only include those made up of conscious individuals that produce their micro behaviors guided by personal interest? Your focus on individual motivations recalls the well-known book by Thomas Schelling entitled Micromotives and Macrobehavior. Schelling also considers the relationships between individual behavior and collective structures that are formed by the behaviors. Is there a connection here with the Theory of Combinatory Systems?

Schelling's work represents a cornerstone for the abundance and provocative nature of the examples it provides of interactions between individual micro behaviors and macro behavior. However, I must first point out that Schelling's analysis, while containing some similarities to CST due to the subject it deals with (as the title of his work implies), offers a different interpretation of such interaction. Above all, his analysis refers to human systems in which the individuals possess the freedom of choice, based on the pursuit of individual interest; however, they make their decisions ignoring the macro consequences of the combination of their behaviors (limited and incomplete information). Second, the individual micro behaviors give rise to various types of structures; that is, to macro effects (Schelling mainly considers those I define as systems of order) that condition the micro behaviors, which in turn serve to maintain or accentuate those structures. The *feedback* is clear, but this mechanism is not seen as the prime engine of the system's dynamics and interpretation. In fact, the attempt to explain such micro and macro interactions focuses on the logical approach provided by game theory as applied to situations of binary choice, in which individual decisions depend solely on the number of subjects that choose a specific alternative. More so than by *feedback*, the forming of social structures is justified by the achievement of various forms of equilibrium, which the collective system tends to reach and maintain even when the individual agents are unaware of it. Game Theory represents a fascinating and effective explanation that is not at odds with CST but, in fact, represents one of the cases foreseen by the latter theory. Combinatory System Theory is more general, in that it does not consider only social systems; in searching for models of the various types of social systems, one has recourse not only to Game Theory but to the theory of combinatory systems as well.

The previous considerations imply that the Theory of Combinatory Systems attributes significant importance to micro-macro circularity or feedback. Is this true?

Yes, fundamental importance is given to this. In fact, "*micro-macro feedback*" is what distinguishes combinatory systems from other types of systems. The micro behaviors produce the macro behavior but are also controlled, conditioned, guided by it. Without the *micro-macro feedback*, every collectivity would merely be a set of elements, or even a different system, not a combinatory one. There is no doubt that the mountain path is the result of the individuals that walk along it, but it is equally clear that those who choose that stretch of slope are conditioned by its existence, which they maintain by choosing to take it (here we must keep in mind that we are dealing with a repetitive system). Recognizing *micro-macro feedback* is indispensable for interpreting collective phenomena as deriving from a combinatory system.

It almost seems as if the elements of the systems are guided by an "invisible hand."

You have touched on a fundamental aspect: the *micro-macro feedback* generates the "invisible hand," which appears to guide the individual behaviors in producing the collective phenomena. Considering the traditional definition of *self-organization* as the emerging phenomenon represented by the formation of ordered structures, or recognizable patterns in the behavior of the agents of a collectivity, as if they were directed, or organized, by an Invisible Hand, Supreme Authority, or Internal Organizer, it is easy to recognize the "synergetic effect" of the *micro-macro feedback* action (or circular causality) between the agents' micro behaviors and the global system's macro behavior or effect. We must not forget that the feedback acts in two directions: on the one hand, the "invisible hand" conditions the micro behaviors, thereby producing the macro behavior, and, on the other, it is precisely the macro behavior that guides the subsequent micro behaviors, thus maintaining the *micro-macro feedback*, that is, the action of the "invisible hand."

How is the micro-macro feedback started and how is it maintained over time? Are there factors which feed it?

This is a strong feature of Combinatory Systems Theory: it is not sufficient to "recognize" the action of the *micro-macro feedback* to interpret the modus operandi of combinatory systems. One must also "identify" two important (sets of) factors: the *necessitating* and *recombining* factors, which represent the "motors," so to speak, of the *micro-macro feedback* action. *Necessitating factors* are all the factors that *force* the agents to *adapt* their micro behavior to the system's macro *behavior* or effect. They may be constraints, rules, laws, convictions, imitative acts, biological or social impulses, *conscious* motivations (necessity, convenience, opportunity, etc.). The stronger the necessitating factors are, the more powerful the micro behavior will be. *Necessitating* factors are necessary but not sufficient. It is also necessary for the system to be able to recombine the *micro* behaviors (and their effects) to produce the dynamics of the *macro* behavior (and the macro effect). The *recombining factors* are the "elements" that *allow* or force the system to *notice* and *recombine* the *micro* behavior (or the *micro* effects) to produce and maintain the *macro* behavior (or the *macro* effect): rules, conventions, algorithms, and so on, which simply follow on from the conditions of the environment or result from the social condition or the culture of the collectivity constituting the system. If the *necessitating* and *recombining* factors are maintained, the *micro-macro feedback* can operate.

There is an excerpt I am not clear about: you state that the micro behavior is determined by the macro behavior, but, at the same time, that the micro behavior shapes the macro behavior. But which comes first? The collectivity must exist to condition the individuals, but the latter must already have produced their micro behaviors in order to form the collectivity. This seems to represent the age-old dilemma: what came first, the chicken or the egg.

If by "egg" we mean micro behaviors and by "chicken" macro behavior, then there is only one possible answer: the egg came first. When the combinatory system initiates, the initial behavior can only be the consequence of the original micro behaviors. Once under way, the macro behavior inevitably conditions the subsequent micro behaviors. I emphasize "inevitably"; otherwise, there would not be any combinatory system.

Your answer raises a question: How large must the collectivity be for us to be able to speak of a combinatory system?

Another feature of combinatory systems theory is that it also considers collectivities composed of a limited number of elements, even two in the limiting case, as occurs in feuds or wars between two factions or nations. It is not necessary for these elements to be so numerous that they cannot be individually observed. However, it is understood that combinatory systems can also be formed by countless numbers of individual components.

A doubt arises: Does a combinatory system correspond to the collectivity or to the group of elements? Or to the macro behavior? Or only to that part of the collectivity that triggers the micro behaviors?

The problem of terminology is, of course, always a delicate one, since it is indispensable in precisely clarifying the concepts used. Combinatory systems concern the collectivity, which is an observable, or even only imaginable, entity whose aim is to produce *macro behavior* or *macro effects*. The macro behavior or effects are also observable, or imaginable, at least in principle. On the other hand, the combinatory system is always "conceptual": it is our way of thinking of collective phenomena caused by the base, of its macro behaviors and macro effects. Above all, let us not forget the *micro-macro feedback*, which is the indispensable condition for the existence of the combinatory system. As a concrete example, we can view the combinatory system as a collectivity "in action" for a certain period; or, generalizing, as a "plurality" of agents (a base) operating in an environment based on self-produced (in the broad sense) information. Every combinatory system has its own unique "base" of agents which often operates in a characteristic environment which must be identified, or specified, in order to observe and describe the system. The "base" can also be open: a fashion, for example, arises in a city with a closed number of inhabitants but then spreads throughout the entire country. A disease breaks out into an epidemic on a continent and subsequently spreads to others.

You've spoken several times about the base operating in an environment. What do you mean by this term? Does the environment in which populations operate impact the behavior of the combinatory system? And, if so, in what way?

Your interest in this concept shows you have understood the importance of the *environment*. In a broad sense, by environment we mean a place—real or imaginary, convenient, limited or unlimited—where the individuals in the base produce their micro behaviors and the system's macro behavior is formed. The environment can condition, even considerably, the micro and macro behaviors. A base, for example, can consist of 50 pairs of dancers, a herd of 500 elephants, 100 members of a choir, or 5000 tractor-trailers on the highway. However, the behavior of these elements differs according to the type and size of the environment. The dancers behave differently depending on whether they are in a narrow or wide, oval, or rectangular dance hall. The elephants behave differently if they are in a broad savannah as opposed to a narrow valley. The tractor-trailers produce different effects if the stretch of highway they are driving along is rectilinear or curved, two-lane or multiple-lane. At times the system has a particular form, in the sense that the elements that make up its base are arranged in an ordered way. For example, combinatory systems whose base is composed of an ant colony, a market, passers-by that write graffiti on walls, or speakers of a certain dialect have no particular order. On the other hand, the Macedonian phalanx and can-can dancers are examples of ordered combinatory systems.

I still don't understand how the initial micro behaviors are generated which give rise to the macro behavior?

This is something we cannot know. Normally, "chance" generates the initial micro behaviors. As Epicurus and Laplace clearly stated, when we have no knowledge of the causes, we introduce chance. Generally, it is sufficient if only several elements in the base start to produce their micro behaviors. How many elements? This depends on the system. At times, only a single element is needed; at other times, the entire collectivity must act. We can define the *minimum activation number* as the number of elements that, at t_0, must contemporaneously produce their micro behavior for a macro behavior to develop, which then activates the *micro-macro feedback*.

So, if I've understood correctly, "chance" plays an essential role in the description of combinatory systems.

Chance intervenes in a great many circumstances to produce micro behaviors from the individuals in a collectivity; however, this does not mean a combinatory system will necessarily be activated. By chance someone inscribes a heart in a tree, but this does not ensure the activation of a combinatory system whereby the tree will be covered in a mass of incisions of all kinds. Combinatory systems start up by chance only when chance intervenes on a *minimum number of elements* that is sufficient to activate the *micro-macro feedback*. This number depends on the nature of the base of the system. If this minimum number is not achieved, then we are not dealing with a "true" combinatory system, but only one whose elements in some way interact.

An example would clarify this important feature

Here is an example from something we have all experienced, which I often provide when presenting the theory of combinatory systems, since it is particularly simple and meaningful. Why does a background buzzing noise occur in a crowded room? Because some of those present begin to speak. Absolute silence is impossible to maintain for long. We cannot know who will be the first to speak, but for the buzzing noise to arise, a certain minimum number of speakers will trigger the *micro-macro feedback*. The buzzing will become louder and, in order to be heard, those in the room will speak in a louder voice, thus raising the level of noise until a deafening din prohibits many from hearing others and being heard by others (the *maximum density* of speakers has been reached). Some who have become deafened by the noise will stop speaking, thus lowering the buzzing volume (the feedback is always at work) until, with calm restored, they once again begin speaking. We all have had dozens of experiences of this type of system, for example, during dinners with friends in a "noisy" restaurant or in classrooms overflowing with students. We often blame others for the buzzing, but we ourselves, by speaking, cause the noise, which then conditions our own way of speaking, perhaps by bringing us to shout "What an uproar! Enough!"

Once started, does the combinatory system continue ad infinitum*?*

A general "metaphysical" law states: "Nothing lasts indefinitely." This law obviously can also be applied to combinatory systems. Once started, the macro behavior will necessarily continue to manifest itself for a certain period, until the elements in the collectivity, by continuing to change their state (micro behavior), modify the state of the collectivity itself (macro behavior). Eventually, the *maximum saturation number* of the system is reached, at which point the macro behavior ceases (or, in general, the state of the collectivity remains unchanged for subsequent periods).

Several times in the last few answers you spoke of necessity. Should we then believe there is a determinism inherent in combinatory systems?

I do not think one can speak of determinism—in the sense of the negation of the individual's freedom to carry out his or her micro behavior—but instead of the *micro-macro feedback* action. I have merely stated that in most cases combinatory systems start up "by chance," but that they then necessarily maintain their macro behavior as long as the *micro-macro feedback* is at work. This represents another feature of combinatory systems: in most cases, they require the joint action of *chance* and *necessity*. Going back to the clarifying example of the crowded room that produces the buzzing noise, we see that there is no determinism in the micro behavior of the speakers, only the need they feel to adjust their voice levels to the background buzz. It is precisely this need that makes it inevitable that the buzzing level will be modified by means of the typical *micro-macro feedback* action. To further understand how the chance–necessity binomial operates, let us consider the combinatory system which explains the development of an epidemic, the flu, for example. This system can be described as follows: a carrier of the flu (an individual) comes into contact with other people (micro behavior), thus spreading the virus (micro effect). In a collectivity (base) of people that interact (macro behavior), the epidemic soon spreads (macro effect). We can easily see that if the first infected individual had not met anyone else, then the combinatory system would not have stared. If "by chance" there are frequent contacts with others, so that the "minimum activation number" is reached, then there is a high probability the system will start up. When the number of infected individuals exceeds the "maximum saturation number," the epidemic will by necessity cease to expand. We must also pay attention to the role of the environment. If the base is made up of 10,000 individuals, then the epidemic effect would be different if those individuals lived in a section of New York City, for example, as opposed to being equally distributed over a 10,000-square kilometer area of Siberia.

There's one thing I don't understand: several times you've talked about the macro effect alongside the macro behavior; likewise, the micro effect alongside the micro behavior. What is the difference?

We know that the collective behavior we attribute to the base (taking the same path or the same stretch of highway, the stadium wave, circular dancing by dancers in a ballroom, sitting on the same bench, masses of graffiti being inscribed, and so on) can be referred to as the system's macro *behavior*. In many cases, this behavior causes clearly observable effects (a path, highway ruts, a wave going around the stadium, a rotating motion, a mass of graffiti, the spread of a fashion, etc.) that can be defined as the system's macro effects. However, how can the buzzing noise form in a room if those present are not speaking to each other at a certain voice level? How can applause break out if the single spectator does not applaud and produce the classic "clap-clap" sound? And how can a fashion spread if the individuals in a collectivity do not show imitative behavior? Here, then, is an important consequence: the *macro* behaviors and correlated *macro* effects derive from the collective action of the base, since each of its elements produces a distinct *micro* behavior that can produce a particular effect: a *micro* effect. Just as the macro behavior derives from the combination of the analogous micro behaviors, the macro effect comes from the combination of the similar micro effects produced by the agents.

A doubt arises: Is the micro behavior determined by the macro behavior or by the macro effect? Does someone who inscribes graffition a park bench do so because he is influenced by the macro behavior or by the mass of graffiti he observes?

In most cases, individual behavior is influenced by the *macro effects* of the collective behavior, which represent *global information*—"self-produced" by the combinatory system—based on which the individual agents decide to change their *present state* by producing its micro behavior, which in turn gives rise to the consequent micro effects. Nevertheless, we cannot give a categorical answer here. In some cases, the distinction between *behavior* and *effect* is clear, in others blurred, and in others still impossible to determine. For example, consider the system that leads to the formation of a path after a certain number of people (micro behavior) have crossed and trampled a grassy field (micro effect), thereby leaving a trace (macro effect) which is then followed by all the other passers-by (macro behavior). The succession of people crossing the field produces the path since, when all the agents that must cross the field observe the path, they think it represents a convenient route, and so decide to take it. Thus, it is simpler and more intuitive to think that the path (the macro effect) conditions the successive micro behaviors (taking the path to cross the field) even if it is the macro behavior, in fact, which derives from the combination of the individual micro behaviors and thus consolidates the path, gradually leading to the disappearance of the grass. The path is the *macro effect*, which represents "self-produced" *global information* guiding the decisions of the successive crossings that reinforce the path. In this example, only the path (macro effect) can be perceived by the agents; the macro behavior (the agents crossing the path in succession) are not observable (except in special cases). The system producing the stadium wave is started when the spectators rise from their seats and then sit down again in rhythmic fashion (micro behaviors), thus creating a type of wave that runs around the entire stadium (macro behavior). We can succinctly state the rule the spectators follow in this way: "if the wave arrives, rise up from your seat, count to three, and then sit down again." The combination of the micro behaviors (rising up-sitting down, rising up-sitting down, etc.) from each agent based on this rule produces a macro behavior that refers to the entire stadium (environment and base), represented by the wave that runs circularly around the stadium several times, sustained by the collective enthusiasm (recombining factor), until a stop signal intervenes. Note that this system does not contain a true macro *effect* distinguishable from the macro *behavior*; the dynamics of the state of the system, represented by a mix of spectators on their feet and spectators sitting down, which form the wave which appears as a continuous motion, directly conditions the micro behaviors. The wave is the "self-produced" *global information* that guides the spectators' behavior.

Your examples appear to show that it is not easy to distinguish between macro behavior and macro effect. For example, is the school of fish a macro behavior or a macro effect? And what about the mass of graffition a wall?

It is not worthwhile focusing too much on this distinction. I am sure that when faced with a phenomenon caused by or attributable to the action of a collectivity, we shall all be able to distinguish between macro and micro *behaviors* and macro and micro *effects*. If we must make a distinction, we can define the macro (micro) effect as that which is observable even when the macro (micro) behavior has ceased. The school of fish breaks up as soon as the fish move off, which makes it appear as a "macro behavior in progress," which we can also interpret as a visible macro effect. A path or an industrial or residential settlement that arises from the macro behaviors of combinatory systems of firms or inhabitants is observable even when the macro behavior has ceased; thus, there are macro effects.

There is another aspect which is not clear. You state that the micro behavior of the individual in a collectivity is determined, conditioned, directed, almost constrained by the macro behavior. But if we consider most of the examples you have presented, we realize that individuals are almost never aware of the macro behavior of the collectivity. How can the micro behaviors be conditioned by the macro behavior if the individuals are not aware of the latter?

This question brings out another extremely delicate aspect for a thorough understanding of the logic of combinatory systems, hitting on an essential problem: why do the individual elements of the system undertake their micro behaviors even though these are destined to lead to undesirable and harmful, at times even catastrophic global effects? Why do most countries continue to speak their own language when English is much more useful? What causes elephant herds in flight to head toward a crevasse, which they will then fall into en masse, with an obvious disastrous outcome? Why do firms continue to increase productivity when this will certainly lead to fewer jobs and social tension? Why does a parent continue a feud even though he knows its effects will fall back on his child? The answer must be sought in the mechanism by which the individual elements of the system carry out their micro behavior. In fact, each agent does not always perceive or realize the system's macro behavior or resulting macro effects, often acting based solely on information about the micro behavior/effect of only "a few" of the other elements that can be directly observed. Combinatory systems with this characteristic operate based on "Incomplete and limited information."

Can combinatory systems be controlled?

Of course! The theory of combinatory systems not only describes the behavior of combinatory systems and the resulting macro effects, it also tries to explain and control them. Let us return to the system leading to the formation of a mass of colored graffiti painted on a wall. If this were left on a surrounding wall in the periphery of a town, then the system would appear to be quite "resistant," in that no weakening interventions would occur. If the graffiti appeared on the façade of a luxury apartment building, then weakening actions would occur, since the custodians would carry out careful surveillance and remove the graffiti that had just been deposited. Even more short-lived would be the graffiti system that developed on a well-guarded monument, for example, a wall of the Sistine Chapel in Rome. In this case, the system would not even arise due to the "No Writing on the Frescoed Walls" notices, the rigorous surveillance, which would quickly lead to the removal of any writing, the deterrent effect of the severe sanctions for those disobeying the prohibition, and, above all, the censure of the other visitors. (Do you notice here a different combinatory system at work?) Have you noticed the state of public monuments visited by students on a school trip due to the lack of control of the students' micro behaviors? With appropriate reinforcing and weakening actions by an external or internal operator, it is possible to control the combinatory system; that is, to ensure the macro behavior conforms to a desired model.

Combinatory System Theory appears to be interesting and even coherent. Does it also have predictive powers?

CST does not in itself contain all the elements to predict collective behavior, even if applied to a collectivity of well-known individuals; chance always determines part of its explanatory power. In fact, most of the simulation models in Chap. 3 make use of the probabilities of a change in state of the agents. Thus, two similar collectivities could produce somewhat different combinatory systems. Even if the laws of probability governing the changes in state of the individuals were known, chaotic macro behavior could occur, in the technical sense of the term. These systems would be part of complex functioning systems, whose dynamics are unpredictable.

Can you tell us how and when this theory arose?

How did I come up with this theory? I only recall that once, during a walk in the mountains, I was observing a valley and asking myself how I could explain the formation of paths in valleys and along mountain slopes. Another time I was at the theater when I had an embarrassing experience. A group of spectators were continuously applauding in an uncoordinated manner. At times, everyone applauded, while at other times the applause stopped. On this occasion, I also wondered what the mechanism was that regulated this phenomenon. A hundred other times I have observed collective phenomena, or the effect produced by collective action, and asked myself why. Then, one summer under the umbrella at an unbelievably crowded beach, I asked myself again why the need for Italian people to go on holiday in August and fully grasped the idea of *micro-macro feedback*. I have worked on and off for the last 30 years on developing the theory presented in this book. I feel I have fully complied with "Thomas Edison's percentages" of scientific research: "*Genius is one percent inspiration, ninety-nine percent perspiration.*" Do you not agree?

The term "combinatory systems" calls to mind combinatory networks. Should these be considered synonymous?

No, they should not. I have used the term "combinatory system" to refer to these systems because it emphasizes the fact their modus operandi is based on the "combination" of analogous behaviors produced by similar elements, not organized according to hierarchical relations or interconnected in network or tree relations (see Definition 1, Sect 1.1.2). This has nothing to do with the combinatory networks, which, based on the normal and traditional definition, are digital electronic circuits that have a commutation function, being capable of automatically calculating a binary function of one or more Boolean variables.

How are combinatory systems different from other systems? As you have written on several occasions, for example, in your book The Magic Ring (Springer 2021; 1st Ed. 2014), in any system (even non-combinatory ones) the functioning of the structure depends on that of its parts; and that of the parts on that of the structure, as part of a whole-parts feedback. Systems are at the same time structured and structuring. Do combinatory systems represent an exception?

No, they are not an exception. There is only one type of systems logic: the structural and functional interaction between whole and parts. Nevertheless, combinatory systems have a feature that distinguishes them from so-called "traditional" systems, whose structure is usually made up of "organs"; that is, *differentiated* elements that carry out *specialized* functions and have a precise *localization* within the structure. They are thus organized systems. Organization creates unity, a consequence of which is the feedback between the micro and macro functioning. Combinatory systems, on the other hand, are non-organized systems whose structure is made up of similar elements that are not interconnected or interacting and carry out analogous functions. The feedback between the micro and macro behavior is not only a consequence but also a true cause; it ensures that a collectivity will become an entity, even though one composed of non-organized individuals.

Then isn't the term "combinatory system" possibly misleading?

The explanation above indicates that the term is not misleading. Nevertheless, I could use other terms for these systems: for example, self-determined systems, chance/necessity systems, non-organized systems, self-produced information systems, and so on. In a 2005 paper entitled *Observing Collectivities as Simplex Systems: The Combinatory Systems Approach* (published in "Nonlinear Dynamics, Psychology and Life Sciences"), I also proposed the term Simplex Systems to differentiate these from Complex Systems. Subsequently, I chose to always use "combinatory systems."

Then Combinatory Systems Theory is part of the theory of dynamical systems?

Since I have defined micro and macro behavior as the dynamics of state variables, I would definitely say so! However, dynamical systems theory—or traditional systems theory—differs from Combinatory System Theory in one significant way: its point of view. Traditional systems theory considers dynamical systems in terms of *input, state* and *output* variables, without, however, focusing on the system or analyzing the nature of the elements forming the structure or the nature of the structural dynamics. It is limited to an external point of view. On the other hand, Combinatory Systems Theory views systems from an internal point of view, trying to understand the nature of the agents that form the collectivity and to describe the structural dynamics in terms of the micro behaviors that interact with the macro behavior and of the micro and macro effects that result from those behaviors.

Are there inputs and outputs in combinatory systems?

The answer is simple: since *micro-macro feedback* is operating, the micro behavior/effect *is* the input for the macro behavior/effect. However, the latter becomes the input that produces the micro behavior/effect. For this reason, I have represented combinatory systems as loops connecting the micro and macro behavior/effects.

Couldn't one use the theory of recursive systems?

Certainly! The combinatory system that turned Pavia, in Italy, into "the town of a hundred towers" (Sect. 2.3.3) could be described using the theory of recursive systems. At each stage, the input is represented by the state of the towered town; that is, by the number of towers erected (macro effect) and the output produced by the new state of the towered city, which depends on the previous state. However, the operating mechanism causing the change in state is represented by the *micro-macro feedback*, which operates on the micro behaviors. Nevertheless, we can take as an input the *probability* that a noble will erect a tower and as an output the new *probability* after the system has acted. The operating mechanism causing the changes in state is represented now by the *micro-macro feedback* operating on the probabilities, which are modified by the macro behavior/effect. I believe Combinatory System Theory once again reveals much more of an explanatory power, since it considers the micro and macro behavior together. Note that there are other systems theories, or other systems models, that can interpret the collective phenomena examined by our theory. However, it is the existence of a plurality of systems models that interpret collective phenomena—models which, mind you, are not mutually exclusive but integrated with each other—that can justify Combinatory System Theory also as a further instrument for gaining knowledge of the fascinating universe of systems and collectivities.

You have referred to Simplex Systems as combinatory systems. However, it seems to me that Combinatory System Theory focuses on the same object of observation as Complex Systems Theory. Is there a difference between the two theories?

Yes, there are two basic differences. The first is the nature of the elements that make up the observed collectivity. CST observes collectivities of similar agents that carry out analogous micro behaviors. In Complex Systems Theory, in particular Complex Evolving Systems Theory, the observed agents are heterogeneous and carry out different behaviors oriented toward survival. They make decisions based on forecasts and expectations and interact and exchange information with their environment to maintain their internal processes over time and develop some form of "cognition." The second difference is that Complex Systems Theory does not focus its explanation on *micro-macro feedback* action or the system's *self-produced global information*. Rather, it studies the dynamics of the agents by using the operational logic of *cellular automata*. The agents produce their micro behavior solely by observing the behavior/effect of the agents that are acting around them and that form their neighborhood. Nigel Gilbert stated: "*If you want to understand why a person acts as she does, it is certainly possible to look around in the immediate environment for an explanation. But often an explanation needs to look also, or perhaps primarily, at events that occurred in the past and at how the present situation developed from previous circumstances.*" Neighborhood, environment, and history dominate the explanation of the dynamics of collectivities in Complex Systems Theory. Nevertheless, CST also considers situations in which the agents operate under conditions of information that is *incomplete* and *limited* to the information deduced from a circumscribed number of other agents. However, this is not a contradiction or a limiting factor: however, agents decide to change their state, this will influence the general state of the combinatory system, so that they will observe neighborhoods whose states are always diverse (and which are part of the macro state of the system) and update their micro state to maintain the *micro-macro feedback.*

Could you further clarify those systems that are self-organized?

Regarding the system of ordered dancing, I refer you to the description of a rotating system, the "Waltz-Spin" System in Sect. 2.5.1. Another typical example of self-organization—one that is usually cited though never interpreted (as far as I know)—involves transforming a "disordered" two-directional flow into two "ordered" mono-directional flows. This system is invariably observed, for example, when many individuals (base) must enter or exit from a door at the same time and in opposite directions (necessitating factor), or walk in the opposite direction on a sidewalk which presents an obstacle to the flow of pedestrians (environment). If, after several disorderly flows of pedestrians, by chance an orderly flow begins to arise—for example, if a sufficient number of people that must leave a room do so by filing out the door from the right or the pedestrians walking along a sidewalk in one direction align themselves so as to get around the obstacle from the right—then by necessity even those people that must enter through a door or walk along a sidewalk will tend to arrange themselves into an ordered flow on the right. Such a shift makes it inevitable that the passers-by that follow will choose the appropriate traffic flow, thereby avoiding moving counter to the main flow so as not to have a physical or verbal confrontation with those that have preceded them (recombining factor). The disorderly flow has been transformed into two orderly and persistent flows (macro behavior) and the initial disorder transformed into order, thanks to the self-organization produced by the *micro-macro feedback*, which makes it necessary to adhere to the orderly arrangement when, by chance, this has occurred.

But how do ants manage to walk in orderly rows, always choosing the shortest route to their destination or back to their nest?

You are perhaps referring to stigmergy (or stigmergie), which is the self-organization noted by Pierre-Paul Grassé when he observed "social insects" and their ordered micro behaviors? The coordination observed in social insects, termites, and ants, in particular, is due to their own micro *behaviors*, which release pheromones as micro *effects* that, by recombining (added up), form a pheromone trail representing the macro effect of the ant colony traveling in that direction. The ants that come later by necessity follow the pheromone trail, strengthening it by releasing additional chemical traces. The feedback is clear here. The explanation provided by stigmergy is not an alternative to that offered by CST. In fact, it represents a special case of the latter.

If I have understood correctly, you are saying that combinatory systems must be viewed mainly as closed systems. But don't they need energy supplies? And if so, don't they then become open systems, at least from the point of view of energy inputs?

It is easy to realize that for many combinatory systems, energy supplies are necessary to activate the micro behaviors and observe the macro behavior. Since the main objective of the Theory of Combinatory Systems is to emphasize the operating logic typical of such systems—by studying the micro-macro feedback and the necessitating and recombining factors—then, to simplify the description of the theory, energy inputs are not usually considered; this is also because the representation of the energy inputs is usually superfluous, if not impossible to achieve in many cases. If we consider the systems characterized by "graffiti on a park bench," "dancers dancing the can-can at the Moulin Rouge," "a herd of elephants in flight," or "soldiers parading," or most of the other combinatory systems where the macro behavior prevails over the macro effect, then it is obvious that taking energy inputs into account is not necessary in order to also consider the operating logic of such systems.

Are there artificial combinatory systems?

As we can see from the examples in the book, combinatory systems appear to be activated spontaneously, which means that the micro and macro rules are already naturally a part of the system's elements. All systems that form thanks to the action of "ad hoc," structured rules to permit the desired macro behavior to be achieved are *artificial* systems. These conditions clearly obtain in the case of dance or parade groups. However, it is legitimate to assume that artificial combinatory systems can be designed. It is necessary to define the base of the system, the micro and macro rules that regulate its dynamics (assuming some type of necessitating and recombining factors) and specify the *micro-macro feedback* action. The design of artificial combinatory systems could lead to the actual start of the planned combinatory system. In this case, the design of the system could substitute for chance in starting the system. Many spontaneous systems can be defined as natural, even in the conventional and relative sense of the term. For example, all systems that form during an epidemic can be referred to as "natural," as well as those that arise from the micro rules inserted into the genetic heritage of living beings comprising the system, as occurs, for example, in social systems based on imitation, rationality, and the sense of trust inherent in the agents in the system.

The reference to systems of accumulation and their efficacy in explaining the formation of urban and industrial settlements and to systems of the diffusion of characteristics or objects reminds me that, to explain such phenomena several years ago, reference was often made to path dependence theory. Do you hold that the Theory of Combinatory Systems has some similarity to this latter theory?

Certainly. The Theory of Combinatory Systems also includes path dependence theory; or, at least, it is capable of better interpreting the collective effects defined by path dependence. But you must understand that there is not only one version of this theory, as I have stated in the text. In any event, all the "serious" versions can be viewed as special cases of combinatory systems. In other words, the Theory of Combinatory Systems is more complete and suitable for interpreting path dependence phenomena. For Combinatory System Theory, the phenomenon of path dependence is not a theory but simply the observation that the dynamics of a collectivity—its macro behavior or macro effect—can be thought to depend on initial chance (dependence from initial conditions) and on the recombining rules of the micro behaviors of the agents. Thus, the individual choices of the agents lead to micro behaviors deriving from history, that is, from the macro behavior (history dependence). In this sense, path dependence is proof of the action of the *micro-macro feedback,* even if path dependence theory does not include this mechanism in the explanation of path dependence.

You have stated that CST allows us not only to understand and describe combinatory systems but also to carry out simulations. Does it adopt any special instrument for this purpose?

All of Chap. 3 deals with simulations. The simulation models are complex in nature since, without the possibility of doing "field" research, I am forced to use probabilities to represent the necessitating factors, making use of the drawing of random numbers to enable the program to decide whether the agents will change their state at every moment of the simulation. The basic instrument used is what I have called the Combinatory Automaton, which is a grid of cells, each of which represents the micro state of an agent in the combinatory system being simulated. At each moment, by using appropriately defined functions, we calculate the "combination" of the states of all the cells, which represents the macro state of the entire combinatory system. Sometimes the combination is a sum, other times some sort of average or the maximum or minimum value of all those in the cells. The micro state of the cells changes depending on the macro state of the system, based on transition of state functions that must be defined, keeping in mind that the transition occurs as a function of the probabilities and of the random numbers drawn. The simulation process is simple in principle; the difficulties arise in having to manage a large amount of data that must be modified at every moment, with respect to a defined time scale, for a certain number of moments within the simulation period.

Note: This interview only considers a brief overview of Combinatory Systems Theory. Only a thorough reading of the entire book, including the examples and simulations, can provide a complete view of the theory.

References

Ahmed, E., Elgazzar, A. S., & Hegazi, A. S. (2005). *An overview of complex adaptive systems*. arXiv preprint nlin/0506059.

Albin, P. S. (1998). In D. K. Foley (Ed.), *Barriers and bounds to rationality: Essays on economic complexity and dynamics in interactive systems*. Princeton University Press.

Aliber, R. Z., Kindleberger, C. P., & McCauley, R. N. (2023). The anatomy of a typical crisis. In *Manias, panics, and crashes*. Palgrave Macmillan.

Andreewsky, E., & Delorme, R. (2006). *Second cybernetics and complexity: Meetings with Heinz Von Foerster (original: Seconde cybernétique et complexité: Rencontres avec Heinz Von Foerster)*. Editions L'Harmattan.

Ashford, N. (2001). Principles for a free society. Jarl Hjalmarson Foundation. Retrieved from: http://www.opportunitiesforafricans.com/wp-content/uploads/2013/04/1.Principles-of-a-FreeSociety.pdf.

Chan, D. (1998). Functional relations among constructs in the same content domain at different levels of analysis: A typology of composition models. *Journal of Applied Psychology, 83*(2), 234.

Csányi, V. (1966). In Khalil & Boulding (Eds.), *Organization, function, and creativity in biological and social systems* (pp. 146–180).

Cupples, B. (1977). Three types of explanation. *Philosophy of Science, 44*(3), 387–408.

Dequech, D. (2001). Bounded rationality, institutions, and uncertainty. *Journal of Economic Issues, 35*(4), 911–929.

Dooley, K. (1996). Complex adaptive systems: A nominal definition. *The Chaos Network, 8*(1), 2–3.

Dooley, K. J. (1997). A complex adaptive systems model of organization change. *Nonlinear Dynamics, Psychology, and Life Sciences, 1*(1), 69–97.

Dupuy, J. P. (2005). Panic and the paradoxes of the social order. Passions in economy, politics, and the media: in discussion with Christian theology, 17, 215–233.

Einstein, A. (1926). Quoted by Heisenberg.

Foley, D. K. (1998). In P. S. Albin (Ed.), *Introduction*.

Foster, J., & Metcalfe, J. S. (Eds.). (2001). *Frontiers of evolutionary economics: Competition, self-organization, and innovation policy*. Edward Elgar Publishing.

Franz, W. (1990). Hysteresis in economic relationships: An overview. In *Hysteresis effects in economic models* (pp. 1–17). Physica-Verlag HD. Retrieved from: https://www.econstor.eu/bitstream/10419/101650/1/733236960.pdf

Gaffeo, E. (1999). In M. Gallegati & A. Kirman (Eds.), *Tutorial on social interaction economics* (pp. 45–73).

Gallegati, M., & Kirman, A. (1999). *Beyond the representative agent*. Edward Elgar Publishing.

Gell-Mann, M. (1994). *The Jaguar and the quark: Adventures in the simple and the complex*. Freeman & Co.

Gell-Mann, M. (1995). What is complexity? *Complexity, 1*(5) Retrieved from: http://complexity.martinsewell.com/Gell95.pdf

Gilbert, N. (1995). Simulation: An emergent perspective. In *Conference on new Technologies in the Social Sciences, October, 27–29*. Retrieved from: http://cress.soc.surrey.ac.uk/resources/emergent.html
Hawking, S. W., & Mlodinow, L. (2010). *The grand design*. Random House Publishing Group.
Heider, F. (1958). *The psychology of interpersonal relations*. Wiley.
Heisenberg, W. (1989). In Salam et al. (Eds.), *Theory, criticism and a philosophy* (pp. 31–56).
Hempel, C. G., & Oppenheim, P. (1948). Studies in the logic of explanation. *Philosophy of Science, 15*(2), 135–175.
Hempel, C. G. (1965). *Aspects of scientific explanation and other essays*. The Free Press.
Holland, J. H. (1995). *Hidden order: How adaptation builds complexity*. Perseus Books.
Jantsch, E. (1980). *The self-organizing universe: Scientific and human implications of the emerging paradigm of evolution*. Pergamon Press.
Keller, E. F., & Lloyd, E. A. (Eds.). (1992). *Keywords in evolutionary biology*. Harvard University Press.
Keynes, J. M. (1937). The general theory of employment. *The Quarterly Journal of Economics, 51*(2), 209–223.
Khalil, E. L., & Boulding, K. E. (Eds.). (1996). *Evolution, order, and complexity*. Routledge.
Le Moigne, J. L. (1994a). La théorie du système général: théorie de la modélisation. Collection Les Classiques du Reseau Intelligence de la Complexite, PUF.
Le Moigne, J. L. (1994b). La théorie du système général: théorie de la modélisation. Collection Les Classiques du Reseau Intelligence de la Complexite. Comments at:: http://archive.mcxapc.org.
Lennox, J. G. (1992). In E. F. Keller & E. A. Lloyd (Eds.), *Teleology* (pp. 323–333).
Maturana, H. R., & Varela, F. J. (1980). *Autopoiesis and cognition. The realization of living*. Reidel Publishing.
Maturana, H. R., & Varela, F. J. (1987). *The tree of knowledge: The biological roots of human understanding*. New Science Library/Shambhala Publications.
Mella, P. (2014). The pillars of learning, understanding, studying and explaining. *Creative Education, 5*(17), 1615–1628.
Mella, P. (Ed.). (2021). *The magic ring. Systems thinking approach to control systems* (Contemporary systems thinking series). Springer.
Mulder, P. (2012). 5 Whys analysis. In Toolshero. https://www.toolshero.com/problem-solving/5-whys-analysis/.
Pelikan, P. (2001). In J. Foster & J. S. Metcalfe (Eds.), *Self-organizing and Darwinian selection in economic and biological evolutions: An enquiry into the sources of organizing information* (pp. 121–151).
Plamondon, A. L. (1979). *Whitehead's organic philosophy of science*. State University of New York Press.
Polanyi, M. (1967). *The tacit dimension*. University of Chicago Press.
Prigogine, I., & Meixner, J. (1976). *L'ordre par fluctuations et le système social* (pp. 7–48). VS Verlag für Sozialwissenschaften.
Reichholf, J. H., & Mattson, P. (1994). Zoology: Biodiversity why are there so many different species? *Universitas, 1*, 42.
Rubinstein, A. (1998). *Modeling bounded rationality*. MIT Press.
Salam, A., Bethe, H. A., Dirac, P., Heisenberg, W., Wigner, E. P., Klein, O., & Lifshitz, E. M. (1989). *From a life of physics*. World Scientific.
Salmon, W. C. (1990). *Four decades of scientific explanation*. University of Minnesota Press.
Sandquist, G. M. (1985). *Introduction to system science*. Prentice-Hall, Inc.
Schelling, T. C. (2006). *Micromotives and macrobehavior*. WW Norton & Company.
Senge, P. M., & Lannon-Kim, C. (1991). *The systems thinking approach. The systems thinker newsletter* (Vol. 2). Kendall Square.
Simon, H. A. (1955). A behavioral model of rational choice. *The Quarterly Journal of Economics, 69*(1), 99–118.

Simon, H. A. (1990). Invariants of human behavior. *Annual Review of Psychology, 41*, 1–19.

Thom, R. (1975). Structural stability and morphogenesis: An outline of a general theory of models. Trans. DH Fowler, Massachusetts: Reading.

Weiner, B. (1994). Integrating social and personal theories of achievement striving. *Educational Research, 64*, 557–573.

von Bertalanffy, L. (1968). *General system theory: Foundations, development, applications*. Braziller.

References

Ahmed, E., Elgazzar, A. S., & Hegazi, A. S. (2005). *An overview of complex adaptive systems*. arXiv preprint nlin/0506059.

Albin, P. S. (1998). In D. K. Foley (Ed.), *Barriers and bounds to rationality: Essays on economic complexity and dynamics in interactive systems*. Princeton University Press.

Albu, M. (1997). *Technological learning and innovation in industrial clusters in the South*. SPRU, Electronic Working Papers Series, (7).

Algorithmic Botany. Website of the biological modeling and visualization research group. http://algorithmicbotany.org

Aliber, R. Z., Kindleberger, C. P., & McCauley, R. N. (2023). The anatomy of a typical crisis. In *Manias, panics, and crashes*. Palgrave Macmillan.

Allen, P. M. (2001). Knowledge, ignorance, and the evolution of complex systems. In J. Foster & J. S. Metcalfe (Eds.), *Frontiers of evolutionary economics: Competition and self-organization and innovation policy*. E. Elgar Pub.

Anderson, P. W., Arrow, K. J., & Pines, D. (1988). *The economy as an evolving complex system*. Addison-Wesley.

Andreewsky, E., & Delorme, R. (2006). *Second Cybernetics and complexity: meetings with Heinz Von Foerster (original: Seconde cybernétique et complexité: rencontres avec Heinz Von Foerster)*. Editions L'Harmattan.

Antonelli, C. (1996). Localized knowledge percolation processes and information networks. *Evolutionary Economics, 6*, 281–295.

Antonelli, C. (1997). The economics of path-dependence in industrial organization. *International Journal of Industrial Oganization, 15*, 643–675.

Ardeni, P. G., & Gallegati, M. (1999). In M. Gallegati & A. Kirman (Eds.), *Fluctuations and growth due to technological innovation and diffusion*. Edward Elgar Publishing.

Arthur, W. B., Durlauf, S. N., & Lane, D. A. (Eds.). (1997). *The economy as an evolving complex system II* (Vol. 28). Perseus Books.

Ashby, R. W. (1957). *An introduction to cybernetics* (2nd ed.). Chapman & Hall.

Ashford, N. (1999). Spontaneous order. *The Freeman: Ideas on Liberty, 49*(7). http://www.fee.org/freeman/99/9907/ashford

Ashford, N. (2001). *Principles for a free society*. Jarl Hjalmarson Foundation. http://www.opportunitiesforafricans.com/wp-content/uploads/2013/04/1.Principles-of-a-FreeSociety.pdf

ASQ. American Society for Quality. (1991). *Quality*. https://asq.org/quality-resources/quality-glossary/q

Atlan, H., Dupuy, J. P., & Kiippel, M. (1986). *Individual alienation and systems intelligence*. IFAC Economics and Artificial Intelligence.

P. Mella, *The Combinatory Systems Theory*, Progress in IS,
https://doi.org/10.1007/978-3-031-86946-4

Axelrod, R. (1997). *The complexity of cooperation. Agent-based models of competition and collaboration*. Princeton University Press.

Aziz-Alaoui, M., & Bertelle, C. (Eds.). (2007). *Emergent Properties in Natural and Artificial Dynamical Systems*. Springer Science & Business Media.

Bak, P. (1994). In G. Cowan, D. Pines, & D. E. Meltzer (Eds.), *Self-organized criticality: A holistic view of nature* (pp. 477–496). Perseus Books.

Bak, P. (1996). *How nature works: The science of self-organized criticality*. Copernicus. http://jasss.soc.surrey.ac.uk/4/4/reviews/bak.html

Band, W. A. (1991). *Creating value for customers: Designing and implementing a total corporate strategy*. Wiley.

Baumol, W., Blackman, S., & Wolff, W. (1989). *Productivity and American leadership: The long view*. MIT Press. Wiley. (1972 first edition).

Beccattini, G., Bellandi, M., & De Propist, L. (2009). *A Handbook of Industrial district*. UK Edward Elgar, Pub.

Beer, S. (1987). In H. R. Maturana & F. J. Varela (Eds.), *Preface*. New Science Library.

Bellandi, M. (1996). On entrepreneurship, region and the constitution of scale and scope economies. *European Planning Studies, 4*(4), 421–438.

Belussi, F., & Pilotti, L. (2002). Knowledge creation, learning and innovation in Italian industrial districts. *Human Geography, B, 84*(2), 125–139.

Ben-Eli, M. U. (1979). Amplifying regulation and variety increase in evolving systems. *Cybernetics and System, 9*(3), 285–296.

Bergman, E. M., & Feser, E. J. (1999). *Industry clusters: A methodology and framework for economic cooperation and development*. OECD, Boosting Innovation. The Cluster Approach.

Berto, F., & Tagliabue, J. (2002, revised in 2023). Cellular automata. *Stanford Encyclopedia of Philosophy*. http://plato.stanford.edu/entries/cellular-automata/

Bikhchandani, S., Hirshleifer, D., & Welch, I. (1992). A theory of fads, fashion, custom, and cultural change as informational cascades. *Journal of Political Economy, 100*(5), 992–1026.

Blank, R. M., & Shapiro, M. D. (2001). *Labour and the sustainability of output and productivity growth*. NBER University Of Michigan.

Bowersox, D. J., Closs, D. J., & Cooper, M. B. (2002). *Supply chain logistics management* (Vol. 2). McGraw-Hill.

Breventano, S. (1570). *History of antiquity, nobility, et notable things of the city of Pavia [original: Istoria della antichità, nobiltà, et delle cose notabili della città di Pauia]*. Hieronimo Bartholi Ed. S. Pietro in Ciel'Aureo. Google Books.

Brockhurst, M. A., Morgan, A. D., Rainey, P. B., & Buckling, A. (2003). Population Mixing Accelerates Coevolution. *Ecology Letters, 6*(11), 975–979.

Brusco, S. (1982). The Emilian model: productive decentralisation and social integration. *Cambridge Journal of Economics, 6*, 167–184.

Brusco, S. (1986). Small firms and industrial districts: the experience of Italy. In Keeble & Wever (Eds.), *New Firms and Regional Developments*. Croom Helm.

Buchanan, K. L. (2000). Stress and the evolution of condition-dependent signals. *Trends in Ecology and Evolution, 15*(4), 156–160.

Buckling, A., & Rainey, P. B. (2002). The role of parasites in sympatric and allopatric host diversification. *Nature, 420*(6915), 496–499.

Burgiel, S. W. (2020). The incident command system: A framework for rapid response to biological invasion. *Biological Invasions, 22*(1), 155–165.

Burkhart, R. (1994). The swarm multi-agent simulation system. In: OOPSLA, *94 Workshop on "The Object Engine"*.

Caldas, J. C., & Coelho, H. (1999). The Origin of Institutions: socio-economic processes, choice, norms and conventions. *Journal of Artificial Societies and Social Simulation, 2*(2), 1.

Camazine, S., et al. (2001). *Self-organization in biological systems*. Princeton University Press.

Campbell, D. T. (1960). Blind variation and selective retentions in creative thought as in other knowledge processes. *Psychological Review, 67*(6), 380.

Campbell, C., & Rozsnyai, C. (2002). Quality assurance and the development of course programmes. Papers on Higher Education.

Capra, F. (1996). *The web of life: A new scientific understanding of living systems*. Anchor Books.

Carlson, S. (2000). Boids of a feather flock together. *Scientific American, 283*(5), 112–114.

CASG, Complex Adaptive Systems Group. (2013). http://www.cas-group.net

Castelfranchi, C. (1998). Through the minds of the agents. *Journal of Artificial Societies and Social Simulation, 1*(1) http://jasss.soc.surrey.ac.uk/1/1/5.html

Casti, J. L. (1985). *Nonlinear system theory*. Academic Press.

Castigli, M. (2022). *Robots are spreading in industry: Here's in which sectors and what they do [original: I robot si diffondono nell'industria: ecco in quali settori e cosa fanno]*. https://www.agendadigitale.eu/industry-4-0/i-robot-si-diffondono-nellindustria-ecco-in-quali-settori-e-cosa-fanno/

Cessna, A. (2015 online). *Protoplanets*. http://www.universetoday.com/37053/protoplanets/

Chan, D. (1998). Functional relations among constructs in the same content domain at different levels of analysis: A typology of composition models. *Journal of Applied Psychology, 83*(2), 234.

Chandler, A. D., Jr. (1990). *Scale and scope: The dynamics of industrial capitalism*. Harvard, University Press.

Chandler, A. D., Hagström, P., & Sölvell, Ö. (Eds.). (1999). *The dynamic firm. The role of technology, strategy, organization, and regions*. Oxford University Press.

Chavalarias, D. (2016). The unlikely encounter between von Foerster and Snowden: When second-order cybernetics sheds light on societal impacts of Big Data. *Big Data & Society, 1*(11) http://bds.sagepub.com/content/3/1/2053951715621086

Cipolla, C. M. (1962). *The economic history of World population*. Penguin Books.

Cook, S. (1995). *Practical Benchmarking: A manager's guide to creating a competitive advantage*. Kogan Page Pub.

Corning, P. A. (1995). Synergy and self-organization in the evolution of complex systems. *Systems Research and Behavioral Science, 12*(2), 89–121.

Coveney, P., & Highfield, R. (1995). *Frontiers of complexity: The search for order in a complex world*. Faber and Faber.

Cowan, G., Pines, D., & Elliott Meltzer, D. (Eds.). (1994). *Complexity: Metaphors, models, and reality* (pp. 477–496). Perseus Books.

Cox, E. (1994). *The fuzzy systems handbook: A practitioner's guide to building, using, and maintaining fuzzy systems* (p. 1994). Academic Press.

Cox, B. (2004). *Swarm. Virtual school*. http://virtualschool.edu/mon/Bionomics/Swarm.html

Coyle, D. (1999). *The weightless world: Strategies for managing the digital economy*. Capstone London and MIT Press.

Creutz, M. (1996). *Self organized criticality*. https://arxiv.org/pdf/hep-lat/9611018.pdf

Crosby, P. B. (1979). *Quality is free: The art of making quality certain*. McGraw-Hill.

Csányi, V. (1966). Organization, function, and creativity in biological and social systems. In E. Khalil & K. Boulding (Eds.), *Evolution, order and complexity* (Vol. 1996, pp. 146–180). Routledge.

Cupples, B. (1977). Three types of explanation. *Philosophy of Science, 44*(3), 387–408.

Darwin, C. (1859). *On the origins of species by means of natural selection*. Murray. http://darwin-online.org.uk/content/frameset?itemID=F373&viewtype=text&pageseq=1

Dawkins, R. (1989). *The selfish gene* (1st ed.). Oxford University Press. 1976, http://pratclif.com/biologie-moleculaire/dawkins.htm

Dawkins, R. (2004). *The ancestor's tale: A pilgrimage to the dawn of evolution*. Houghton Mifflin.

De Vries, H. (1905). *Species and varieties: Their origin by mutation: Lectures delivered at the University of California by Hugo de Vries*. Open Court Publishing Co, Edward Elgar.

Delorme, R. (2001). *Theorizing complexity*, in Foster, J. and Metcalfe, J. S. (2001b). Edward Elgar.

Demartini, C., & Mella, P. (2011). Time competition. The new strategic frontier. *iBusiness*, 136–145.

Deneubourg, J. L., & Goss, S. (1989). Collective patterns and decision-making. *Ethology Ecology & Evolution, 1*(4), 295–311.

Dequech, D. (2001). Bounded rationality, institutions, and uncertainty. *Journal of Economic Issues, 35*(4), 911–929.

Dettre, M. (2022). *Just 24 wild rabbits started Australia's most devastating plague*. https://www.queenslandcountrylife.com.au/story/7873584/how-the-rabbit-plague-took-over-australia/

Devaney, R. (1989). *An introduction to chaotic dynamical systems* (2nd ed.). Addison-Wesley.

Dewdney, A. K. (1990). *The magic machine: a handbook of computer sorcery*. WH Freeman &.

Dolan, A. (1998). *Floys. New members in the Artificial Life zoo*. http://www.aridolan.com/eFloys.html [the site is no longer online].

Dooley, K. (1996). Complex adaptive systems: A nominal definition. *The Chaos Network, 8*(1), 2–3.

Dooley, K. J. (1997). A complex adaptive systems model of organization change. *Nonlinear Dynamics, Psychology, and Life Sciences, 1*(1), 69–97.

Dopfer, K., & Potts, J. (2007). *The general theory of economic evolution*. Routledge.

Dorigo, M., Di Caro, G., & Gambardella, L. M. (1999). Ant algorithms for discrete optimization. *Artificial Life, 5*(2), 137–172.

Dorigo, M., Bonabeau, E., & Theraulaz, G. (2000). Ant algorithms and stigmergy. *Future Generation Computer Systems, 16*(8), 851–871.

Drogoul, A., & Ferber, J. (2018). Multi-agent simulation as a tool for studying emergent processes in societies. In *Simulating societies* (pp. 127–142). Routledge.

Drogoul, A., Ferber, J., & Cambier, C. (1994). Multi-agent simulation as a tool for analysing emergent processes in societies. In *Proc. Simulating Societies Symposium: the Computer Simulation of Social Phenomena* (pp. 127–142). Routledge.

Drucker, P. F. (1977). *People and performance: The best of Peter Drucker on management*. Routledge.

Dupuy, J. P. (2005). Panic and the paradoxes of the social order. *Passions in Economy, Politics, and the Media: In Discussion with Christian Theology, 17*, 215–233.

Dupuy, J. P. (2006). In E. Andreewsky & R. Delorme (Eds.), *Que reste-t-il de la Cybernetique a` l'ère des sciences cognitives*.

Dyer, J. H. (1997). Effective interfirm collaboration: How firms minimize transaction costs and maximise transaction value. *Strategic Management Journal, 18*(7), 535–556.

Edwards, M. (2003). *A brief history of Holons*. http://www.integralworld.net/edwards13.html

Einstein, A. (1926). Quoted by Heisenberg, (1989).

Eldredge, N., & Gould, S. J. (1972). Punctuated equilibria: an alternative to phyletic gradualism. *Models in Paleobiology, 82*, 115. https://www.researchgate.net/publication/221958736_Punctuated_Equilibria_An_Alternative_to_Phyletic_Gradualism

EMK, Complexity Group. (2014). *Complexity Lexicon*. http://emk-complexity.org/guide/lexicon.html

Epstein, M. J., & Axtell, R. (1996). *Growing artificial societies, social science from the bottom up*. Brookings Institution Press.

Fedoriaka, D. (2010). *L-systems list*. https://fedimser.github.io/l-systems.html

Feigenbaum, A. V. (1951). *Quality control: Principles, practice, and administration: An industrial management tool for improving product quality and design and for reducing operating costs and losses*. McGraw-Hill.

Fennell, J. (2016). *Planetesimal: Definition, theory & hypothesis*. http://study.com/academy/lesson/planetesimal-definition-theory-hypothesis.html

Flake, G. W. (1998). *The computational beauty of nature: Computer explorations of fractals, chaos, complex systems, and adaptation*. The MIT Press.

Flood, R. L. (1993). *Beyond TQM*. Wiley.

Floreano, D., & Keller, L. (2010). Evolution of adaptive behaviour in robots by means of Darwinian selection. *PLoS Biology, 8*(1), e1000292. https://journals.plos.org/plosbiology/article?id=10.1371/journal.pbio.1000292

Fogel, D. B. (1995). *Evolutionary computation: Toward a new philosophy of machine intelligence*. IEEE Press.

Foley, D. K. (1998). Introduction. In P. S. Albin (Ed.).

Forrest, S., & Jones, T. (1994). Modeling complex adaptive systems with Echo. *Complex Systems: Mechanism of Adaptation, 2*, 3–21.

Forrester, J. W. (1961). *Industrial dynamics*. Pegasus Communications.

Forrester, J. W. (1999). *System dynamics: The foundation under systems thinking*. Sloan School of Management, MIT. http://clexchange.org/ftp/documents/systemdynamics/SD2011-01SDFoundationunderST.pdf

Foster, J., & Metcalfe, J. S. (2001a). In J. Foster & J. S. Metcalfe (Eds.), *Modern evolutionary economic perspectives: an overview* (pp. 1–16). Edward Elgar.

Foster, J., & Metcalfe, J. S. (Eds.). (2001b). *Frontiers of evolutionary economics: competition, self-organization, and innovation policy*. Edward Elgar Publishing.

Franz, W. (1990). Hysteresis in economic relationships: An overview. In *Hysteresis effects in economic models* (pp. 1–17). Physica-Verlag HD. https://www.econstor.eu/bitstream/10419/101650/1/733236960.pdf

Gabriel, R. A. (2010). *Philip II of Macedonia: Greater than Alexander*. Potomac Books, Inc.

Gaffeo, E. (1999). In M. Gallegati & A. Kirman (Eds.), *Tutorial on social interaction economics* (pp. 45–73). Edward Elgar Publishing.

Galangashi, R. B., Kohansal, A. R., & Adnani, S. S. (2017). Designing a rotary converter for absorbing sea wave energy. *Research in Marine Sciences, 2*(4), 188–193.

Gallegati, M., & Kirman, A. (1999). *Beyond the representative agent*. Edward Elgar Publishing.

Gardner, M. (1970). Mathematical games: The fantastic combinations of John Conway's new solitaire game "life". *Scientific American, 223*(4), 120–123.

Garvin, D. A., & March, A. (1986). *The views of Deming, Juran, and Crosby*. Harvard Business School. Harvard Business Publishing.

Geertz, C. (1973). *The interpretation of culture*. Basic Books.

Gell-Mann, M. (1992). Complexity and complex adaptive systems. In *Santa Fe Institute Studies in the Sciences of Complexity-Proceedings* (Vol. 11, pp. 177–177). Addison-Wesley Publishing Co.

Gell-Mann, M. (1994). *The Jaguar and the Quark: Adventures in the simple and the complex*. Freeman & Co.

Gell-Mann, M. (1995). What is complexity? *Complexity, 1*(5) http://complexity.martinsewell.com/Gell95.pdf

Gilbert, N. (1995). Simulation: An emergent perspective. In *Conference on new technologies in the social sciences* (pp. 27–29).

Gilbert, N. (1999). Simulation: A new way of doing social science. *The American Behavioral Scientist, 42*(10), 1485.

Gilbert, N., & Doran, J. (1994). *Simulating societies. The computer simulation of social phenomena*. UCL Press.

Gilbert, N., & Troitzsch, K. G. (1999). *Simulation for the social scientist*. Open University Press.

Glass, L. (1975). Classification of biological networks by their qualitative dynamics. *Journal of Theoretical Biology, 54*(1), 85–107.

Gleick, J. (1987). *Chaos theory: Making a new science*. Viking.

Goldberg, D. E. (1989). *Genetic algorithms in search, optimization, and machine learning*. Addison-Wesley.

Goldspink, C. (2000). Modelling social systems as complex: Towards a social simulation meta-model. *Journal of Artificial Societies and Social Simulation, 3*(2), 1–23.

Gordon, I. (1999). Internationalisation and urban competition. *Urban Studies, 36*(5/6), 1001–1016.

Gould, S. (2000). *The theory of options: A new theory of the evolution of human behavior*. Universal Publishers/Upublish.com.

Gould, S. (2001). *Introduction to the theory of options*. http://www.oocities.org/athens/acropolis/1628/A01into.htm

Gould, S. J. (2002). *The structure of evolutionary theory*. Harvard University Press.

Gould, S. J., & Eldredge, N. (1977). Punctuated equilibria: The tempo and mode of evolution reconsidered. *Paleobiology, 3*(2), 115–151.

Grassé, P. P. (1959). La reconstrucion du nid et les coordinations inteindividuelles chez Bellicositermes Natalensis et Cubitermes. La théorie de la stigmergie: essai d'interpretation du comportement des termites constructeurs. *Insect Sociaux, 6*(1), 41–80.
Grassé, P. P. (1960). The automatic regulations of collective behavior of social insect and "stigmergy". *Journal de Psychologie Normale et Pathologique, 57*, 1–10.
Grassé, P. P. (1967). Nouvelles experiences sur le termite de Müller (Macrotermes mülleri) et considerations sur la théorie de la stigmergie. *Insectes Sociaux, 14*(1), 73–101.
Green, D. G., & Green, D. (1993). *L-SYSTEMS. Environmental and information sciences*. Charles Sturt University. https://www-archiv.fdm.uni-hamburg.de/b-online/e28_3/lsys.html
Gregory, R. L. (1987). *The Oxford Companion to the Mind*. Oxford University Press.
Grimmett, G. R. (1999a). Inequalities and entanglements for percolation and random-cluster models. In *Perplexing problems in probability* (pp. 91–105). Birkhäuser.
Grimmett, G. R. (1999b). *Percolation*. Springer-Verlag.
Guizzo, E. (2010). *World robot population*. https://spectrum.ieee.org/automaton/robotics/industrial-robots/041410-world-robot-population
Haak, T. (2020).·*15 Examples of workplace nudging*. https://hrtrendinstitute.com/2020/11/09/15-examples-of-workplace-nudging/
Hakansson, H., & Snehota, I. (1994). *Developing relationships in business networks*. Routledge.
Haken, H. (1977a). *Synergetics. An introduction*. Springer-Verlag.
Haken, H. (1977b). Some aspects of synergetics. In *Synergetics* (pp. 2–17). Springer-Verlag.
Haken, H. (Ed.). (1982). *Evolution of order and chaos: In physics, chemistry, and biology*. Springer-Verlag.
Haken, H. (1983). *Synergetics. Introduction and advanced topics*. Springer-Verlag.
Harvey, L., & Green, D. (1993). Defining quality. *Assessment & Evaluation in Higher Education, 18*(1), 9–34.
Hawking, S. W., & Mlodinow, L. (2010). *The Grand Design*. Random House Publishing Group.
Heider, F. (1958). *The psychology of interpersonal relations*. Wiley.
Heisenberg, W. (1989). In Salam et al. (Eds.), *Theory, criticism and a philosophy* (pp. 31–56). World Scientific Publishing Co Pte Ltd.
Hempel, C. G. (1965). *Aspects of scientific explanation and other essays*. The Free Press.
Hempel, C. G., & Oppenheim, P. (1948). Studies in the logic of explanation. *Philosophy of Science, 15*(2), 135–175.
Henderson, P. (Ed.). (2000). *Systems engineering for business process change: Collected papers from the EPSRC research programme*. Springer Science & Business Media.
Heylighen, F. (1989). Self-organization, emergence and the architecture of complexity. In *Proceedings of the 1st European conference on System Science* (Vol. 18, pp. 23–32). AFCET.
Heylighen, F. (1993). Blind variation and selective retention. *Principia cybernetica web*. http://pespmc1.vub.ac.be/bvsr.html
Heylighen, F. (2007). *Why is open access development so successful? Stigmergic organization and the economics of information*. https://arxiv.org/pdf/cs/0612071.pdf
Heylighen, F. (2009 on line). *Self-organization. Principia Cybernetica Web*. http://pespmc1.vub.ac.be/SELFORG.html
Highfield, R. (2002). *Lionesses don't prefer blonds - in the mane*. https://www.telegraph.co.uk/news/science/science-news/4768860/Lionesses-dont-prefer-blonds-in-the-mane.html
Hofstadter, D. (1979). *Godel, Escher, Bach: an Eternal Golden Braid*. Basic Books, Inc.. http://themindi.blogspot.it/2007/02/chapter-11-prelude-ant-fugue.html
Holland, J. H. (1975). *Adaptation in natural and artificial systems: an introductory analysis with applications to biology, control, and artificial intelligence*. University of Michigan Press.
Holland, J. H. (1992). Complex adaptive systems. *Daedalus, 121*(1), 17–30.
Holland, J. H. (1995). *Hidden order: How adaptation builds complexity*. Perseus Books.
Holland, J. H. (1998). *Emergence: From chaos to order*. Addison-Wesley.
Hölldobler, B., & Wilson, E. O. (1990). *The ants*. Harvard University Press.

Honavar, V. (2006). *Complex adaptive systems group at Iowa State University*. https://web.mit.edu/esd.83/www/notebook/Complex%20Adaptive%20Systems.pdf

House, R., Rousseau, D. M., & Thomas-Hunt, M. (1995). The Meso Paradigm: A framework for the integration of micro and macro organizational behavior. *Research in Organizational Behavior, 17*, 71–114.

Hubble, E. P. (1936). *The realm of the nebulae* (Vol. 25). Yale University Press.

Huxley, J. (1942). *Evolution: The modern synthesis*. George Allen and Unwin.

IEP, Internet Encyclopedia of Philosophy. (2014). *Theories of explanation*. http://www.iep.utm.edu/explanat/

IFR. International Federation of Robotics. (2019 online). *Industrial and service Robots*. https://ifr.org/downloads/press2018/IFR%20World%20Robotics%20Presentation%20-%2018%20Sept%202019.pdf

Illiyas, F. T., Mani, S. K., Pradeepkumar, A. P., & Mohan, K. (2013). Human stampedes during religious festivals: A comparative review of mass gathering emergencies in India. *International Journal of Disaster Risk Reduction, 5*, 10–18.

ISAL, International Society for Artificial Life. (2016). http://alife.org/index.php?page=alife&context=alife

Jaffe, A., Trajtenberg, M., & Henderson, R. (1993). Geographic localisation of knowledge spillovers as evidenced by patent citations. *Quarterly Journal of Economics, 108*, 577–598.

James, W. (1907). Pragmatism's conception of truth. In *Lecture 6 in Pragmatism: A new name for some old ways of thinking* (pp. 76–91). Longman Green and Co. https://brocku.ca/MeadProject/James/James_1907/James_1907_06.html

Jantsch, E. (1980). *The self-organizing universe: Scientific and human implications of the emerging paradigm of evolution*. Pergamon Press.

Johnson, N. L., & Kotz, S. (1977). *Urn models and their application; an approach to modern discrete probability theory*. John Wiley & Sons.

Johnson-Laird, P. N. (1983). *Mental models: Towards a cognitive science of language, inference, and consciousness* (Vol. 6). Harvard University Press.

Kareiva, P., Mullen, A., & Southwood, R. (1990). Population dynamics in spatially complex environments: Theory and data. *Philosophical Transactions of the Royal Society B: Biological Sciences, 330*(1257), 175–190.

Kauffman, S. A. (1993). *The origins of order: Self organization and selection in evolution*. Oxford University Press.

Kefan, X., Song, Y., Liu, S., & Liu, J. (2019). Analysis of crowd stampede risk mechanism: A systems thinking perspective. *Kybernetes, 48*(1), 124–142.

Keller, E. F., & Lloyd, E. A. (Eds.). (1992). *Keywords in evolutionary biology*. Harvard University Press.

Kellert, S. H. (1994). *In the wake of chaos: Unpredictable order in dynamical systems*. University of Chicago Press.

Keynes, J. M. (1937). The general theory of employment. *The Quarterly Journal of Economics, 51*(2), 209–223.

Khalil, E. L., & Boulding, K. E. (Eds.). (1996). *Evolution, order, and complexity*. Routledge.

Klir, G. J. (2013). *Facets of systems science* (Vol. 7). Springer Science & Business Media.

Koestler, A. (1967). *The ghost in the machine*. Arkana.

Koestler, A. (1972). *The roots of coincidence* (2nd ed.). Hutchinson. (Paperback, 1973).

Koestler, A. (1978). *Janus; a summing up*. Random House.

Koza, J. R. (1992). *Genetic programming: on the programming of computers by means of natural selection* (Vol. 1). The MIT Press.

Krizna, M., Tahara, T., Tainaka, K., Ito, H., Morita, S., Ichinose, G., Togashi, T., Nagatani, T., & Yoshimura, J. (2018). Multi-species coexistence in Lotka-Volterra competitive systems with crowding effects. *Scientific Reports, 8*(1), 1–8.

Kuhn, T. S. (1962). *The Structure of Scientific Revolutions*. 2nd edn. 1970. University of Chicago.

Lakatos, I. (1978). *The methodology of scientific research programmes*. Cambridge University Press.

Lamarck, J. B. (1809). *Philosophie zoologique, ou, Exposition des considérations relative à l'histoire naturelle des animaux*. Hachette Livre, BNF (reprint, 2012).

Langton, C. G. (1989). *Artificial life* (pp. 1–48). Addison-Wesley Publishing Company.

Laprie, J. C. (2008). From dependability to resilience. In *38th IEEE/IFIP int. conf. on dependable systems and networks* (pp. G8–G9). https://users.ece.cmu.edu/~koopman/dsn08/fastabs/dsn08fastabs_laprie.pdf

Lazerson, M. H., & Lorenzoni, G. (1999). The firms that feed industrial districts: A return to the Italian source. *Industrial and Corporate Change, 8*(2), 235–266.

Le Moigne, J. L. (1994). *La théorie du système général: théorie de la modélisation*. Collection Les Classiques du Reseau Intelligence de la Complexite. http://archive.mcxapc.org

Lennox, J. G. (1992). In E. F. Keller & E. A. Lloyd (Eds.), *Teleology* (pp. 323–333). Scientific Research Publishing Inc.

Levy, S. (1992). *Artificial life: The quest for a new creation*. Penguin Books.

Lewin, R. (1999). *Complexity: Life at the edge of chaos*. University of Chicago Press.

Lewis, M. P., Simons, G. F., & Fennig, C. D. (Eds.). (2013). *Ethnologue: Languages of the world*. SIL International. http://www.ethnologue.com/ethno_docs/distribution.asp?by=

Liebowitz, S. J., & Margolis, S. E. (1995). Path dependence, lock-in, and history. *Journal of Law, Economics and Organization, 11*(1), 205–226.

Liebowitz, S. J., & Margolis, S. E. (1998 online). *Path dependence*. https://www.utdallas.edu/~liebowit/palgrave/palpd.html

Liekens, A. (2000). From individuals to populations, approaches to the study of biological emergent phenomena. In *Artificial life. Proceedings of the seventh international conference on artificial life*. The MIT Press.

Lindenmayer, A., & Rozenberg, G. (1972). Developmental systems and languages. In *Proceedings of the fourth annual ACM symposium on Theory of computing* (pp. 214–221). ACM.

Loeb, A. L. (1977). *Preface to Fuller & Applewhite, 1979*. Synergetics.

Loeb, A. L. (1987). Synergetics Dictionary ed. by EJ Applewhite. *Leonardo, 20*(1), 98–99.

Lorenzoni, G., & Lipparini, A. (1999). The leveraging of interfirm relationships as a distinctive organizational capability: a longitudinal study. *Strategic Management Journal, 20*(4), 317–338.

Lotka, A. J. (1925). *Elements of physical biology*. Williams & Wilkins.

Lundvall, B. Å. (1992). Introduction. In B. Å. Lundvall (Ed.), *National systems of innovation. towards a theory of innovation and interactive learning*. Pinter.

Lundvall, B. Å., & Johnson, B. (1994). The learning economy. *Journal of Industry Studies, 1*(2), 23–42.

Lustick, S. (2000). Agent-based modelling of collective identity: Testing constructivist theory. *Journal of Artificial Societies and Social Simulation, 3*(1), 1.

Maillat, D. (1998). Interaction between urban systems and localized productive systems: An approach to endogenous regional development in terms of innovative milieu. *European Planning Studies, 6*, 117–129.

Malthus, T. R. (1798). *An essay on the principle of population, as it affects the future improvement of society, with remarks on the speculations of Mr. Godwin, M. Condorcet, and other writers*. J. Johnson. Recent Edition (2007), The Lawbook Exchange.

Manzoni, A. (1985). *The Betrothed [I promessi sposi, about 1840]*. Mondadori. https://moodle2.units.it/pluginfile.php/441330/mod_resource/content/1/MANZONI-PROMESSI%20SPOSI.pdf

Marshall, A. (1890). *Principles of economics*. Macmillan and Co. Preface. https://www.econlib.org/library/Marshall/marP.html

Marshall, A. (1919). *Industry and trade. A Study of industrial technique and business organization; and of their influences on the condition of various classes and nations*. Macmillan and Co.

Martelli, M. (1999). *Introduction to discrete dynamical systems and chaos*. J. Wiley & Sons.

Maslow, A. H. (1943). A theory of human motivation. *Psychological Review, 50*, 370–396.

Maslow, A. H. (1954). *Motivation and personality*. Harper and Row.

Maturana, H. R., & Guiloff, G. D. (1980). The quest for the intelligence of intelligence. *Journal of Social and Biological Structures, 3*(2), 135–148.

Maturana, H. R., & Varela, F. J. (1980 1st ed. 1972). *Autopoiesis and cognition. The Realization of living*. Reidel Publishing.

Maturana, H. R., & Varela, F. J. (1987). *The tree of knowledge: The biological roots of human understanding*. New Science Library/Shambhala Publications.

McLeod, S. (2007 online). *Maslow's hierarchy of needs*. http://www.simplypsychology.org/maslow.html

Megacities online. https://www.statista.com/chart/1826/population-growth-in-the-worlds-megacities/

Mella, P. (2005). Observing collectivities as simplex systems: The combinatory systems approach. *Nonlinear Dynamics, Psychology and Life Sciences, 9*(2), 121–153.

Mella, P. (2006). Spatial co-localisation of firms and entrepreneurial dynamics. The combinatory system view. *The International Entrepreneurship and Management Journal, 2*, 391–412.

Mella, P. (2009). *The Holonic Revolution. Holons, Holarchies and Holonic Networks. The ghost in the production machine*. University Press.

Mella, P. (2011). The quality policy in value based management. *The Journal of American Academy of Business, Cambridge, 17*, 68–180.

Mella, P. (2012). *Systems thinking. Intelligence in action*. Springer Verlag.

Mella, P. (2014). The pillars of learning, understanding, studying and explaining. *Creative Education, 5*(17), 1615–1628.

Mella, P. (2017). The unexpected cybernetics life of collectivities. The combinatory systems approach. *Kybernetes, 46*(7), 1086–1111.

Mella, P. (2018). The law of increasing productivity. *International Journal of Markets and Business Systems, 3*(4), 297–316.

Mella, P. (2021, 1st Ed. 2014), *The magic ring. Systems thinking approach to control systems. Contemporary systems thinking series*. Springer International Publishing.

Mitleton-Kelly, E. (1997a). In P. Henderson (Ed.), (2000) *Complexity: Partial support for BPR* (pp. 24–37). Springer.

Mitleton-Kelly, E. (1997b). *Complex adaptive systems in an organisational context: organisations as co-evolving complex adaptive systems*. ESRC Business Processes Resource Centre, University of Warwick.

Mitleton-Kelly, E. (Ed.). (2003). *Complex systems and evolutionary perspectives on organisations: the application of complexity theory to organisations. Advanced series in management*. Elsevier.

Monod, J. (1970). *Chance and necessity: Essay on the natural philosophy of modern biology*. New York: Alfred A. Knopf, 1971 (Original, *Le hazard et la nécéssité: Essai sur la philosophie naturelle de la biologie moderne*. Paris: Seuil).

Morgeson, F. P., & Hofmann, D. A. (1999). The structure and function of collective constructs: Implications for multilevel research and theory development. *Academy of Management Review, 24*(2), 249–265.

Mulder, P. (2012). 5 whys analysis. In *Toolshero*. https://www.toolshero.com/problem-solving/5-whys-analysis/

NCB, New Catholic Bible. (2019). *Genesis* 11:1–9. https://bible.usccb.org/bible/genesis/11

NECSI. (online). *New England complex systems institute*. Cambridge, MA USA. http://www.necsi.org

Negoita, C. V. (1981). *Fuzzy systems (Cybernetics and systems)*. Abacus Press.

Negoita, C., Zadeh, L. A., & Zimmermann, H. (1978). Fuzzy sets as a basis for a theory of possibility. *Fuzzy Sets and Systems, 1*(3), 61–72.

Nelsen, D., & Daniels, S. E. (2007). Quality glossary. *Quality Progress, 40*(6), 39.

Nelson, R. (1995). Theorizing about economic change. *Journal of Economic Literature, 33*, 48–90.

Nelson, R., & Winter, S. (1982). *An evolutionary theory of economic change*. The Belknap Press of Harvard University Press.

Ngai, K. M., Burkle, F. M., Hsu, A., & Hsu, E. B. (2009). Human stampedes: a systematic review of historical and peer-reviewed sources. *Disaster Medicine and Public Health Preparedness, 3*(4), 191–195.

Nguyen, D. H., & Yin, G. (2016). Coexistence and exclusion of stochastic competitive Lotka-Volterra models. *Journal of Differential Equations, 262*(3), 1192–1225.

Nonaka, I. (1994). A dynamic theory of organizational knowledge creation. *Organization Science, 5*(1), 14–37.

Nonaka, I., & Konno, N. (1998). The concept of "ba": Building a foundation for knowledge creation. *California Management Review, 40*(3), 40–54.

Nooteboom, B. (1999). Innovation, learning and industrial organization. *Cambridge Journal of Economics, 23*(2), 127–150.

Ohno, T. (1988). *The Toyota production system; beyond large-scale production.* Productivity Press.

Olsen, M. (2019). Stigmergy for biological spatial modeling. *Complex Adaptive Systems: Views from the Physical, Natural, and Social Sciences*, 169–197.

Otter, H., van der Veen, A., & de Vriend, H. J. (2001). ABLOoM: Location behaviour, spatial patterns, and agent-based modelling. *Journal of Artificial Societies and Social Simulation, 4*(4) http://jasss.soc.surrey.ac.uk/4/4/2.html

Packard, A. S. (1901). *Lamarck, the founder of evolution. His life and work with translations.* Longmans, Green.

Pelikan, P. (2001). In J. Foster & J. S. Metcalfe (Eds.), *Self-organizing and Darwinian selection in economic and biological evolutions: an enquiry into the sources of organizing information* (pp. 121–151). Elgar.

PestSmart Connect (online). https://pestsmart.org.au

Pirsig, R. (1974). *Zen and the Art of Motorcycle Maintenance: An Inquiry into Values.* Bantam Books.

Plamondon, A. L. (1979). *Whitehead's organic philosophy of science.* State University of New York Press.

Polanyi, M. (1967). *The tacit dimension.* University of Chicago Press.

Porter, M. E., & Sölvell, Ö. (1998). In A. D. Chandler, P. Hagström, & Ö. Sölvell (Eds.), *The role of geography in the process of innovation and the sustainable competitive advantage of firms.* Oxford Academic Press. (1999).

Potts, J. (2001). Knowledge and markets. *Journal of Evolutionary Economics, 11*(4), 413–431.

Pray, L. (2008). Discovery of DNA structure and function: Watson and Crick. *Nature Education, 1*(1), 100.

Prigogine, I. (1985). *New perspectives on complexity* (pp. 483–492). United Nations University. Reprinted in: Klir G. J. (2013).

Prigogine, I., & Meixner, J. (1976). *L'ordre par fluctuations et le système social* (pp. 7–48). VS Verlag für Sozialwissenschaften.

Prigogine, I., & Nicolis, G. (1989). *Exploring complexity: An introduction.* Freeman & Co.

Prigogine, I., & Stengers, I. (1984). *Order out of chaos. Man's new dialogue with nature.* Bantam Books. (Original: La Nouvelle Alliance. Paris: Gallimard 1989).

Prusinkiewicz, P., & Lindenmayer, A. (1990). *The algorithmic beauty of plants.* Springer-Verlag.

Rabellotti, R. (1997). *External economies and cooperation in industrial districts.* Mc Millan Press.

Rangone, A., & Mella, P. (2019). Obstacles to managing dynamic systems. The systems thinking approach. *International Journal of Business and Social Science, 10*(8), 24–41.

Reichholf, J. H., & Mattson, P. (1994). Zoology: Biodiversity why are there so many different species? *Universitas, 1*, 42.

Reynolds, C. W. (1987). Flocks, herds and schools: A distributed behavioral model. In *Proceedings of the 14th annual conference on Computer graphics and interactive techniques* (pp. 25–34). http://www.macs.hw.ac.uk/~dwcorne/Teaching/Craig%20Reynolds%20Flocks,%20Herds,%20and%20Schools%20A%20Distributed%20Behavioral%20Model.htm

Richmond, B. (1993). Systems thinking: Critical thinking skills for the 1990s and beyond. *System Dynamics Review, 9*(2), 113–133.

Rigopoulou, I. D., Chaniotakis, I. E., Lymperopoulos, C., & Siomkos, G. I. (2008). After-sales service quality as an antecedent of customer satisfaction: The case of electronic appliances. *Managing Service Quality: An International Journal, 18*(5), 512–527.

Ringwood, A. E. (1979). *Origin of the Earth and Moon*. Springer-Verlag.

Riva, A., & Pilotti, L. (2019). Sassuolo's ceramic district and strategic changes in the sector during the time. *Economia Aziendale Online, Special Issue, 10*(3), 455–482.

Rondo-Brovetto, P., & Saliterer, I. (2007). Comparing regions, cities, and communities: Local government benchmarking as an instrument for improving performance and competitiveness. *The Innovation Journal: The Public Sector Innovation Journal, 12*(3), 1–18.

Rotary wave. (2014, online). Webpage: https://www.rotarywave.com/en/company/

Rousseau, D. M. (1985). Issues of level in organizational research: Multi-level and cross-level perspectives. *Research in Organizational Behavior, 7*(1), 1–37.

Rubinstein, A. (1998). *Modeling bounded rationality*. MIT Press.

Russell, S. J., & Norvig, P. (2016). *Artificial intelligence: A modern approach*. Pearson.

Sakoda, J. M. (1971). The checkerboard model of social interaction. *The Journal of Mathematical Sociology, 1*(1), 119–132.

Salam, A., Bethe, H. A., Dirac, P., Heisenberg, W., Wigner, E. P., Klein, O., & Lifshitz, E. M. (1989). *From a life of Physics*. World Scientific.

Salmon, W. C. (1990). *Four decades of scientific explanation*. University of Minnesota Press.

Sandquist, G. M. (1985). *Introduction to system science*. Prentice-Hall, Inc.

Sapir, E. (1968). *Selected Writings of Edward Sapir*. Univ of California Press.

Schatten, A. (2004). *Modelling on the edge of chaos: Cellular automata and agents*. http://www.schatten.info/publications/presentations/modell_edge_chaos-ca.pdf

Schelling, T. C. (1971). Dynamic models of segregation. *Journal of Mathematical Sociology, 1*(2), 143–186.

Schelling, T. C. (2006 first ed. 1978). *Micromotives and macrobehavior*. WW Norton & Company.

Schlenker, P. (2004). Conditionals as definite descriptions. *Research on Language and Computation, 2*(3), 417–462.

Schluter, D., Price, T. D., & Grant, P. R. (1985). Ecological character displacement in Darwin's finches. *Science, 227*(4690), 1056–1059.

Schmitz, J. A., Jr. (2005). What determines productivity? Lessons from the dramatic recovery of the US and Canadian iron ore industries following their early 1980s crisis. *Journal of Political Economy, 113*(3), 582–625.

Schoener, T. W. (2011). The newest synthesis: Understanding the interplay of evolutionary and ecological dynamics. *Science, 331*(6016), 426–429.

Schowalter, T. D. (2016). *Ecosystem Structure and Function* (Chap. 11). In *Insect Ecology. (2022)*. Elsevier, Academic Press.

Senge, P. M. (1990). *The fifth discipline: The art & practice of the learning organization*. Currency Doubleday.

Senge, P. M., & Lannon-Kim, C. (1991). *The systems thinking approach. The Systems Thinker Newsletter* (Vol. 2(5)). Kendall Square.

Senge, P. M., Kleiner, A., Roberts, C., Ross, R., & Smith, B. (1994). *The fifth discipline fieldbook*. Doubleday, Random House Inc.

Serra, R., & Zanarini, G. (1990). *Complex systems and cognitive processes*. Springer-Verlag.

Shimizu, H. (1987). A general approach to complex systems in bioholonics. In *Lasers and synergetics: A colloquium on coherence and self-organization in nature* (pp. 204–223). Berlin, Springer Berlin Heidelberg.

Simon, H. A. (1955). A behavioral model of rational choice. *The Quarterly Journal of Economics, 69*(1), 99–118.

Simon, H. A. (1990). Invariants of human behavior. *Annual Review of Psychology, 41*, 1–19.

Simon, H. A., & Kaplan, C. A. (1989). *Foundations of cognitive science*. MIT Press.

Singer, P. W. (2009). Isaac Asimov's laws of robotics are wrong. *Brookings Institute, 18*, 2009.

Smith, A. (1776). *An inquiry into the nature and causes of the wealth of nations*. Edited with an Introduction, Notes, Marginal Summary and an Enlarged Index by Edwin Cannan. London: Methuen, 1904). Vol. 1.

Spelta, A. M. (1603). In P. Bartoli (Ed.), *History written by Antonio Maria Spelta, citizen of Pavia [original: Historia d'Antonio Maria Spelta, Cittadino Pavese]*. Google books.

Stacey, R. D. (1995). The science of complexity: An alternative perspective for strategic change processes. *Strategic Management Journal, 16*(6), 477–495.

Staddon, J. E. (2016, 1st ed. 1983). *Adaptive behavior and learning*. Cambridge.

Sterman, J. D. (2001). System dynamics modeling: Tools for learning in a complex world. *California Management Review, 43*(4), 8–25.

Stonier, R. J., & Yu, X. H. (1994). *Complex systems: Mechanism of adaptation*. IOS Press.

Sugden, R. (1989). Spontaneous order. *Journal of Economic Perspectives, 3*(4), 85–97.

SWARM online. (2024). Main page at: www.swarm.org/wiki/Main_Page

Swenson, R. (2000). Spontaneous order, autocatakinetic closure, and the development of space-time. *Annals of the New York Academy of Sciences, 901*(1), 311–319.

Szeliski, R. (2022). *Computer vision: Algorithms and applications*. Springer Nature.

Takeuchi, Y. (1995). *Global dynamical properties of Lotka-Volterra systems*. World Scientific.

Tansley, A. G. (1935). The use and abuse of vegetational concepts and terms. *Ecology, 16*(3), 284–307.

Thaler, R., & Sunstein, C. (2008). *Nudge: Improving decisions about health, wealth and happiness*. Yale University Press.

Thaxton, C., Bradley, W., & Olson, R. (1984). *The mystery of life's origin*. Philosophical Library.

Thom, R. (1975). *Structural stability and morphogenesis: An outline of a general theory of models* (D. H. Fowler, Trans.). Reading.

Thorelli, H. B. (1986). Networks: Between markets and hierarchies. *Strategic Management Journal, 7*, 37–51.

Toffoli, T., & Margolus, N. (1987). *Cellular automata machines: A new environment for modeling*. The MIT Press.

Turing, A. M. (1950). Computing machinery and intelligence. *Mind, 59*, 433–460.

Tylor, E. B. (1871). *Primitive culture: Researches into the development of mythology, philosophy, religion, art, and custom (book in 2 Volumes)*. J. Murray.

Uhrmacher, A. M., & Weyns, D. (Eds.). (2009). *Multi-agent systems: Simulation and applications*. CRC press (Taylor & Francis).

Ulam, S. M. (1986). *Science, computers, and people: from the tree of mathematics*. Birkhauser.

Ulam, S. M. (1991). *Adventures of a mathematician*. Univ of California Press.

Varela, F. (1979). *Principles of biological autonomy*. Elsevier.

Varela, F. (1981). *Describing the logic of the Living. The adequacy and limitations of the idea of Autopoiesis,* in Zeleny.

Volterra, V. (1926). Variations and fluctuations of the number of individuals in animal species living together. In *Animal Ecology*. McGraw-Hill. (1931) [Original: Variazioni e fluttuazioni del numero d'individui in specie animali conviventi. Atti dell'Accademia Nazionale dei Lincei, Italy, 31-113].

Volterra, V. (1931). *Lessons on the mathematical theory of struggle for life [original: Leçons sur la théorie mathématique de la Lutte pour la vie]*. Gauthier-Villars.

von Bertalanffy, L. (1968). *General system theory: Foundations, development, applications*. Braziller.

von der Pütten, J. C. (2008). *Moral issues and concerns about China's one-child policy: A cosmopolitan perspective*. GRIN Verlag.

von Foerster, H. (1960). In M. C. Yovits & S. Cameron (Eds.), *On self-organising systems and their environments. Self-organising systems* (pp. 31–50). Pergamon Press.

von Foerster, H. (1972). Perception of the future and the future of perception. *Instructional Science, 1*(1), 31–43.

von Hayek, F. (1942). Scientism and the study of society. Part I. *Economica, 9*(35), 267–291.

Von Mises, L. (1998). *Human action: The scholar's edition*. The Mises Institute.

Von Neumann, J., & Burks, A. W. (1966). Theory of self-reproducing automata. *IEEE Transactions on Neural Networks, 5*(1), 3–14. https://cba.mit.edu/events/03.11.ASE/docs/VonNeumann.pdf

Waddington, C. H. (1957). *The strategy of the genes*. Allen and Unwin. (Reprinted 2014).

Waldrop, M. M. (1993). *Complexity: The emerging science at the edge of order and chaos*. Simon and Schuster.

Watson, J. D., & Crick, F. H. (1953). Molecular structure of nucleic acids. *Nature, 171*(4356), 737–738.

Weidenschilling, S. J. (1991). Formation of planetesimals. *Planetary Sciences: American and Soviet Research, 1*, 82–97.

Weidenschilling, S. J. (2000). Formation of planetesimals and accretion of the terrestrial planets. *Space Science Reviews, 92*(1–2), 295–310.

Weiner, B. (1994). Integrating social and personal theories of achievement striving. *Educational Research, 64*, 557–573.

Weinshall, T. D., & Raveh, Y. A. (1983). *Managing growing organizations: A new approach*. John Wiley & Sons.

Weis, E. (2008). *Fundamentals of complex evolving systems: A primer*. Inst. of Social Ecology, IFF-Fac. for Interdisciplinary Studies, Klagenfurt Univ.

West, P. M., & Packer, C. (2002). Sexual selection, temperature and the lion's mane. *Science, 297*(5585), 1339–1343.

Whitaker, R. (2003). *The observer web: Autopoiesis and enaction*. ENCYCLOPAEDIA AUTOPOIETICA WEB.

Whorf, B. L. (1956). *Language thought, and reality*. The MIT Press.

Widhiarso, W. (2009). The influence of language on thought study Benjamin Whorf hypothesis and Edward Sapir. SSRN. https://ssrn.com/abstract=1496642

Wiener, N. (1961). *Cybernetics or control and communication in the animal and the machine* (1st ed.). MIT Press. Google Books. 1948.

Wikipedia. (2024a). *Al-Aimmah Bridge disaster*. https://en.wikipedia.org/wiki/2005_Al-Aimmah_Bridge_disaster

Wikipedia. (2024b). *Hillsborough disaster*. https://en.wikipedia.org/wiki/Hillsborough_disaster

Wikipedia. (2024c). *List of fatal crowd crushes*. https://en.wikipedia.org/wiki/List_of_fatal_crowd_crushes

Wikipedia. (2024d). *Megacity*. https://en.wikipedia.org/wiki/Megacity

Wikipedia. (2024e). *Mina stampede*. https://en.wikipedia.org/wiki/2015_Mina_stampede

Wikipedia. (2024f). *Punctuated equilibrium*. https://en.wikipedia.org/wiki/Punctuated_equilibrium

Wikipedia. (2024g). *Turin, Italy, Stampede* (2017). https://en.wikipedia.org/wiki/2017_Turin_stampede

Wikipedia. (2024h). *Isaac Asimov*. https://en.wikipedia.org/wiki/Isaac_Asimov

Wilber, K. (2000). *Sex, ecology, spirituality: The spirit of evolution* (1st ed.). Shambhala. 1995.

Wilber, K. (2001). *A brief history of everything* (1st ed.). Shambhala. 1996.

Williamson, O. E. (1993). The evolving science of organization. *Journal of Institutional and Theoretical Economics, 149*, 36–63.

Willis, A. J. (1997). The ecosystem: An evolving concept viewed historically. *Functional Ecology*, 268–271.

Wolfram, S. (1984). Cellular automata as models of complexity. *Nature, 311*(5985), 419–424.

Wolfram, S. (2004). *A new kind of science and the future of mathematics*. Joint Mathematics Meetings, Phoenix, Arizona. http://www.stephenwolfram.com/publications/a-new-kind-of-science-future-mathematics/

Yager, R. R., & Zadeh, L. A. (Eds.). (1991). *An introduction to fuzzy logic applications in intelligent systems*. Kluwer Academic Publishers.

YouTube. (2022). *Massive Mexican Wave! - London 2012 Olympic Stadium*. https://www.youtube.com/watch?v=hq387m0PRac

Zadeh, L. A. (1978). Fuzzy sets as a basis for a theory of possibility. *Fuzzy Sets and Systems, 1*(1), 3–28.

Zeleny, M. (1981). *Autopoiesis, a theory of living organization*. Elsevier.

Zollo, G., Iandoli, L. & De Maio, L. (2001). An application of an Ant Colony system to the analysis of organizational learning processes. In *VIII SIGEF Congress proceedings, new logics for the new economy, Naples*.

Zuradelli, C. (1888). *The towers of Pavia [original: Le torri di Pavia]*. Forni.

Index

P. Mella, *The Combinatory Systems Theory*, Progress in IS,
https://doi.org/10.1007/978-3-031-86946-4

I

K

L

M

N

The manufacturer's authorised representative in the EU is Springer Nature Customer Service Centre GmbH, Europaplatz 3, 69115 Heidelberg, Germany. If you have any concerns regarding our products, please contact ProductSafety@springernature.com

Printed and bound by CPI Group (UK) Ltd, Croydon, CR0 4YY
01/10/2026
02225949-0013